Julius Adler.
U. of Wis
Biochemistry
Rm. 75
Sensory Reception

21040
43376-S
8w
(N)

D0948277

ANNALS OF THE NEW YORK ACADEMY OF SCIENCES

Volume 510

EDITORIAL STAFF

Executive Editor
BILL BOLAND

Managing Editor
JUSTINE CULLINAN

Associate Editor
LINDA H. MEHTA

The New York Academy of Sciences
2 East 63rd Street
New York, New York 10021

THE NEW YORK ACADEMY OF SCIENCES
(Founded in 1817)
BOARD OF GOVERNORS, 1987

FLEUR L. STRAND, *President*
WILLIAM T. GOLDEN, *President-Elect*

Honorary Life Governors

SERGE A. KORFF   H. CHRISTINE REILLY   IRVING J. SELIKOFF

Vice-Presidents

PIERRE C. HOHENBERG   DENNIS D. KELLY   JACQUELINE MESSITE
JAMES G. WETMUR   VICTOR WOUK

ALAN PATRICOF, *Secretary-Treasurer*

Elected Governors-at-Large

CYRIL M. HARRIS   PETER D. LAX   NEAL E. MILLER
ERIC J. SIMON   JOSEPH F. TRAUB

KURT SALZINGER   *Past Presidents (Governors)*   WILLIAM S. CAIN

HEINZ R. PAGELS, *Executive Director*

OLFACTION AND TASTE IX

ANNALS OF THE NEW YORK ACADEMY OF SCIENCES
Volume 510

OLFACTION AND TASTE IX

Edited by Stephen D. Roper and Jelle Atema

The New York Academy of Sciences
New York, New York
1987

Copyright © 1987 by the New York Academy of Sciences. All rights reserved. Under the provisions of the United States Copyright Act of 1976, individual readers of the Annals *are permitted to make fair use of the material in them for teaching or research. Permission is granted to quote from the* Annals *provided that the customary acknowledgment is made of the source. Material in the* Annals *may be republished only by permission of the Academy. Address inquiries to the Executive Editor at the New York Academy of Sciences.*

Copying fees: *For each copy of an article made beyond the free copying permitted under Section 107 or 108 of the 1976 Copyright Act, a fee should be paid through the Copyright Clearance Center, 21 Congress Street, Salem, MA 01970. For articles of more than 3 pages the copying fee is $1.75.*

Library of Congress Cataloging-in-Publication Data

Main entry under title:

Olfaction and taste IX.

(Annals of the New York Academy of Sciences ;
v. 510)
    Based on the Ninth International Symposium on Olfaction
and Taste held at the Snowmass Village, Colorado, July
20-24, 1986 by the International Commission on Olfaction
and Taste.
    Includes bibliographies and index.
    1. Smell—Physiological aspects—Congresses.
2. Taste—Physiological aspects—Congresses.
I. Roper, Stephen D.   II. Atema, Jelle.
III. International Symposium on Olfaction and
Taste (9th : 1986 : Snowmass Village, Colo.)
IV. International Commission on Olfaction and Taste.
Q11.N5 vol. 510   500 s   [612'.86]   87-34753
[QP458]
ISBN 0-89766-416-7
ISBN 0-89766-415-9 (pbk.)

PCP
Printed in the United States of America
ISBN 0-89766-415-9 (paper)
ISBN 0-89766-416-7 (cloth)

microvilli at the site of stimulus reception. Teleost fishes,[11] such as the trout, have both microvillar and ciliated olfactory receptors (FIGS. 7 and 8). Mammals, it seems, generally utilize modified cilia at the transducer sites of their olfactory receptors—and have microvillar receptor cells in the vomeronasal organs of their accessory olfactory systems.[12] Some birds, such as the duck, have olfactory receptor neurons that have both cilia and microvilli on the tip of the same dendrite.[13] Gustatory receptors tend to follow the same broad phylogenetic pattern as olfactory receptors; invertebrate contact chemoreceptors tend to be ciliary derivatives, whereas vertebrate taste cells deploy microvilli at the receptor site.

DISCUSSION

It is becoming increasingly apparent that the plasma membranes of chemosensory neurons contain specific macromolecular receptors that participate in the detection of chemical stimulants.[14-16] Cilia and microvilli, being long, slender extensions of the cell surface, provide considerable amplification of the area of the plasma membrane and glycocalyx available for stimulant-receptor interactions. Since cells, left to passively bow to minimum-energy consideration, would be spheres, it takes special intracellular structures to support axial extensions of the cell surface. Consequently, both cilia and microvilli are generated and supported by the polymerization of proteinaceous subunits into cytoskeletal structures. Cilia are supported by axonemes; axonemes are made of microtubules; and microtubules are polymers of tubulin, an ancient protein[17] as old as the mitotic spindle itself. Microvilli are supported by core filaments made of actin; actin, too, is an ancient protein, common to the vast majority of cells that can change their shape or move organelles about within their cytoplasm.

Throughout the course of evolution, sensation and motility have been closely allied to one another. The adaptive advantage of this alliance is clear; if an organism senses a change in its surroundings, its sensitivity will be selected for only if it can respond to the change by moving itself or a part of itself in response. As mentioned above, sensory cells have made widespread use of these two ancient proteins, tubulin and actin, to generate axial extensions of the cell surface that promote the molecular interactions central to sensation. Evolution tends to be conservative, and it is certainly no accident that tubulin and actin are also the basic building blocks of many major motile systems such as motile cilia, flagella, and muscle. Consequently, sensation and motility are inextricably linked together not only at the behavioral and physiological levels, but at the molecular level as well.

FIGURE 7. Scanning electron micrograph of trout olfactory epithelium showing the surfaces of ciliated (C) and microvillar (M) olfactory receptor neurons. Magnification × 11,000.

FIGURE 8. High-voltage electron micrograph of trout olfactory epithelium showing the apical poles of ciliated (C) and microvillar (M) olfactory receptor neurons. Magnification × 20,000.

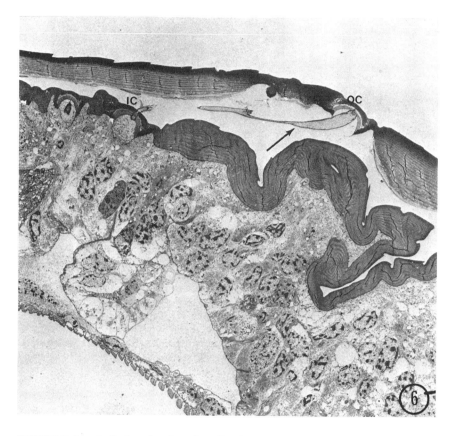

FIGURE 6. Longitudinal section through a ciliated mechanoreceptor (campaniform sensillum) of an insect before molting. Note extension of cilium (arrow) that interconnects cuticular cap in new, inner cuticle (IC) with that in old, outer cuticle (OC). Magnification × 800.

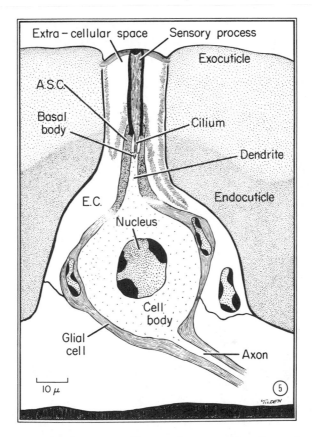

FIGURE 5. Drawing of an insect campaniform sensillum, a proprioceptive mechanoreceptor in the cuticle.

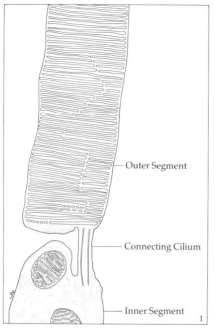

Outer Segment

Connecting Cilium

Inner Segment

1

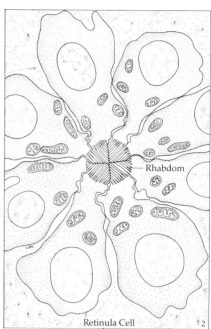

Rhabdom

Retinula Cell

2

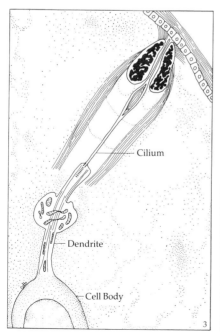

Cilium

Dendrite

Cell Body

3

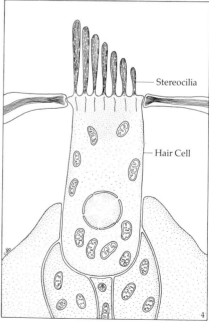

Stereocilia

Hair Cell

4

orthogonal orientation of the microvilli is central to the bee's capacity to detect polarized light. It is interesting to note that other Metazoa, such as the annelid *Nereis,* have both microvilli and (rudimentary) cilia in their photoreceptors.[6]

MECHANORECEPTORS

A variety of ciliated and microvillar sensory cells are present in metazoan mechanoreceptors. Insect acoustic receptors, such as the ear of the noctuid moth that can detect the ultrasonic cries of bats, are innervated by chordotonal sensilla. Chordotonal sensilla, such as the one illustrated in FIGURE 3, center their function around one or two bipolar neurons, each of which sends a modified cilium to the site of stimulus reception.[7] This stands in sharp contrast to the vertebrate inner ear, in which "hair cells" are present that have an orderly array of stiff microvilli at the site of stimulus reception (FIG. 4). These microvilli, misleadingly named "stereocilia," are the cellular elements responsible for sensory transduction in inner ear hair cells.[8]

Other invertebrate mechanoreceptors, such as the insect campaniform sensillum, employ cilia that not only play a role in sensory transduction, but in development as well.[9] Each campaniform sensillum has a bipolar neuron that sends a modified cilium to a cap in the cuticle that is the site of stimulus reception (FIG. 5). Before moulting, heterometabolous insects such as cockroaches have two cuticles: an inner, "new" cuticle, which will serve as its exoskeleton after the moult, and an outer, "old" cuticle, destined to be shed at ecdysis. At this time (FIG. 6), the cilium of the campaniform sensillum undergoes tremendous elongation. It still originates at the dendrite tip of the bipolar neuron, but passes through the cap in the new cuticle, traverses the moulting space, and makes a physiologically functional connection with the cap in the old cuticle.

CHEMORECEPTORS

A variety of chemosensory cells are equipped with cilia, microvilli, or both. Insect olfactory receptors, for example, are innervated by bipolar neurons whose dendrites bear modified cilia.[10] The olfactory receptors of sharks and rays, however, employ

---

FIGURE 1. Vertebrate photoreceptor with cilium; longitudinal section through part of retinal rod.

FIGURE 2. Invertebrate photoreceptor with microvilli; cross section through ommatidium of bee eye.

FIGURE 3. Invertebrate acoustic receptor with cilium; chordotonal sensillum of insect ear.

FIGURE 4. Vertebrate acoustic receptor with microvilli (stereocilia); longitudinal section through hair cell of inner ear.

Evolutionary Patterns in Sensory Receptors

An Exercise in Ultrastructural Paleontology[a]

DAVID T. MORAN

Rocky Mountain Taste & Smell Center
Department of Cellular and Structural Biology
University of Colorado School of Medicine
Denver, Colorado 80262

In order to survive and reproduce, all living things great and small, be they elephants or amoebae, must respond to changes in their immediate surroundings. Organisms, then, tend to be sensitive; both sensation and motility are necessary for responsiveness. Among the Protozoa, an amoeba, for example, will detect and glide away from a region of unfavorable pH. *Euglena,* armed with its photosensitive eyespot, will move toward the sunlight so essential for photosynthesis. The agile and fast-moving *Paramecium* is equipped with many motile cilia that not only serve as mechanoreceptors, but are covered by excitable membranes.[1] Among the Metazoa, an impressive array of sensory cells exists that respond to a wide range of stimuli. Investigation of the ultrastructure of metazoan sensory cells reveals a pattern of organization common to most of them: many sensory cells employ cilia, microvilli, or both at the site of stimulus reception.[2,3]

PHOTORECEPTORS

The structure of the outer segment of a "typical" vertebrate photoreceptor is drawn in FIGURE 1. Here, the stacked membranes that bear the photopigments are attached to the inner segment by a connecting cilium. Vertebrate photoreceptors, then, like many other photoreceptors, are ciliary derivatives.[4] Insect photoreceptors, on the other hand, are constructed quite differently. A cross section through a photoreceptive unit, or ommatidium, of the honeybee retina is drawn in FIGURE 2. Here, a series of retinula cells, arranged in a circle, send out microvilli—slender tubular extensions of the apical cell surface—into the center of the ommatidium.[5] Here the microvilli, which contain the photopigments, interdigitate to form the light-sensitive rhabdom. The

[a]Supported by National Institutes of Health Program Project Grant No. NS-20486 and National Science Foundation Research Grant No. BNS-821037.

Planning an international conference involves such a vast number of helpful associates that this preface could easily become an endless litany of acknowledgments. While it is not possible to cite everyone, we would be remiss if we did not express our gratitude to Ms. Jan Chase of the University of Colorado Health Sciences Center, and Ms. Eileen Banks of the Department of Anatomy and Neurobiology, Colorado State University, for their administrative assistance in organizing the meeting. We would also like to thank Linda Mehta of the New York Academy of Sciences for her cheerful and professional assistance in seeing this book through the press.

For the organizing committee,
STEPHEN D. ROPER
JELLE ATEMA

Preface

Every three years the International Commission on Olfaction and Taste designates a host country for ISOT, the International Symposium on Olfaction and Taste. Until 1986, the United States had not hosted this event for nearly two decades. Prior symposia had been held in Melbourne (1983), Noordwijkerhout, the Netherlands (1980), Paris (1977), Melbourne (1974), Seewiesen, West Germany (1971), New York (1968), Tokyo (1965), and Stockholm (1962). Early in 1985, the International Commission invited the Association for Chemoreception Sciences (AChemS) to host this triennial symposium in North America. Dr. Maxwell Mozell, at that time Chairman of the International Commission on Olfaction and Taste, and Dr. David Smith, then Executive Chair of AChemS, asked us whether we would be willing to organize ISOT at a convenient site in the United States. We were delighted to help out in this all-important symposium. After some investigation, we suggested that the clean, crisp air and exceptional beauty of the Colorado Rocky Mountains would be an ideal backdrop for an intensive working conference such as ISOT. An organizing committee was formed, chaired by Stephen Roper and consisting of Jelle Atema, John Caprio, Thomas Finger, Robert Gesteland, Bruce Halpern, John C. Kinnamon, Maxwell Mozell, David Smith, and Gordon Shepherd. We decided to hold the Ninth International Symposium on Olfaction and Taste at the Snowmass Village Resort, deep in the Maroon Bell Mountain Range of Colorado, July 20-24, 1986.

We felt that informal, personal interactions are the most critical components for a successful conference. We attempted to design the meeting to maximize these interactions by programming small breakfast workshops, held outdoors each day on the plaza; by including poster and slide sessions; and by leaving afternoons free for informal discussions. Even small details, such as arranging seats to be in a semicircular array in the larger conferences to facilitate discussions among the audience, spreading posters as far apart as possible to reduce interference between adjacent presentations, arranging for plentiful hot coffee and numerous comfortable chairs, were examined closely and deemed important. We were pleased to see that the 400 participants found the meeting a success and that novel research ideas and new collaborative programs have resulted directly from the interactions experienced at ISOT IX.

The organizing committee decided to divide the field of chemosensory biology into three global, but not necessarily all-encompassing questions: Where do the molecular events of chemosensory transduction take place? How is peripheral input processed in the central nervous system? Do responses to mixtures of chemical stimuli differ fundamentally from responses to pure stimuli? The symposia for the first three days focused on each of these questions in turn. The final symposium, on the fourth day, was entitled "From Reception to Perception: Summary and Synthesis."

One of the themes underlying the four-day conference was an attempt to integrate modern biophysical and molecular techniques into the study of chemosensory processes. A number of experiments utilizing patch recording techniques to study individual ion channels in receptor cell membranes were introduced at this conference. Exciting new data on molecular mechanisms of chemosensory transduction were presented. At the other extreme, novel holistic approaches to the chemical senses were presented: Two new "languages" for defining olfactory responses were described in the final symposium. Given this wide range of topics and experimental approaches, participants were exposed to an enormous menu of new ideas and new approaches to solving the questions confronting investigators. This was particularly stimulating for students and for scientists entering the field of chemical senses for the first time.

Financial assistance was received from:

- NATIONAL INSTITUTES OF HEALTH
- NATIONAL SCIENCE FOUNDATION
- BROWN & WILLIAMSON TOBACCO CORPORATION
- COCA-COLA COMPANY
- COLORADO STATE UNIVERSITY
- GENERAL FOODS CORPORATION
- GIVAUDAN CORPORATION
- PHILIP MORRIS INCORPORATED
- PROCTOR & GAMBLE COMPANY
- R. J. REYNOLDS COMPANY
- TAKASAGO CORPORATION

The New York Academy of Sciences believes that it has a responsibility to provide an open forum for discussion of scientific questions. The positions taken by the participants in the reported conferences are their own and not necessarily those of the Academy. The Academy has no intent to influence legislation by providing such forums.

ANNALS OF THE NEW YORK ACADEMY OF SCIENCES

Volume 510
November 30, 1987

OLFACTION AND TASTE IX [a]

Editors
STEPHEN D. ROPER AND JELLE ATEMA

CONTENTS

[a] This volume is the result of a symposium entitled Ninth International Symposium on Olfaction and Taste, held in Snowmass Village, Colorado on July 20-24, 1986, by the International Commission on Olfaction and Taste.

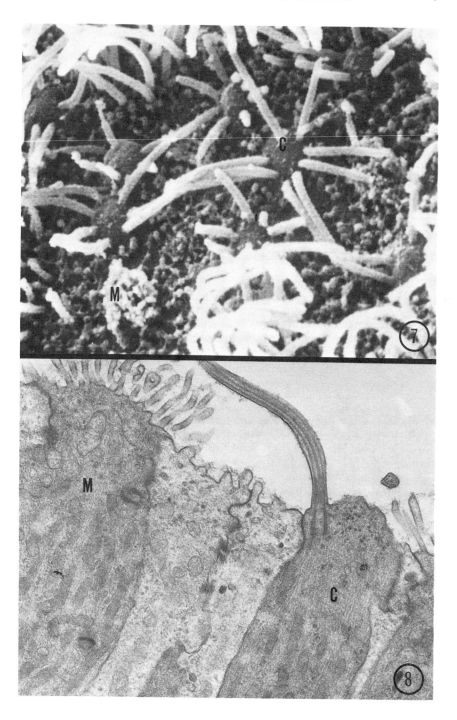

ACKNOWLEDGMENTS

The author thanks J. Carter Rowley for his collaboration during all phases of this project, Cecile Duray-Bito for making the drawings (FIGS. 1-4), Kathy Duran for assistance with photography, and George Dickel for spirited discussion.

REFERENCES

1. VAN HOUTEN, J. 1987. Eukaryotic unicells: How useful in studying chemoreception? Ann. N.Y. Acad. Sci. This volume.
2. VINNIKOV, YA. A. 1975. The evolution of olfaction and taste. *In* Olfaction and Taste. D. A. Denton & J. P. Coghlan, Eds. Vol. V: 175-187.
3. MORAN, D. T. & J. C. ROWLEY III. 1983. The structure and function of sensory cilia. J. Submicrosc. Cytol. **15**: 157-162.
4. EAKIN, R. M. 1965. Evolution of photoreceptors. Cold Spring Harbor Symp. Quant. Biol. **XXX**: 363-370.
5. VARELA, F. G. & K. R. PORTER. 1969. Fine structure of the visual system of the honeybee (*Apis mellifera*). I. The retina. J. Ultrastruct. Res. **29**: 236-259.
6. EAKIN, R. M. & J. L. BRANDENBERGER. 1985. Effects of light and dark on photoreceptors in the polychaete annelid *Nereis limnicola*. Cell Tissue Res. **242**: 613-622.
7. MORAN, D. T., J. C. ROWLEY III & F. G. VARELA. 1975. Ultrastructure of the grasshopper proximal femoral chordotonal organ. Cell Tissue Res. **161**: 445-457.
8. HUDSPETH, A. J. 1983. The hair cells of the inner ear. Sci. Am. **248**: 54-64.
9. MORAN, D. T., J. C. ROWLEY III, S. N. ZILL & F. G. VARELA. 1976. The mechanism of sensory transduction in a mechanoreceptor: Functional stages in campaniform sensilla during the molting cycle. J. Cell Biol. **71**: 832-847.
10. KEIL, T. A. 1986. Lectin binding sites in olfactory sensilla of the silkmoth, *Antheraea polyphemus*. Ann. N.Y. Acad. Sci. This volume.
11. HARA, T. J. 1975. Olfaction in fish. Prog. Neurobiol. **5**: 271-335.
12. GRAZIADEI, P. P. C. 1973. The ultrastructure of vertebrates olfactory mucosa. *In* The Ultrastructure of Sensory Organs. I. Friedmann, Ed.: 267-305. Elsevier. New York.
13. GRAZIADEI, P. P. C. & L. H. BANNISTER. 1967. Some observations on the fine structure of the olfactory epithelium in the domestic duck. Z. Zellforsch. Mikrosk. Anat. **80**: 220-232.
14. VINNIKOV, YA. A. 1986. Glycocalyx of receptor cell membranes. Chem. Senses **11**: 243-258.
15. LANCET, D. 1986. Vertebrate olfactory reception. Ann. Rev. Neurosci. **9**: 239-355.
16. LANCET, D. 1987. Toward a comprehensive analysis of olfactory transduction. Ann. N.Y. Acad. Sci. This volume.
17. LITTLE, M., G. KRAMMER, M. SINGHOFER-WOWRA & R. F. LUDUENA. 1986. Evolution of tubulin structure. Ann. N.Y. Acad. Sci. **406**: 8-12.

Sensory Transduction in Flagellate Bacteria

JUDITH P. ARMITAGE

Microbiology Unit
Department of Biochemistry
University of Oxford
Oxford OX1 3QU, England

R. ELIZABETH SOCKETT[a]

Department of Botany and Microbiology
University College London
London W. C. 1E 6BT, England

INTRODUCTION

Many bacterial species are motile by means of flagella. It takes about 30 genes to make a flagellum, a significant percentage of the genome; it would be unlikely, therefore, that bacteria would waste the metabolic energy required to replicate, transcribe, and translate all those genes just to swim randomly about their environment. Even though the mean generation time is longer, motile bacteria overgrow nonmotile mutants very quickly in all except the best mixed chemostat, suggesting a survival advantage for many motile bacteria in their natural environment. Flagellate bacteria use their motility to reach their optimum environments, for example, the gut wall for *Vibrio cholera* or *Salmonella typhi,* the root of a specific legume for a nitrogen-fixing *Rhizobium.* In general bacteria must balance information about the whole of their environment, oxygen levels, light, useful and harmful chemicals, temperature, and even, in some cases, pheromones to maintain themselves in and move toward regions where their growth rates can be sustained or increased. The different mechanisms used by different bacterial species to respond to different stimuli, and the way in which these signals may be integrated will be the subject of this very brief review of what is now probably the best understood of all behavioral systems. There have been several recent detailed reviews.[1-4]

[a] R. E. S. holds a Science and Engineering Research Council (UK) studentship.

9

MOTILITY

Bacteria use the gradient of protons formed across the cytoplasmic membrane to rotate passive, self-assembling, semi-rigid helical flagella: the only rotary motors known in nature.

The rotation of two protein rings associated with the cytoplasmic membrane is transferred to the flagellum. The total number of proteins involved in flagellar rotation is not certain, nor is the role of each protein in that rotation. Within the protein complex there are proton pores, switches, rotational proteins, and structural proteins. The driving force for rotation is, unlike eukaryotic systems, not ATP directly, but the unidirectional, electrochemical proton gradient. The mechanism by which an electrochemical gradient can be transformed into mechanical, rotational energy is a matter of great speculation.[5–8]

Despite the unidirectional nature of the proton gradient, most bacteria rotate their flagella in both a clockwise (CW) and counterclockwise (CCW) direction, with the exception of bacteria such as *Rhodobacter sphaeroides. Escherichia coli* and *Salmonella typhimurium* rotate their flagella about 95% CCW, 5% CW, whereas *Pseudomonas aeruginosa* and *Rhodospirillum rubrum* rotate theirs 50% in each direction. Switching between CCW and CW rotation causes the cell to change direction, or briefly tumble, every few seconds. The frequency of switching is altered if an environmental stimulus is encountered, biasing the overall direction of swim toward an attractant or away from a repellent. The basic mechanism of bacterial taxis does not therefore involve orientation of the cell toward an attractant, but rather a biasing of its usually random motility in a favorable direction.

SENSORY RECEPTION

This can be divided into two main categories, (1) sensing of any change in the energy status of the cell, and (2) a more sophisticated chemotaxis system, biasing cells toward an environment that may support a faster growth rate. FIGURE 1 illustrates some of the mechanisms involved in bacterial response to stimulus.

Taxis Involving the Energy Level

The basic requirement of all organisms is that their overall energy level does not fall dangerously low. The overriding taxis system in bacteria is designed to move them as quickly as possible out of areas where the electrochemical proton gradient (PMF) is threatened, or maintain them where electron transport is maximal. Bacteria show a repellent response to ionophores that cause a fall in PMF, changing direction more often if the concentration is increasing, until the PMF falls below a critical level at which all direction changing stops and the cells swim smoothly, increasing the chances of leaving a very dangerous environment as quickly as possible.[9]

Under natural conditions the aerobic and anaerobic responses to terminal electron acceptors (O_2, NO_3, fumarate) in *E. coli,* and *S. typhimurium,*[10,11] and the aerotactic

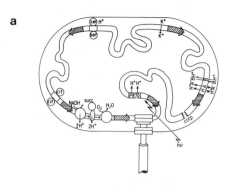

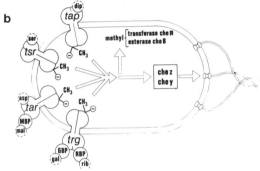

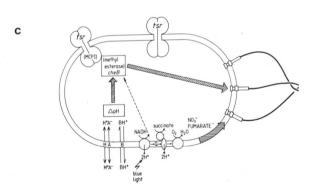

FIGURE 1. Diagrammatic representation of some of the mechanisms involved in bacterial taxis. (a) PMF-dependent taxis in *R. sphaeroides;* (b) MCP-dependent taxis in enteric bacteria; and (c) pH- and PMF-dependent taxis in enteric bacteria.

HA = weak acid; B = weak base; P870 = photosynthetic reaction center; CCCP = carbonyl cyanide *m*-chlorophenyl hydrazone; cit = citrate; ser = serine; MBP = maltose binding protein; GBP = galactose binding protein; RBP = ribose binding protein; dip = dipeptide; asp = aspartate.

(FIG. 1b copyright © [1986] The Open University Press, programme 532518; used with permission.)

and phototactic responses in *R. sphaeroides*[12,13] have all been shown to depend directly on changes in the size of the PMF, caused by changes in the rate of electron transport.

The interaction of aerotaxis and phototaxis in *R. sphaeroides* is a good illustration of how PMF can be used to maintain bacteria in their optimum environment. Photosynthetically grown cells possess both photosynthetic and respiratory electron-transport chains, the two chains sharing some components. The pathway to the high-affinity terminal oxidase from the shared cytochromes is, however, not coupled to any increase in PMF. The binding of oxygen to the terminal cytochrome of a cell in the light results in a loss of electrons from the photosynthetic electron transport chain and a consequent fall in PMF, and a repellent response (−ve aerotaxis). In the dark, however, when only the respiratory electron transport chain is operating, there is an increase in PMF when oxygen binds to the terminal acceptor, and therefore a +ve aerotactic response. Fluctuations in PMF can prevent cells swimming up an attractant light gradient into a toxic aerobic region, but cause the reversed response in the dark.[13]

The PMF sensor is unknown, but it seems likely that as the flagellar motor is itself driven by PMF, it may be able to sense the PMF level directly. It is not possible at the moment to exclude a redox-sensing small molecule within the cell.

Receptor-Dependent Taxis

If the energy level of the cell is high, it may then respond to changes in its environment causing it to move to an area that could allow an increased growth rate. Once in the improved environment, the cell must adapt to the new conditions to be able to respond to any future stimuli. The best studied system is that involving methyl-accepting chemotaxis proteins (MCPs). Four different sensory transducing proteins, MCPs, have been identified in *E. coli* and *S. typhimurium* each sending signals about the extracellular concentration of a range of chemicals to control flagellar switching frequency. Each MCP responds to the binding of a few chemoeffector chemicals, mainly a limited range of sugars and amino acids.[2,14]

There are about 2,000 of each MCP per cell, randomly distributed over the cell surface. Examination of the gene sequence suggests some relationship between each type, possibly the result of tandem gene duplication. The proteins, which function as tetramers, are about 55 kDa in size, with a variable periplasmic receptor structure, a membrane-spanning region, and a conserved intracellular signaling region.[14] Antibodies raised against the *E. coli* trg (MCP 111) protein cross-react with membrane proteins from species as diverse as the gram-positive *Bacillus subtilis,* the stalked bacterium *Caulobacter crescentus* and *Spirochaeta aurentia*, suggesting an ancient ancestry for the sensory system.[15] When a chemoeffector binds to the receptor, a signal is sent from the receptor to the flagella, changing the frequency of switching between CW and CCW, biasing the direction of swim. An initial conformational change within the MCP, caused by the binding of the chemoeffector, may cause the release of the sensory signal.[16] The response times, ($\simeq$200 msec) and the range and decay times of the signal in filamentous cells, have suggested that the signal must be a diffusible compound in the size range 10-80 kDa.[17,18] Using an *in vitro* system of flagellate cell envelopes, from wild-type and behavioral mutants, into which different possible sensory proteins were added back, it has been shown that the signal is probably the protein product of the *cheY* gene.[19] CheY causes CW rotation, the concentration available to bind to the motor switch is probably controlled by the amount bound to the MCPs, a reduction of chemoeffector binding releasing CheY to diffuse to the switch. The structure of CheY and the nature of the binding and release are unknown. The binding-release and activation-deactivation probably also involve other cytoplasmic gene products known to be required for normal chemotaxis, CheZ, CheW, and CheA, as well as ATP and possibly Ca^+.[19]

The MCPs are also responsible for the adaptation of the cells to the chemoeffector sensed through them. Specific glutamyl residues on the cytoplasmic side of the MCP are methylated or demethylated by two specific enzymes, a methyl transferase (CheR) and a methyl esterase (CheB), in response to chemoeffector binding or release. The sensory signal is blocked when there is some stoichiometric relationship between the number of bound sites and the number of methylated sites. The time course for adaptation can be minutes, depending on the concentration of chemoeffector added.

Enteric bacteria also respond to a range of stimuli for which there are no recognizable membrane receptors, but the response to which apparently involves the MCPs, specifically MCP1 (Tsr). These include weak acid repellents, weak base attractants, and changes in temperature. It seems that a transient change in intracellular pH caused by the rapid, direct permeation of weak acids or bases into the cell alters the interaction of the CheB protein (methyl esterase) with MCP1 and causes a repellent or attractant signal, respectively, to be sent to the flagellum.[20]

Membrane Potential Dependent and MCP-Independent Taxes

The photosynthetic bacteria *Rhodospirillum rubrum* and *R. sphaeroides* have no MCPs but can still respond to chemical changes in their environment (in preparation). Unlike studies using *E. coli*, where all attempts to identify membrane potential dependent, or surface charge dependent signaling systems have failed in anything other than the aerotaxis system, changes in membrane potential have been shown to be involved in the chemotactic responses of *Spirillum serpens, Spirochaeta aurantia, R. rubrum,* and *R. sphaeroides.* At least one of these species, *S. aurantia,* also possesses an MCP-type sensory system.[21] In all these species chemotaxis is inhibited if the membrane potential is clamped in such a way that transient changes are inhibited.

The potential dependent absorption spectrum of the carotenoid pigments in *R. sphaeroides* made it possible to examine the interrelationship between the chemotactic response and the change in the membrane potential. The primary attractants in these organisms are weak acids, major repellents in enteric bacteria. The response to the strong enteric attractants, for example serine, is very poor and dependent on the size of the baseline potential; cells incubated anaerobically in the dark showed a much stronger tactic response than those incubated in the light. Examination of the change in carotenoid absorption in the same cells revealed a much larger potential change on chemoeffector addition to the dark cells with the lower baseline potential addition than in cells incubated in the light. The response to weak acids is independent of pH, and not directly related to the oxidation state of the compound; therefore, neither changes in intracellular pH nor metabolism are directly involved (in preparation). Possibly the binding of some chemicals to the cytoplasmic membrane before uptake causes a transient change in surface charge, which is directly transmitted to the flagellar motor, comparable to PMF taxis. Under conditions where electron transport is high, for example, high light intensity, the high membrane potential would swamp any small change in charge, responses would therefore be expected to be lower under these conditions, or when the membrane potential is clamped. Photosynthetic bacteria can, however, still adapt to stimuli, but rather faster than the enteric bacteria, and adaptation does not depend on methionine.

Mutants of *E. coli* in the adaptation enzymes (CheBR) or in MCPI behave very much like photosynthetic bacteria, that is, reversed responses to weak acids and weak responses to amino acids, and show adaptation in the absence of the methylating enzymes.[22,23] It has been suggested that this represents a relationship between the net charge on MCP1 and the motor, the switching of which is altered if the MCP1-*cheB* interaction is changed.[24] It is possible that this more basic tactic system underlies the dominant MCP system in *E. coli* and is related to that identified in photosynthetic bacteria, where the complex photosynthetic intracellular membrane system may make a diffusing signal inefficient.

SUMMARY

Flagellate bacteria can respond to a wide range of environmental chemicals and a variety of physical parameters, and integrate those responses. The most important thing for a cell is to maintain its energy level; bacteria therefore respond directly to any changes in their PMF. This has been likened to higher organisms responding to a physiological change, for example, a fall in blood glucose.[1]

In addition, if the PMF is high, the cell is free to respond to a limited range of metabolites and possibly move to an area that will allow an increased growth rate. Bacteria do not sense all amino acids, as the space available on the cytoplasmic membrane is limited, and a change in a few important metabolites is probably a good measure of the general environment around the cell. The sensory response does not require either transport into the cell or metabolism of the chemical, only the binding to the specific MCP. The cell could have a mutation in the pathway metabolizing the chemoeffector, but it would still respond to changes in the concentration of that compound. This taken with the ability of the cells to adapt to the stimulus has been considered to be the prokaryotic equivalent of smell and taste.

REFERENCES

1. MACNAB, R. M. 1982. Soc. Exp. Biol. Symp. **35:** 77-104.
2. BOYD, A. & M. SIMON. 1982. Ann. Rev. Physiol. **44:** 501-517.
3. TAYLOR, B. L. 1982. Ann. Rev. Microbiol. **37:** 551-573.
4. MACNAB, R. M. 1984. Ann. Rev. Biophys. Bioeng. **13:** 51-83.
5. MACNAB, R. M. 1983. *In* Biological Structures and Coupled Flow. A. Oplatka & M. Balaban, Eds.: 147-160. Academic Press. New York.
6. BERG, H. C. & S. KHAN. 1983. *In* Mobility and Recognition in Cell Biology. H. Sund & C. Veeger, Eds.: 485-497. Walter de Guyter. Berlin, Germany.
7. MITCHELL, P. 1984. FEBS Lett. **176:** 287-294.
8. WAGENKNECKT, T. 1986. FEBS Lett. **196:** 193-197.
9. KHAN, S. & R. M. MACNAB. 1980. J. Mol. Biol. **138:** 563-597.
10. LAZLO, D. J. & B. L. TAYLOR. 1981. J. Bacteriol. **145:** 990-1001.
11. LAZLO, D. L., M. NIWANO, W. W. GORAL & B. L. TAYLOR. 1984. J. Bacteriol. **159:** 663-667.

12. ARMITAGE, J. P. & M. C. W. EVANS. 1981. FEMS Microbiol. Lett. **11:** 89-92.
13. ARMITAGE, J. P., C. INGHAM & M. C. W. EVANS. 1985. J. Bacteriol. **161:** 967-972.
14. BOYD, A., A. KRIKOS, N. MUTOH & M. SIMON. 1983. *In* Mobility and Recognition in Cell Biology. H. Sund & C. Veeger, Eds.: 551-562. Walter de Gruyter. Berlin, Germany.
15. NOWLIN, R., D. O. NETTLETON, G. W. ORDAL & G. L. HAZELBAUER. 1985. J. Bacteriol. **163:** 262-266.
16. ZUKIN, R. S., P. R. HARTIG & D. E. KOSHLAND, JR. 1979. Biochemistry **18:** 5599-5605.
17. BLOCK, S. M., J. E. SEGALL & H. C. BERG. 1982. Cell **31:** 215-226.
18. SEGALL, J. E., A. ISHIHARA & H. C. BERG. 1985. J. Bacteriol. **161:** 51-59.
19. EISENBACH, M., Y. MARGOLIN & S. RAVID. 1985. *In* Sensing and Response in Microorganisms. M. Eisenbach & M. Balaban, Eds.: 43-61. Elsevier. Amsterdam, the Netherlands.
20. SLONCZEWSKI, J. L., R. M. MACNAB, J. R. ALGER & A. M. CASTLE. 1982. J. Bacteriol. **152:** 384-399.
21. GOULBOURNE, E. A., JR. & E. P. GREENBERG. 1983. J. Bacteriol. **155:** 1443-1445.
22. REPASKE, D. R. & J. ADLER. 1981. J. Bacteriol. **145:** 1196-1208.
23. STOCK, J., A. BORCZUK, F. CHIOI & J. E. B. BURCHENAL. 1985. Proc. Natl. Acad. Sci. USA **82:** 8364-8368.
24. DANG, C. V., M. NIWANO, J-I. RYU & B. L. TAYLOR. 1986. J. Bacteriol. **166:** 275-280.

Eukaryotic Unicells: How Useful in Studying Chemoreception?

JUDITH VAN HOUTEN AND ROBIN R. PRESTON

Department of Zoology
University of Vermont
Burlington, Vermont 05405

INTRODUCTION

There are compelling reasons to use unicellular organisms in the study of chemoreception, particularly when examining receptor cell function in taste and olfaction. To a student of receptor cell function, it would be useful to have a system characterized by large receptor cell size for electrophysiology, homogeneous receptor cell populations that can be grown in quantity sufficient for biochemistry, and short generation time for genetic analysis and manipulation. Unicellular organisms provide all of these characteristics and more, such as rapid, clear-cut assays of chemoresponse. Armitage,[1] Adler,[2] and Kleene[3] in recent reviews describe how bacteria can provide large amounts of material and mutants for an elegant dissection of chemosensory transduction, and these authors make the point that bacteria can clearly be of use in the study of chemoreception. There are limits to the study of sensory transduction in bacteria because, while giant bacteria can be penetrated with electrodes,[4] they are not routinely studied with conventional electrophysiology. Eukaryotic unicells, on the other hand, not only can provide a large cell for penetration by multiple electrodes, but they also have excitable plasma membranes and display neuronal properties,[5] like those of a primary neuron in chemoreception.

Eukaryotic unicellular organisms can provide useful advantages for the study of chemoreception and, while they will not replace metazoan systems in the studies of taste and smell, they can contribute information that is likely to be common to all chemoreceptor cells. Many eukaryotic unicellular chemoresponses have been described, but the most commonly studied are the chemotaxis of the slime mold *Dictyostelium*[6] in its amoeboid stage; the chemotaxis of leukocytes[7] in their response to wound and infection, and chemokineses of ciliates. Among the ciliates, *Tetrahymena* responds to peptides and amino acids,[8] which might indicate food; *Blepharisma* responds to pheromones to become mating reactive and to locate potential mates.[9] The most systematically studied ciliate chemoresponse, however, is that of *Paramecium*. The following is a description of one unicellular system, *P. tetraurelia,* to demonstrate its usefulness in studying receptor cell function.

16

METHODOLOGY AVAILABLE TO STUDY *PARAMECIUM*

Paramecia are grown inexpensively in mass culture axenically[10] or in bacterized rye extracts.[11] Kilogram amounts of cells can be harvested,[10] but generally there is sufficient material in one to three liters of culture to study membrane proteins.

Paramecia of several species have been studied using standard electrophysiological techniques[12] including voltage clamping.[13] Patch clamping is now being applied to the study of ion channels in the *Paramecium* membrane.[14]

The surface membrane of *Paramecium* is almost evenly divided between that covering the cilia and that covering the rest of the cell body. Cilia and their membrane can be harvested separately from the cell body membrane[15] and, likewise, the cell body membrane can be harvested free of most of the cilia.[16]

Calcium-sensitive fluorescent dyes, such as Quin-2/AM,[17] are taken into the cells and the ester bond cleaved to trap the dye in its impermeable, calcium-sensitive form inside the cell. Likewise, diacetyl-carboxyfluorescein, a pH indicating fluorescent dye,[18] is taken into the cells and deesterified.

Paramecia can be mutated using chemicals or radiation as mutagens.[19] Kung has shown that a large variety of mutants can be isolated on the basis of their altered swimming behavior and many of these mutants have altered ion channels or gates.[20] It is a remarkable feat for one laboratory to mutate and study such a large number of genes for membrane functions. This is possible in part because the components of *Paramecium* swimming behavior (frequency of turning and speed of forward swimming) are very clear and easily observed, and these components happen to be under membrane electrical control. Frequency of turning is a function of the frequency of calcium action potentials,[12] which transiently reverse the ciliary beat and cause a jerky turn, and speed of swimming is a function of frequency and angle of ciliary beating, which is controlled by the membrane potential.[21]

It has been possible to isolate mutants with specific deficits in chemoresponse behavior. For example, there are three complementation groups of mutants that specifically cannot respond to folate but that are normal in response to other chemical stimuli.[22] These mutants can be genetically characterized and show assortment typical of mutations in single genes. This is significant because the loss of single gene products that are components of the chemosensory transduction pathway makes it possible to dissect the pathway component by component. Revertants of the original mutants have also been isolated.[23] Mutants that have reverted their phenotype back to the wild type by a second-site mutation allow an analysis of a complex pathway that very often is not possible by other means. The second mutation may have no discernable phenotype by itself, but because it alters a chemoreception mutant phenotype, it is certain that the second gene product somehow participates in or affects the chemosensory pathway. (See Huang *et al.*[24] for an example of the use of revertants in *Chlamydomonas.*)

It is possible to microinject cytoplasm, RNA, or DNA into paramecia and to follow up with electrophysiological or behavioral tests in order to test the "curing" effects of the injected component on mutants or to perform quasi-complementation tests.[25] Microinjection will make it possible to transform paramecia for manipulation of cloned genes. The molecular genetics of *Paramecium* is still relatively new, and it will be necessary to meet some special problems with codon usage as its development proceeds.[26]

REVIEW OF CHEMORECEPTION IN *PARAMECIUM*

Paramecia are attracted to folate, acetate, cAMP, and other chemicals, which probably signify food. The cells respond by a kinetic mechanism that results in fast, smooth swimming up gradients of attractant.[27] Through the use of mutants, we have shown that paramecia modulate both frequency of turning and swimming speed, which add up to the net population attraction or dispersal.[27,28] From the elegant work of Jennings, Eckert, Naitoh, Kung, Machemer, and others, we know that turning and speed depend upon membrane potential.[20] An understanding of the physiology of these individual components of *Paramecium* behavior allowed us to make predictions about the effect of chemical stimuli on membrane potential during the complex swimming behavior of chemoresponse. These predictions were borne out by direct electrophysiological measurements.[29] Put simply, attractants generally hyperpolarize and repellents generally depolarize; therefore, even a complex behavior can be explained in terms of relatively simple components of *Paramecium* physiology.

How then does an external chemical stimulus affect membrane potential, which controls swimming behavior? One accepted paradigm is that receptors bind the stimulus and that this binding is transduced into internal chemical and electrical information, that is, the second and third messengers. We set about identifying receptors by first examining the number and affinities of surface binding sites, which should include the receptor. For example, binding studies with one of the available stimuli, [^{3}H]folate, indicated that there are saturable, specific binding sites on the *Paramecium* cell. Binding to these sites is reduced to a low-level, nonspecific binding in a chemoreception mutant.[30] Fluorescein-folate can also be used to study whole cell binding.[31] When dyed with fluorescein-folate, normal cells show intense fluorescence that is specific for folate, while mutant cells show little discernable fluorescence above autofluorescence.[31] Revertants that have recovered the wild-type phenotype also have recovered fluorescence in fluorescein-folate.[32] The binding of the folate conjugate is most likely to be to the outside of the cell for the following reasons: (1) the mutant has lost surface [^{3}H]folate binding capacity and (2) cells preincubated with detergent Triton X-100 to purposely permeabilize cells to the dye show an increase in fluorescence.[32] The use of fluorescein-folate has been adapted to a small scale so that individual clones in microtiter wells can be screened for binding mutants using a dissecting, epifluorescence microscope.[31]

Our next step in receptor characterization is the identification of binding proteins from the cell membrane. The *Paramecium* folate binding proteins of interest are to be found primarily on the cell body and not the ciliary membrane[30,33] and these binding sites are not likely to be evenly distributed down the cell as indicated by local perfusion during electrophysiological recording.[33] The dearth of specific binding sites on cilia does not invalidate *Paramecium* as a model for chemoreception. Cilia may still play a role in *Paramecium* chemoreception because they have an adenylate cyclase that may be part of the chemosensory pathway involved after the change in membrane potential is elicited by the stimulus (see below), and not all chemoreceptor systems, even in olfaction systems, involve cilia.[3]

We have identified folate binding proteins from the cell body membrane by affinity chromatography and determined that five of these are surface exposed, as expected for a receptor.[34] Like many other external chemoreceptor systems, the binding proteins involved in chemoreception should be of relatively low affinity; therefore, to circumvent problems of affinity chromatography of weakly binding ligands, we turned to the sensitive method of immunodetection to identify folate binding proteins.[35] The rationale was that the chemoreceptor would be among the membrane proteins to which we could cross-link folate because cross-linking folate onto whole cells specifically inhibited attraction to folate, but not to acetate[36] (see TABLE 1).

Folate binding proteins from isolated cell body membrane that was cross-linked with folate were identified by electroblotting proteins from polyacrylamide gels onto nitrocellulose and immunodetection of the nitrocellulose with anti-folate antibodies.[35] Currently, we are cataloguing cross-linked proteins, which compare well with proteins identified by affinity chromatography, examining cross-linked proteins from chemoreception mutants, and cross-linking folate to intact cells instead of isolated membranes.

Folate is not the only stimulus that can be used in binding studies and for which there should be a receptor. Cyclic AMP is an attractant that shows saturable, specific binding to whole cells.[37] Affinity chromatography has identified only one protein that is a specific cAMP binding protein (approximately 48,000 molecular weight) from the cell body membrane (Gagnon and Van Houten, unpublished results). Nucleotides present the advantage of using [^{32}P]azido compounds that are photoreactive and will covalently cross-link cAMP to its binding site *in situ* and thereby alleviate some of the problems associated with affinity chromatography. Preliminary [^{32}P]azido-cAMP data identify one integral membrane protein of 48,000 molecular weight that is labeled

TABLE 1. Immunodetection for the Identification of Folate Binding Proteins

| | I_{che} | |
Stimulus	Cross-Linked Cells	Control Cells
2.5 mM K$_2$ folate	0.54 ± 0.06	0.72 ± 0.06
5 mM K-OAc	0.61 ± 0.06	0.67 ± 0.06

[a] Data are averages of 12 experiments for folate tests and six experiments for acetate tests ± one standard deviation. I_{che} is an index of chemoattraction. $I_{che} > 0.5$ indicates attraction; < 0.5 indicates repulsion.

on intact cells. We are pursuing the identification of the cAMP binding proteins using both normal cells and mutants that are not normally attracted to cAMP.

Binding to receptor is transduced into a change in membrane potential. The attractant-induced hyperpolarization has been studied using conventional electrophysiology, including voltage clamping. There is no obvious reversal potential and no dependence on external K or Na.[38] There are permeability studies and work with a mutant[39] (restless, courtesy of E. Richard) that point to a role for internal calcium that is not voltage dependent. Temperature affects the size of the hyperpolarization slightly, and we are inclined to consider electrogenic pumps in the production of the hyperpolarization. Studies of external pH rule out the attractant entering the cell, dissociating and activating the Na-H antiporter or other H exchange. Amiloride, a diuretic that blocks the Na-H antiporter, does block chemoresponse in Na solutions,[40] but a direct role for a Na-H exchange in the chemosensory transduction pathway needs further examination.

Receptor binding should elicit second and third messengers, which in the case of *Paramecium* must account for the change in membrane potential, change in ciliary beating, and adaptation. The first messengers to come to mind are: internal calcium, internal pH, cyclic nucleotides, and IP$_3$. Indirect evidence exists for a role for calcium in chemoreception, and we are examining internal calcium movements using not only electrophysiology but also calcium-sensitive fluorescent dyes.[17] Quin-2, for example, can be loaded into the cells and gives a fluorescence signal large enough to be useful without causing extreme buffering of internal calcium with the consequent destruction of the chemoresponse. The fluorescence signal is not from dye leaking out of the cell or trapped between cilia. Quin-2-loaded cells are centrifuged to remove them from their surrounding buffer and resuspended in fresh buffer. The fluorescence signal from the supernatant is insignificant, but the cells show a large fluorescent signal. Cells incubated in Quin-2/AM for < 1 minute and not given a chance to take up dye show little fluorescence, indicating that the fluorescent signal from cells loaded with Quin-2/AM is from internalized dye and there is little trapping of dye between cilia.

Like calcium, internal pH can be examined with permeant fluorescent dyes,[18] which cells convert to a pH-sensitive form by de-esterifying the compounds, such as diacetyl-carboxy fluorescein (Van Houten and Preston, unpublished results). These dyes promise to be useful in sorting out second messenger functions particularly since changes in internal Ca are likely to be accompanied by concomitant changes in pH_i.

As pointed out by Moran,[41] the sensing of chemical stimuli must subsequently be translated into response. Cyclic nucleotides may be the second messengers that carry out this translational function for *Paramecium*. Cyclic nucleotides have been implicated in ciliary beating control.[42-44] In particular, increased levels of cAMP are associated with hyperpolarization;[44] therefore, the attractant-induced hyperpolarization may elicit a change in cyclic nucleotide levels, which, in turn, affect ciliary beating and behavior. Levels of cyclic nucleotides can be examined by RIA and HPLC,[37] among other methods.[44] It is almost certain that cyclic nucleotides will have a role somewhere in the *Paramecium* chemosensory-response pathway and the study of this role has added significance in light of the recently recognized role of adenylate cyclase in olfaction and taste.[45]

There is a recently renewed appreciation for the role of inositol phospholipids in receptor function.[46] Phosphoinositol lipids can be labeled and quantified in protozoa.[32] Lithium, which inhibits the recycling of inositol into phospholipids, profoundly affects chemoresponse;[36] however, in our hands, there is little change in phosphoinositol lipids with chemoreception stimulation, but direct measurement of IP_3 levels over time will be necessary to determine whether there is a role for IP_3 and its subsequent calcium release in *Paramecium* chemoreception. Another product of phospholipid degradation is diacylglycerol, which phorbol esters mimic in the activation of protein kinase C.[46] Phorbol esters do not affect chemoresponse, casting doubt on protein kinase C as a component of the chemosensory pathway.

There are other aspects of *Paramecium* chemoreception, such as adaptation and the effects of methylation,[47] that have not entered into this review. It is not our intent to be exhaustive but rather to give a feeling for the kinds of studies that can be accomplished with *Paramecium*. It is now time to move the *Paramecium* system into molecular genetics and cloning of genes for receptors and other pathway components as they are identified. This need to move into the molecular level is an appropriate and welcome development for chemoreception science in general.[48]

SUMMARY

The description of the chemoreception pathway in *Paramecium* is incomplete, but the technical means are available to study these pathways at the molecular level. The hallmark of ciliates is their versatility and their most important attribute is the availability of useful mutants. It is just this versatility and amenability to genetic manipulation that will move the study of *Paramecium* chemoreception forward and provide useful information for chemoreceptor cell function in general.

REFERENCES

1. ARMITAGE, J. P. 1987. Sensory transduction in flagellate bacteria. Ann. N.Y. Acad. Sci. This volume.
2. ADLER, J. 1987. How motile bacteria sense and respond to chemicals. Ann. N.Y. Acad. Sci. This volume.
3. KLEENE, S. J. 1986. Bacterial chemotaxis in vertebrate olfaction. Experientia **42:** 241-250.
4. FELLE, H., J. S. PORTER, C. L. SLAYMAN & H. R. KABACK. 1980. Quantitative measurements of membrane potential in *E. coli.* Biochemistry **19:** 3585-3590.
5. NAITOH, Y. 1982. Protozoa. *In* Electrical conduction and behavior in "simple" invertebrates.: 1-48. Clarendon Press. Oxford, England.
6. GERISCH, G. 1982. Chemotaxis in *Dictyostelium.* Ann. Rev. Physiol. **44:** 535-552.
7. SCHIFFMANN, E. 1982. Leukocyte chemotaxis. Ann. Rev. Physiol. **44:** 553-568.
8. LEVANDOWSKY, M., T. CHENG, A. KEHR, J. KIM, L. GARDNER, L. SILVERN, L. TSANG, G. KAI, C. CHUNG & E. PRAKASH. 1984. Chemosensory responses to amino acids and certain amines by the ciliate *Tetrahymena:* A flat capillary assay. Biol. Bull. **167:** 322-330.
9. MIYAKE, A. 1981. Physiology and biochemistry of conjugation in ciliates. *In* Biochemistry and Physiology of Protozoa. 2nd edition. S. H. Hutner & M. Levandowsky, Eds. **4:** 126-198. Academic Press. New York.
10. SCHÖNEFELD, U., A. ALFERMANN & J. E. SCHULTZ. Economic mass culturing of *Paramecium tetraurelia* on a 200-liter scale. J. Protozool. **33:** 222-225.
11. SONNEBORN, T. M. 1974. *In* Handbook of genetics. J. King, Ed. **4:** 469-592. Plenum Press. New York.
12. ECKERT, R. 1972. Bioelectric control of ciliary activity. Science **176:** 473-481.
13. SAIMI, Y., R. HINRICHSEN, M. FORTE & C. KUNG. 1983. Mutant analysis shows that the calcium-induced K current shuts off one type of excitation in *Paramecium.* Proc. Natl. Acad. Sci. **80:** 5112-5116.
14. MARTINAC, B., Y. SAIMI, M. C. GUSTIN & C. KUNG. 1986. Single-channel recording in *Paramecium.* Biophys. J. **49:** 167a.
15. ADOUTTE, A., R. RAMANATHAN, R. LEWIS, R. DUTE, K-Y. LING, C. KUNG & D. L. NELSON. 1980. Biochemical studies of the excitable membrane of *P. tetraurelia.* J. Cell Biol. **84:** 717-738.
16. BILINSKI, M., H. PLATTNER & R. TIGGEMANN. 1981. Isolation of surface membranes from normal and excytotic mutant strains of *P. tetraurelia.* Eur. J. Cell Biol. **24:** 108-115.
17. TSIEN, R., T. POZZAN & T. RINK. 1982. Calcium homeostasis in intact lymphocytes: cytoplasmic free calcium monitored with a new, intracellularly mapped fluorescent indicator. J. Cell Biol. **94:** 325-334.
18. RINK, T., R. TSIEN & T. POZZAN. 1982. Cytoplasmic pH and free Mg^{2+} in lymphocytes. J. Cell Biol. **95:** 189-196.
19. CHANG, S., J. VAN HOUTEN, L. ROBLES, S. LUI & C. KUNG. 1974. An extensive behavioral and genetic analysis of Pawn mutants of *P. aurelia.* Gen. Res. Camb. **23:** 165-173.
20. KUNG, C. & Y. SAIMI. 1982. The physiological basis of taxes in *Paramecium.* Ann. Rev. Physiol. **44:** 519-534.
21. MACHEMER, H. & J. dePEYER. 1977. Swimming sensory cells: Electrical membrane parameters, receptor properties and motor control in ciliated protozoa. Verch. Desch. Zool. Ges. **1977:** 86-110.
22. DiNALLO, M., M. WOHLFORD & J. VAN HOUTEN. 1982. Mutants of *Paramecium* defective in chemokinesis to folate. Genetics **102:** 149-158.
23. WHITE, M. & J. VAN HOUTEN. Unpublished results.

24. HUANG, B., Z. RAMANIS & D. J. L. LUCK. 1982. Suppressor mutations in Chlamydomonas reveal a regulatory mechanism for flagellar function. Cell **28**: 115-124.
25. HAGA, N., M. FORTE, R. RAMANATHAN, T. HENNESSEY, M. TAKAHASHI & C. KUNG. 1984. Characterization and purification of a soluble protein controlling Ca channel activity in *Paramecium*. Cell **39**: 71-78.
26. PREER, J., L. PREER, B. RUDMAN & A. BARNETT. 1985. Deviation from the universal code shown by the gene for surface protein 51A in *Paramecium*. Nature **314**: 188-190.
27. VAN HOUTEN, J. 1978. Two mechanisms of chemotaxis in *Paramecium*. J. Comp. Physiol. **127**: 167-174.
28. VAN HOUTEN, J. 1977. A mutant of *Paramecium* defective in chemotaxis. Science **198**: 746-748.
29. VAN HOUTEN, J. 1979. Membrane potential changes during chemotaxis in *Paramecium*. Science **204**: 1100-1103.
30. SCHULZ, S., M. DENARO, A. XYPOLYTA-BULLOCH & J. VAN HOUTEN. 1984. Relationship of folate binding to chemoreception in *Paramecium*. J. Comp. Physiol. **155A**: 113-119.
31. VAN HOUTEN, J., R. SMITH, J. WYMER, B. PALMER & M. DENARO. 1985. Fluorescein-conjugated folate as an indicator of specific folate binding to *Paramecium*. J. Protozool. **32**: 613-616.
32. VAN HOUTEN, J. Unpublished results.
33. PRESTON, R. R. & J. VAN HOUTEN. 1987. Localization of the chemoreceptive properties of the surface membrane of *Paramecium tetraurelia*. J. Comp. Physiol. In press.
34. SCHULZ, S., J. M. SASNER & J. VAN HOUTEN. 1987. Progress on the identification of the folate chemoreceptor of *Paramecium*. Ann. N.Y. Acad. Sci. This volume.
35. SASNER, J. M., S. SCHULZ & J. VAN HOUTEN. 1986. Anti-ligand antibody as a probe for a chemoreceptor in *Paramecium*. J. Cell Biol. **103**: 212a.
36. VAN HOUTEN, J. & R. R. PRESTON. 1987. Chemoreception: *Paramecium* as a chemoreceptor cell. *In* Advances in Experimental Biology and Medicine. Y. Ehrlich, Ed. Plenum Press. New York. In press.
37. SMITH, R., R. R. PRESTON, S. SCHULZ, M. L. GAGNON & J. VAN HOUTEN. 1987. Biochim. Biophys. Acta, submitted.
38. PRESTON, R. R. & J. VAN HOUTEN. 1987. Chemoreception in *Paramecium tetraurelia:* Acetate and folate-induced membrane hyperpolarization. J. Comp. Physiol. In press.
39. RICHARD, E., R. HINRICHSEN & C. KUNG. 1986. Single gene mutation that affects the K conductance and resting membrane potential in *Paramecium*. J. Neurogenet. **2**: 239-252.
40. VAN HOUTEN, J. & R. R. PRESTON. 1985. Effect of amiloride on *Paramecium* chemoreception. Chem. Senses **10**: 466.
41. MORAN, D. 1987. Evolutionary patterns in sensory receptors: An exercise in ultrastructural paleontology. Ann. N.Y. Acad. Sci. This volume.
42. SCHULTZ, J., R. GRÜNEMUND, R. VON HIRSCHHAUSEN & U. SCHÖNEFELD. 1984. Ionic regulation of cyclic AMP levels in *Paramecium tetraurelia in vivo*. FEBS Lett. **167**: 113-116.
43. MAJIMA, T., T. HAMASAKI & T. ARAI. 1986. Increase in cellular cyclic GMP level by potassium stimulation and its relation to ciliary orientation in *Paramecium*. Experientia **42**: 62-64.
44. GUSTIN, M., N. BONINI & D. NELSON. 1983. Membrane potential regulation of cAMP: Control mechanism for swimming behavior in the ciliate *Paramecium*. Soc. Neurosci. Abstr. **9**: 167.
45. LANCET, D. 1987. Toward a comprehensive analysis of olfactory transduction. Ann. N.Y. Acad. Sci. This volume.
46. NISHIZUKA, Y. 1984. Turnover in inositol phospholipids and signal transduction. Science **225**: 1365-1370.
47. VAN HOUTEN, J., J. WYMER, M. CUSHMAN & R. PRESTON. 1984. Effects of S-adenosyl-L-methionine on chemoreception in *Paramecium tetraurelia*. J. Cell Biol. **99**: 242a.
48. MARGOLIS, F., N. SYDOR, Z. TEITELBAUM, R. BLACHER, M. GRILLO, K. RODGERS, R. SUN & U. GUBLER. Molecular biological approaches to the olfactory system. Chem. Senses **10**: 163-174.

Ionic Mechanism of Generation of Receptor Potential in Frog Taste Cells[a]

TOSHIHIDE SATO,[b] YUKIO OKADA, AND
TAKENORI MIYAMOTO

Department of Physiology
Nagasaki University School of Dentistry
Nagasaki 852, Japan

A taste cell membrane is divided into two parts: the taste-receptor membrane covered with a superficial fluid (SF) and the basolateral membrane surrounded by an interstitial fluid (ISF).[1] It has been considered that both ionic permeability of receptor and basolateral membranes[2-5] and phase boundary potential[5,6] occurring at the receptor membrane surface may play important roles in generating the receptor potentials in response to taste stimuli;[7,8] however, an exact understanding of generation of the receptor potential in a taste cell is still lacking.

The purpose of this study is to examine the mechanisms underlying the generation of receptor potentials in a taste cell in response to salt, acid, and bitter stimuli by replacing SF and ISF with various modified salines.

METHODS

Bullfrogs anesthetized with urethane were used. Receptor potentials were recorded intracellularly from taste cells in the *in situ* tongue with 3 M KCl-filled microelectrodes. The effects of a particular ion in SF and ISF on the receptor potential were evaluated by comparing the control and test values obtained from different populations of taste cells. Perfusion of the tongue through its artery was carried out to exchange the normal ISF with various modified saline solutions. The tongue was adapted to frog saline (115 mM NaCl, 2.5 mM KCl, 1.8 mM CaCl$_2$, and 5 mM HEPES, pH 7.2), which was regarded as standard SF.

[a] This study was supported in part by Grants-in-Aid for Scientific Research Nos. 57480345, 59223011, and 60480402 from the Ministry of Education, Science and Culture of Japan.

[b] Address for correspondence: Toshihide Sato, Department of Physiology, Nagasaki University School of Dentistry, Nagasaki, 852, Japan.

23

RESULTS AND DISCUSSION

Salt Stimulation

After the normal SF covering the receptor membrane was replaced with 1 mM amiloride saline, the amplitude of receptor potential in response to 0.5 M NaCl was reduced by 30%. As has already been reported,[9,10] replacing the normal ISF with Na$^+$-free saline reduced the receptor potential by 40%. Interstitial 5 mM Co^{2+} saline had no effect on the receptor potential. These findings suggest that NaCl-induced receptor potential is concerned with a large contribution of ionic permeabilities of the

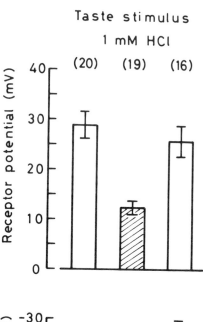

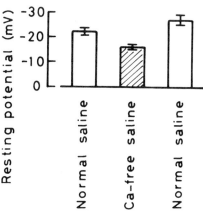

FIGURE 1. Effect of superficial Ca-free (1.8 mM Mg) saline on receptor potential (upper) in response to 1 mM HCl and resting potential (lower) in frog taste cells. Right columns denote the after-control under normal saline. In this and in FIGURE 2, numerals in parentheses mean the number of taste cells sampled.

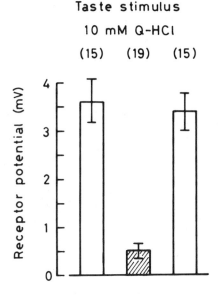

Taste stimulus

10 mM Q-HCl

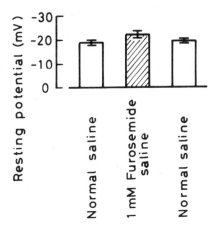

FIGURE 2. Effect of interstitial 1 m*M* furosemide saline on receptor potential (upper) in response to 10 m*M* Q-HCl and resting potential (lower). Right columns denote the after-control.

receptor and basolateral membranes and a small contribution of the phase boundary potential.

Acid Stimulation

As has been mentioned previously,[11] receptor potentials elicited by 1 m*M* HCl did not change even after any ion component in ISF was removed. The amplitude of the HCl-induced receptor potential was reduced by 65% and 50%, respectively, after both Na^+ and Ca^{2+} or Ca^{2+} alone were removed from the normal SF (FIG. 1). Elevating Ca^{2+} concentration in SF increased the response amplitude. Either Co^{2+} or Cd^{2+} in SF suppressed the receptor potential. Removing all ions from SF reduced the taste cell response by 65%. From these results, it may be concluded that HCl-induced receptor potential is concerned with a 65% contribution of receptor membrane Ca channel and a 35% contribution of phase boundary potential.

Bitter Stimulation

Removal of Ca^{2+} and Na^+ in normal SF did not affect the receptor potential induced by 10 mM quinine-HCl (Q-HCl), while decrease of Cl^- concentration in SF augmented it. Removing Na^+ or Cl^- in ISF reduced the response by 75%, as has been mentioned previously.[12] Furosemide (1 mM) in ISF greatly reduced the receptor response (FIG. 2). These findings suggest that 75% of the Q-HCl response may be concerned with a release of Cl^- through the receptor membrane, but the remaining 25% probably with the phase boundary potential.

REFERENCES

1. SATO, T., K. SUGIMOTO, Y. OKADA & T. MIYAMOTO. 1984. Ionic basis of resting membrane potential in frog taste cells. Jpn. J. Physiol. **34:** 973-983.
2. BEIDLER, L. M. 1967. Anion influences on taste receptor response. *In* Olfaction and Taste. T. Hayashi, Ed. Vol. II: 509-534. Pergamon Press. Oxford and New York.
3. BEIDLER, L. M. 1971. Taste receptor stimulation with salts and acids. *In* Handbook of Sensory Physiology. Vol. IV, Chemical Senses. L. M. Beidler, Ed. Part **2:** 200-220. Springer-Verlag. Berlin, Heidelberg, and New York.
4. AKAIKE, N. & M. SATO. 1976. Role of anions and cations in frog taste cell stimulation. Comp. Biochem. Physiol. **55A:** 383-391.
5. DeSIMONE, J. A. & S. PRICE. 1976. A model for the stimulation of taste receptor cells by salt. J. Gen. Physiol. **16:** 869-881.
6. KURIHARA, K., N. KAMO & Y. KOBATAKE. 1978. Transduction mechanism in chemoreception. Adv. Biophys. **10:** 27-95.
7. SATO, T. 1980. Recent advances in the physiology of taste cells. Prog. Neurobiol. **14:** 25-67.
8. SATO, T. 1986. Receptor potential in rat taste cells. *In* Progress in Sensory Physiology. H. Autrum, D. Ottoson, E. R. Perl, R. F. Schmidt, H. Shimazu & W. D. Willis, Eds. Vol. **6:** 1-37. Springer-Verlag. Berlin, Heidelberg, New York, and Tokyo.
9. SATO, T., K. SUGIMOTO & Y. OKADA. 1982. Ionic basis of receptor potential in frog taste cell in response to salt stimuli. Jpn. J. Physiol. **32:** 459-462.
10. MIYAMOTO, T., Y. OKADA & T. SATO. 1985. Ionic basis of salt-induced receptor potential in frog taste cells.: 185-188. Proc. 19th Jpn. Symp. Taste Smell.
11. OKADA, Y. & T. SATO. 1983. The role of Ca ions in receptor potential of frog taste cell in response to acid stimuli. J. Physiol. Soc. Jpn. **45:** 495.
12. OKADA, Y., T. MIYAMOTO & T. SATO. 1984. Receptor potential of frog taste cell in response to bitter substance.: 41-44. Proc. 18th Jpn. Symp. Taste Smell.

Toward a Comprehensive Molecular Analysis of Olfactory Transduction

DORON LANCET,[a] ZEHAVA CHEN,[a] ADINA
CIOBOTARIU,[a] FRITZ ECKSTEIN,[c] MIRIAM KHEN,[a]
JUDITH HELDMAN,[a] DOV OPHIR,[b] IRIS SHAFIR,[a]
AND UMBERTO PACE[a]

[a]Department of Membrane Research
The Weizmann Institute of Science
Rehovot, Israel

[b]Department of Otolaryngology
Kaplan Hospital
Rehovot, Israel

[c]Abteilung Chemie
Max Planck Institut für Experimentelle Medizin
D-3400 Göttingen, West Germany

INTRODUCTION

The last two decades of olfactory research have led to a much better understanding of the molecular mechanisms of odor recognition. Olfactory biochemistry was transformed from a largely speculative discipline into an established body of knowledge related to cellular transduction mechanisms. This development has its root in the early theories of Moncrief, Amoore, Beets, and others, suggesting that odorants are recognized by cell-surface receptors on the basis of their molecular stereochemistry.[1] Similar mechanisms are now known to be common to all aspects of biological recognition, from bacterial chemotaxis to cell-cell contacts in multicellular organisms.

CYCLIC AMP: AN OLFACTORY SECOND MESSENGER

It is now rather well demonstrated that odorants, which interact with yet unidentified olfactory receptor (OR) proteins in the membrane of olfactory cilia, activate an enzyme cascade that leads to the production of cyclic AMP (cAMP).[2-4] A GTP-binding protein (G-protein) of the stimulatory type (G_s) appears to mediate coupling between OR molecules and the enzyme adenylate cyclase,[2,5,6] the latter producing cAMP from ATP. cAMP is then believed to interact, directly or indirectly, with the ion channels responsible for membrane depolarization.[4,7] Such mechanisms are homologous to those found in visual transduction, where light activates the photoreceptor

27

protein rhodopsin. Rhodopsin then interacts with a G-protein (transducin) to activate a phosphodiesterase, an enzyme that breaks down cyclic GMP (cGMP), the visual second messenger.[8] Similarly, many hormones and neurotransmitters exert their action by modulating the activity of cyclic nucleotide-processing enzymes;[9] thus a rewarding aspect of current olfactory research is the notion that chemoreception shares molecular details, beyond stereospecific receptors, with other cellular mechanisms involved in transmembrane signaling. As in other systems, notably photoreception,[8] the adenylate cyclase-G-protein cascade could play a role in signal amplification. This may explain the extreme sensitivity of odorant detection in some species (cf. Pace and Lancet[6]).

INTERPLAY OF BIOCHEMISTRY AND ELECTROPHYSIOLOGY

A complete understanding of olfactory transduction can only be reached through a combined effort of biochemists and electrophysiologists. The first indication of a role of cAMP in olfactory transduction came through electrophysiological experiments. The second messenger cAMP, or its membrane penetrable analogue dibutyryl cAMP, can produce odorant-like electroolfactogram responses recorded under fluid.[10] Pretreatment with such compounds or with phosphodiesterase inhibitors (that increase intracellular cAMP) diminishes the odorant responses, presumably by saturating the transduction system.[10,11]

More recently, biochemical evidence has been obtained that confirms the role of cAMP in vertebrate olfactory transduction. Preparation of isolated olfactory cilia were shown to be highly enriched in the enzyme adenylate cyclase.[2,3,6] Importantly, various odorants were found to specifically activate the cAMP-producing ciliary enzyme, and they do so in a GTP-dependent manner,[2,3,6] ruling out nonspecific activation. The odorant concentration range required for adenylate cyclase activation corresponded to that reported in electrophysiological recordings.[4]

Additional corroboration is expected to arise from electrophysiological experiments, using the whole array of reagents provided by the emerging biochemical picture of olfactory transduction. These reagents can be applied when monitoring summated responses (electroolfactogram, olfactory nerve responses, bulbar potentials, transepithelial short-circuit currents), single cell responses (extracellular, intracellular) in intact epithelium or in isolated cells, as well as single-ion channel recordings in natural or reconstituted membrane patches. The potential modulatory reagents include: (a) cyclic AMP and its analogues; (b) guanine nucleotides, such as GTPγS and GDPβS, that respectively activate or inhibit the signal-coupling GTP-binding protein; (c) forskolin, that directly activates the adenylate cyclase catalytic subunit; (d) compounds that inhibit adenylate cyclase, such as di-deoxyadenosine; (e) phosphodiesterase inhibitors, such as isobutyl methyl xantine (IBMX) or theophylline; and (f) the catalytic subunit of cAMP-dependent protein kinase, or its specific inhibitor (PKI). Attention has to be paid to the question of membrane penetrability of these reagents.[13]

OLFACTORY G-PROTEIN

A defined polypeptide component of the proposed chemosensory transduction mechanism is olfactory G-protein.[2,4-6] This guanine nucleotide binding protein is enriched in olfactory cilia,[2,5] and has properties similar to those of G_s, the stimulatory G-protein of hormone reception.[9] Its α (heavy) subunit is a 42-kDa polypeptide, which is specifically modified (ADP-ribosylated) by cholera toxin and NAD.[2,6] It reacts with G_α-specific antisera,[5] and we recently found that it is capable of reconstituting G_s activity in the G_s-negative cell line S49 cyc⁻. We are currently examining the possibility that olfactory G_s, although similar to G_s in other membranes, is a distinct species. One of the approaches taken is to use novel analogues of GTP, synthesized by F. Eckstein and colleagues, which have thio substitutions at the α and β position (cf. Yamanaka *et al.*[12]). Each of these analogues has two optical isomers, and together they allow us to probe the stereochemistry of GTP-binding sites. We find that among the two α stereoisomers, the R configuration activates olfactory cilia adenylate cyclase much more efficiently than the S configuration. This is in contrast to brain or liver membranes, where both isomers appear to be equally weak activators. This difference may point to a unique configuration of the GTP binding site of olfactory G-protein.

OLFACTORY RECEPTOR PROTEIN(S)

The family of protein receptors believed to underlie odor recognition at the surface of different chemosensory neurons is yet to be identified and characterized (cf. refs. 4,7,13,14). Our strategy has been an "indirect" one: to approach the receptors by their general protein-chemical properties, and by learning about the transduction mechanism to which they are coupled.[4,15] It was argued that well-characterized transductory components could be used as receptor probes.[6,13]

Mapping the specific proteins in the frog isolated olfactory cilia preparation,[15,16] we have identified a surface glycoprotein of 95 kDa (gp95), which has several unique properties conforming with those expected for olfactory receptor (OR) molecules.[15,18] Polypeptide gp95 is specific to olfactory cilia with respect to respiratory cilia, brain membranes, and several other control membrane preparation.[15–18] It is also enriched in the cilia compared to deciliated epithelial membranes, suggesting some mode of segregation into the sensory organelle.[17] The abundance of gp95 (but not of many other, more minor ciliary proteins)[15] agrees with that expected for the specific proteins seen as freeze-fracture intramembranous particles in olfactory cilia.[19] Furthermore, gp95 is the only major ciliary polypeptide that behaves as a transmembrane (integral membrane) protein,[17] again in conformity with its being part of a signal transfer machinery, and with its corresponding to intramembranous particles. gp95 appears to have glycoprotein homologues in the 85-95-kDa range in several other vertebrate species.[17]

Recently we have been able to obtain results suggesting a functional role for gp95. We used the lectin wheat germ agglutinin (WGA), which binds specifically to N-acetyl glucosamine (GlcNAc) and sialic acid carbohydrate residues on glycoproteins. gp95 is highly enriched in complex-type oligosaccharides that contain such sugar moieties, and in this respect it is also unique among the glycoproteins of the frog isolated olfactory cilia preparation.[15,18] When the lectin WGA is applied to the surface of frog olfactory epithelium, it is found to inhibit the electroolfactogram responses to several odorants, and this inhibition is reversed by the soluble monosaccharide GlcNAc.

In order to correlate this *in vivo* result with the biochemistry of transduction, we also examined the effect of WGA on odorant activation of adenylate cyclase in the test tube. Here, we used the odorant-sensitive enzyme as the first *in vitro* assay for olfactory activation. The lectin was found to inhibit the *in vitro* adenylate cyclase odorant response as well, and the inhibition was again reversed by GlcNAc. The correspondence of the *in vivo* and *in vitro* results is encouraging, both in terms of information on gp95, and regarding the validity of adenylate cyclase measurements as a functional olfactory-related assay.

Additional functional clues are obtained through studying the effect of 18.1, a monoclonal antibody specific to gp95.[17] This antibody, which immunofluorescently stains the olfactory epithelial ciliary surface[17] and immunoprecipitates gp95 from nonionic detergent extracts of frog olfactory cilia, is also found to quantitatively immunoprecipitate the ciliary adenylate cyclase activity, suggesting a connection between the major ciliary glycoprotein and an enzyme involved in chemosensory transduction.

FUTURE RECEPTOR STUDIES

Studying olfactory receptor (OR) molecules is probably the most important future challenge to olfactory research. The identification, isolation, and eventual molecular cloning of OR proteins could open up exciting possibilities for understanding animal and human olfaction; thus, it may become possible to identify all the genes that code for the different OR proteins. Specific anosmias, genetic differences in olfactory sensitivity, and even individual odor preference could find their basis in population polymorphisms through DNA mapping, as has recently been done for the genes coding for the color vision photoreceptor proteins.[20] Judging by the findings in color vision, it is anticipated that since there may be at least several dozen OR genes,[1,4] individual differences should be the rule rather than the exception. We anticipate that this complexity of olfactory genetics will lend itself to scrutiny through molecular genetic tools.

Our efforts to isolate and clone OR protein proceed along two routes. One approach is based on the notion that gp95 is a promising OR candidate, hence is worth further study. We have purified gp95 to apparent homogeneity by a procedure that includes Triton X-114 extraction of the frog olfactory cilia preparation, Triton X-114 phase separation (that isolates integral membrane proteins), followed by WGA-Sepharose chromatography and elution with GlcNAc. Cyanogen bromide fragments have been prepared and are currently analyzed by gel chromatography and by reverse-phase high-performance liquid chromatography. Partial amino-acid sequence of such fragments is expected to direct the synthesis of oligonucleotide probes, to be used for identifying cDNA clones coding for gp95 sequence(s). If gp95 corresponds to OR proteins, it is expected that many gp95-related homologous cDNA clones will be found. This is because OR molecules are anticipated to have multiplicity corresponding to diverse odorant specificities.[4,15,21]

The second approach assumes no knowledge of OR molecules, and seeks to identify them through their ability to interact with the known transduction components: adenylate cyclase and G-protein. Reconstitution methods are used to study the interaction of ciliary components with G-protein and adenylate cyclase from olfactory cilia or from other membranes. One assay established in our laboratory used G_s from turkey erythrocytes, which is known to bind GTPγS only when activated by agonist-bound β-adrenergic receptor. It is anticipated that if heterologous interactions occur, only odorant-activated OR will induce GTPγS incorporation into G_s. The activated, solubilized turkey erythrocyte or olfactory cilia G_s are assayed for their ability to activate adenylate cyclase in membranes from the G_s-negative cell line cyc⁻. Activation of adenylate cyclase by G_s can thus be used as an assay to identify OR molecules in detergent extracts of the olfactory cilia preparation.

FROM CYCLIC AMP TO ION CHANNELS

Little is known about the nature of the ion channels responsible for the odorant-induced membrane depolarization in the chemosensory neurons. There are two possible ways by which cyclic nucleotides could affect such channels: by initiating protein phosphorylation[23] or through a direct modulation effect.[8] Aspects of the first route have been studied by us in some detail.[22] We have shown that cAMP-dependent protein kinase is present in the olfactory cilia preparation, measured through the phosphorylation of an exogenous substrate, histone. In addition, we identified several ciliary polypeptides, notably a 24-kDa phosphoprotein, pp24, whose phosphorylation is modulated by cAMP. These phosphoproteins could be ion-channel components or could interact with such functional structures.

Direct cyclic nucleotide modulation of ion channels has only been described in one case—vertebrate photoreception.[8] The intriguing possibility that cAMP directly affects the conductance of olfactory ion channels can be examined by single-ion channel recordings and other electrophysiological techniques.

CONCLUSION

The next few years of olfactory research should lead to a comprehensive molecular analysis of the mechanisms of odor recognition and possibly also of taste reception.[24] This may open the way to a rigorous and detailed description of the selectivity and diversity attributes of the chemosensory process. An understanding of the molecular transduction machinery could allow one to control molecular amplification parameters, hence could lead to reagents that diminish or enhance chemosensory sensitivity. Finally, a better understanding of the molecules that define the individuality of olfactory sensory neurons may help us to understand the cell-cell recognition processes that allow chemosensory axons to find their appropriate synaptic targets during development and regeneration.

REFERENCES

1. AMOORE, J. E. 1982. Odor theory and odor classification. *In* Fragrance chemistry: The science of the sense of smell. E. T. Theimer, Ed.: 27-76. Academic Press. New York, London.
2. PACE, U., E. HANSKY, Y. SALOMON & D. LANCET. 1985. Odorant sensitive adenylate cyclase may mediate olfactory reception. Nature **316**: 255-258.

3. SKLAR, P. B., R. H. ANHOLT & S. H. SNYDER. 1986. The odorant sensitive adenylate cyclase of olfactory receptor cells: Differential stimulation by distinct classes of odorants. J. Biol. Chem. **261:** 15538-15543.

4. LANCET, D. 1986. Vertebrate olfactory receptor. Ann. Rev. Neurosci. **9:** 329-355.

5. ANHOLT, R. R. H., S. M. MUMBY, D. A. STOFFERS, P. R. GIRARD, J. F. KUO & S. H. SNYDER. 1986. Transductory proteins of olfactory receptor cells: Identification of guanosine nucleotide binding proteins and protein kinase C. Biochemistry **26:** 788-795.

6. PACE, U. & D. LANCET. 1986. Olfactory GTP-binding protein: Signal transducing polypeptide of vertebrate chemosensory neurons. Proc. Natl. Acad. Sci. USA **83:** 4947-4951.

7. GETCHELL, T. V. 1986. Functional properties of vertebrate olfactory receptor neurons. Physiol. Rev. **66:** 772-817.

8. STRYER, L. 1986. Cyclic GMP cascade of vision. Ann. Rev. Neurosci. **9:** 87-119.

9. SCHRAMM, M. & Z. SELINGER. 1984. Message transmission: Receptor controlled adenylate cyclase system. Science **225:** 1350-1356.

10. MINOR, A. V. & N. L. SAKINA. 1973. Role of cyclic adenosine-3',5'-monophosphate in olfactory reception. Neurofysiologya **5:** 415-422.

11. MENEVSE, A., G. DODD & T. M. POYNDER. 1977. Evidence for the specific involvement of cAMP in the olfactory transduction. Biochem. Biophys. Res. Commun. **77:** 671-677.

12. YAMANAKA, G., F. ECKSTEIN & L. STRYER. 1985. Stereochemistry of the guanyl nucleotide binding site of transducin probed by phosphorothioate analogues of GTP and GDP. Biochemistry **24:** 8094-8101.

13. LANCET, D. 1986. Molecular components of olfactory reception and transduction. *In* Molecular Neurobiology of the Olfactory System. F. L. Margolis & T. V. Getchell, Eds. In press. Plenum Press. New York.

14. SNYDER, S. H., P. B. SKLAR & J. PEVSNER. 1986. Olfactory receptor mechanisms: Odorant binding protein and adenylate cyclase. *In* Molecular Neurobiology of the Olfactory System. F. L. Margolis & T. V. Getchell, Eds. In press. Plenum Press. New York.

15. CHEN, Z. & D. LANCET. 1984. Membrane proteins unique to olfactory cilia: Candidates for sensory receptor molecules. Proc. Natl. Acad. Sci. USA **81:** 1859-1863.

16. CHEN, Z., U. PACE, J. HELDMAN, A. SHAPIRA & D. LANCET. 1986. Isolated frog olfactory cilia: A preparation of dendritic membranes from chemosensory neurons. J. Neurosci. **6:** 2146-2154.

17. CHEN, Z., U. PACE, D. RONEN & D. LANCET. 1986. Polypeptide gp95: A unique glycoprotein of olfactory cilia with transmembrane receptor properties. J. Biol. Chem. **261:** 1299-1305.

18. CHEN, Z., D. OPHIR & D. LANCET. 1986. Monoclonal antibodies to ciliary glycoproteins of frog olfactory neurons. Brain Res. **368:** 329-338.

19. MENCO, B. PH. M. 1980. Qualitative and quantitative freeze-fracture studies on olfactory and nasal respiratory epithelial surfaces of frog, ox, rat and dog. II Cell apices, cilia and microvilli. Cell Tissue Res. **211:** 5-29.

20. NATHANS, J., D. THOMAS & D. S. HOGNESS. 1986. Molecular genetics of human color vision: The genes encoding blue, green, and red pigments. Science **232:** 193-210.

21. BOYSE, E. A., G. K. BEAUCHAMP, K. YAMAZAKI, J. BARD & L. THOMAS. 1982. Chemosensory communication: A new aspect of the major histocompatibility complex and other genes in the mouse. Oncodev. Biol. Med. **4:** 101-116.

22. HELDMAN, J. & D. LANCET. 1986. Cyclic AMP dependent protein phosphorylation in chemosensory neurons: Identification of cyclic nucleotide regulated phosphoproteins in olfactory cilia. J. Neurochem. **47:** 1527-1533.

23. LEVITAN, I. B. 1985. Phosphorylation of ion channels. J. Membr. Biol. **87:** 177-190.

24. STRIEM, B. J., U. PACE, U. ZEHAVI, M. NAIM & D. LANCET. 1986. Is adenylate cyclase involved in sweet taste transduction? Chem. Senses. **11:** 669.

Neural Derivation of Sound Source Location in the Barn Owl

An Example of a Computational Map

ERIC I. KNUDSEN

Department of Neurobiology
Stanford University of School of Medicine
Sherman Fairchild Science Building
Stanford, California 94305-5401

INTRODUCTION

The central nervous system must derive information about many features of the environment that are not represented topographically in the sensory periphery. The analysis of such features requires the evaluation of spatiotemporal patterns of activity from across receptor arrays. Often this analysis involves hierarchies of serial and parallel processing that give rise to higher order neurons tuned for the derived feature. In some instances, the brain creates a new topography based on a systematic ordering of neurons according to their tuning for the computed feature, an organization that I will refer to as a computational map.[1]

The location of a sound source in space is an example of a stimulus feature that must be computed from the spatiotemporal pattern of receptor activity: Sound frequency, and not source location, is represented topographically across the sensory epithelium of the cochlea. The auditory system derives information about the location of a stimulus from a variety of cues that are only indirectly related to the location of the sound source. The most important of these cues are binaural differences in the timing and intensity of sound and monaural cues resulting from direction-dependent spectral shaping by the head and external ears.

The spatial significance of an auditory cue is highly frequency-specific. For example, an interaural intensity difference of 6 dB at 2 kHz indicates different locations in space than the same interaural intensity difference at 8 kHz;[2,3] the same principle applies to interaural delay cues.[4,5] For this reason, the auditory system must evaluate localization cues in a frequency-specific manner. In addition, the spatial significance of any single cue value is ambiguous, since the same value can arise from many different locations. To resolve spatial ambiguity, the auditory system must evaluate several cues (the more the better) simultaneously and determine the location in space that is common to all of them. The process of sound localization thus involves two steps: (1) the determination of frequency-specific cue values, and (2) the association of particular sets of cue values with appropriate locations in space. Neural correlates of this process are found in the responses of high-order neurons that are tuned for sound source location and are organized into an auditory map of space. This paper describes briefly the characteristics of this map, the final anatomical and physiological steps that lead to it, and the properties that it shares with other computational maps in the brain.

A BIMODAL MAP OF SPACE IN THE OPTIC TECTUM

High-order neurons involved in the spatial analysis of sounds are found in the owl's optic tectum, the homologue of the mammalian superior colliculus.[6,7] The optic tectum is a sensorimotor structure that receives input from nearly all sensory modalities and commands movements that orient an animal to stimuli of interest. Because the optic tectum is involved with spatial orientation, it is reasonable that its sensory input should convey spatial information. When tested with a free-field acoustic stimulus, units in the optic tectum respond only when the source is located within a limited region of space, or receptive field, and respond best when the source is located within a particular portion of the receptive field called the best area. The best area of a unit does not change with the intensity or the spectral or temporal properties of the sound; however, the best areas of different units vary systematically with the positions of the units in the tectum, thereby forming a map of auditory space. The map, based on best areas, encompasses most of the contralateral hemifield, extending from approximately 115° contralateral to 20° ipsilateral and from 40° above to 80° below the horizontal plane. Regions of space directly in front of the animal are represented in a disproportionately large portion of the map. Thus, by the level of the optic tectum, the representation of auditory space, which at the periphery is based on the relative timing and intensity of binaural input, has been transformed into neuronal best areas and a map of space.

In addition to being sharply tuned for sound source location, most of these neurons also respond to visual stimuli.[6] Visual receptive fields are smaller than auditory fields, but they align reliably with auditory best areas, indicating that the auditory and visual maps of space are in register.

AN AUDITORY MAP OF SPACE IN THE INFERIOR COLLICULUS

The auditory map of space that appears in the optic tectum is already present in a lower order nucleus, the inferior colliculus. Small injections of HRP into the optic tectum[8] reveal a point-to-point projection from the external nucleus of the inferior colliculus (ICx). Electrophysiological recordings in the ICx demonstrate that units there are tuned for space and are organized into a map of space, properties that are remarkably similar to those observed in the optic tectum.[9,10] The notable exception is that units in the ICx do not respond to visual stimuli. Experiments that employ two sound sources (one to drive the unit and the other to test the influence of sound sources outside of the receptive field) reveal further that the receptive fields of ICx units have an antagonistic, inhibitory surround reminiscent of visual and somatosensory receptive fields in many parts of the brain.[11]

The binaural bases for the spatial selectivity of these units has been studied by Moiseff and Konishi[12] who, after measuring a unit's receptive field properties with a free-field source, inserted loudspeakers into the owl's ear canals and tested the unit's sensitivity to variations in individual interaural parameters. These experiments demonstrate that the space specificity of ICx units is derived from their absolute requirements for the simultaneous occurrence of particular interaural delays *and* interaural intensity differences. Interaural differences that deviate by more than a few tens of microseconds or by a few decibels from these values inhibit the unit. The tuning of a unit to interaural timing largely determines the azimuthal properties of its receptive field: the optimal interaural delay for driving a unit predicts the azimuthal location of its best area, and the selectivity of the unit for that delay corresponds with the width of its receptive field. The elevation and vertical dimension of a unit's receptive field are determined by the sensitivity of the unit to interaural intensity differences.[1] These data indicate that the spatial tuning of ICx units results from a convergence of timing and intensity information from the two ears (which we know is processed in parallel ascending pathways),[14] and that the map of space is established on the basis of a systematic progression in the tuning for interaural delay along the rostrocaudal axis and a variation in the tuning for interaural intensity difference along the dorsoventral axis of the ICx.

In contrast to their sharp tuning for source location, ICx units tend to be broadly tuned for frequency.[9] Moreover, these units are not organized tonotopically, in marked contrast to the properties of the units from which they derive their inputs.

INTEGRATION OF INFORMATION FROM DIFFERENT FREQUENCIES

The source of input to the ICx is the central nucleus of the inferior colliculus (ICc), a nucleus containing units organized according to frequency tuning instead of spatial tuning. A small injection of HRP into the ICx labels several groups of neurons in the ipsilateral ICc.[15] Most of the neurons are located near the border of the ICc with the ICx, but a few are scattered throughout the nucleus. The dispersion of sources of input from the tonotopically organized ICc indicates that the pathway from the ICc to the ICx converges information from a wide range of frequencies onto single neurons.

Electrophysiological measurements agree with this interpretation.[9] Units in the ICc respond well to tones and are narrowly tuned for frequency. In addition, the ICc is clearly tonotopically organized: Units sensitive to low frequencies are located dorsally, while those sensitive to high frequencies are located ventrally. A progression of frequency tuning along the dorsoventral axis of the ICc is gradual and unerring, and represents the most conspicuous (though not the only) determinant of functional organization in the ICc.

The receptive fields of units in the ICc are not space-specific.[9] Instead, these units either (1) respond to all possible source locations (omnidirectional units); (2) respond to sounds from a certain location, but have receptive fields that expand dramatically with stimulus intensity (space-preferring units); or (3) have two to four excitatory fields separated by areas from which sounds either are inhibitory or are less effective in driving the unit (multifield units).

In the region of the ICc that borders on the ICx and provides most of the input to the ICx, only space-preferring and multifield units are found.[9] The representation of space by these response types is ambiguous at the unit level in that the receptive fields of space-preferring units vary with intensity, and multifield units respond to several discrete directions in space. Thus, auditory space is not mapped in the ICc.

These units do, however, respond differentially to sound location in ways that cannot be explained simply by the directionality of the ears. Rather, these units must be sensitive to certain binaural parameters of the stimulus. Binaural sensitivity is manifested most clearly in the responses of multifield units. Among these units, the angular separation of the excitatory receptive fields correlates with the characteristic frequency of each unit: the excitatory fields represent those directions in space that give rise to equivalent interaural phases of the unit's characteristic frequency.[16] For example, a multifield unit with a characteristic frequency of 8 kHz may be excited by stimuli in up to three discrete regions of space. The period at 8 kHz is 125 microseconds. It is known from direct measurements that for every degree of change in azimuth of the sound source there is about a 2-microsecond change in interaural delay.[12] Therefore, a 62.5° change in azimuth results in a 125-microsecond change in interaural delay, exactly one period of an 8-kHz signal. Thus, directions in space separated by 62.5° in azimuth will give rise to the same interaural phase at 8 kHz. The stimulus locations that elicit peak excitatory responses from a multifield unit with an 8-kHz characteristic frequency are spaced approximately 60° apart. This correlation, which holds for all multifield units studied, indicates that these units respond to a particular interaural phase difference at their characteristic frequencies. The spatial ambiguity of their responses arises because a given phase difference can result from different locations. Thus, multifield and space-preferring units in the ICc apparently select for particular frequency-specific binaural cues, but they do not code for space per se.

NEURAL COMPUTATION OF SPACE

On theoretical grounds, the evaluation of spatial cues must be carried out in frequency-specific channels. This is because the head and ears alter sound intensities and phases differently depending on frequency. Consequently, the correlation of interaural phase or intensity difference cues with directions in space varies greatly with frequency; however, a frequency-specific interaural difference cue, by itself, is spatially ambiguous. At best, it restricts the source to a curved surface in space (known to psychophysicists as a "cone of confusion"); at worst, it limits the source to one of a number of possible curved surfaces. To eliminate spatial ambiguity, several cues from different frequency bands must be combined. When enough cues are evaluated simultaneously, only one region in space will be consistent with all of them.

This theoretical sequence is manifested in the transformation in the unit-response types from the ICc to the ICx. Units in the ICc are narrowly tuned for frequency and are sensitive to one or possibly two interaural difference cues. As a consequence, the receptive fields are spatially ambiguous. In the ICx, the integration of multiple binaural cues from a wide range of frequencies eliminates the spatial ambiguities inherent in individual cues.

The map of space generated in the ICx manifests a transition from the encoding of binaural cues within a framework of tonotopic organization to a representation of sound source location within a newly derived, spatiotopic organization. Once spatial information is encoded in this form, it can be readily integrated with spatial information from other sensory systems. This integration occurs in the optic tectum. Auditory and visual inputs sensitive to the same regions in space converge onto single cells, creating a map of space in which neuronal activation signifies not a particular sensory modality, but the location of a stimulus independent of modality. This physiological merger of modalities in the optic tectum indicates that the transformation of auditory spatial information from timing and intensity differences into a topographic representation of space has been completed.

CONCLUSION

As described in this paper, the brain can create topographic representations of sensory features that are derived from spatiotemporal patterns of receptor activity. Such computational maps have been observed frequently in the auditory and visual pathways, and they may occur commonly in the brain (for a review see Knudsen et al.[1]). Properties of the auditory space map that are shared by other computational maps include the following:

1. Parameters that are mapped are biologically important to the species and are ones that must be evaluated quantitatively with high precision. The maps are most differentiated in species that rely heavily on the evaluation of the parameter for survival.
2. The tuning of neurons for a computed parameter is broad relative to the total range of the map. As a consequence, a stimulus will activate neurons across a large portion of the map. However, because the tuning curves of neurons are peaked, high-resolution information about the value of the parameter is contained in the relative response rates of neurons and is represented by the location of peak activity within the map.
3. The relative magnification of the representation of parameter values can vary within computational maps. Where such anisotropies occur, they correlate with behavioral performance. (For example, the greatly expanded representation of frontal space in the owl's auditory space map corresponds to a region of exceptionally high localization accuracy and precision; Knudsen et al.[1]). Such anisotropies do not arise as a result of variations in receptor densities (as do analogous anisotropies in maps of the sensory epithelium), but instead are created by the central nervous system through a differential scaling of the algorithm that generates the map. The appearance of such highly ordered representations of derived information in various neural circuits suggests that computational maps offer a strategy for information processing that is needed in many situations.

REFERENCES

1. KNUDSEN, E. I., S. DU LAC & S. D. ESTERLY. 1987. Computational maps in the brain. Ann. Rev. Neurosci. **10:** 41-65.

2. SHAW, E. A. G. 1974. Transformation of sound pressure level from the free field to the eardrum in the horizontal plane. J. Acoust. Soc. Am. **56:** 1848-1861.
3. KNUDSEN, E. I. 1980. Sound localization in birds. *In* Comparative Studies of Hearing in Vertebrates. A. N. Popper & R. R. Fay, Eds.: 287-322. Springer. New York.
4. KUHN, G. F. 1977. Model for the interaural time differences in the azimuthal plane. J. Acoust. Soc. Am. **62:** 157-167.
5. ROTH, G. L., R. K. KOCHHAR & J. E. HIND. 1980. Interaural time differences: Implications regarding the neurophysiology of sound localization. J. Acoust. Soc. Am. **68:** 1643-1651.
6. KNUDSEN, E. I. 1982. Auditory and visual maps of space in the optic tectum of the owl. J. Neurosci. **2:** 1177-1194.
7. KNUDSEN, E. I. 1984. Auditory properties of space-tuned units in owl's optic tectum. J. Neurophysiol. **52:** 709-723.
8. KNUDSEN, E. I. & P. F. KNUDSEN. 1983. Space-mapped auditory projections from the inferior colliculus to the optic tectum in the barn owl (*Tyto alba*). J. Comp. Neurol. **218:** 187-196.
9. KNUDSEN, E. I. & M. KONISHI. 1978. Space and frequency are represented separately in the auditory midbrain of the owl. J. Neurophysiol. **41:** 870-884.
10. KNUDSEN, E. I. & M. KONISHI. 1978. A neural map of auditory space in the owl. Science **200:** 795-797.
11. KNUDSEN, E. I. & M. KONISHI. 1978. Center-surround organization of auditory receptive fields in the owl. Science **202:** 778-780.
12. MOISEFF, A. & M. KONISHI. 1983. Binaural characteristics of units in the owl's brainstem auditory pathway: Precursors of restricted spatial receptive fields. J. Neurosci. **3:** 2553-2562.
13. KNUDSEN, E. I. & M. KONISHI. 1980. Monaural occlusion shifts receptive-field locations of auditory midbrain units in the owl. J. Neurophysiol. **44:** 687-695.
14. TAKAHASHI, T., A. MOISEFF & M. KONISHI. 1984. Time and intensity cues are processed independently in the auditory system of the owl. J. Neurosci. **4:** 1781-1786.
15. KNUDSEN, E. I. 1983. Subdivisions of the inferior colliculus in the barn owl (*Tyto alba*). **218:** 174-186.
16. KNUDSEN, E. I. 1984. Synthesis of a neural map of auditory space in the owl. *In* Dynamic Aspects of Neocortical Function. G. M. Edelman, W. M. Cowan & W. E. Gall, Eds.: 375-396. John Wiley and Sons. New York.

Neurophysiology and Neuroanatomy of the Olfactory Pathway in the Cockroach

J. BOECKH, K.-D. ERNST, AND P. SELSAM

Institute for Zoology
University of Regensburg
D-8400 Regensburg, West Germany

As an example of the "identified unit" approach to the olfactory pathway, we present a condensed summary of our attempts at stepwise analysis of the olfactory pathway of the American cockroach from the receptor cell level to descending neurons (cf. also Boeckh et al.[1]; Hildebrand and Montague[2]; Light[3]). As is the case with other insects, olfactory receptor cells of *Periplaneta* are found in sensilla with perforated hair walls on the antennae and other body appendages. According to their odor spectra, about 20 different types of cells have been recognized, some comprising 10^3, others $4 \cdot 10^4$ individuals. Their representatives come in standard sets beneath certain sensillum types. Two types respond to components of the female sexual pheromone (FIG. 1a, 4 and 2). The two others from the same sensillum (c.f. FIG. 1c) respond invariably to terpenoid compounds such as eugenol, while other substances constitute a variable "side" spectrum. Fifteen other types are characterized by wide and partially overlapping spectra including series of homologous alcohols, esters, aldehydes, and hydrocarbonic acids. Each type responds maximally to one or two characteristic compound(s). Another population of "nongrouped" cells respond with "individualistic" spectra. Except the distance-attractant component of the female pheromone (i.e. periplanone B), every odorant affects more than one receptor cell type and elicits a characteristic across-fiber pattern within the receptor population (FIG. 1a). Naturally occurring odors contain many compounds in certain ratios, and the receptors pick up "their" compounds according to their spectra. Because of overlap of receptor spectra and of compounds in the odors, several cell types respond in a rank order that is characteristic for each of these natural complex odors (FIG. 1a). By means of dye infusion into sockets of antennal sensilla, processes of the according receptor cells can be followed up to their terminals in certain glomeruli of the antennal lobes (FIG. 1c). Each of the 160 glomeruli is a characteristic unit, defined by its location, the "dendritic" tree of its output neurons, and probably a certain set of receptor terminals. It is separately represented in other parts of the brain via the projection of its output neuron(s). Local interneurons interconnect between glomeruli. They receive the majority of the receptor terminals (FIG. 1b).

Individual glomeruli have been successfully labeled for their location but only a few for their inputs and outputs. The sexually dimorphic macroglomerulus receives inputs from about 40,000 receptor cells for each, periplanone B and periplanone A. About 20 output neurons end up in about 1-2,000 terminals in the protocerebrum (FIGS. 1c, 2). This input convergence is probably the source of the extreme sensitivity

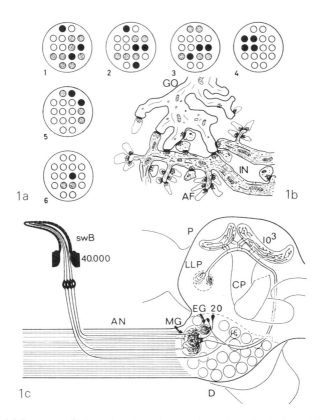

FIGURE 1. (a) Responses of 14 reaction types of antennal receptor cells to odors of (1) banana, (2) orange, (3) "ripe" meat, (4) calling female, (5) n-hexanol-1, (6) n-hexanoic acid-1 (after data of Sass[4] and Selzer[5]. (b) Receptor axons (AF), labeled by degeneration, terminate at identified deutocerebral interneuron (IN) or identified glomerular output neuron (GO). (c) Receptor cells underneath antennal swB sensillum terminate in macroglomerulus (MG) with 20 output neurons, in "eucalyptol" glomerulus (EG) with one output neuron, and in anonymous glomerulus. Output neurons reach calyces of corpora pedunculata (CP) with altogether 10^3 terminals, and lateral protocerebral lobe (LLP), AN. antennal nerve; D, deutocerebrum; P, protocerebrum.

of the neurons, which are activated if only a fraction of the receptor cells receive single molecules (FIG. 3a). Another glomerulus with inputs from the swB sensillum has an output neuron sensitive to terpenoid molecules ("eucalyptol" glomerulus, FIG. 1c). Output neurons of a few other glomeruli react rather specifically and constantly to only one aroma from our selection, for example, to "cheese" odor. The output neurons of the majority of identified glomeruli display unpredictable broad spectra with wide overlap, aroma-characteristic response profiles across this population are not prominent. An unexpected result was that output neurons of the very same glomeruli in different specimens showed odor spectra that were not only not identical but showed no closer relationship to one another than the spectra of nonhomologous glomeruli. It remains open whether this inconsistency is caused by differences in experimental conditions of the recorded cells or by variations between neurons' specificities from one specimen to the other.

Multimodal output neurons of the deutocerebrum react to various odors and to wind-induced fluttering movements of the antennal flagellum that are transmitted by campaniform sensilla at the margin of the annules.

While no "order" of deutocerebral projections is visible in the calyces of the corpora pedunculata (FIGS. 1c, 2), pheromone-sensitive neurons, mechanosensory (multimodal?) neurons, and neurons sensitive to "other" odors show different location and arrangements of terminals in the lateral lobe (LLP). Some "loop" neurons, leaving the CP and entering LLP or the CP calyces, and also "descending" neurons with

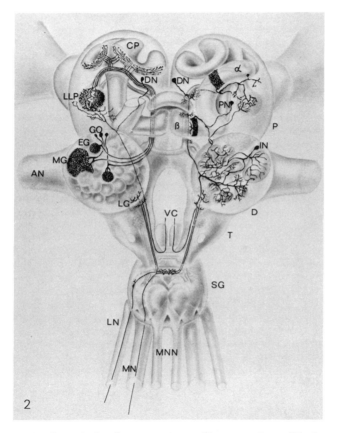

FIGURE 2. Brain of a male *Periplaneta americana* with protocerebrum (P), deutocerebrum (D), tritocerebrum (T), and suboesophageal ganglion (SG). AN, antennal nerve; α, α-lobe; β, β-lobe; CP, calyx of mushroom body; DN, descending neuron; EG, "eucalyptol" glomerulus; GO, glomerular output neuron with tract into P. IN, deutocerebral interneuron without axon; LLP, lateral lobe of P; LN, labial nerve and MN, maxillary nerve with chemosensory afferents terminating bilaterally in lobus glomeratus (LG); MG "macroglomerulus"; MNN, mandibular nerves; PN, protocerebral interneuron; VC, ventral cord.

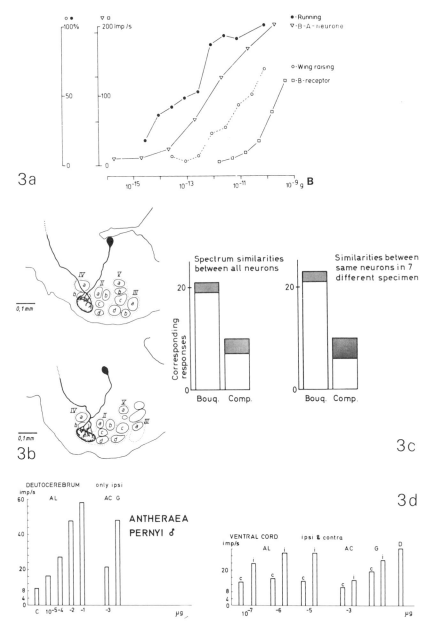

FIGURE 3. (a) Dosage-response characteristics for distant-attractant (periplanone B) in female sexual pheromone of receptor cells, central neurons and pheromone-aroused behavior of male *Periplaneta* (after Boeckh *et al.*[1]). Stimulus strength given in nanograms synthetic material at odor source. (b) Same deutocerebral glomerulus in two specimens of *Periplaneta* males. Camera lucida drawing of cobalt-injected, identified output neurons. Labels denominate identified glomeruli. (c) Similarity of spectra in numbers of similar responses to series of complex odors

(Bouquets) and single compounds (Comp.) between any pairs of glomeruli (left columns, $n = 70$) and homologous glomeruli in seven specimens (right columns). Blank areas, excitatory responses; shaded areas, inhibitory responses. (c) Responses of deutocerebral neuron and thoracic axon (probably of descending neuron) in a male saturniid moth to female pheromone. AC, hexadecadienylacetate; AL, hexadecadienal: synthetic pheromone components in μg at stimulus source. G, fresh female gland. C, in left diagram, response to control stimulus; c,i, in right diagram, stimulation at contralateral, ipsilateral antenna; D, rapid darkening of the visual field.

arbors in "olfactory" or "antennal" areas of the proto- and the deutocerebrum respond to odor stimulation at the antenna (FIG. 2). The combined anatomical and physiological identification of the latter has not yet been accomplished in *Periplaneta*. Large (probably descending) axons in the thoracic cord of a saturniid moth responded very sensitively to female pheromone and, different from deutocerebral units, react to bilateral olfactory inputs and to visual inputs (moving stripe pattern). They rapidly reach saturation with increased pheromone concentrations (FIG. 3C) and might account for simple turning movements toward the odor source.

REFERENCES

1. BOECKH, J., K. D. ERNST, H. SASS & U. WALDOW. 1984. Anatomical and physiological characteristics of individual neurones in the central antennal pathway of insects. J. Insect. Physiol. **30:** 15-26.
2. HILDEBRAND, J. G. & R. A. MONTAGUE. 1986. Functional organization of olfactory pathways in the central nervous system of *Manduca sexta. In* Mechanisms in Insect Olfaction. T. L. Payne, M. C. Birch & C. E. J. Kennedy, Eds.: 91-109. Clarendon Press. Oxford, England.
3. LIGHT, D. M. 1986. Central integration of sensory signals: An exploration of processing of pheromonal and multimodal information in *Lepidopteran* brains. *In* Mechanisms in Insect Olfaction. T. L. Payne, M. C. Birch & C. E. J. Kennedy, Eds.: 287-301. Clarendon Press. Oxford, England.
4. SASS, H. 1978. Olfactory receptors on the antenna of *Periplaneta americana*: Response constellations that encode food odors. J. Comp. Physiol. **128:** 227-233.
5. SELZER, R. 1983. On the specificities of antennal olfactory receptor cells of *Periplaneta americana.* Chem. Senses **8:** 375-395.

Organization of Olfactory Bulb Output Cells and Their Local Circuits

JOHN W. SCOTT

Department of Anatomy and Cell Biology
Emory University School of Medicine
Atlanta, Georgia 30322

Observations with Golgi methods, peptide neurohistochemistry, and horseradish peroxidase (HRP) transport demonstrated several classes of olfactory bulb mitral and tufted cells.[1,2] Some superficial tufted cells are interneurons lacking the long basal dendrites characteristic of other mitral and tufted cells.[1] The long basal dendrites of output cells have extensive interaction with inhibitory granule cells.[1,2,5] I will emphasize work relating output cells to the sublayers of olfactory bulb external plexiform layer (EPL).

These cells are traditionally divided into a group of mitral cells, with cell bodies in a single layer, and three groups of tufted cells (external, middle, and internal), with cell bodies in the EPL or periglomerular region. Instead of the traditional terminology, I will discuss the output mitral and tufted cells in terms of the EPL sublayers. Reconstruction of labeled mitral and tufted cells after small, extracellular injections of HRP into the EPL[3] suggested a subdivision of the EPL into three zones (superficial, intermediate, and deep). Each zone contained basal dendrites from separate sets of mitral and tufted cells. We saw two types of mitral cells and three groups of tufted cells. Type I mitral cells had basal dendrites exclusively in the deep zone, while type II mitral cell basal dendrites entered the intermediate zone. Tufted cells in the superficial zone and at the edge of the glomeruli distributed their basal dendrites only in the superficial zone. The remaining tufted cells had somata either in the intermediate zone or the deep zone, but distributed basal dendrites only in the intermediate zone.

Other investigators also support a sublaminar arrangement of the EPL. Staining with enkephalin antibodies and antibodies to glutamic acid, decarboxylase showed a division within the EPL that reflected staining of different populations of granule cells. Immunohistochemical staining for tyrosine hydroxylase and for cholecystokinin-labeled tufted cells concentrated in the outer part of the EPL and in the periglomerular region.[1,2] These observations were not directly compared to the laminar patterns of mitral and tufted cell dendrites.

We compared another marker, cytochrome oxidase (CO), with distributions of mitral and tufted dendrites.[4] The EPL stains more darkly for CO than the mitral cell layer or the periglomerular region, but there is a central band of darkest staining that corresponds to the intermediate zone. We confirmed this relation with small HRP injections in the EPL of olfactory bulbs processed for CO. Both CO and HRP can be seen in the same tissue. Reconstructions of labeled cells show that labeled mitral and tufted dendrites respect the boundaries marked out by the CO stain. The correspondence between the CO banding and dendritic layering remains even when the size of the bulb is reduced by neonatal occlusion of one nostril.[9] The CO staining of the bulb ipsilateral to the occlusion is less distinct with relatively thinner intermediate and superficial layers. HRP injections show that the type II mitral cells and tufted cells normally contributing basal dendrites to the intermediate zone continue to do so under this condition. These sublaminar basal dendritic patterns are important because they provide the potential that output cells in the different layers interact with different populations of interneurons. Existing data indicate that mitral and tufted cell basal dendrites and cell bodies synapse only with granule cell gemmules. While some granule cells innervate the whole EPL, two granule cell populations restricted spine distributions in the EPL.[1,3] Superficial granule cells extend dendritic shafts across the deep zone of the EPL with few or no branches and spines but show profuse branching and spines in the intermediate and superficial zones. These superficial granule cells probably do not synapse extensively with type I mitral cells. The deepest granule cells have many branches and spines in the deep and intermediate zones but do not reach the superficial zone; therefore they cannot synapse with the tufted cell basal dendrites of that zone. Laminarly organized output cell axon collaterals, anterior commissure axons, and centrifugal axons may differentially activate these granule cell populations and have selective influences on the output cells of the three EPL zones.[1-3,6]

CENTRAL PROJECTIONS

There is no strict topographic organization of the olfactory bulb output to olfactory cortex,[1,2] although the projection is not random. For example, olfactory tubercle projections arise preferentially from more ventral parts of the olfactory bulb. The presence of cells with different degrees of axon branching[8] (FIG. 1) makes studies of the projection pattern by retrograde transport difficult to interpret and may require other techniques. Retrograde transport studies also showed a stronger correlation of projection pattern with the cell type (laminar position of labeled cells) than with the bulb sector in which the cell lay. Mitral cells have long projections that include posterior piriform cortex while tufted cells project only to more rostral regions. It appears that the more superficial the tufted cell somata, the shorter the axon. This does not mean that the projections of mitral and tufted cells are completely isolated. For example, both mitral and tufted cells participate in the projection to the pars externa of the anterior olfactory nucleus (AON pe).[8,10] We reconstructed axons from 22 mitral and tufted cells filled by extracellular injection of the phaseolus vulgaris lectin (FIG. 1). Mitral cells had few axon collaterals in the rostral piriform cortex, which receives heavy tufted cell inputs. Type I and type II mitral cells both had wide ranges of axon diameters, numbers of collaterals, and distributions of axons. Our best-filled tufted cell had extensive fine collaterals in the rostral piriform cortex. More observations on tufted cells and cells from different regions of the bulb are needed to understand the organization of these projections; however, it is likely that the mitral cells have substantially less influence on rostral parts of the olfactory cortex than do tufted cells. The mitral cell axons were not fully filled, but, in their rostral regions, the two types of mitral cells overlap extensively in their projections, indicating that the zones of dendritic ramification do not correlate perfectly with the length of projection.

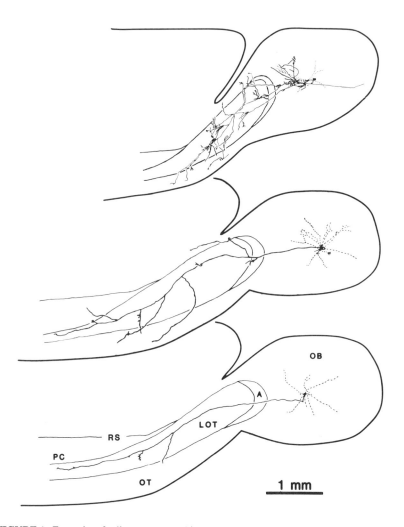

FIGURE 1. Examples of cells reconstructed in a parasagittal view from *Phaseolus vulgaris* lectin injections. The top cell is a tufted cell of the intermediate zone of the EPL. Its dendrites (represented by dotted lines) lie in a nearly radial distribution, and it had extensive axon collaterals within the olfactory bulb. It also had many fine collaterals in the rostral piriform cortex just deep to the LOT. The two type I mitral cells below were filled back to the caudal edge of the olfactory tubercle. The more caudal parts of most mitral cells were not stained. These cells had longer, radially distributed basal dendrites. They indicate typical mitral cell branching patterns. Some, like the example of the middle panel, had a moderate number of branches in this rostral region. Others, like the cell of the bottom panel, had very few rostral branches. Type II mitral cells had a similar range of branching patterns. Abbreviations: OB—olfactory bulb; A—anterior olfactory nucleus pars externa; LOT—lateral olfactory tract; OT—olfactory tubercle; RS—rhinal sulcus; PC—piriform cortex.

PHYSIOLOGICAL STUDIES

The longer axons of mitral cells are useful for distinguishing them from tufted cells for electrophysiology. Although mitral cells have larger axon diameters than tufted cells, we cannot use conduction velocity to indicate cell type[7] because of the extensive branching in the lateral olfactory tract; thus we are denied a powerful tool used by retinal physiologists for studying ganglion cell types. HRP injections indicate no tufted cell axons caudal to the olfactory tubercle. We therefore define tufted cells electrophysiologically as cells projecting into the lateral olfactory tract but not to the posterior piriform cortex.[7,6]

While previous studies showed that cells recorded in the superficial EPL respond to electrical stimulation of the olfactory nerve with more spikes than cells deep in the EPL, the cell types were not identified. Identifying mitral and tufted cells by their antidromic activation pattern[7] showed that tufted cells had lower thresholds, more multiple spikes, greater probability of activation from the olfactory nerve and a greater distribution of effective stimulus sites on the olfactory nerve layer. The antidromic activation pattern correlated more strongly with these responses than did the laminar position of the recordings. This method could not distinguish between the two types of mitral cells. More precise methods, such as intracellular marks, are being tried to identify these mitral cells.

Several authors found a variety of temporal patterns of unit response to odor stimulation (see Harrison and Scott[6] for review), but these patterns often change with different odors or different stimulus concentrations and do not correspond to cell types. The morphology of the cells was not known. Correlating odor responses with cell types will require a good series of odor concentrations and a method for comparing response magnitude relative to response variability. We devised an index of dissimilarity between odor response patterns that allows measurement of the change in response induced by a given stimulus change.[6,11] This makes possible comparison of stimulus-response functions for different cells. This type of approach should promote comparisons across the range of stimulus concentration for a battery of odors.

CONCLUSIONS

While there have been many studies of the odor responses of neurons in the olfactory bulb and the central olfactory pathway, interpretation has been hampered by problems in specification of stimuli, of responses and of the particular cells being observed. The rapidly accumulating knowledge on the olfactory pathways suggests good reasons to expect physiological differences among the mitral and tufted cell classes. The recent literature has emphasized the correspondence of output cell type and laminar position to its projection pattern. Our work supports this correspondence, but there are also substantial overlaps in projection. The two types of mitral cells may turn out to have very similar projections. This suggests that processing in the different zones of the EPL is not just related to the length of output cell axons. Potentially there is parallel transmission over different channels, similar to the on- and off-center paths of vision, that are useful for neural computation in posterior cortical areas. The presence of important differences in processing by cells with dendrites in the three zones can only be conclusively demonstrated with intracellular marks, unless more reliable extracellular methods are found to discriminate these cells. Future progress will also require careful control of stimulus intensity, recording site in the bulb, new ways of measuring the response, and attention to the spatial and temporal aspects of the stimulus.

REFERENCES

1. MACRIDES, F., T. A. SCHOENFELD, J. E. MARCHAND & A. N. CLANCY. 1985. Evidence for morphologically, neurochemically and functionally heterogenous classes of mitral and tufted cells in the olfactory bulb. Chem. Senses **10:** 175-202.
2. SCOTT, J. W. 1986. The olfactory bulb and central pathways. Experientia **42:** 223-232.
3. ORONA, E., E. C. RAINER & J. W. SCOTT. 1984. Dendritic and axonal populations of mitral and tufted cells innervate the rat olfactory bulb. J. Comp. Neurol. **226:** 346-356.
4. MOURADIAN, L. E. & J. W. SCOTT. 1985. Demonstration of the sublaminar pattern of the olfactory bulb external plexiform layer with cytochrome oxidase. Chem. Senses **10:** 401.
5. SCOTT, J. W. & T. A. HARRISON. 1987. The olfactory bulb: Anatomy and Physiology. *In* Neurobiology of Taste and Smell. T. Finger & W. Silver, Eds.: 151-178. J. Wiley. New York.
6. HARRISON, T. A. & J. W. SCOTT. 1986. Olfactory bulb responses to odor stimulation: Analysis of response pattern and intensity relationships. J. Neurophysiol. **56:** 1571-1589.
7. SCHNEIDER, S. P. & J. W. SCOTT. 1983. Orthodromic response properties of rat olfactory bulb mitral and tufted cells correlate with their projection patterns. J. Neurophysiol. **50:** 358-378.
8. OJIMA, H., K. MORI & K. KISHI. 1984. The trajectory of mitral cell axons in the rabbit olfactory cortex revealed by intracellular HRP injection. J. Comp. Neurol. **230:** 77-87.
9. BRUNJES, P. C. 1985. Unilateral odor deprivation: Time course of changes in laminar volume. Brain Res. Bull.: 233-237.
10. SCOTT, J. W., E. C. RAINER, J. L. PEMBERTON, E. ORONA & L. E. MOURADIAN. 1985. Pattern of rat olfactory bulb mitral and tufted cell connections to the anterior olfactory nucleus pars externa. J. Comp. Neurol. **242:** 415-424.
11. HARRISON, T. A. & J. W. SCOTT. 1986. Analysis of olfactory neural responses by a method of spike train matching. Ann. N.Y. Acad. Sci. This volume.

Cortical Organization in Gustatory Perception[a]

TAKASHI YAMAMOTO

Department of Oral Physiology
Faculty of Dentistry
Osaka University
1-8 Yamadaoka, Suita
Osaka 565, Japan

Perception of taste involves highly integrated brain functions such as emotional reactions, memory, experience, and learning as well as discrimination of quality and intensity of taste stimuli. Although it is generally accepted that the cerebral cortex is important in sensory reception, the role of the cortical taste area in perception of taste still remains to be studied. The functional significance and neuronal organization of the cortical taste area would be better understood through the analyses of unit responses of cortical neurons during ingestive behavior. Our recent experimental results will be presented here with some introductory remarks on neuroanatomical and functional aspects of the cortical taste area.

NEUROANATOMY

Projection areas for the tongue sensory modalities are separately represented in the cortex.[1-3] Neurons responding to tongue tactile stimulation are located in the granular cortex (or in the ventralmost somatic sensory area I). Neurons responding to taste stimulation of the tongue are found most ventrally near the rhinal sulcus. Neurons responsive to thermal stimulation of the tongue exist with some overlap between the tongue tactile area and the taste area. According to the cytoarchitectonic divisions of the insular cortex,[4] the gustatory area belongs to the posterior agranular insular cortex. Thermoresponsive neurons are located in the transitional zone between the granular and agranular cortices, which may be called "granular insular cortex" or "dysgranular cortex." Some taste-responsive neurons exist in this transitional zone, especially in the rostral part of the taste area. The insular cortex receives inputs of different sensory modalities such as olfaction, vision, somatosensation, visceral sense as well as taste; therefore, in this article, I will use a term "IGC" (insular gustatory cortex) for the cortical taste area.

Gustatory inputs reach the IGC via the different paths. According to our recent anatomical study by anterograde and retrograde WGA-HRP methods in rats and

[a]Supported by Grants-in-Aid for Scientific Research (No. 61570822) and Special Project Research (No. 61134026) from the Ministry of Education, Science and Culture of Japan.

49

hamsters in addition to the similar work by Yasui et al.[5-7] in cats, thalamic afferents terminate to layers 1 and 6, axons from the pontine parabrachial nuclei project to layer 6. Layer 5 receives inputs from the contralateral IGC. On the other hand, IGC neurons send axons to different brain areas, that is, from layer 6 to the thalamic taste area, from layer 5 to the contralateral IGC and to the reticular formation, and from layer 3 to the amygdala.

FUNCTION

Primates with brain damages including the cortical taste area show hypogeusia or ageusia. In rodents, however, bilateral cortical lesions including the IGC exerted little effect on acceptance and rejection thresholds. One of the feature lesion effects is disruption of acquisition or retention of the conditioned taste aversion.[8] Therefore, it is assumed that the IGC plays an important role in some cognitive processes including discrimination of taste quality and intensity, association between conditioned and unconditioned stimuli, memory, learning, and so forth.

As already described above, the IGC is a part of the insular cortex, which is often called "viscero-autonomic cortex" by some investigators. The IGC may not be homogeneous in function, but the cells in this area mediate functions other than gustation, such as vagal visceral sensibility, olfaction, respiration, and blood circulation.

Corticofugal fibers from the IGC to the reticular formation may be responsible for orolingual movements and salivary secretion; our latest study has shown that repetitive electrical stimulation applied to the IGC induces rhythmic discharges in the mylohyoid, masseter, and hypoglossal nerves, and in both the parasympathetic and sympathetic secretory nerves to the submandibular gland.

NEURON TYPES

To investigate neural organization of the IGC, unit activity in the somatosensory cortex and IGC was recorded through chronically implanted electrodes, consisting of five or six 35-μm o.d. teflon-coated platinum-iridium wires, while Wistar male rats were freely licking liquids from a drinking spout and eating dry pellets in a test box. Rats could lick the fluids when two shutters, a black shutter and a transparent shutter, which were interposed between the animal and the spout, were opened. Stimulation consisted of: 10°C, 25°C, and 40°C distilled water, sucrose, NaCl, HCl, quinine hydrochloride at various concentrations, and other sapid solutions. The licking pattern and electromyographic activity of the temporalis muscle were also monitored. After chronic recordings, the animals were lightly anesthetized, then the adequate stimulus and receptive field were examined for several neurons. The animals were deprived of water for about 20 hours before a recording session. Fifty-six neurons that were histologically identified in the somatosensory cortex and IGC were classified into the following seven groups according to their response properties. Each of these neurons changed its neural activity (impulses/5 sec) during ingestive behavior more than 2.5 standard deviations of the spontaneous rate/5 sec for the neuron. Surface location of each neuron is represented in FIGURE 1. Some of the neuron types share the common characteristics with those for the parabrachial neurons in conscious rabbits.[9]

"Phasic-Tactile" Neurons (n = 13)

These neurons showed rhythmic phasic activities at different phases of the licking cycle during licking of liquids regardless of their qualities. The firing pattern was

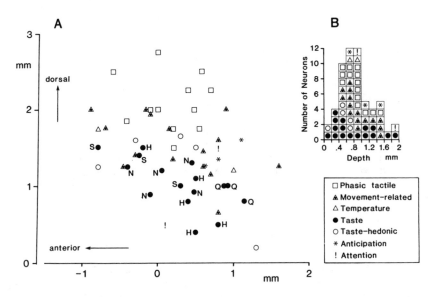

FIGURE 1. Surface localization of electrophysiologically isolated single neurons grouped into each of the seven types (**A**), and depth distribution of each neuron from the cortical surface (**B**). Neurons were mapped on a coordinate system projected upon the lateral side of the hemisphere of the cortex. Abscissa indicates the anteroposterior distance from the standard zero point, which is the rostralmost corpus callosum. Ordinate indicates the dorsoventral distance from the standard line, which is the rhinal sulcus. Letters S, N, H, and Q indicate that the taste neurons near each of these letters belong to either of sucrose-best, NaCl-best, HCl-best, and quinine-best neurons, respectively.

dependent on the location of receptive field on the orofacial area. For example, a neuron that had a confined tactile receptive field on the tip of the tongue discharged only when the tongue made contact with the drinking spout, while a neuron with a receptive field on the upper lip responded when the tongue retracted into the oral cavity. The mean tongue pressure against the spout (0.6 cm o.d.) during licking was 1.8 g/0.28 cm^2, when measured by a pressure-sensitive transducer attached to the spout. These neurons were located within the somatosensory area I (granular cortex).

"Movement-Related" Neurons (n = 15)

These neurons increased or decreased their activities during licking, chewing, or grooming without correlations with any cycle of jaw or tongue movements. This type of neurons was called "lick-responsive" neurons in a previous paper.[10] These neurons showed a tonic discharge pattern during orolingual movements, in contrast to the phasic response in phasic-tactile neurons. Although receptive fields and adequate stimuli for these neurons could not be identified, judging from the response properties, it is assumed that they receive inputs from mechanoreceptors situated in a fairly large oral region. Alternatively, considering the fact that the IGC is responsible for orolingual as well as salivatory activities, some of these neurons might be related to these motor and autonomic functions. Histologically these neurons were mostly situated in the dysgranular cortex between the dorsal somatosensory cortex and the ventral IGC.

"Temperature" Neurons (n = 2)

One of these neurons increased its spontaneous firing rate to licking of cold water and decreased to hot water. The other neurons showed the reversal of the responsiveness to low and high temperatures of water. We obtained only two neurons of this type, while many neurons were found to respond to cold stimulation of the tongue in anesthetized rats.[1] These neurons were also located in the dysgranular cortex.

"Anticipation" Neurons (n = 3)

Neurons that seemed to be related to anticipation of ingestion were found in the dysgranular cortex or IGC. These neurons showed a large increase in firing rate when the black shutter was opened: during this period the rat could see the spout through the transparent shutter. When the transparent shutter was opened, the rat began to lick and the increased activity decreased slightly. When the black shutter was closed, the activity returned to its original background level. If the animal was satiated and had no intention of taking liquid, opening the black shutter did not elicit any change in activity of these neurons.

"Attention" Neurons (n = 2)

These neurons responded only slightly during licking of solutions, but strong responses were elicited by startle stimuli such as sound, light flash, body touch, or an air-puff to the body. These neurons were also found in dysgranular cortex or IGC. The function of these and the previous anticipation neurons might not be directly related to taste recognition.

"Taste" Neurons (n = 16)

These neurons increased or decreased their discharge rates during licking of particular taste solutions. The responsiveness of these neurons differed greatly depending on the quality of solutions the animals ingested. Some neurons responded vigorously during licking of a solution of rat food pellets. When the taste neurons were classified into "best-stimulus" categories, depending on their best sensitivity to any one of the four basic stimuli, sucrose-best, NaCl-best, HCl-best, and quinine-best neurons were found to be located in this order from anterior to posterior within the IGC (see the letter marks near the taste neurons in FIG. 1). Such a relative chemotopic organization of taste responsiveness was by and large in agreement with the results obtained in anesthetized rats.[11] Some taste neurons received convergent inputs from other sensory stimuli such as temperature (cold), olfaction (odors of almond or acetic acid), or visceral discomfort (intraperitoneal injection of LiCl).

Two taste neurons showed enhanced responses, either in excitatory or inhibitory directions, to a conditioned taste stimulus after acquisition of aversion by pairing the conditioned stimulus with an intraperitoneal injection of 0.15 M LiCl (2% of body weight).

"Taste-Hedonic" Neurons (n = 5)

These neurons were also responsive to taste stimuli. The characteristic difference of these neurons from the taste neurons is that these neurons showed differential response patterns that differentiated the palatable (including water) and unpalatable liquids. The unpalatable liquids were HCl, quinine, and high concentrations (0.5 M and 1.0 M) of NaCl in our experiments. Taste responses were modified in one neuron by establishment of the conditioned taste aversion. In this case, the inhibitory response of a taste-hedonic neuron to 0.1 M NaCl changed to the excitatory response to this solution as observed during the licking of quinine. After extinction of the conditioned aversion, the responsiveness of the neuron to 0.1 M NaCl returned to the preconditioning (or inhibitory) response pattern. Histological survey has suggested that these hedonic-type neurons exist around the taste area where taste neurons are packed (see FIG. 1).

CONCLUSIONS

Neurons in the cortical taste and orolingual sensory areas possess different response characteristics corresponding to each aspect of ingestive behavior as demonstrated by seven distinct neuron types. Some of these functionally different neurons are organized anatomically separately within the cortex. Cortical neurons with these modality-specific and behavior-dependent responsiveness together with their spatial arrangement in the cortex may serve for perception of taste, integration of orolingual sensory inputs, and control of ingestive behavior.

REFERENCES

1. YAMAMOTO, T., N. YUYAMA & Y. KAWAMURA. 1981. Brain Res. **221:** 202-206.
2. NORGREN, R., E. KOSAR & H. J. GRILL. 1982. Soc. Neurosci. Abstr. **8:** 201.
3. YAMAMOTO, T., N. YUYAMA, T. KATO & Y. KAWAMURA. 1984. J. Neurophysiol. **51:** 616-635.
4. KRETTEK, J. E. & J. L. PRICE. 1977. J. Comp. Neurol. **171:** 157-192.
5. YASUI, Y., K. ITOH & N. MIZUNO. 1984. Brain Res. **292:** 151-155.
6. YASUI, Y., K. ITOH, M. TAKATA, A. MITANI, T. KANEKO & N. MIZUNO. 1985. J. Comp. Neurol. **234:** 77-86.
7. YASUI, Y., K. ITOH, A. MITANI, K. TAKADA & N. MIZUNO. 1985. J. Comp. Neurol. **241:** 348-356.
8. BRAUN, J. J., P. S. LASITER & S. W. KIEFER. 1982. Physiol. Psychol. **10:** 13-45.
9. SCHWARTZBAUM, J. 1983. Brain Res. Bull. **11:** 61-89.
10. YAMAMOTO, T., K. ASAI & Y. KAWAMURA. 1986. *In* Emotions. Y. Oomura, Ed.: 127-135. Japan Scientific Society Press. Tokyo.
11. YAMAMOTO, T., N. YUYAMA, T. KATO & Y. KAWAMURA. 1985. J. Neurophysiol. **53:** 1356-1369.

Effects of Odorant Mixtures on Olfactory Receptor Cells[a]

STEVEN PRICE

Department of Physiology and Biophysics
Medical College of Virginia
Virginia Commonwealth University
Richmond, Virginia 23298

The overall question addressed by this symposium is, "Do responses to mixtures of stimuli differ fundamentally from responses to pure stimuli?" This paper will be an attempt to explore this at the level of receptor cells. There are several obstacles to providing an answer. One is that our knowledge of how olfactory receptor cells respond to pure stimuli is primitive. Another is that odorants probably always present the cells with more than one compound despite our efforts to avoid this. A third is that there is nearly no published information on cellular responses to mixtures presented intentionally. The approach taken will be to briefly review the mechanisms by which odorants might influence receptor cells, then to discuss the basis for questioning whether mixtures can be avoided in any situation, and finally discuss the kinds of data that would be needed in order to approach the original question.

HOW DO OLFACTORY RECEPTOR CELLS RESPOND TO ODORANTS?

Olfactory receptor cells are bipolar neurons in the olfactory epithelium. Their dendrites are bathed in an overlying layer of mucus and their axons project through the cribriform plate to the olfactory bulb where they make the first synapse. The responses that these cells can make that are of significance from the standpoint of olfaction are changes in the frequency with which they discharge action potentials. This is directly related to their state of polarization; the more nearly they are depolarized, the greater their discharge frequency and vice versa; thus, the question of how odorants cause responses is basically that of how they affect the resting potentials. There are two broad categories of mechanisms for this: the specific effects, probably mediated by receptor molecules, and nonspecific effects of interactions with various cellular components or the extracellular milieu.

It is generally agreed that the first step in the specific mechanisms is reversible binding of the odorant to a cell membrane protein with a high affinity for it. Several candidate olfactory receptor proteins have been extracted.[1-9] The evidence that each

[a] Supported by the United States Army Research Office.

55

is a receptor protein has been reviewed.[10] There are several proposals for the mechanism by which membrane potential is altered when a receptor protein binds an odorant. One is that configurational changes in the proteins open ion-conducting channels in the membrane.[11] Another is that the protein initiates a cyclic nucleotide cascade leading to depolarization, probably via phosphorylation of channel proteins.[8,12–14] The important feature from the standpoint of our discussion is that the effect is due to a rather specific interaction of the odorant with some constituent of the cell that is specialized for transduction.

The second type of mechanism of stimulation of olfactory receptor cells is mediation by nonspecific effects of the odorant; that is, there can be effects similar to those the compound might have on any other cell where the existence of specific receptor proteins is so unlikely that it can be ignored. It has been suggested that some odorants may stimulate olfactory receptor cells only by such mechanisms.[15–18] There is little direct evidence bearing on their role in olfaction and they will be mentioned here only briefly.

To the extent that the resting potential of a neuron depends on the extracellular milieu, any effect of an odorant upon the composition of the extracellular fluid or mucus can influence it. Simple cases might be direct effects as odorants dissolve in the fluid. For example, organic acids probably alter the mucus pH and, perhaps, that of the extracellular fluid as well by dissolving in it and dissociating. It is well known that resting potential depends on the pH of the medium. Possible indirect effects include alterations secondary to local blood flow changes from vasoactive odorants or from effects on cells that secrete mucus.

To the extent that the resting potential of a neuron depends upon the cell's metabolism, there can be effects from odorants that alter the metabolic rate. Many odorants are lipophilic and therefore readily penetrate cells and have access to intracellular metabolic systems.

To the extent that the resting potential depends upon the physical properties of the cell membrane, odorants that partition into the membrane will influence firing frequency by this route. Surface charge density contributes to resting potentials, although the magnitude of the contribution varies in different cell types.[19–23] Anionic or cationic odorants will alter the surface charge density in opposite directions; uncharged odorants will reduce the surface charge density by expanding the membrane and thus increasing the mean distance between intrinsic charged groups. Electrostatic interactions of membrane charges with ionic odorants can also influence surface charge density.[21]

IS OLFACTORY TISSUE EVER PRESENTED WITH JUST A SINGLE ODORANT?

It was asserted that olfactory stimuli are probably almost always mixtures of compounds at the receptor cell. There are two reasons for this. One is that any compound used as an olfactory stimulus is likely to be contaminated. Some chemicals are available in very high states of purity and hence are less likely to have their effects confounded by those of contaminants, especially when low concentrations are used. It must be remembered, however, that olfactory stimuli are delivered in the gas phase but must partition into the aqueous mucus before reaching the receptor cells; therefore, the relative concentrations of an odorant and a contaminant depend not only on their gas phase concentrations but on their relative partition coefficients. Let us consider a hypothetical example, in which butanol (a commonly used odorant) is contaminated with small amounts of butyric acid (an oxidation product of butanol). Butanol partitions into aqueous solutions from air, becoming about 25 times more concentrated in the aqueous phase at equilibrium. Butyric acid partitions into the aqueous phase until it is totally extracted from the air, and is a potent odorant itself. It is easy to see how what would appear to be a minor level of contamination might result in a significant proportion of the total response as being due to the contaminant.

A second reason that purified odorants probably present receptor cells with mixtures is that many odorants can be metabolized in olfactory epithelium. In such cases the "odorant" will be present as a mixture with its metabolites, and the composition of the mixture will vary with time. Olfactory epithelium is the second richest source of the cytochrome P-450 dependent oxygenases,[24] also known as drug-metabolizing enzymes because of their ability to metabolize exogenous compounds. Such compounds are transformed into readily excretable forms by these enzymes, which thus have an important protective function. No doubt there are other enzymes that alter odorant structures in mucus and therefore contribute to the formation of mixtures even from highly purified odorants. It is sometimes suggested that metabolism of odorants serves to terminate stimulation of receptor cells; however, in at least one case the metabolic product is as potent a stimulus as the parent odorant.[25] Anisole, a commonly used odorant, is metabolized to phenol and formaldehyde,[24] both of which are extremely irritating to neurons. A few workers have hypothesized that particular odorant metabolic products are stimuli to olfactory receptor cells.[15,26,27]

HOW CAN ODORANT MIXTURES AFFECT OLFACTORY RECEPTOR CELLS?

We can consider that all effects of odorants on olfactory neurons are mediated by specific receptor molecules, various nonspecific actions, or some combination of the two. The nonspecific actions may be direct or indirect. This will be true whether an odorant is homogeneous and not subject to metabolism in olfactory tissue, a situation that is probably rare, or if other compounds are present with it at the receptor cell.

Unless the compounds in a mixture react with each other, each can be expected to exert its effects independently. Membrane potential and firing frequency will change as the net effect of all of the actions of the constituents of the mixture. This is presumably true for every olfactory cell response that has ever been measured. How would the cell's firing frequency change in the presence of a mixture compared to what would happen with the individual pure compounds? There are no published data on this question with vertebrate material. In the one report of receptor cell responses to odorant mixtures,[28] the parameter measured was the electroolfactogram (EOG), the extracellular potential change resulting from changing currents across the tissue. The EOG of rabbits was measured in response to mixtures made from equipotent concentrations of cyclohexanol and benzaldehyde. The EOG elicited by the mixtures was always smaller than that elicited by either pure compound. This was interpreted as evidence for reciprocal masking of responses. As will be shown, this presumes certain characteristics of the relation of concentration to response for each compound.

HOW CAN RESPONSES TO MIXTURES BE ANALYZED?

Even relatively simple questions about interactions of pairs of odorants require careful experimental design. Analysis of interactions between mixtures of biologically active compounds is common. The problem of analyzing effects of odorant mixtures is simply a specific example of this. The question of whether mixtures have the effects expected from those of the individual components is not as simple and straightforward as it might seem.

Ignoring the likelihood that more than just a single compound is being presented to the receptor cell, consider some typical relations between concentration and response. The simplest is a straight line passing through the origin. In such cases, increments of concentration are purely additive. Such relations are rare. They can only occur over a limited concentration range and include no maximum response at high concentrations. More common is the situation in which there is some maximum response that cannot be exceeded at any concentration. Such relations are consistent with models in which reversible interaction of the stimulus with some cellular constituent initiates the response, the maximum response occurring when that constituent is saturated. Such data can be approximated by an adsorption isotherm (for example, the Michaelis-Menten equation), even when the model is incorrect. At low concentrations the relation of concentration to response is approximately linear, passing through the origin. At increasing concentrations, increments elicit progressively smaller increments of response. As the maximum response is approached, they elicit no experimentally measurable increments at all.

The most common relation of odorant concentration to receptor cell response is that of a sigmoid curve. The response increments increase more than linearly with concentration at low concentrations. At intermediate concentrations there is a region of approximate linearity (not passing through the origin). At high concentrations there is a region in which concentration increments elicit progressively smaller response increments as a maximum response is approached. A model that generates such a curve is one in which binding of one stimulus molecule to a receptor lowers the probability of another being bound. This is the so-called allosteric model. Another is one in which more than one molecule of the stimulus must be bound in order for a response to be elicited. The equations formally describing the curves derived from these models do not include a threshold concentration below which no response occurs; however, at low concentrations there will be responses that do not exceed the noise level in experimental systems, and this will be the "subthreshold" range.

We can consider how effects might be analyzed with this as background. First, it should be obvious that absence of interaction between the components of a mixture need not lead to simple additivity in their effects. Indeed, except in the cases where the response and concentration are related by a line passing through the origin, the effect of an increment of a compound added to itself (an increase in concentration) will not be additive. Clearly, the complete curve relating concentration to response must be determined for each compound alone. Only then can the effect of one or more concentrations of one of the responses to the other be subjected to analysis for interaction. Even then, the results must be interpreted with care. Every biological response has some maximum that cannot be exceeded; therefore, subtracting the incremental response of one component from that of the mixtures will result in lowering the apparent maximal response to the second component. The situation might be analogized to measuring the speed at which an animal runs in response to each of two stimuli, such as food reward and a shock. Providing both stimuli simultaneously cannot result in a speed in excess of the maximum, and it would be an error to conclude from such an experiment that the stimuli are mutually suppressive.

A protocol that has been used widely in analyzing effects of drug mixtures is the so-called method of isoboles.[29] Two drugs are adjusted to equipotent concentrations and then mixed in various proportions, mutually diluting each other. It is presumed that the level of potency of every such mixture will be equal, the increased concentration of one component offsetting the decreased concentration of the other, unless the drugs interact. This reasoning is sound only when there is formal identity of the relation between concentration and response for the two drugs.[30] In all other cases, the reduction in potency in reducing the concentration of one component is not offset by an equally increased potency of the elevated concentration of the other. The isobole method does avoid the difficulty encountered by the existence of a maximum to the possible response noted above, however.

The only published study of odorant mixtures in which a vertebrate receptor cell response, the EOG, was measured essentially followed the isobole protocol.[28] This is also true for most studies of odorant mixtures in which magnitude estimation or bulbar responses were measured.[31] Equipotent concentrations of cyclohexanol and benzaldehyde were mixed in various proportions. The responses were always below those of the individual odorants, and it was concluded that they were mutually suppressive. As with mixtures of drugs, this conclusion would be warranted only if the relation of concentration to response was identical in form for each odorant.

SUMMARY AND CONCLUSIONS

The general mechanisms by which chemical stimuli may influence the firing frequency of olfactory neurons were briefly described. They include specific mechanisms mediated by receptor molecules and nonspecific mechanisms involving general properties of the chemicals and of cells. It is difficult to imagine that odorant mixtures influence receptor cells by mechanisms that are fundamentally different from those by which homogeneous chemicals act.

It is argued that even under the best experimental conditions the presentation of odorants usually or always involves exposing the receptor cells to more than one additional molecular species compared to the unstimulated condition. This is because odorants invariably have contaminants that may be of potency such that their contribution to the odor is large even though their contribution to the number of molecules in the stimulus stream is small. Furthermore, the partition coefficients of the major and minor components are unlikely to be identical; therefore, their relative concentrations in the aqueous environment of the receptor cells can differ greatly from that in the gas phase. Finally, metabolic transformations of odorants in the olfactory mucosa can result in the exposure of receptor cells to mixtures of odorant and metabolites, with the mixture composition varying with time.

Finally, some pitfalls in analyzing the effects of odorant mixtures are discussed. At the very least, it is necessary to determine the relation of concentration to response for each odorant in the mixture in order to interpret results in terms of interactions. Even with such data caution must be used, especially in attaching significance to reductions in the apparent maximal responses to one odorant induced by the presence of the other.

ACKNOWLEDGMENTS

I am grateful to Dr. John A. DeSimone for stimulating discussions.

REFERENCES

1. GENNINGS, J. N., D. B. GOWER & L. H. BANNISTER. 1977. Biochim. Biophys. Acta **496:** 547-566.
2. PRICE, S. 1978. Chem. Senses **3:** 51-55.
3. FESENKO, E. E., V. I. NOVOSELOV, N. F. MJASOEDOV & G. V. SIDOROV. 1978. Stud. Biophys. **73:** 71-84.
4. CAGAN, R. H. & W. N. ZEIGER. 1978. Proc. Natl. Acad. Sci. **75:** 4679-4683.
5. RHEIN, L. D. & R. H. CAGAN. 1981. In Biochemistry of Taste and Olfaction. R. H. Cagan & M. R. Kare, Eds.: 47-68. Academic Press. New York.
6. BROWN, S. B. & T. J. HARA. 1982. In Chemoreception in Fishes. T. J. Hara, Ed.: 159-180. Elsevier Scientific Publishing Co. Amsterdam, the Netherlands.
7. PELOSI, P., N. E. BALDACCINI & A. H. PISANELLI. 1982. Biochem J. **201:** 245-248.
8. CHEN, Z. & D. LANCET. 1984. Proc. Natl. Acad. Sci. **81:** 1859-1863.
9. PRICE, S. & A. WILLEY. 1987. Ann. N.Y. Acad. Sci. This volume.
10. PRICE, S. 1981. In Biochemistry of Taste and Olfaction. R. H. Cagan & M. R. Kare, Eds.: 69-84. Academic Press. New York.
11. VODYANOY, V. & R. B. MURPHY. 1983. Science **220:** 717-719.
12. KURIHARA, K. & N. KOYAMA. 1972. Biochem. Biophys. Res. Commun. **48:** 30-34.
13. MINOR, A. V. & N. L. SAKINA. 1973. Neurofiziol. **5:** 415-422.
14. MENEVSE, A., G. DODD & T. M. POYNDER. 1977. Biochem. Biophys. Res. Commun. **77:** 671-677.
15. PRICE, S. 1984. Chem. Senses **8:** 341-354.
16. ARVANITAKI, A., H. TAKEUCHI & N. CHALAZONITIS. 1967. In Olfaction and Taste. II. Y. Zotterman, Ed.: 573-598. Pergamon Press. New York.
17. MOZELL, M. M. 1970. J. Gen. Physiol. **56:** 46-63.
18. KASHIWAYANAGI, M. & K. KURIHARA. 1985. Brain Res. **359:** 97-103.
19. OHKI, S. 1981. Physiol. Chem. Phys. **13:** 195-210.
20. MIYAKE, M. & K. KURIHARA. 1983. Biochim. Biophys. Acta **762:** 256-264.
21. CHAN, D. S. & H. H. WANG. 1984. Biochim. Biophys. Acta **770:** 55-64.
22. KASHIWAYANAGI, M. & K. KURIHARA. 1984. Brain Res. **293:** 251-258.
23. OHSHIMA, H. & S. OHKI. 1985. Biophys. J. **47:** 673-678.
24. DAHL, A. R., W. M. HADLEY, F. F. HAHN, J. M. BENSON & R. O. MCCLELLAN. 1982. Science **216:** 57-58.
25. GOWER, D. B., M. R. HANCOCK & L. H. BANNISTER. 1981. In Biochemistry of Taste and Olfaction. R. H. Cagan & M. R. Kare, Eds.: 8-31. Academic Press. New York.
26. VINNIKOV, YA. A., G. A. PYATKINA, E. I. SHAKHMATOVA & YU. V. NATOCHIN. 1979. Doklad. Akad. Nauk USSR **245:** 750-753.
27. HORNUNG, D. E. & M. M. MOZELL. 1981. In Biochemistry of Taste & Olfaction. R. H. Cagan & M. R. Kare, Eds.: 33-45. Academic Press. New York.
28. MACLEOD, P. 1968. Cah. Oto-Rhino-Laryngol. Suppl. **1:** 25-27.
29. LOEWE, S. 1953. Arzneim. Forsch. **3:** 285-307.
30. BERENBAUM, M. C. 1985. J. Theor. Biol. **114:** 413-431.
31. CAIN, W. S. 1975. Chem. Senses **1:** 339-352.

Coding of Chemosensory Stimulus Mixtures

DAVID G. LAING[a]

CSIRO Division of Food Research
Food Research Laboratory
North Ryde, N.S.W., Australia 2113

INTRODUCTION

Perception of odors in the environment by humans, animals, or insects invariably involves complex mixtures that often contain dozens of odorous components. Odor perception, therefore, usually depends on the reception and neural processing of many components.

Little is known about how and where mixtures are processed by the olfactory system, what factors determine how many odorants will be perceived, and what influence the physicochemical features of the stimuli have on the perception of mixtures.

Perhaps the most important problem to resolve initially concerns the role of peripheral and central components of the olfactory system in the processing of mixtures. If substantial processing occurs at the odor receptors through competitive or noncompetitive binding, an understanding of additivity, masking, and synergism, the most common consequences of mixing odors, is likely to be gained through structure-activity studies. On the other hand, if processing is primarily a central phenomenon, an intimate knowledge of neural circuitry and response properties of cells in different regions of the olfactory system, for example, olfactory bulb, will be required.

This article describes evidence for two mechanisms involved in the coding of odor mixtures at the vertebrate olfactory receptor epithelium. The first mechanism involves the separation of mixture constituents by the olfactory mucus and their adsorption in epithelial regions where receptor cells respond best to them; the second focuses on the time taken by odorants to stimulate receptor cells once they have been adsorbed by the mucus.

[a] Address for correspondence: Dr. David G. Laing, CSIRO Division of Food Research, P.O. Box 52, North Ryde NSW, Australia 2113.

DIFFERENTIAL ADSORPTION OF MIXTURE CONSTITUENTS
AND RECEPTOR SELECTIVITY

In vertebrates, the nose contains olfactory receptor cells that are located in an epithelium that lines sections of the nasal cavity and is covered by a thin layer of aqueous mucus.[1] The mucus provides a means of selectively adsorbing odorants from air passing over it. Data from studies with the frog indicate that, during a sniff, odorants are selectively adsorbed by the mucus according to their polarity in a chromatographic-like process and diffuse through the latter to the dendrites of receptor cells.[2] Polar odorants, such as alcohols and acids, are adsorbed in the anterior epithelium, while nonpolar hydrocarbons, such as octane, are poorly but evenly adsorbed over most of the epithelium. This adsorptive behavior is typical of polar stationary phases in chromatography.

So far, measurements of differential adsorption have been confined to single odorants in the frog,[2] an animal that has a relatively flat olfactory epithelium lining the dorsal and ventral surfaces of the nasal cavity. However, the location of the epithelia in most land-based vertebrates is more complex and generally includes the septum and several turbinates. A question of fundamental importance is whether each of these structures exhibits differential adsorption similar to that observed with the frog. Given that the surface of each structure is comparatively flat, and the gaps between them small, differential adsorption is possible. Therefore, as inspired air flows across the surfaces of the septum and turbinates in higher animals, the constituents of a mixture may be differentially adsorbed by the mucus covering each of the structures.

If differential adsorption occurs, how does the olfactory system take advantage of the separation process? An answer can be found from studies of the responses of salamander receptor cells. Like the frog, the salamander has a flattish olfactory epithelium; however, of special significance is the fact that cells of similar responsiveness are located in similar regions of the epithelium.[3] Cells that are most responsive to polar odorants are located in anterior regions near the external nares, while cells that are most responsive to hydrocarbons, like limonene or pinene, are found in the posterior region near the internal nares. Although the differential adsorption of odorants by salamander epithelium has not been reported, the response data suggest that receptor cells are located in regions that are most likely to be contacted by the odor molecules that stimulate them "best." If differential adsorption is a feature of the olfactory detection system, then it is possible that a similar arrangement of receptor cells occurs across the septum and turbinates of species with these structures. Cells on a turbinate, for example, would be aligned in the direction of odor flow to maximize their contact with and their response to odors adsorbed by the mucus. As yet, a systematic study of cell responses across a turbinate or septum has not been reported, but is clearly needed. In lieu of such information, studies in this laboratory with 2-deoxyglucose and cytochrome oxidase are currently aimed at marking the location of receptor cells stimulated by odorants of different polarity.

Differential adsorption of odorants, however, has special implications for mixture perception. Thus, if a binary mixture consisting of a polar and a nonpolar odorant is passed over an epithelium, differential adsorption may not only separate the two, but in doing so could prevent the polar odorant from reaching posterior regions that contain cells that are most responsive to nonpolar odorants. In effect, the polar odorant would not be able to interfere with the stimulation of nonpolar cells and therefore could not suppress the responses of receptor cells to the nonpolar odorant. This proposition was recently tested in studies with humans in this laboratory[4] using mixtures containing the nonpolar odorants limonene and pinene and the polar odorant propionic acid. As predicted, propionic acid had no effect on the perception of the hydrocarbons. In contrast, limonene and pinene strongly suppressed perception of the acid. Since cells in the anterior regions of the septum and turbinates were predicted to have low responsivity to hydrocarbons, there was no *a priori* reason for predicting this latter event. However, [³H]2-deoxyglucose (2-DG) studies with the rat using the same mixture that produced suppression of the acid showed there was a significant decrease in uptake of the tracer in glomeruli in the olfactory bulb that normally were very active when only propionic acid was the stimulant.[5] Since the terminals of olfactory nerve axons are considered responsible for most of the uptake of 2-DG in glomeruli,[6] the result indicated limonene was not stimulating the cells normally excited by propionic acid. On the contrary, the result suggested limonene was inhibiting responses of receptor cells to the acid. In hindsight, the result is not so surprising, since hydrocarbons are adsorbed evenly over the epithelial surface in the frog,[2] and therefore are capable of reaching receptor cells that are most responsive to the acid. Since the structures of the hydrocarbons are very different from the structure of the acid, it is unlikely the hydrocarbons would compete directly for acid sites. Inhibition, because of noncompetitive binding, would account for the result.

Further evidence for differential adsorption influencing mixture events was obtained from human studies with mixtures of limonene and pinene.[4] Here it was proposed that suppression of both components would occur since cells that were most responsive to both odorants would be located in similar regions of the anterior epithelium, thus increasing the probability of competition for cells or inhibition of the cells best stimulated by the other odor. As predicted, each of the components was able to suppress the other. The factor determining which odor predominated in a mixture was the perceived intensity of each in the unmixed state. The higher intensity component always suppressed the other to a greater degree.

The evidence, though limited at present, clearly supports the idea of a chromatographic-like process separating mixture components and concentrating them in regions of the epithelium where cells are most responsive to them. By preventing odors from reaching regions stimulated by others, the separation process can stop odorants from interacting with each other. On the other hand, interaction is likely to occur when odors are adsorbed in similar regions of the epithelium. The separation process may occur in a similar manner across different structures in the nose with both spatial and temporal information passing from each structure to glomeruli in the bulb, where it can be integrated or modified according to the situation.

DIFFUSION AND RECEPTION OF ODORANTS FROM MIXTURES

The second factor that may play an important role in the coding of mixtures is the time taken by odorants to diffuse through the mucus layer and stimulate receptor cells.

It has been reported that the characteristic latency of the excitatory response of vertebrate receptor cells to odorants is determined by the diffusion coefficient of the odorant, depth of receptor sites, and threshold concentration of odorant needed to initiate a cell response.[7] Observed latencies range from 45 to 2,000 msec and are far greater than the calculated time (1.4 msec) for molecules to diffuse from the gas phase above the mucus into the mucus layer.[7] Furthermore, latency changes markedly with odor concentration. For example, response latencies of a salamander receptor cell to safrole decreased from 425 to 150 msec when the concentration of the odorant in the gas phase above the mucus was increased threefold from the threshold level. In the case of another cell, latency changed from 1,977 to 517 msec in response to a similar concentration change. Since increasing the concentration of odorant in the mucus does not change the time taken for molecules to diffuse to a receptor site, the reduced latencies seen with increased concentration must be due to increases in the rate that receptor sites are occupied. In other words, increasing the odor concentration in the vicinity of a receptor cell increases the probability of site occupation. Response latencies of a cell also vary with odor type,[7] which suggests that a cell has different receptor-site types and that the time taken for different odorants to reach the threshold concentration is different. Thus, if two odorants are present in similar concentrations in the vicinity of a cell that has different numbers of the two site types and requires the same number of sites to be occupied in either case for stimulation to occur, then the odor that fits the predominant site type will have the greater probability of occupying the requisite number of sites and record the shorter latency. The cell will therefore exhibit different response latencies to the two odorants.

The important point is that these studies showed that latencies are large, and that differences between odorants or concentrations are of the order of hundreds of milliseconds.

To appreciate fully how latency differences might affect the perception of mixture components, a second feature of the olfactory system needs to be considered, namely the rapidity with which suppression of an odor can be induced. For example, human psychophysical studies have shown that a very small difference in time (1.0 msec) or in concentration (10%) are enough during birhinal stimulation to suppress perception of the stimulus in either the later or weaker stimulated nostril.[8] Similarly, an electro-physiological study with the rabbit has shown that a time difference of less than 3 msec is sufficient to halve the responses of glomeruli in the contralateral bulb.[9] Although concentration differences affected glomerular activity, the outcome was more complex, and it is difficult to draw a conclusion. The important finding of both studies, however, was that a time difference of only a few milliseconds was sufficient to significantly reduce perception of cell responses in the side receiving the later stimulation. Clearly, three milliseconds is vastly less than the hundreds of milliseconds that characterized latency differences between different odorants or different concentrations of an odorant[7] described above. The finding indicates that if significant interbulbar effects can operate with a time difference of a few milliseconds, then intrabulbar effects are likely to be just as efficient given their shorter nerve pathways.

Thus, when a mixture of two odorants is presented to the nose (i.e., in the normal mode of presentation), and one has a stimulation time that is significantly shorter than that of the other, it is very likely that perception of the second odorant will be affected either through competition for receptor cells or the modulating actions of bulbar cells triggered by the first odorant.

Evidence for this proposal was recently obtained in studies with humans where two odorants were simultaneously presented to the same nostril (physical mixture) or to separate nostrils (dichorhinic mixture).[10] The odorants were the same as used in the study described earlier,[4] namely limonene, pinene, and propionic acid. It was predicted from studies of reciprocal bulbar inhibition[8,9] that, during dichorhinic stimulation, suppression would depend on odor intensity and be independent of odor quality. Accordingly, propionic acid and the hydrocarbons were expected to exhibit reciprocal suppression. The result, however, was very different than that predicted. The hydrocarbons suppressed the acid during both modes of presentation, but the acid had no effect on the perception of the hydrocarbons.

In addition to suggesting that suppression of the acid is primarily a peripheral event, the results indicated that the failure of the acid to suppress the hydrocarbons was due to another factor. It is proposed that this factor is the time taken to stimulate receptor cells. Thus, it is suggested that the time taken by the acid to stimulate receptor cells is far greater than that taken by the hydrocarbons. Accordingly, when the odorants are presented separately to the nostrils, the hydrocarbons suppress the acid via interbulbar inhibition of the contralateral bulb. When presented as a physical mixture to the same nostril, the acid is suppressed via noncompetitive binding at the receptor epithelium as proposed earlier,[4] and possibly by inhibition within the ipsilateral bulb.

SUMMARY

It appears likely that substantial processing of odor mixtures is done by peripheral olfactory structures. Both physiological and psychophysical data suggest that the first step is differential adsorption of constituents by the olfactory mucus, which separates and concentrates constituents in epithelial regions where neurons are most responsive to them. The second step involves the diffusion of adsorbed odorants through the mucus and activation of the receptor neurons. In contrast to the minute times involved in the first step, the time taken to diffuse through the layer and stimulate neurons is large, varies markedly between odorants, and is concentration dependent. The second step, therefore, provides another process for separating the actions of individual odorants. Both processes either separately or combined can account for mixture phenomena such as suppression and masking. Defining the role of each process with different mixtures of odorants that have different molecular properties is clearly a goal for future studies.

REFERENCES

1. GRAZIADEI, P. P. C. 1971. The olfactory mucus of vertebrates. *In* Handbook of Sensory Physiology. L. M. Beidler, Ed. **4:** 29-58. Springer-Verlag. Berlin, Germany.
2. MOZELL, M. M. & M. JAGODOWICZ. 1973. Chromatographic separation of odorants by the nose: Retention times measured across *in vivo* olfactory mucosa. Science **181:** 1247-1249.
3. MACKAY-SIM, A., P. SHAMAN & D. G. MOULTON. 1982. Topographic coding of olfactory quality: Odorant-specific patterns of epithelial responsivity in the salamander. J. Neurophysiol. **48:** 584-596.
4. LAING, D. G. 1985. Human perception of odours. Neurosci. Lett. Suppl. 19. Abstract S11.
5. BELL, G. A. 1984. Application of the radioactive 2-deoxyglucose technique to the study of the sense of smell. Proceedings of the Australian Society of Biophysics, 9th Annual Meeting. Wollongong, Australia.
6. BENSON, T. E., G. D. BURD, D. M. D. LANDIS & G. M. SHEPHERD. 1985. High resolution 2-deoxyglucose in quick-frozen slabs of neonatal rat olfactory bulb. Brain Res. **339:** 67-78.
7. GETCHELL, T. V., G. L. HECK, J. A. DeSIMONE & S. PRICE. 1980. The location of olfactory receptor sites. Inferences from latency measurements. Biophys. J. **29:** 397-412.

8. VON BEKESY, G. 1964. Olfactory analogue to directional hearing. J. Appl. Physiol. **19:** 369-373.
9. LEVETEAU, J. & P. MACLEOD. 1969. Reciprocal inhibition at glomerular level during bilateral stimulation. *In* Proceedings of the International Symposium on Olfaction and Taste. C. P. Pfaffman, Ed. **3:** 212-215. Rockefeller University Press. New York.
10. LAING, D. G. 1987. An investigation of the mechanisms of odor suppression using physical and dichorhinic mixtures. Behav. Brain Res. In press.

Psychophysical Models for Mixtures of Tastants and Mixtures of Odorants

JAN E. R. FRIJTERS

Department of Market Research
Department of Food Science
The Netherlands' Agricultural University
6703 BC Wageningen, the Netherlands

INTRODUCTION

The relation between one or more perceived attributes of compounds and one or more attributes of mixtures of these substances can be influenced by various factors. As several authors (e.g. Berglund *et al.*,[1] Gillan,[2] and Lawless[3]) have pointed out, there are many opportunities at different levels in the organism for interaction between a pair of tastants or between a pair of odorants in a mixture, before some kind of perceptual interaction at the central level may occur. The first interaction(s) may already take place at the presensory level as a result of a chemical reaction, mutual inhibition of physical activity,[4] or change of the structure of the medium.[5] Molecules of different compounds can also interact at a peripheral-sensory level, where they may limit each other's accessibility to the receptor(s).[6,7] These interactions should be distinguished from electrophysiological interactions in the neural structures, such as the medulla, hypothalamus, chorda tympani, or bulbus olfactorius. Central interactions occur at the level of the sensory stores, short-term memory, and stored percepts. At this stage factors like selective attention, coding, classification, arousal, and habituation operate to influence odor and taste perceptions. Psychological contrast and assimilation effects may potentially change the intensities and qualities of the perceived component tastes or odors compared to their perceived attributes when tasted or smelled outside the mixture. The psychophysical and sensory psychological study of mixture phenomena is directed at the central level of interaction. This paper deals with the models that are concerned with this field.

DEFINITIONS AND NOMENCLATURE

Psychophysical knowledge about chemosensory mixture phenomena can only be obtained by comparing perceived olfactory or gustatory attributes of single chemicals to such attribute(s) of mixtures of these compounds. This is a basic rule. In all reported mixture studies, the qualities and/or intensities of perceived tastes or odors of two or more chemicals (tastants or odorants) of known concentrations are compared to some perceived attribute of a mixture of these compounds. Such a mixture attribute may be the quality (or qualities) of the taste(s) or odor(s): It can be the intensity of a specific taste or odor quality, or it may be the total intensity of the mixture's taste or smell. Sometimes the inquiry concerns the relationship between the intensities of different taste or odor sensations as elements of a more complex mixture impression. All these studies have in common that some operational definition is used for the concept of "taste" or "odor" quality, that some scale or unit is used to express "intensity," and that a particular comparison rule is followed to assess the kind and / or magnitude of the interaction.

Which perceived attribute(s) of single compounds can be meaningfully compared to such attribute(s) of a tastant-tastant or an odorant-odorant mixture? First of all this depends on the kind of percept that is associated with the mixture. Berglund *et al.*[1] have tried to systematize the relations between attributes of components and attributes of mixtures toward the development of a nomenclature. They distinguish between homogeneous and heterogeneous percepts of smell, a distinction that can be applied to taste as well. A homogeneous percept is formed when two odors (not odorants) blend completely into a new integrated percept when mixed (quality α plus quality β results in a mixture's quality γ). According to this definition, a homogeneous percept is formed when a new quality is obtained, and if the mixture's percept consists of one quality. In order to obtain such a percept, the system blends, fuses, or synthesizes the original qualities into a new one. I have found that the emergence of a new quality is not essential. One can also speak of a homogeneous percept if the same odor (or taste) quality is added to itself without changing the perceived complexity. A heterogeneous percept is obtained, according to Berglund *et al.*,[1] when two odors (or tastes) do not blend or fuse in the mixture but remain distinct (quality α plus quality β results in a mixture's percept having both quality α and quality β).

These definitions imply that homogeneity is identical to singularity and heterogeneity to complexity. Furthermore, they presuppose that sensory qualities are readily discriminable and classifiable, and that isolated qualities function as basic elements ("building bricks") for a percept. This view originates from the structuralistic postulate that a "content of consciousness" can be reconstructed from mental elements of which sensations are one kind. Empirical research on singularity-complexity has generated a continuing debate, especially in taste research, and remains a matter of controversy.[8-13] This issue is not pursued further here. It is sufficient to conclude that single tastants do not always result in single perceived tastes, and that mixtures of tastants do not always lead to heterogeneous percepts, although O'Mahoney *et al.*[12] found that perceived complexity tends to increase with actual physical complexity of a taste mixture.

In addition to qualitative interactions, quantitative interactions may occur between tastants or between odorants in a mixture. Conclusions about quantitative interactions are usually drawn by comparing the intensities of the qualities of the components as assessed outside the mixture to the intensity of the mixture's taste or odor, or to the intensity of a particular quality in that mixture. Berglund *et al.*'s terminology[1] distinguishes two primary types of comparisons depending on the nature of the mixture's percept.

A homogeneous percept is formed by a single quality, and this may smell or taste stronger, equal, or weaker than the sum of the intensities of the single qualities as assessed outside the mixture. These interactions are called hyper-addition ($\Psi_\gamma > \psi_\alpha + \psi_\beta$), complete addition ($\Psi_\gamma = \psi_\alpha + \psi_\beta$) and hypo-addition ($\Psi_\gamma < \psi_\alpha + \psi_\beta$), respectively. (The symbol Ψ is used throughout this paper to denote a sensory attribute of a mixture, or a sensory attribute of a component in a mixture. The symbol ψ stands for a sensory attribute of a component outside a mixture. A Greek subscript stands for a taste or odor quality, a Roman subscript refers to a chemical). However, it is perfectly legitimate to compare the mixture's homogeneous intensity to each of the single qualities outside the mixture. If the mixture quality intensity is greater than one of the single intensities ($\Psi_\gamma > \psi_\alpha$ or ψ_β), intermediate ($\psi_\alpha < \Psi_\gamma < \psi_\beta$), or smaller ($\Psi_\gamma < \psi_\alpha$ or ψ_β), the terms partial addition, compromise, and subtraction are used.

In case of a heterogeneous mixture percept, the relationship between the intensities of the tastes or odors outside the mixture and the mixture's total intensity may be the most obvious issue but not the most important one. In this case it seems that the first question to be answered concerns the relationship between the intensity of a taste or odor outside the mixture and the intensity of the same attribute in the mixture. According to the definitions of Berglund et al.[1] one should distinguish between synergism ($\Psi_\alpha > \psi_\alpha$ or $\Psi_\beta > \psi_\beta$), independence ($\Psi_\alpha = \psi_\alpha$ or $\Psi_\beta = \psi_\beta$), and antagonism ($\Psi_\alpha < \psi_\alpha$ or $\Psi_\beta < \psi_\beta$). Comparing the intensity of a taste or odor component when tasted alone to its intensity when included in a mixture provides information on the effect of the intensity of that taste or odor on the presence of another one. Obviously, a complete description of the relation between the intensities of odors and tastes outside a mixture and the total intensity (to be denoted as $\Psi_{\alpha\beta}$) of a heterogeneous mixture percept involves two discrete steps. The first is the establishment of a psychological relationship between ψ_α and Ψ_α (or/and between ψ_β and Ψ_β). The second is the development of a psychological integration rule that describes the relation between Ψ_α and $\Psi_{\alpha\beta}$ (or/and Ψ_β and $\Psi_{\alpha\beta}$).

Conclusions about hyper-addition, complete addition, and hypo-addition require that the intensity of a taste or odor sensation be assessed with a ratio scale, since these types of comparisons include an arithmetic additive operation (called *summated response comparison*[14]). On the contrary, conclusions about partial addition, compromise, subtraction, synergism, independence, and antagonism require assessment at least on an ordinal scale, but not necessarily on an interval or ratio scale. The issue of most adequate scaling of odors and tastes (and other sensations) has been discussed extensively in numerous papers. No unanimous conclusions have been reached, especially not about the metric properties to be assigned to the various types of scales that can be used. I do not intend to deal with scaling issues, for example, whether category scales or visual analogue scales have interval-scale properties, or whether scales obtained by "ratio scaling" techniques have ratio properties, although there is sufficient evidence now to conclude that they have not.[15] The most objective judgment of claims about the properties of a scale used in a particular experiment evolves from an additional analysis,[16] which shows the internal consistency of the scale.[17]

MIXTURE MODELS

Most mixture studies are restricted to the use of one or a few concentrations of two compounds for mixture construction. In a minority of studies, the effect of a fixed concentration of one compound on the psychophysical function of another substance has been determined.[18–22] No models are available for the prediction of a mixture's quality on the basis of the quality of the components of the mixture. Also, a paucity of models exist that describe or predict the degree of complexity of a mixture. The models developed so far are quantitative expressions for the prediction of a quantitative response to a mixture's taste or odor intensity. Five models are briefly discussed: the addition model, the substitution model, the vector summation model, the equiratio model, and Beidler's mixture equation.

Addition Model

This model states that the intensity of a taste mixture is simply the sum of the intensities of its components when tasted in isolation.[23-25] The basic assumption is that the gustatory system treats the two substances in a binary mixture separately, and that the two perceived intensities of the components are then added at a central neural level. There is no peripheral or neural interaction and the psychological integration is additive. This model was specified in conjunction with the psychophysical power law, and it assumes that magnitude estimation as a scaling procedure results in a true ratio scale. The model makes no distinction between homogeneous and heterogeneous mixture percepts (hence in this section the notation Ψ_{mix} is used instead of Ψ_γ or $\Psi_{\alpha\beta}$). If the psychophysical functions of the substances A and B are given by, respectively,

$$\psi_{\alpha \cdot i} = k_a \phi_{a \cdot i}^m \text{ and } \psi_{\beta \cdot j} = k_b \phi_{b \cdot j}^n$$

then the predicted mixture intensity $\Psi_{mix \cdot ij}$ of the mixture $\Phi_{ab \cdot ij}$ ($= \phi_{a \cdot i} + \phi_{b \cdot j}$) is given by:

$$\Psi_{mix \cdot ij} = \Psi_{\alpha \cdot i} + \Psi_{\beta \cdot j} \qquad (1)$$

under the assumption that $\psi_{\alpha \cdot i} = \Psi_{\alpha \cdot i}$ and $\psi_{\beta \cdot j} = \Psi_{\beta \cdot j}$.

From Equation 1 it follows that:

$$\Psi_{mix \cdot ij} = k_a \phi_{a \cdot i}^m + k_b \phi_{b \cdot j}^n \qquad (2)$$

Although the model was designed to describe the mixture interaction between substances that are processed independently at peripheral and neural levels, Moskowitz used acids[25] and sugars[23,24] for mixture composition in experiments designed to test this model. These substances are not independently processed since they show cross-adaptation to various degrees.[26,27] It has been pointed out that the addition model must lead to erroneous predictions in such a case.[28,29]

Substitution Model

In contrast to the addition model, the substitution model has been formulated especially for peripherally dependent substances. Moskowitz[24] expressed the basic idea as follows: "The sensory system adds together the concentrations of the mixture components, treats the sum as a higher concentration of the reference chemical, and then transforms the concentration of the reference chemical into subjective magnitude according to an intensity function appropriate for the reference chemical." This model thus assumes that there is no peripheral competition between molecules of the different substances in the mixture, and also, that the (unknown) psychophysical function of the mixture becomes equal to that of the reference chemical. Like the addition model, this one is also based on the assumption that magnitude estimation results in a true ratio scale. It makes no distinction between homogeneous and heterogeneous percepts. Again, let the psychophysical functions of the substances A and B be given by:

$$\psi_{\alpha \cdot i} = k_a \phi_{a \cdot i}^m \text{ and } \psi_{\beta \cdot j} = k_b \phi_{b \cdot j}^n$$

then on setting ψ_α equal to ψ_β one obtains:

$$k_a \phi_{a \cdot i}^m = k_b \phi_{b \cdot j}^n \text{ and hence } \phi_{a \cdot i} = (k_b \phi_{b \cdot j}^n / k_a)^{1/m}$$

For a binary mixture of A and B of concentration $\Phi_{ab \cdot ij}$ $(= \phi_{a \cdot i} + \phi_{b \cdot j})$, the substitution model predicts the mixture intensity $\Psi_{mix \cdot ij}$ by:

$$\Psi_{mix \cdot ij} = k_a (\phi_{a \cdot i} + (k_b \phi_{b \cdot j}^n / k_a)^{1/m})^m \tag{3}$$

In Equation 3, the concentration $\phi_{b \cdot j}$ has been expressed in a perceptually equally intense concentration of the "reference" chemical A. It has been shown[28] by derivation that the predicted mixture intensity of a particular binary mixture depends on which of the two substances is chosen as the reference chemical. From this observation it was concluded that this model is invalid because of lack of internal consistency. This conclusion is irrespective of the plausibility of the assumptions of the model (see also De Graaf and Frijters.[30])

Vector Summation Model

The vector summation model is a psychological model developed in odor research[31] for the prediction of a homogeneous odor intensity of a mixture (Ψ_γ) on the basis of the intensities of the odors of the unmixed components (ψ_α and ψ_β in case of a binary mixture). These intensities assessed by magnitude estimation (although category scales have also been used[32]) are represented as vectors in a plane. The predicted intensity of a binary mixture is given by:

$$\Psi_\gamma = (\psi_\alpha^2 + \psi_\beta^2 + 2\psi_\alpha \psi_\beta \cos x_{\alpha\beta})^{0.5} \tag{4}$$

The angle $x_{\alpha\beta}$ is equal to the angle between the vectors ψ_α and ψ_β. It reflects the degree of fusion or blending of odors within a mixture (that is how Ψ_α and Ψ_β interact, not ψ_α and ψ_β). Implicit in this model are the assumptions that an odor of low perceived intensity has a weak influence on the odor intensity of the mixture, and also, that two odors of equal perceived intensity interact maximally.[1]

The value $\cos x_{\alpha\beta}$ can never be larger than one, so that the model excludes potential hyperaddition. Since virtually all odorants investigated produced compressed psychophysical power functions, this seems not to be a serious deficiency, because these observations make it highly improbable that the (unknown) psychophysical mixture function would have a different shape. If it is correct that the psychophysical function for a binary mixture is also a compressed one, hyperaddition can never occur. (See, however, the section: Some Generalizations Reconsidered.)

The interactions that can be described by the model are complete addition ($x = 0°$) or hypoaddition (maximally when $x = 180°$). As noted already by Jones and Woskow[33] interactions of this nature are predicted by the general rule that some degree of hypoaddition usually occurs in this situation. Cain[34] has noted in reviewing the literature that $x_{\alpha\beta}$ always lies between 105-130 degrees,[1,32,35-37] so that the magnitude of this angle can no longer be considered specific for (the qualities of) a pair of odorants, or related to the similarities between the odor qualities. He concluded that the invariance of the angle over different pairs of odors suggests that this parameter probably represents a stable feature of the psychological integration process. Although this may be the case, it should be borne in mind that integration would occur at the level of Ψ_α and Ψ_β, not at the level of ψ_α and ψ_β.

Since the model predicts Ψ_γ on the basis of ψ_α and ψ_β, it is clear that interactions at receptor and neural levels occurring in perceiving a mixture of ϕ_a and ϕ_b are also not accounted for by the model.

Some authors[38,39] have proposed modifications of the model; others[39-41] generalized it to more than two component mixtures. The main conclusion from these studies is that the original model, or its modifications, underestimate the higher order mixture intensities.

Laing et al.[32] summarized the findings obtained by binary mixture studies, most of which employed the vector summation model, as follows:

> The intensity of odorants not their quality determines the contribution of each to the quality of a mixture. (2) In mixtures of unequal (unmixed) intensity odorants; (a) the stronger odorant predominates, or is the only one perceived; (b) suppression exerted by the stronger odorant is far greater than by a weaker one. (3) In mixtures of equal (unmixed) intensity odorants: (a) both odorants are perceived: both odorants suppress each other equally, with maximum interaction occurring between two high-intensity odorants and very little when both are weak. (4) Dissimilar odorants do not blend to form a new odor. (5) The total intensity of a mixture is less than the sum of the intensities of the components, but never less than the intensity of the weaker component. (6) Suppression is the most common result of interaction between components, and its magnitude is greatest with components of high intensity. (7) Synergism is rare but may occur when both components are weak.

Note that these authors use the term odorant for the odor of a component and not for the substance, as was the case in the previous part of this paper.

Some Generalizations Reconsidered

Pfaffmann et al.[42] and later Moskowitz[23-25] draw some general conclusions about mixture interactions in taste. They generalized that mixtures of substances of the same quality have a perceived taste intensity that is equal to or greater than the simple sum of the perceived intensities of the components. They also concluded that in mixtures of substances with different taste qualities, the perceived intensity of a mixture is less than the simple sum of the perceived intensities of the components. Bartoshuk[2] challenged these generalizations and postulated instead that the similarity or dissimilarity of the taste qualities of the unmixed components is a dimension that is unrelated to the kind of interaction that occurs in a mixture. According to her view, the essential characteristics of the components dictating the kind of interaction that will occur are the shapes of the psychophysical functions of the components outside the mixture. Because Meiselman[44] had been able to demonstrate that these forms can be manipulated by changing the stimulus delivery procedure, Bartoshuk and Cleveland[45] were in a position to produce the experimental evidence required to support this hypothesis. In their study with mixtures of sugars and mixtures of acids, they showed that there exists an (unquantified) association between the shapes of the functions of the mixture's components and the intensity of a mixture's taste. Because they used substances with similar tastes, the operation of two possible confounding factors was prevented. The first is heterogeneity of the mixture's percept. Substances with similar taste have a greater likelihood of producing a homogeneous percept when mixed, so that the subject's taste response is less likely to be affected by the psychological integration process of combining two (or more) different taste qualities to obtain the total intensity of the mixture's taste. The second potential confounder is central mixture suppression.[46,47] Substances with similar tastes are not likely to cause central mixture suppression, whereas substances with different taste qualities that are not near threshold concentrations may produce suppression.[3,48-51]

Bartoshuk's view that the psychophysical functions of the components of a mixture determine the mixture's taste intensity has changed in the course of time. Initially[43,45] she held the position that this rule can be applied to mixtures of components with similar and also to such mixtures with dissimilar tasting constituents. Her later position[52] is adequately reflected by the following quotation: "We can easily imagine two chemically different substances that are identical to the taste system because they share a structure that is an effective taste stimulus. Such stimuli would be expected to produce psychophysical functions of the same shape and they would, of course, add along that common function. However, there is no obvious reason why mixtures of substances with different tastes should be related to the shapes of their psychophysical function." In this outlook the emphasis is still being put on *the* shape of the psychophysical function, despite the fact that the form of such function can be influenced by the experimental procedure; expanded functions can even be changed into compressed functions. Moreover, some subjects may have expanded functions while others may have compressed ones for the same substance under identical experimental conditions (e.g., Frijters[53]). It seems, therefore, that the idea of a substance-specific shape is dubious.

Frijters and Ophuis[28] postulated that there is a direct relationship between the psychophysical functions of the components of a mixture and a different psychophysical function of a particular mixture "type" (see next section) if each of these is determined under the same experimental conditions. The mixture function is not a "common" function of the first two, but a separate one. Its identity depends on the mixing ratio of the components. According to Frijters and Ophuis,[28] for mixtures of substances with similar tastes, the relationship between the psychophysical functions of the components and a particular mixture function can expected to be simple, because it is not confounded with effects resulting from central mixture suppression or the influence of psychological integration due to heterogeneity of the mixture's percept. Following this line of reasoning, we see that the main problem to be resolved is not the determination of a specific kind of interaction that occurs in a particular case, because, as discussed earlier, interaction phenomena are not specific for the substances or taste qualities only, but depend also on the stimulus procedure, the scaling method used, the concentration levels of the substances,[54] the degree of adaptation,[3] and the organism's state of habituation.[46] The main task to be accomplished is proper modeling of the relation between the psychophysical functions of the components and the psychophysical mixture function. Once a valid model of this kind is available, the question as to whether hyper-addition, complete addition, or hypo-addition has occurred in a particular mixture is a matter of secondary concern, since any comparison that one would wish to make can simply be done on the basis of these three known functions.

Equiratio Mixture Model

The purpose of the equiratio mixture model is to predict the sensory response to the intensity of a mixture (with a homogeneous percept) on the basis of the concentrations and mixing ratio of the components (having a similar taste quality) in a binary mixture. The model incorporates a specific relationship between the psychophysical taste functions of the substances to be used for mixture composition, and a psychophysical equiratio mixture function.

Since the concept of "equiratio mixture" is relatively new, it should first be explained. Any mixture of two substances is characterized in a physical sense by two factors. One of these is the total concentration of substance in the mixture, and the other is the mixing ratio of the substances. For example, a mixture of 1 *M* glucose and 1 *M* fructose contains 2 *M* of substance and has a mixing ratio of 0.50:0.50. Using the same substances and the same mixing ratio, a mixture of a total molarity of 1 *M* contains 0.5 *M* glucose plus 0.5 *M* fructose. When the dilution is continued using the same factor, the next mixture has a molarity of 0.50 *M* and so forth. An equiratio mixture type is therefore defined as a series of mixtures of different concentrations, in which the mixing ratio is held constant.[28] By changing the mixture ratio systematically, an unlimited number of different "types" of equiratio mixtures for each pair of substances can be constructed. Taking the total molarity as an independent variable, a psychophysical power function can be established in the same way as for an unmixed substance. The general equation is given by:

$$R_{ab \cdot ij \cdot pq} = k_{ab \cdot pq} \Phi^u_{ab \cdot ij \cdot pq} \qquad (5)$$

where R is the sensory response, Φ is the total molar concentration of a mixture, and k and u are constants to be estimated. The subscripts a and b refer to the substances, i and j to the concentrations of A and B, respectively, and p and q to the mixture proportions. It should be noted that Frijters and Ophuis[28] made a formal distinction between sweetness response (R) and perceived taste intensity (Ψ or ψ); the sensory response may not be a linear behavioral representation of the perceived taste intensity. The relationship between the perceived taste intensity and the sensory response is described by a response output function.

In case of a homogeneous mixture percept with similar tasting substances that are sensory dependent (either at the receptor site or in the afferent channel), a particular equiratio mixture function can be predicted on the basis of the psychophysical functions of the constituent mixture components. Under these conditions, the exponent of the mixture function can be estimated on the basis of the exponents of the psychophysical power functions of the substances A and B (being m and n, respectively) as follows:

$$\hat{u} = pm + qn \qquad (6)$$

The constant $k_{ab \cdot pq}$ in Equation 5 can be obtained from:

$$\hat{k}_{ab \cdot pq} = (pk_a/\phi_{a \cdot s} + qk_b/\phi_{a \cdot s})/(p/\phi_{a \cdot s} + q/\phi_{b \cdot s}) \qquad (7)$$

where k_a and k_b are the constants of the power functions of A and B, and $\phi_{a \cdot s}$ and $\phi_{b \cdot s}$ are the concentrations of the substances A and B, respectively, which give rise to a sensory response of the same magnitude as the response given to the "standard stimulus" in a magnitude estimation experiment. Since any psychophysical equiratio mixture function can also be determined experimentally, it is possible to test the model. Such a test consists of a comparison of the function as predicted on the basis of the Equations 6 and 7 to the function established on the basis of experimental data. In three of such tests, two[28,55] with sweet substances and one[56] with sour compounds, the equiratio mixture model appeared to predict the psychophysical functions of the mixtures with great precision.

Beidler's Mixture Equation

In his theory of taste stimulation, Beidler[57] postulated that adsorption of stimulus molecules to receptor sites elicits a neural response of a magnitude proportional to the number of occupied receptor sites. Because the number of sites is limited, the response magnitude (in most electrophysiological studies, the integrated whole-nerve chorda tympani response) approaches asymptotically to a maximum response at high concentrations. In addition to the equation for single compounds, Beidler[58,59] proposed a mixture equation to describe the peripheral interaction of two taste substances under the condition that the stimulus molecules of both substances compete for adsorption at the same receptor sites. According to this equation, the magnitude of the response to a mixture of concentration i of substance A ($= \phi_{a \cdot j}$) plus the concentration j of substance B ($= \phi_{b \cdot j}$) is given by:

$$R_{ab \cdot ij} = \frac{k_a \phi_{a \cdot i} R_{s \cdot a} + k_b \phi_{b \cdot j} R_{s \cdot b}}{1 + k_a \phi_{a \cdot i} + k_b \phi_{b \cdot j}} \tag{8}$$

where $R_{ab \cdot ij}$ = response to mixture of A and B; $R_{s \cdot a}$, $R_{s \cdot b}$ = responses to A and B, respectively; and k_a, k_b = association constants of A and B, respectively.

This equation was generalized onto a perceptual level by Curtis et al.[60] This means that $R_{ab \cdot ij}$ was set equal to $\Psi_{\gamma \cdot ij}$ and $R_{s \cdot a}$ and $R_{s \cdot b}$ were replaced by $\psi_{s \cdot \alpha}$ and $\psi_{s \cdot \beta}$. These investigators evaluated the mixture equation in a psychophysical experiment using the method of magnitude estimation. They concluded that the equation provides "an excellent description of the psychophysical relation for mixture data, if it is assumed that a nonlinear response transformation [between perceived intensity and sensory response] is introduced in judgement." Unfortunately, the shape of such output transformation is generally not known.[61-63] A nonlinear response output function implies that the sensory taste responses are nonlinear with perceived taste intensities; even if the relationship between a mixture's composition and concentration on one hand, and its perceived intensity on the other hand, could be described in principle by Beidler's mixture equation, a psychophysical power function obtained by magnitude estimation would obscure this relationship. Another problem encountered when attempting to determine the equation in a psychophysical experiment is the impracticability of finding the maximum response to a particular substance.

In a study with glucose-fructose mixtures[64] both problems were bypassed by using an indirect psychophysical method (which does not include a numerical response output function) and elimination of $\psi_{s \cdot \alpha}$ and $\psi_{s \cdot \beta}$ through substitution, so that the equation could be tested at five specified taste intensity levels. The results showed that at the lowest sweetness level (0.125 M glucose), the equation predicted with great precision. At the highest level (2 M glucose), the prediction error was about 9%. In the same study two variations of a mixture equation based on Heck and Erickson's rate theory[65] were derived. These appeared to result in the same predictions as Beidler's mixture equation.

CONCLUSION

Five models were reviewed. The addition and substitution model are now merely of historical interest, because these are either internally inconsistent or the description of mixture interaction phenomena is inadequate in the light of empirical results. The vector summation model describes the psychological integration of component sensations into a mixture's percept. It is limited in its explanatory power because it cannot describe hyper-addition. Empirical evidence shows that for more complex mixtures it results in underestimation of the mixture's intensity and shows also that it has little discriminative ability. Psychological integration seems to be very similar for different pairs of odors. The equiratio model accounts for the sensory response to the intensity of a mixture of sensory-dependent substances having similar taste qualities, and having a homogeneous percept. The model does not contain parameters related to central mixture suppression or heterogeneity of percepts. Beidler's mixture equation can only be used in the same circumstances as the equiratio model. Explicit is the assumption of mutual competition between molecules of the substances used for mixture composition.

A few aspects of mixture interaction deserve further investigation. These are the effects caused by the state of the organism, such as degree of adaptation and state of habituation. Studies should also be carried out on attention processes, rules of integration, and the relation between perceived intensity or quality and sensory response. An explicit separation of psychophysical relationships (stimulus-sensation) and psychometrical ones (sensation-response) deserves further study.[17,54]

REFERENCES

1. BERGLUND, B., U. BERGLUND & T. LINDVALL. 1976. Psychological processing of odor mixtures. Psychol. Rev. **83:** 432-441.
2. GILLAN, D. 1982. Mixture suppression: The effect of spatial separation between sucrose and NaCl. Percept. Psychophys. **32:** 504-510.
3. LAWLESS, H. T. 1979. Evidence for neural inhibition in bittersweet taste mixtures. J. Comp. Physiol. Psychol. **93:** 538-547.
4. BARROW, G. M. 1974. Physical Chemistry for the Life Sciences. McGraw-Hill, New York.
5. MATHLOUTI, M., A. SEUVRE & G. G. BIRCH. 1986. Relationship between the structure and the properties of carbohydrates in aqueous solutions: Sweetness of chlorinated sugars. Carbohydr. Res. **152:** 47-61.
6. BEIDLER, L. M. 1978. Biophysics and chemistry of taste. *In* Handbook of Perception, Vol. VIA.: Tasting and Smelling. E. C. Carterette & M. P. Friedman, Eds.: 21-49. Academic Press. New York.
7. HOLLEY, A., A. DUCHAMP, M. REVIAL, A. JUGE & P. MACLEOD. 1974. Qualitative and quantitative discrmination in the frog olfactory receptors: Analysis from electrophysiological data. Ann. N. Y. Acad. Sci. **237:** 102-114.
8. ERICKSON, R. P. 1982. Studies on the perception of taste: Do primaries exist? Physiol. Behav. **28:** 57-62.
9. KUZNICKI, J. T. & N. ASHBAUGH. 1979. Taste quality differences within sweet and salty taste categories. Sens. Proc. **3:** 157-182.
10. KUZNICKI, J. T. & N. ASHBAUGH. 1982. Space and time separation of taste mixture components. Chem. Senses **7:** 39-62.
11. KUZNICKI, J. T., M. HAYWARD & J. SCHULTZ. 1983. Perceptual processing of taste quality. Chem. Senses **7:** 273-292.

12. O'MAHONY, M., S. ATASSI-SHELDON, L. ROTHMAN & T. MURPHY-ELLISON. 1983. Relative singularity/mixedness judgements for selected taste stimuli. Physiol. Behav. **31:** 749-755.
13. SCHIFFMAN, S. S. & R. P. ERICKSON. 1980. The issue of primary tastes versus a taste continuum. Neurosci. Biobehav. Rev. **4:** 109-117.
14. DE GRAAF, C. & J. E. R. FRIJTERS. 1986. Sweetness intensity of binary mixtures lies in between intensities of equimolar concentration of the mixture's constituents. Submitted for publication.
15. BIRNBAUM, M. H. 1982. Problems with so-called "direct" scaling. *In* Sensory Methods: Problems and Approaches to Hedonics. J. T. Kuznicki, R. A. Johnson & A. F. Rutkiewic, Eds.: 34-48. ASTM Philadelphia, PA.
16. TORGERSON, W. S. 1961. Distances and ratios in psychophysical scaling. Acta Psychol. **19:** 201-205.
17. DE GRAAF, C., J. E. R. FRIJTERS & H. C. M. VAN TRIJP. 1987. Taste interaction between glucose and fructose assessed by functional measurement. Percept. Psychophys. In press.
18. CAIN, W. S. 1975. Odor intensity: Mixtures and masking. Chem. Senses Flavor **1:** 339-352.
19. MOSKOWITZ, H. R. 1971. Intensity scales for pure tastes and for taste mixtures. Percept. Psychophys. **9:** 51-57.
20. MOSKOWITZ, H. R. 1972. Perceptual changes in taste mixtures. Percept. Psychoph. **11:** 257-262.
21. STONE, H. & S. M. OLIVER. 1969. Measurement of the relative sweetness of selected sweeteners and sweetener mixtures. J. Food Sci. **34:** 215-222.
22. STONE, H., S. OLIVER & J. KLOEHN. 1969. Temperature and pH effects on the relative sweetness of suprathreshold mixtures of dextrose fructose. Percept. Psychophys. **5:** 257-260.
23. MOSKOWITZ, H. R. 1973. Models of sweetness additivity. J. Exp. Psychol. **99:** 88-98.
24. MOSKOWITZ, H. R. 1974. Models of additivity for sugar sweetness. *In* Sensation and Measurement: Papers in Honor of S. S. Stevens. H. R. Moskowitz, B. Scharf. & J. C. Stevens, Eds.: 378-388. Reidel. Dordrecht, the Netherlands.
25. MOSKOWITZ, H. R. 1974. The sourness of acid mixtures. J. Exp. Psychol. **102:** 640-647.
26. MCBURNEY, D. H. 1972. Gustatory cross adaptation between sweet-tasting compounds. Percept. Psychophys. **11:** 225-227.
27. MCBURNEY, D. H., D. V. SMITH & T. F. SHICK. 1972. Gustatory cross adaptation: Sourness and bitterness. Percept. Psychophys. **11:** 228-232.
28. FRIJTERS, J. E. R. & P. A. M. OUDE OPHUIS. 1983. The construction and prediction of psychophysical power functions for the sweetness of equiratio sugar mixtures. Perception **12:** 753-767.
29. BARTOSHUK, L. M. 1977. Psychophysical studies of taste mixtures. *In* Olfaction and Taste VI. J. Le Magnen & P. MacLeod, Eds.: 377-391. I.R.L. London, England.
30. DE GRAAF, C. & J. E. R. FRIJTERS. 1985. A note on the inconsistency of the substitution model for taste mixtures. Perception **14:** 499-500.
31. BERGLUND, B., U. BERGLUND & T. LINDVALL. 1971. On the principle of odor interaction. Acta Psychol. **35:** 255-268.
32. LAING, D. G., H. PANHUBER, M. E. WILLCOX & E. A. PITTMAN. 1984. Quality and intensity in binary odor mixtures. Physiol. Behav. **33:** 309-319.
33. JONES, F. N. & M. H. WOSKOW. 1964. On the intensity of odor mixtures. Ann. N.Y. Acad. Sci. **116:** 484-494.
34. CAIN, W. S. 1986. Olfaction. To be published in the Handbook of Experimental Psychology.
35. BERGLUND, B., U. BERGLUND, T. LINDVALL & L. T. SVENSSON. 1973. A quantitative principle of perceived intensity summation in odor mixtures. J. Exp. Psychol. **100:** 29-38.
36. LAING, D. G. & M. E. WILLCOX. 1983. Perception of components in binary odor mixtures. Chem. Senses **7:** 249-264.
37. MOSKOWITZ, H. R. & C. D. BARBE. 1977. Profiling of odor components and their mixtures. Senses Proc. **1:** 212-226.
38. LAFFORT, P. & A. DRAVNIEKS. 1982. Several models of suprathreshold quantitative olfactory interaction in humans, applied to binary, ternary, and quaternary mixtures. Chem. Senses **7:** 153-174.

39. PATTE, F. & P. LAFFORT. 1979. An alternative model of olfactory quantitative in binary mixtures. Chem. Senses Flavors **4**: 707-719.

40. BERGLUND, B. 1974. Quantitative and qualitative analysis of industrial odors with human observers. Ann. N.Y. Acad. Sci. **237**: 427-439.

41. CAIN, W. S. & M. DREXLER. 1974. Scope and evaluation of odor counteraction Ann. N.Y. Acad. Sci. **237**: 427-439.

42. PFAFFMANN, C. M., L. M. BARTOSHUK & D. H. MCBURNEY. 1971. Taste psychophysics. *In* Handbook of Sensory Physiology, Vol IV, Part 2. L. M. Beidler, Ed.: 75-101. Springer. New York.

43. BARTOSHUK, L. M. 1975. Taste mixtures: Is mixture suppression related to compression? Physiol. Behav. **14**: 643-649.

44. MEISELMAN, H. L. 1971. Effect of presentation procedure on taste intensity functions. Percept. Psychophys **10**: 15-18.

45. BARTOSHUK, L. M. & C. T. CLEVELAND. 1977. Mixtures of substances with similar tastes. Senses Proc. **1**: 177-186.

46. KROEZE, J. H. A. 1982. After repetitious sucrose stimulation saltiness in NaCl-sucrose mixtures is diminished: Implications for a central mixture-suppression mechanism. Chem. Senses **7**: 81-92.

47. KROEZE, J. H. A. & L. M. BARTOSHUK. 1985. Bitterness suppression as revealed by split-tongue taste stimulation in humans. Physiol. Behav. **35**: 779-783.

48. FABIAN, F. W. & H. B. BLUM. 1943. Relative taste potency for some basic food constituents and their competitive and compensatory action. Food Res. **8**: 179-193.

49. INDOW, T. 1969. An application of the τ scale of taste. Interactions among the four qualities of taste. Percept. Psychophys. **5**: 347-351.

50. KROEZE, J. H. A. 1978. The taste of sodium chloride: Masking and adaptation. Chem. Senses Flavors **3**: 443-449.

51. PANGBORN, R. M. 1962. Taste interrelationships III: Suprathreshold solutions of sucrose and sodium chloride. J. Food Sci. **27**: 495-500.

52. BARTOSHUK, L. M. & J. F. GENT. 1985. Taste mixtures. An analysis of synthesis. *In* Taste, Olfaction and the Central Nervous System. W. Pfaff, Ed. The Rockefeller Press. New York.

53. FRIJTERS, J. E. R. 1984. Sweetness intensity perception and sweetness pleasantness in women varying in reported restraint of eating. Appetite **5**: 103-108.

54. MCBRIDE, R. L. 1986. The sweetness of binary mixtures of sucrose, fructose and glucose. J. Exp. Psychol.: Hum. Percept. Perf. **12**: 584-592.

55. FRIJTERS, J. E. R., C. DE GRAAF & H. C. M. KOOLEN. 1984. The validity of the equiratio taste mixture model investigated with sorbitol-sucrose mixtures. Chem. Senses **9**: 241-248.

56. FRIJTERS, J. E. R. & D. A. STEVENS. 1985. Psychophysical functions of equiratio acid mixtures. Internal Report, Department Psychology. Clark University. Worcester, MA.

57. BEIDLER, L. M. 1954. A theory of taste stimulation. J. Gen. Physiol. **83**: 133-139.

58. BEIDLER, L. M. 1962. Taste receptor stimulation. Prog. Biophys. Biophys. Chem. **12**: 107-151.

59. BEIDLER, L. M. 1971. Taste receptor stimulation with salts and acids. *In* Handbook of Sensory Physiology. Vol IV. Chemical Senses, Part 2. L. M. Beidler, Ed.: 200-220. Springer-Verlag. New York.

60. CURTIS, D. W., D. A. STEVENS & H. T. LAWLESS. 1984. Perceived intensity of the taste of sugar mixtures and acid mixtures. Chem. Senses **9**: 107-120.

61. BIRNBAUM, M. H. 1980. Comparison of two theories of "ratio" and "difference" judgments. J. Exp. Psychol.: Gen. **109**: 304-319.

62. RULE, S. J. & D. W. CURTIS. 1980. Ordinal properties of subjective ratios and differences. Comment on Veit. J. Exp. Psychol.: Gen. **109**: 206-300.

63. VEIT, C. T. 1978. Ratio and substractive processes in psychophysical judgement. J. Exp. Psychol.: Gen. **107**: 81-107.

64. DE GRAAF, C. & J. E. R. FRIJTERS. 1986. A psychophysical investigation of Beidler's mixture equation. Chem. Senses **11**: 295-314.

65. HECK, G. L. & R. P. ERICKSON. 1973. A rate theory of gustatory stimulation. Behav. Biol. **8**: 687-712.

Electrophysiological Responses of Olfactory Receptor Neurons to Stimulation with Mixtures of Individual Pheromone Components[a]

ROBERT J. O'CONNELL AND ALAN J. GRANT

The Worcester Foundation for Experimental Biology
Shrewsbury, Massachusetts 01545

INTRODUCTION

Living organisms are inevitably awash in a sea of volatile organic compounds. Within this melange one usually finds, for any one species, a set of biologically relevant odorants only a fraction of which are products of the animal's own communication system. These latter substances, when considered together, may constitute a single behaviorally important multicomponent chemical signal. Often the relative amounts of the individual components have important signal value to the receiving animal as do their absolute concentrations. Many of the remaining chemical compounds may function, either singly or collectively, as chemical signals in some other context or may appear to be of interest to other individuals. Thus, it is almost always true that the multicomponent pheromonal signals that are of interest to an organism must be discriminated against a chemical background containing very large amounts of many other volatiles. In many cases we know that these compounds signal vital information to the organism upon which various life and death decisions are ultimately to be made.

The task of detecting, transmitting, filtering, and interpreting these complex messages, in the face of such a vast volatile background, is the special provenance of the olfactory system. When viewed from the perspective of individual olfactory receptors, there are two related tasks that must be undertaken. First the relevant compounds must be detected and discriminated from among the background odors, and second their relative amounts must be determined. One relatively straightforward mechanism for insuring that the resolving power required of the chemical communication system will be adequate for the initial discrimination is to build into the system a high degree of chemical specificity. That is, the enormity of the background against which the relevant discriminations must be made can be effectively reduced to what appears to be reasonably manageable levels by insuring that the primary chemical detectors in the initial stages of the system are insensitive to most of the background array of behaviorally irrelevant compounds. In those cases where several different compounds

[a] Supported by The Alden Trust and by National Institute of Neurological and Communicative Disorders and Stroke Grant NS 14453.

are integral components of an important biological signal, one may provide additional specific receptor neurons, each exclusively sensitive to one of the components in order to detect the appropriate message. When the chemical signal in question is used for long-distance communication, this range of specificity is coupled to a high degree of sensitivity. With this type of solution to the problem, there is little need for much "central" processing as most of the discrimination is handled in the periphery.

There are several obvious disadvantages to this strategy. In the first case, it requires an additional specialized receptor neuron, one for each of the discriminable stimuli in the mixture. As the number of discriminable odors in a multicomponent blend and the number of multicomponent blends themselves grows, so too does the number of specialized receptor neurons that are required. Coupled to this is the concomitant limitation on the discriminative range available to the organism. That is, only those odor compounds that have been designed into the organism's sensory capabilities are available for the signaling of additional information. Thus the chemical cues emanating from a new predator go undetected if they happen to fall outside the sensory capabilities available to the organism. Such odor blindness is relatively nonadaptive. As has been noted many times, this deficiency is easily corrected by simply increasing the bandwidth of the individual receptor neurons in the sensory system. Efficient detection and encoding of a very large number of different odor compounds can be accomplished by a relatively small number of different olfactory receptor neuron types if each group is uniquely responsive to a range of different odors. The identity of a particular stimulus may then be unambiguously decoded by the central nervous system if it evaluates, in parallel, the relative amounts of neural activity elicited across the whole population of receptor neuron types. It is important to note that the relative breadth of tuning among the different classes of olfactory receptor neurons need not be constant in this type of encoding mechanism. Thus, some fraction of the available input channels could be more narrowly tuned than others. For that matter, it should be obvious that the parallel processing of the individual components of a multicomponent blend could be abandoned altogether by designing a type of receptor neuron that is activated uniquely only by the appropriate mixture of components in a blend and that is insensitive to some or all of the individual chemicals that make up the signal. In a sense this is simply a "higher order" of chemical specificity that requires a specific mixture of odorants for activation rather than an individual compound. This leads us directly to the question under consideration here. Are the responses elicited in primary olfactory receptor neurons by behaviorally relevant mixtures of chemical stimuli fundamentally different from their responses to pure stimuli?

REVIEW

As is usual in biology, the answer to this question depends in part on the particular species of which it is asked. From nearly the earliest single-unit recordings obtained from vertebrate olfactory receptor neurons,[1] it was clear that multiple active compounds were the norm. Although the upper boundary on the number of different chemical compounds that will influence the electrical activity of an individual olfactory receptor neuron has never been firmly established, it is likely more than ten and less than one hundred. It was also first determined in the frog that mixtures of odorants often failed to produce responses that could be predicted from a knowledge of the responses elicited by their individual components.[1,2] In many cases, the responses elicited by a mixture were smaller than the sum of the responses elicited by the individual components in the blend (FIG. 1A). This suggests that each of these compounds may be competing for a common site of interaction with the neuron. However, in other receptor neurons (FIG. 1B), the responses elicited by mixtures were significantly larger than those expected from the responses elicited by the individual components. The magnitude of the response to mixtures in this latter type of receptor neuron was often several hundred percent greater than expected. This indicates the involvement of distinctly nonadditive mechanisms.

Unfortunately, recordings from vertebrates are very difficult to obtain and maintain for the time required to evaluate the interactions among several stimuli. As a consequence, little detailed information concerning the processing of multicomponent odor stimuli is available. In invertebrates, where the recording conditions are often more favorable, several studies have examined the responses of olfactory receptor neurons to mixtures of odorants.[3] In most cases these experiments involved the processing of behaviorally relevant pheromone blends. Because much more is known about the chemical composition of insect pheromone blends, it is possible to state that mixture interactions in these species may normally be quite complex. In the cabbage looper (CL) moth (*Trichoplusia ni*) the earliest component identified from female glands[4-5] and the major volatile component of the gland is (Z)7-dodecenyl acetate (Z7-12:AC). This compound is essential for flight initiation in males and was originally thought to be entirely responsible for the biological activity obtained in bioassays with female gland extracts. Subsequently, (Z)7-dodecenol (Z7-12:OH) was shown to significantly arrest upwind flight in males exposed to either calling females or to sources of synthetic Z7-12:AC.[6-8] Although this compound has occasionally been found in gland extracts, it is generally considered to arise artificially in the process of chemical separation or analysis.[9] It is thus classified, in common with other behaviorally active compounds that are not components of a female's pheromone blend, as an interspecific compound, perhaps one involved in some speciation mechanism that includes the CL. A saturated 12-carbon acetate, dodecyl acetate (12:AC), was next identified in female glands and was shown to modulate several of the close-range search behaviors of males including landing frequency and total time spent on the pheromone source.[10] Recently, four additional compounds have been found in the female gland. They have been identified chemically and implicated behaviorally as important components of the pheromone blend.[11] These compounds include: (Z)5-dodecenyl acetate (Z5-12:AC); 11-dodecenyl acetate (11-12:AC); (Z)7-tetradecenyl acetate (Z7-14:AC); and (Z)9-tetradecenyl acetate (Z9-14:AC). Although the exact roles of these latter four compounds in modulating the normal behavior of the male is still being evaluated, there is little doubt that, in wind-tunnel assays, the two previously identified CL female components (Z7-12:AC and 12:AC) elicit only about 25% of the close-to-source behaviors observed when animals are exposed to the total complement of the six female-produced components.[9,11] Moreover the more complex six-component synthetic blend elicits behavioral responses that are quantitatively equivalent to those elicited by excised virgin female pheromone glands both in terms of the number of males responding and the amount of time required for the full behavioral sequence. There can be little doubt that the whole pheromone blend of this animal elicits behavioral responses in the male that are quantitatively and qualitatively different from the sum of the behavioral responses elicited by its single components. What then is the outcome when these substances are used singly and in various combinations to evaluate the response properties of individual olfactory receptor neurons?

Extracellular action potentials have been recorded from sensilla (HS) on the male

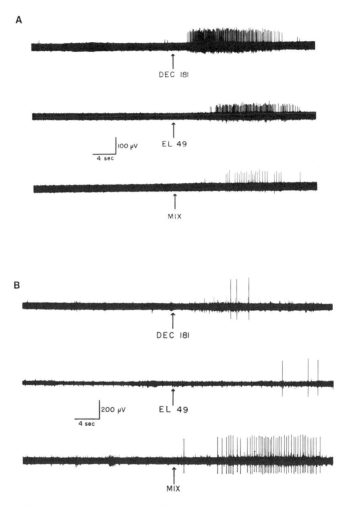

FIGURE 1. (**A**) Responses of a single frog olfactory receptor neuron to stimulation with one cm^3 of the indicated concentrations (in $\mu M/1$) of two odorants. The gas-phase mixture of these two odorants actually contained slightly less of the diethyl carbonate (DEC, 170 μM) and slightly more of the ethyl lactate (EL, 62 μM) than the single-component stimuli because of the small changes in flow rate engendered by adding the two odor streams together in the gas-sampling valve that was used to deliver odors to the exposed olfactory epithelium. (**B**) Responses of a second receptor neuron to the same three stimuli. Action potentials have been retouched for clarity.

CL antenna that are known to contain olfactory receptor neurons that are highly sensitive to the major pheromone component, Z7-12:AC. The neurophysiological and morphological characteristics of this class of sensilla has been described and contrasted with the properties of the other class of pheromone-sensitive sensilla on the antenna.[12,13] Each HS sensillum is actually innervated by two spontaneously active receptor neurons that produce action potentials that are reliably differentiated from each other by their amplitudes and waveforms. By convention, the receptor neuron producing the larger amplitude action potential is designated the A neuron and that producing the smaller is designated the B neuron. Individual HS sensilla are characterized by their length, shape, and responsiveness to low doses (0.00005-0.005 μg) of Z7-12:AC and Z7-12:OH. Five of the seven behaviorally relevant compounds have been evaluated as individual stimuli. In each case their intensities were adjusted to approximate those measured in female pheromone glands except for Z7-12:OH, which is not produced by the female and thus was evaluated at an intensity equal to that used for Z7-12:AC (0.005μg). As demonstrated in previous studies,[12-17] the average A receptor neuron in HS sensilla is reliably responsive to low doses (0.005 μg) of Z7-12:AC whereas the average B receptor neuron is responsive to comparable amounts of Z7-12:OH. Neither receptor neuron is particularly responsive to 12:AC even when larger intensities are evaluated. Neither Z5-12:AC (0.0005 μg) nor Z7-14:AC (0.00005 μg) alone, elicited consistent amounts of electrical activity in either of the HS receptor neurons. Only a few of the potential multicomponent blends for this animal have yet been evaluated. In general, addition of only 10% 12:AC to Z7-12:AC enhances the amount of neural activity elicited in A receptor neurons especially when compared to the amount of activity ascribable to the presence of Z7-12:AC alone. Although the magnitude of this increase has varied from sample to sample,[15-17] it remains impressive in the light of the relatively narrow tuning of A receptor neurons and the general failure of 12:AC to be an effective single stimulus for either of the two receptor neurons in HS sensilla.[12] This ability to synergize the responses elicited in A receptor neurons seems unique to 12:AC since the addition of the remaining four female-produced compounds to the blend failed to elicit an additional increase in the amount of discharge obtained when compared to that evoked by stimulation with the binary mixture alone.[17] However these additional materials are not without impact on neural activity because they do prevent the significant reduction in discharge expected when Z7-12:OH is added to the total blend. That is, the addition of Z7-12:OH to the binary mixture containing Z7-12:AC and 12:AC normally results in a net reduction in the response evoked by the trinary mixture. However, the response elicited by stimulation with the total blend plus Z7-12:OH remains equivalent to that obtained with the total blend alone.[17]

From this brief summary, it should be apparent that individual olfactory receptors may produce responses to odor mixtures that are greater than the sum of the responses elicited by the mixtures' components. It should be obvious from discussions elsewhere in this volume that there are a great number of mechanistic possibilities, largely drawn from the enzymological literature, that could account for both the positive and negative interactions observed. It is not possible to construct a simple explanation for these observations that involves only one unitary stimulus interaction and accounts for both the observed responses to single pheromone components and the various interaction effects observed with blends.

Recent studies in our laboratory have revealed additional, time-varying alterations in the physiological properties of the receptor neurons contained within HS sensilla. These include alterations in spontaneous activity, absolute sensitivity to pheromone components and the extent to which neural discharges are modulated by blends.

SUMMARY

Multicomponent pheromone systems are the rule in many species. As our knowledge about the number and kinds of different chemical compounds actually employed in the communication system of a particular species has increased, so too has our appreciation for the level of neurobiological complexity that must underlie these capabilities. The supposition that mixtures are differentially processed in the nervous system arises most easily when biologically relevant materials are evaluated, either singly or in multicomponent blends, with modern behavioral assay techniques. It is becoming increasingly clear that this increase in the chemical and behavioral complexity of a particular communication system must be paralleled by an increase in the efficiency of the physiological mechanisms employed for the neural encoding of behaviorally relevant odor compounds and blends.

Here we review several studies that have examined the electrical activity elicited in primary olfactory receptor neurons when they are stimulated with mixtures of odorants. Particular attention is given to the responses elicited in a subset of the individual pheromone-sensitive sensilla on the antennae of male cabbage looper moths (*Trichoplusia ni*). Electrophysiological responses to single- and multiple-component stimuli, each drawn from among the seven known behaviorally active compounds for this insect, were obtained at several different stimulus intensities. Both (Z)7-dodecenyl acetate and (Z)7-dodecenol were effective stimuli for both of the receptor neurons found in one of the two classes of pheromone-sensitive sensilla, even at relatively low stimulus intensities (0.0005 μg). Dodecyl acetate, although behaviorally active, did not significantly excite either of these receptor neurons. However, when mixed with either of the unsaturated components, it significantly enhanced the receptor neuron's response to its appropriate parent compound only in the middle range of stimulus intensities. A mixture of all three components did not show this enhancement and at the middle range of intensities actually elicited reduced responses when compared to those elicited by appropriate amounts of any of the one- and two-component stimuli evaluated.

Thus, some blends elicited electrical responses from primary olfactory receptor neurons that were not readily predicted from a knowledge of the receptor neurons' response to individual components.

REFERENCES

1. GESTELAND, R. C., J. Y. LETTVIN, W. H. PITTS & S. H. CHUNG. 1968. *In* Cybernetic Problems in Bionics. H. L. Oestreicher & D. R. Moore, Eds.: 313-322. Gordon & Breach. New York.
2. O'CONNELL, R. J. 1967. Quantitative Stimulation of Single Olfactory Receptors. Ph.D. Thesis. SUNY Upstate Medical Center. Syracuse, NY.
3. DERBY, C. D. & B. W. ACHE. 1984. J. Neurophysiol. **51:** 906-924.
4. IGNOFFO, C. M., R. S. BERGER, H. M. GRAHAM & D. F. MARTIN. 1963. Science **141:** 902-903.
5. BERGER, R. S. 1966. Ann. Entomol. Soc. Am. **59:** 767-771.
6. TOBA, H. H., N. GREEN, A. N. KISHABA, M. JACOBSON & J. W. DEBOLT. 1970. J. Econ. Entomol. **63:** 1048-1051.
7. TUMLINSON, J. H., E. R. MITCHELL, S. M. BROWNER, M. S. MAYER, N. GREEN, R. HINES & D. A. LINDQUIST. 1972. Environ. Entomol. **1:** 354-358.

8. McLaughlin, J. R., E. R. Mitchell, D. L. Chambers & J. H. Tumlinson. 1974. Environ. Entomol. **3:** 677-680.
9. Linn, C. E. Jr., L. B. Bjostad, J. W. Du & W. L. Roelofs. 1984. J. Chem. Ecol. **10:** 1635-1658.
10. Bjostad, L. B., L. K. Gaston, L. L. Noble, J. H. Moyer & H. H. Shorey. 1980. J. Chem. Ecol. **6:** 727-734.
11. Bjostad, L. B., C. E. Linn Jr., J. W. Du & W. L. Roelofs. 1984. J. Chem. Ecol. **10:** 1309-1323.
12. O'Connell, R. J., A. J. Grant, M. S. Mayer & R. W. Mankin. 1983. Science **220:** 1408-1410.
13. Grant, A. J. & R. J. O'Connell. 1986. J. Insect Physiol. **32:** 503-515.
14. O'Connell, R. J. 1985. Experientia **42:** 232-241.
15. O'Connell, R. J. 1985. *In* Mechanisms of Perception and Orientation to Insect Olfactory Signals. T. L. Payne, M. C. Birch & C. E. J. Kennedy, Eds.: 217-224. Oxford University Press. Oxford, England.
16. O'Connell, R. J. 1985. J. Comp. Physiol. A **156:** 747-761.
17. O'Connell, R. J., J. T. Beauchamp & A. J. Grant. 1986. J. Chem. Ecol. **12:** 451-467.

Odor-Taste Mixtures

DAVID E. HORNUNG [a] AND MELVIN P. ENNS [b]

[a]Department of Biology and
[b]Department of Psychology
St. Lawrence University
Canton, New York 13617

The papers of this symposium have so far considered the question of mixtures within a single modality. The purpose of this paper is to consider the perception that arises when the sensation of smell is combined with the sensation of taste, that is, to consider mixtures that involve both modalities. This discussion will focus on two primary issues. The first concerns the relationship between the intensity of flavor and the intensity of the component parts, smell and taste. The "curiously unitary character" of perceived flavors[1] may cause one to question the appropriateness of studying the relationship between the taste and smell components of complex mixtures.[2] However, the assumption made in this paper is that by analyzing some of the elements of flavor, it may be possible to better understand the manner in which this perceptual process is achieved.

The second issue concerns the effect that taste has on the perception of smell and vice versa. Early textbooks on sensation[1,3] noted that, when a complex food was sampled, subjects often mistakenly labeled sensations that were derived from odorants as tastes, and thus taste was often perceived as more important. However, only recently have investigators attempted to unravel the influences that each modality has on the other.

THE ADDITION OF TASTE AND SMELL

In recent attempts to describe and evaluate the "anatomy" of flavor, one general theme seems to emerge. Estimates of the intensity of olfaction and gustation add together in some way to produce the sensation of flavor. Whether the additivity is complete or incomplete appears to depend on how the intensities of the odorants and tastants have been determined. On the one hand, Murphy and her coworkers[4,5] demonstrated that the estimate of the "overall intensity" of a smell-taste mixture was equal to the sum of the overall intensity estimates of the odorant and the tastant when each was unmixed. This result has been demonstrated when the odorant was placed in the mouth[4,5] or at the external nares.[6]

At the other extreme, complete additivity was not observed when the estimate of overall intensity of the taste-smell mixture was compared to the sum of the intensity estimates of the smell and taste of the unmixed components.[7-9] This lack of complete additivity has also been reported when the odorant was presented to the external nares or in the mouth.[6]

The conclusion from these studies seems to be that when subjects are asked to estimate the "overall intensity" of a particular concentration of an odorant or tastant, on the average, the stimuli are rated as less intense than when the subjects are asked to estimate the "intensity of the smell" or the "intensity of the taste" of the same concentration of odorant or tastant, respectively.[6,9]

Hornung and Enns[6] have recently proposed a mathematical model to describe the relationship between the estimates of the overall intensity of a taste-smell mixture to the smell and taste of the unmixed components. They suggest that when estimating overall intensity, the psychophysical functions of taste and smell are reduced by a constant amount. This relationship is:

$$\Psi_{\text{Overall Intensity}} = k_s (\Psi_{\text{Smell}}) + k_t (\Psi_{\text{Taste}})$$

The model is conceived such that overall intensity of a complex mixture can be predicted from the "corrected" intensities of the taste and smell components. These corrections represent a measure of the amount by which smell and taste are reduced when the subject estimates overall intensity. The product of k_s and the intensity of the smell is equal to the overall intensity that one would assign to the smell when the taste was equal to zero. Likewise, the product of k_t and the intensity of the taste is equal to the overall intensity that one would assign to the taste when smell was equal to zero.

As the model predicts, and as the experimental data confirm,[6] compared to taste alone, there is a reduction in the perceived overall intensity even when only water vapor is in the nose. Likewise, compared to smell alone, there is a reduction in the overall intensity when water is present in the mouth. These reductions occur even though there is no apparent opportunity for cross-modal suppression, masking, or interference. These observations thus suggest that the reductions occur as a result of a central (cognitive) mechanism.

THE INFLUENCES OF SMELL ON TASTE AND TASTE ON SMELL

In 1824 Chevreul[10] reported that when the nostrils were closed, all that could be observed as the result of placing a piece of camphor gum on the tongue was a peculiar pricking sensation of touch, similar to that produced by various other substances. Thus, Chevreul[10] concluded that although camphor was thought to have a distinct taste, the sensation produced by the camphor was not a taste at all, but a fusion of odor and touch.

Expanding on this thesis, Hollingworth and Poffenberger[11] suggested that a "very great number" of what are commonly perceived as tastes are not tastes at all, but are really odors. In support of this position, Hollingworth and Poffenberger[11] reported that when various substances were reduced to a uniform consistency (so that they could not be recognized by the tactile sense), and when the nostrils were pinched, it was impossible to distinguish between quinine and coffee or between apple and onion. In a more practical example they asked, "how often has the nasty taste of medicine been softened by Chevreul's simple technique of holding the nose?"

These authors also suggested that the reverse may be true. That is, there may be some cases in which volatile substances entering the mouth through the nostrils may stimulate the taste buds in the upper and back part of the mouth. In such "rare" cases the real taste is mistakenly interpreted as an odor. As an example, Hollingworth and Poffenberger[11] suggest that chloroform seemed to have the characteristic odor that was in all probability a sweet taste due to stimulation of the taste buds by the chloroform vapor.

The first experiment designed to study directly the effect of smell on taste and taste on smell was conducted by Murphy, Cain, and Bartoshuk.[4] These authors reported that the patency of the nostrils did not influence the taste magnitude of solutions containing sodium saccharin (a stimulus with an intense taste but minimal smell). However, nasal patency had a substantial influence on the taste of solutions of ethyl butyrate (a stimulus with an intense smell and low taste). When compared to estimates with the nostrils open, the magnitude of the taste of ethyl butyrate was reduced when the nostrils were closed, and this reduction was directly related to the concentration of the solution.

In a second study, Murphy and Cain[5] reported that the addition of the odorant citral to solutions of either sucrose or sodium chloride resulted in an increase in the estimates of the taste intensity even though citral had been shown not to have a taste component. These effects were labeled "smell-taste confusions."

Subsequent studies have reported that smell-taste confusions may be stimulus specific and that some tastants may affect the intensity ratings of smell.[8,9]

Another test for detecting the influences of smell on taste and taste on smell is when subjects are asked to estimate the intensity of the smell of distilled water when paired with a tastant in the mouth, or the intensity of the taste of distilled water when paired with a sniff of an odorant. In this situation, the smell of almond extract[9] and ethyl butyrate[6] increased the intensity estimates of the taste of distilled water. Likewise, the taste of almond increased the intensity estimates of the smell of distilled water.[9] However, this latter effect was not observed when the taste was sucrose.[6]

Gillan[12] found influences of smell on taste and taste on smell when subjects rated the component qualities of a smell-taste mixture. That is, subjects judged the intensity of sweet, salt, lemon, and licorice after they had tasted solutions of sucrose and sodium chloride, smelled vapors of citral and anethole, and tasted and smelled the four stimuli in combination. The intensity estimates of the component qualities were lower when the taste and smell were combined than when either was presented alone. The data from this study also showed that an odor stimulus that strongly suppressed the perception of another odor had less effect on a taste stimulus. Likewise, a taste stimulus that strongly suppressed a concurrent taste perception had a weaker effect on concurrent odor perception. These findings suggest that cross-modality mixture suppression is perhaps less dramatic than intra-modality suppression.

Although not specifically designed to study the effects of smell on taste and vice versa, the work of Mozell et al.[13] demonstrated the importance of olfaction in flavor identification. In this experiment, when a stream of humidified air was forced through the external nares (and so prevented odorant molecules from reaching the olfactory receptors), human volunteers were unable to identify most common foods. When the airstream was odorized, the foods could be identified. These results again illustrate that what is commonly perceived as taste may actually be a smell-taste confusion.

However, this does not necessarily imply that smell alone is required for flavor identification. As Desor and Beauchamp[14] demonstrated, even using odor cues, subjects had a surprisingly low level of identification of common food items. Clearly, the relative role of smell and taste in flavor identification requires more work.

Perhaps part of the mechanism by which taste can influence smell is by the diffusion of odorant molecules from the pharynx, through the posterior nares, to the olfactory receptors. This method of stimulating the olfactory receptors is called "retronasal olfaction" by Burdach, Kroeze, and Koster.[15] These investigators reported that the presentation of solutions either nasally or retronasally produced similar detection thresholds. The addition of sucrose to the test solution increased the detection threshold for stimuli that were presented retronasally but not for stimuli presented nasally. The results of this study again suggest that taste can influence the perception of odor.

It is intriguing to question the relationship between nasal and retronasal olfaction. In studies of the contributions of smell and/or taste to the identification of complex foods, several investigators make the assumption that the two methods of stimulating the olfactory receptors are at least qualitatively similar.[16,17] Rozin,[18] however, suggested that the two methods of olfactory stimulation might be qualitatively different. Certainly, the limited available data do not yet allow a definitive conclusion on this matter.

SUMMARY

A solution with both an odor and a taste may be considered to be a mixture that involves two sensory modalities. Estimates of the intensity of such mixtures appear to be additive. If the overall intensity of each of the unmixed components is compared with the overall intensity of the mixture, the additivity approaches 100%. If the intensities of the smell and taste of the unmixed components are compared with the overall intensity of the mixture, the additivity is less than 100%. Thus, the specific question that is given to the subjects influences the magnitude of the estimations. This suggests that the additive process involves a central (cognitive) mechanism. Considering that the perception of complex flavors also involves sensory information of touch, temperature, and possibly vision and hearing,[2] a central interpretation seems appropriate.

The influences of smell on the perception of taste also appear to involve a cognitive mechanism. These smell-taste confusions appear to be stimulus specific and are usually resolved in favor of taste. This may be true because the sensations of pressure, movement, and resistance are usually localized in the mouth. These accompanying sensations then suggest that the taste organs are active in determining the result even when no true taste is present.

The influences of taste on the perception of smell are most pronounced when the tastant contains an odor. This suggests that the effect may be peripheral. That is, odorant molecules may be moving from the pharynx, through the posterior nares, to the olfactory receptors. If this interpretation is correct, the influences of taste on smell may be an odor-odor mixture involving "retronasal" and "nasal" olfaction.

REFERENCES

1. GIBSON, J. J. 1966. The Senses Considered as Perceptual Systems. Houghton Mifflin Company. Boston, MA.

2. MCBURNEY, D. H. 1986. Taste, smell, and flavor terminology: Taking the confusion out of fusion. In Clinical Measurements of Taste and Smell. H. L. Meiselman & R. S. Rivlin, Eds.: 117-125. Macmillan. New York.

3. GELDARD, F. A. 1972. The Human Senses. 2nd edit. John Wiley & Sons, Inc. New York.
4. MURPHY, C., W. S. CAIN & L. M. BARTOSHUK. 1977. Mutual action of taste and olfaction. Sensory Proc. 1: 204-211.
5. MURPHY, C. & W. S. CAIN. 1980. Taste and olfaction: Independence vs. interaction. Physiol. Behav. 24: 601-605.
6. HORNUNG, D. E. & M. P. ENNS. 1986. The contribution of smell and taste to overall intensity: A model. Percept. Psychophys. 39: 385-391.
7. GARCIA-MEDINA, M. A. 1981. Flavor-odor taste interactions in solutions of acetic acid and coffee. Chem. Senses 6: 13-22.
8. HORNUNG, D. E. & M. P. ENNS. 1984. The independence and integration of olfaction and taste. Chem. Senses 9: 97-106.
9. ENNS, M. P. & D. E. HORNUNG. 1985. Contributions of smell and taste to overall intensity. Chem. Senses 10: 357-366.
10. CHEVREUL, M. E. 1824. Considerations generales sur l'analyse organique et sur ses applica'tions. Chez F.-G. Levrault, Libaraire Paris. Paris, France.
11. HOLLINGWORTH, H. L. & A. T. POFFENBERGER, JR. 1917. The Sense of Taste. Moffat, Yard and Company. New York.
12. GILLAN, D. J. 1983. Taste-taste, odor-odor, and taste-odor mixtures: Greater suppression within than between modalities. Percept. Psychophys. 33: 183-185.
13. MOZELL, M. M., B. P. SMITH, P. E. SMITH, R. L. SULLIVAN, JR. & P. SWENDER. 1969. Nasal chemoreception in flavor identification. Arch. Otolaryngol. 90: 367-373.
14. DESOR, J. A. & G. K. BEAUCHAMP. 1974. The human capacity to transmit olfactory information. Percept. Psychophys. 16: 551-556.
15. BURDACH, K. J., J. H. A. KROEZE & E. P. KOSTER. 1984. Nasal, retronasal, and gustatory perception: An experimental comparison. Percept. Psychophys. 36: 205-208.
16. SCHIFFMAN, S. 1977. Food recognition in the elderly. J. Gerontol. 32: 586-592.
17. STEVENS, D. A. & H. T. LAWLESS. 1981. Age-related changes in flavor perception. Appetite 2: 127-136.
18. ROZIN, P. 1982. Taste-smell confusions and the duality of the olfactory sense. Percept. Psychophys. 31: 397-401.

The Mixture Symposium

Summary and Perspectives[a]

MAXWELL M. MOZELL

Clinical Olfactory Research Center
State University of New York
Health Science Center
Syracuse, New York 13210

I have never actively entered the field of trying to decipher the effects of chemical mixtures on chemoreceptive responses, although some years ago I became interested in a related topic, namely, looking at the relative contributions of the nose and tongue to the identification of flavors.[1] However, a summary by someone not immediately involved in the area perhaps could give a more global view. Even if I do not come up with any ingenious insights, maybe I can at least be an irritative focus for stimulating discussion.

Each of the contributors has demonstrated how, from his particular approach to the field, the response to a mixture of chemicals differs from the responses to the chemicals given alone; but in describing these approaches each contributor has drawn attention to the formidable task we face in trying to unravel how stimulus mixtures are processed by the chemical senses. As Price chidingly noted, even before we address how single stimuli and mixtures of stimuli are differentially processed, we must first give some thought to whether we have ever recorded electrophysiological discharges or observed behavioral responses to anything but mixtures. This is especially true in olfaction. Recall that even a slight contaminant can become major if its partitioning between the air phase and the mucosa phase markedly favors the mucosa phase. Even if absolutely pure on entering the nose, a single stimulant could possibly become a mixture of stimulants if it is metabolized. It may be, therefore, that we cannot compare mixture responses to single-stimulus responses because we have not yet seen any of the latter.

Frijter's discussion did not make me feel any easier. He emphasized the problems involved in trying to derive a psychophysical model for taste-taste and smell-smell mixtures. He pointed out that even the most successful models are not universally applicable, each having some significant limitation. This was further emphasized by Hornung and Enns who pointed out that the specifics of the model might even depend upon the semantics of the judgments required. That is, they proposed that when subjects are instructed to judge the "intensities," they give different estimates than when instructed to judge the "*overall* intensities." This is all somewhat disappointing, but with all due respect to those working diligently in this field, the big disappointment for a nonparticipant like myself is that all the models for predicting the effect of

[a] Prepared with the support of National Institutes of Health Grant NS 19658.

chemosensory mixtures bear heavily upon the intensity of the perception rather than its quality. It is not that the final intensity of a mixture is not important, but it is, after all, the separation of qualities that appears to be the hallmark of the chemical senses. To disregard quality in understanding chemosensory mixture effects would be tantamount to developing models for color mixture that predict the brightness of the result but not the hue. The model may be correct and have its uses both in practice and in theory, but it would seem to miss the pertinent point. Likewise, it is, of course, interesting that in the mixing of a taste and a smell the combined intensity is less (to a degree depending upon experimental conditions) than is the sum of the taste and smell taken separately. Certainly, if, as was done by Hornung and Enns, such results are cleverly analyzed with appropriate follow-up experiments, the pursuit of the mixture effect upon perceived intensity may indeed shed light upon a number of important mechanisms underlying chemosensory processes including those determining the qualities of mixtures. Nevertheless, as an outsider looking in, I am concerned that in the effort to develop models for the prediction of the flavor intensities of mixtures the quest to understand the generation of their qualities may get lost.

Let me become even more of an irritating focus. I wonder whether we can even reasonably expect to develop an all-inclusive model to predict chemosensory mixing, either quantitatively or qualitatively. Think of all the phenomena that such a model must take into account and somehow incorporate. Consider first the inputs of several involved cranial nerves each with its own set of receptor cells, loci in the stimulation path, and projection into the central nervous system. These inputs include the olfactory, trigeminal, facial, glossopharyngeal, and vagus nerves. We already know that at least some of these inputs can affect each other. For instance, Cain and his coworkers[2,3] have documented that the trigeminal input from the nose adds to an odorant's perceived smell intensity even when the odorant is not itself perceptually "trigeminal" in character, and they have also shown that there can be a mutual inhibition of pungency on odor and odor on pungency. To make matters even more complex, this interplay between the olfactory and trigeminal nerve inputs is both concentration dependent and time dependent.

Even when one considers just one cranial nerve input, there are still many complicating factors. Laing and his coworkers[4,5] showed that if two odorants are at similar perceptual intensities, both odorants, when mixed, will be perceived, although some of their perceptual qualitative components may be altered. On the other hand, if the two odorants are at different perceptual intensities, the weaker one will not be perceived at all when they are mixed. Thus, the relative intensities of the odorants can produce perceptual variations that change precipitously rather than progressively. Add to this Price's admonition that the presented chemicals and their concentrations need not be predictably related to the stimuli actually contacting the receptors and one must begin to worry whether any model can be adequately formulated to account for all the permutations of interplaying events upon which chemosensory mixture perception may be based. Still other complicating factors may be mentioned: differences in the access of the molecules of different chemicals to the receptors, their differential topographic and temporal distributions across the mucosa, and differences in the neural interactions within and across modalities.

Another approach to understanding mixture processing, somewhat more direct than searching for models, is to begin with known chemosensory phenomena that could conceivably play a role in mixture perception and then test whether they do play such a role. This approach was represented by Laing. He began by citing two related sets of data[6,7] taken from amphibians: One set suggested that the molecules of different odorants are sorbed across the mucosa in different spatial concentration gradients depending upon their air-mucosa partition coefficients, and the other set suggested that receptors of like sensitivity are aggregated into those regions of the mucosa where the molecules to which they are particularly sensitive are likely to be sorbed. Laing reasoned from these earlier data that if the mucosal sorption region of odorant A totally encompassed that of odorant B but not vice versa, odorant A could possibly mask the response to B but B would less likely mask the response to A. Laing tested this hypothesis and reported to us a positive result. Thus, we may have our first really testable proposed mechanism for smell mixture effects based upon manipulatable physical events. It depends upon how concordant the molecular distribution patterns of different odorants are across the mucosa, which in turn depends upon the similarity of the odorant air-mucosa partition coefficients. Similarly, reasoning from other data also already in the literature,[8] Laing suggested that the odorant first exciting the receptors (in part determined by its relative diffusion rate through the mucus) will inhibit responses to later odorants both by competition at receptor sites and by intra-and interbulbar effects.

The approach of O'Connell and Grant has become even more basic, carrying the mixture question into intracellular molecular events. Recall the rather unexpected finding[9] that a behaviorally relevant pheromone need not be processed by its own particularly tuned receptor neuron but may instead modify the responses of receptor neurons excited by other members of the mixture blend. Thus, already knowing which odorants interplay with each other both in terms of behavior and electrophysiology, O'Connell and Grant have begun to ask about the molecular events at the intracellular level that can explain these observations.

Because of the multi-component, multi-stage complexity of the mixture problem as outlined above, it is going to be difficult if not impossible to understand its mechanisms without breaking it down into smaller bits, and strategies to do so must be used more extensively. For instance, dichorhinic stimulation (i.e., presenting different stimuli to each side of the nose), as already used by Cain[10] and von Bekesy,[11] *may* permit a separation of receptor-level masking events from more central masking events. However, as discussed by Laing, the use of dichorhinic mixture experiments and their interpretation requires a new measure of caution. That is, the differential speeds with which the molecules of the different odorants on the two sides of the nose diffuse through the mucus could complicate any "central" versus "peripheral" masking effects. A test of this concept appears feasible by using odorants with differing mucus diffusivities and by presenting odorants to the two nostrils with different time disparities. Similarly, stimulation of different sides of the tongue (as for instance practiced at the Connecticut Chemosensory Clinical Research Center[12]) may help separate central and peripheral events in taste. By taking advantage of approaches that could help identify mainly olfactory and mainly trigeminal odorants,[13,14] we may be able to separate the mixture effects of the first and fifth cranial nerves.

Certainly we can look forward to the wider use of newer techniques. Patch clamping[15] will allow us to better compare transduction events of single-component stimuli with multiple-component stimuli. The use of voltage-sensitive dyes[16] offers a possible chance to compare the mixture-generated activity across the olfactory bulb (and perhaps mucosa) to that generated by one-component stimuli. Many other techniques hold promise for helping us to understand mixture mechanisms and could be mentioned but certainly the immunological and molecular techniques cited in many of the contributions to this symposium and so well represented by the work of Doron Lancet[17] must be given special mention. The closer we get to understanding the stimulus-receptor interface and the chain of molecular events to which it gives rise, the more will the special case of mixture effects become clear.

REFERENCES

1. MOZELL, M. M., B. P. SMITH, P. E. SMITH, R. L. SULLIVAN, JR. & P. SWENDER. 1969. Nasal chemoreception in flavor identification. Arch. Otolaryngol. **90:** 131-137.
2. CAIN, W. S. 1974. Contribution of the trigeminal nerve to perceived odor magnitude. Ann. N.Y. Acad. Sci. **237:** 28-34.
3. CAIN, W. S. & C. L. MURPHY. 1980. Interaction between chemoreceptive modalities of odour and irritation. Nature **284:** 255-257.
4. LAING, D. G., H. PANHUBER, M. E. WILLCOX & E. A. PITTMAN. 1984. Quality and intensity of binary odor mixtures. Physiol. Behav. **33:** 309-319.
5. LAING, D. G. & M. E. WILLCOX. 1983. Perception of components in binary odour mixtures. Chem. Senses **7:** 249-264.
6. MOZELL, M. M. & M. JAGODOWICZ. 1973. Chromatographic separation of odorants by the nose: Retention times measured across *in vivo* olfactory mucosa. Science **181:** 1247-1249.
7. MACKAY-SIM, A., P. SHAMAN & D. G. MOULTON. 1982. Topographic coding of olfactory quality: Odorant-specific patterns of epithelial responsivity in the salamander. J. Neurophysiol. **48:** 584-596.
8. GETCHELL, T. V., G. L. HECK, J. A. DeSIMONE & S. PRICE. 1980. The location of olfactory receptor sites. Inferences from latency measurements. Biophys. J. **29:** 397-412.
9. O'CONNELL, R. J. 1985. Responses to pheromone blends in insect olfactory receptor neurons. J. Comp. Physiol. A. **156:** 747-761.
10. CAIN, W. S. 1975. Odor intensity: Mixtures and masking. Chem. Senses Flavors **1:** 339-352.
11. VON BEKESY, G. 1964. Olfactory analogue to directional hearing. J. Appl. Physiol. **19:** 369-373.
12. GENT, J., W. S. CAIN & L. BARTOSHUK. 1986. Taste smell measurement in a clinical setting. *In* Clinical Measurement of Taste and Smell. H. L. Meiselman and R. S. Rivlin, Eds.: 107-116. Macmillan Publishing Co. New York.
13. DOTY, R. L., W. E. BRUGGER, P. L. JURS, M. A. ORNDORFF, P. F. SNYDER & L. D. LOWRY. 1978. Intranasal trigeminal stimulation from odorous volatiles: Psychometric responses from anosmic and normal humans. Physiol. Behav. **20:** 175-187.
14. SILVER, W. L. & J. A. MARUNIAK. 1981. Trigeminal chemoreception in the nasal and oral cavities. Chem. Senses **6:** 295-305.
15. SUZUKI, N. 1986. Voltage-dependent ionic currents in isolated olfactory receptor cells. ISOT IX. Abstract S46.
16. ORBACH, H. S. & L. B. COHEN. 1983. Optical monitoring of activity from many areas of the *in vitro* and *in vivo* salamander olfactory bulb: A new method for studying functional organization in the vertebrate central nervous system. J. Neurophysiol. **3:** 2251-2262.
17. PACE, V., E. HANSKI, Y. SALOMON & D. LANCET. 1985. Odorant sensitive adenylate cyclase may mediate olfactory reception. Nature **316:** 255-258.

How Motile Bacteria Sense and Respond to Chemicals

JULIUS ADLER

Departments of Biochemistry and Genetics
University of Wisconsin
Madison, Wisconsin 53706

Bacteria are attracted by certain chemicals and repelled by others; this is chemotaxis. The work of Wilhelm Pfeffer established this behavior 100 years ago.

How do bacteria sense the chemicals that attract or repel them? How are the flagella told what to do? How do flagella work and how are they coordinated? Can bacteria "learn" to change their chemotactic behavior? We have applied the tools of biochemistry and genetics to provide answers to these questions.

It used to be believed that bacteria sense an attractant by measuring the energy produced during its metabolism. We showed that metabolism of a chemical was neither required nor sufficient for it to be an attractant; instead the bacteria have sensors (or "receptors" or "signalers" or "transducers") that detect the chemical itself, not any product formed from it. By now some twenty different chemosensors are known for various attractants and repellents.

I discovered that methionine is required for chemotaxis, and this led to our discovery of a novel sensor, the methyl-accepting chemotaxis protein. Four distinct methyl-accepting proteins are now known, each serving a different set of attractants and repellents. (See FIG. 1.)

In an as yet unknown manner, these methyl-accepting chemotaxis proteins signal to the flagella to tell them whether to rotate counterclockwise, leading to swimming in a straight line and thus movement toward attractants, or whether to rotate clockwise, leading to tumbling and thus avoidance of repellents: This is called "excitation." The mechanism of excitation is completely unknown, and it is one of our major current goals to discover this mechanism.

Once initiated, excitation is terminated, even though the stimulus is still present. This "adaptation" consists of increased methylation of methyl-accepting chemotaxis protein in the case of attractants or decreased methylation for repellents. By now much is known about the adaptation mechanism, but much remains to be learned.

The route of discovery of this novel sensor was to first isolate behavioral mutants that fail in each of the steps indicated, and then the biochemistry missing in each mutant was determined. (See FIG. 2 for a summary.)

It appears that bacteria can "learn" to change their chemotactic behavior. We are currently working out the genetics and biochemistry of "learning" in bacteria.

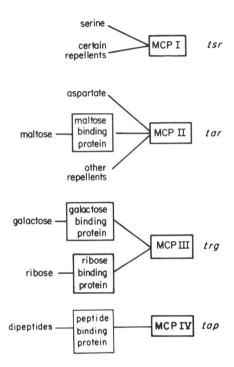

FIGURE 1. Sensors in bacterial chemotaxis. MCP = methyl-accepting chemotaxis protein. The names of the genes for the MCPs are shown on the right. (From the laboratories of Adler, Simon, Manson, and Higgins.)

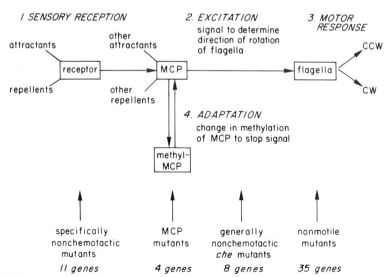

FIGURE 2. The mechanism of bacterial chemotaxis as it is understood today. Mutant types are indicated on the bottom. CCW = counterclockwise rotation of the flagella; CW = clockwise rotation; MCP = methyl-accepting chemotaxis protein. The methyl donor is S-adenosylmethionine, derived from L-methionine; the product of demethylation is methanol.

REFERENCE

1. ADLER, JULIUS. 1983. The behavior of rats, bacteria, and man. *In* Biochemistry of Metabolic Processes. D. L. F. Lennon, F. W. Stratman, and R. N. Zahlten, Eds.: 367-383. Elsevier. New York.

A Molecular Vocabulary for Olfaction[a]

GORDON M. SHEPHERD

Section of Neuroanatomy
Yale University School of Medicine
New Haven, CT 06510

The introduction of methods of molecular biology to olfactory research in the past few years has brought with it two kinds of benefits. Precise information about mechanisms of sensory transduction and the biochemical individuality of olfactory neurons are becoming available, bringing olfaction more into line with other sensory systems. An additional benefit arises from the fact that olfactory information is fundamentally molecular; the new level of inquiry thus gives direct insight into the nature of the sensory information itself. Although we are still at an early stage, it has begun to be apparent that we need new terms to characterize adequately the mechanisms whereby the information contained in the odor molecules is transformed into a neural code.

This review will summarize briefly our present understanding of the functional organization of the olfactory epithelium at the cellular and molecular level. We will focus on several areas where recent advances strain the old terminology, and several suggestions will be made that point the way toward a molecular vocabulary for olfaction.

THE OLFACTORY STIMULUS

There is widespread agreement that olfactory transduction involves an initial interaction between the adequate stimulus (odor molecules) and receptor sites in the cilia of the olfactory receptor neuron. This has in fact been suspected for many years (cf. Moulton and Beidler[1] and Ottoson and Shepherd[2]). One of the chief merits of the new work is to recast this general scheme into specifically molecular terms, each aspect of which has need of an appropriate terminology.

To begin with, the adequate stimulus is traditionally referred to as an "odor molecule." This term arose from psychophysical studies, where it is appropriate ("odor" is a perception, and an "odor molecule" is a molecular type that gives rise to perception of an odor). However, it loses its precision when applied to analysis of the transduction mechanism itself; modern analysis typically involves biochemical preparations of isolated cilia or electrophysiological recordings of membrane events, situations in which the stimulating molecules can hardly be considered to give rise to

[a] This work has been supported by the National Institute for Nervous and Communicative Disorders and Stroke.

a perceived odor. As an alternative, the term "odorant" is often used. It too has psychophysical connotations, as well as carrying the additional burdens of being phonetically unpleasant and being defined as something you cannot smell when you use a deodorant.

For analysis of transduction at the molecular level, it would be useful to have a term defined specifically as a molecule that binds to or otherwise stimulates a receptor to give rise to a sensory response in the olfactory receptor neuron. It will be noted that this paraphrases the standard textbook definition of an antigen in the immune system (cf. Alberts et al.[3]). By analogy, the odor molecule may be more precisely designated as an olfactory neuron response-generating molecule, or, more conveniently, an odogen (cf. Shepherd[4]).

What does an odogen bind to or otherwise stimulate? "Olfactory receptor" is the traditional term. This gives rise to confusion, however, because anatomists and physiologists have used this to refer to the whole olfactory receptor neuron (alternatively, olfactory receptor cell), whereas, to biochemists, olfactory receptor more naturally means the protein that binds the ligand. A convenient way to avoid this confusion is to designate the olfactory receptor cell as the O cell, and its receptor molecule as the O-cell receptor, in analogy with the T cell and T-cell receptor of the immune system.

How does the odogen interact with the O-cell receptor? Two main concepts have been that stimulation depends on the stereochemical fit between molecule and receptor (Amoore), and on the interactions between functional groups and receptor sites (Beets). In the immune system, the antigenic determinants on the antigen molecule are called epitopes. By analogy, the odogenic determinants on an odogen molecule may be referred to as odotopes. The advantage of such a term is that it is not tied to one particular kind of determinant, and may refer to any aspect of the odogen molecule (its stereochemical conformation, types of active groups, length of carbon chain, presence of hydrophobic (lipophilic) groups) that contributes to its ability to interact with receptor sites in the O cell.

OLFACTORY RECEPTORS

One may postulate, therefore, that odotopes on the odogen bind to or interact with O-cell receptors. A candidate for a receptor is gp95,[5] a 95-kDa integral membrane glycoprotein that is a dominant surface component of olfactory cilia. Further characterization of this and related membrane-bound proteins has high priority for the near future.

Evidence for soluble protein receptors has also been obtained. In insects, an "odor binding protein" (OBP) is present in the endolymph around the tip of the sensory cilium; it is believed that this binds the odogen and conveys it to receptors in the cilia (cf. Vogt[6]). An "odorant binding protein" has also been found in the vertebrate, localized to nasal glands and secretions (cf. Pevsner et al.[7]). Another possible function of such a protein is to inactivate or remove the odogen from the O-cell receptor. Rate constants for receptor activation and inactivation have already been postulated to account for O-cell impulse frequencies during constant odogen stimulation.[8]

A convenient term for these types of proteins secreted into the mucus would be odobody (cf. Shepherd[4]). Presumably odobodies in the viscous mucus are secreted by supporting cells (S cells), and in the outer layer of watery mucus by gland cells.

SECOND MESSENGERS

There is now considerable evidence that, following binding of odogen to receptor, there is activation of a G-binding protein and stimulation of adenylate cyclase to make the second messenger cyclic AMP.[9,10] This is at present probably the best characterized step in olfactory transduction at the molecular level (cf. Anholt et al.[11] and Sklar et al.[12]). Other types of second messengers (e. g. inositol lipids) are presently under study (cf. Anholt et al.[11]).

RECEPTOR POTENTIAL

The next step in olfactory transduction is a change in membrane conductance that generates the receptor potential. Intracellular and patch clamp recordings have thus far been in agreement that the receptor potential is depolarizing, and that it is accompanied by an increase in membrane conductance (cf. Getchell,[13] Suzuki,[14] and Trotier and MacLeod[15]). This suggests that the conductance channels pass cations. It remains for further experiments to determine whether the channels are opened by direct binding of some odogens in addition to action of second messengers.

VOLTAGE-GATED CHANNELS

The final step in signal transduction is spread of the receptor potential to sites of voltage-gated channels, where the impulse discharge is generated that transmits the sensory response to the central nervous system. In patch recordings, the voltage-gated channels are more readily encountered than odogen-gated channels. Several types of depolarizing and hyperpolarizing voltage-gated conductances have thus far been characterized (cf. Dionne,[16] Firestein and Werblin,[17] Suzuki,[18] and Anderson and Ache[19]). These could be involved in generating the impulse response in the soma and axon, and in boosting the receptor potential in the dendrite and cilia.

The impulse discharges elicited by odor stimuli have traditionally been the main means by which physiologists have characterized the encoding of odor information. Although the problem of olfactory encoding goes beyond the aim of the present review, the discussion thus far makes it clear that the encoding process needs to be characterized in molecular terms at each step, beginning with odotope-odoreceptor interactions; activation of one of several possible second messengers; opening or closing of conductance channels (possibly involving phosphorylation); spread, boosting, and integration of receptor potentials in different cilia; and the final generation of impulse discharges in the entire array of differentially tuned neurons.

INTRAEPITHELIAL MESSENGERS

It remains to consider mechanisms within the olfactory epithelium involved in controlling differentiation of new O cells from stem cells. It now seems well established that in the vertebrate, new O cells arise from stem (basal) cells during adult life, and that transection of the adult olfactory nerve elicits a massive degeneration of the transected neurons followed by differentiation of a new set of neurons from stem cells (cf. Graziadei and Monti Graziadei[20]).

Thus far, at least four molecular signals have been postulated to take part in these events. First, expression of olfactory marker protein (OMP) appears to depend on an axonal maturation signal (A signal) from synaptic terminals (Margolis et al.[21]). Second, a degeneration (D) signal must also be transported retrogradely, from the site of nerve transection, to trigger degeneration. Third, there must be an extracellular mitogenic (M) signal between O cells and stem cells that modulates stem cell mitosis during normal O-cell turnover, and that is released from degenerating cells following nerve transection. Fourth, evidence has been obtained for a raised input resistance in some S cells following nerve transection, suggesting an uncoupling of gap junctions (Masukawa et al.[22]). It is thus possible that there is an uncoupling (U) signal from degenerating O cells to S cells.

A final type of molecular messenger to be considered is the odogen itself. In addition to their sensory roles, odogens may have actions that modulate the cellular environment of the olfactory epithelium. Of particular interest is the possibility that they may contribute to control of stem cell mitosis and subsequent differentiation and maturation of O cells. Molecules accessing the olfactory epithelium may therefore act as odomitogens as well as odogens. Since the terminal web at the surface of the epithelium limits penetration from the mucus, odomitogenic actions would presumably be mediated via O or S cells.

SUMMARY

Olfactory research is entering a new phase, in which molecular mechanisms are being revealed that go considerably beyond traditional concepts. New ways of characterizing these mechanisms are needed, and some suggestions toward that goal have been made in this review. These suggestions recognize that, whereas formerly our terms and concepts regarding olfactory stimulus-response characteristics came mainly from organic chemists and psychophysicists, the main impetus at present comes from molecular biology. A desirable terminology, therefore, is one that is familiar to molecular biologists and can facilitate comparisons with other systems—immune, endocrine, nervous—where similar methods and terms are in use.

The suggestions made here for the olfactory system could also be adapted for the taste system. Taste stimulation could be characterized, for example, in terms of gustagens interacting with G-cell receptors, stimulation being determined by the gustatope of a particular ion or molecule. It should be emphasized that such terms and mechanisms may not need to be invoked in studies at behavioral or psychophysical levels. However, the need for them at the receptor level may well be an accurate reflection of our progress in applying methods of molecular biology to these systems.

ACKNOWLEDGMENTS

I am grateful to my colleagues, Drs. Leona Masukawa, Charles Greer, Patricia Pederson, and William Stewart; to Shridar Ganesan and Tom Woolf; and to Drs. John Hildebrand, Jurgen Boeckh, Doron Lancet, James Schwob, and Tom Morton for valuable discussions.

REFERENCES

1. MOULTON, D. & L. M. BEIDLER. 1967. Structure and function in the peripheral olfactory system. Physiol. Rev. 47: 1-52.
2. OTTOSON, D. & G. M. SHEPHERD. 1967. Experiments and concepts in olfactory physiology. *In* Progress in Brain Research, Vol. 23. Sensory Mechanisms. Y. Zotterman, Ed.: 83-138. Elsevier. Amsterdam, The Netherlands.
3. ALBERTS, B. 1983. Molecular Biology of the Cell. Garland Publisher. New York.
4. SHEPHERD, G. M. 1985. Welcome whiff of biochemistry. Nature 316: 214-215.
5. CHEN, Z. & D. LANCET. 1984. Membrane proteins unique to vertebrate olfactory cilia: Candidates for sensory receptor molecules. Proc. Natl. Acad. Sci. USA 81: 1859-1863.
6. VOGT, R. G. 1987. Variation in olfactory proteins: A conceptual poster on the evolution of behavior. Ann. N. Y. Acad. Sci. This volume.
7. PEVSNER, J., P. B. SKLAR & S. H. SNYDER. 1987. Localization of odorant binding protein (OBP) to nasal glands and secretions. Ann. N. Y. Acad. Sci. This volume.
8. GETCHELL, T. V. & G. M. SHEPHERD. 1978. Responses of olfactory receptor cells to step pulses of odour at different concentrations in the salamander. J. Physiol. 282: 521-540.
9. PACE, U., E. HANSKI, Y. SALOMON, & D. LANCET. 1985. Odorant sensitive adenylate cyclase may mediate olfactory reception. Nature 316: 255-258.
10. LANCET, D. 1987. Toward a comprehensive analysis of olfactory transduction. Ann. N. Y. Acad. Sci. This volume.
11. ANHOLT, R. R. H., S. M. MUMBY, D. A. STOFFERS, P. R. GIRARD, J. F. KUO, A. G. GILMAN & S. H. SNYDER. 1987. Transductory proteins of olfactory receptor cells: Identification of guanosine nucleotide binding proteins and protein kinase C. Ann. N. Y. Acad. Sci. This volume.
12. SKLAR, P. B., R. R. H. ANHOLDT & R. R. H. SNYDER. 1987. The odorant-sensitive adenylate cyclase of olfactory receptor cells: Differential stimulation of distinct classes of odorants. Ann. N. Y. Acad. Sci. This volume.
13. GETCHELL, T. V. 1977. Analysis of intracellular recordings from salamander olfactory epithelium. Brain Res. 123: 275-280.
14. SUZUKI, N. 1977. Intracellular responses of lamprey olfactory receptors to current and chemical stimulation. *In* Food Intake and Chemical Senses. Y. Katsuki, M. Sato, S. Takagi, and Y. Oomura, Eds.: 12-22. University of Tokyo Press. Tokyo, Japan.
15. TROTIER, D. & P. MACLEOD. 1983. Intracellular recordings from salamander olfactory receptor cells. Brain Res. 268: 225-237.
16. DIONNE, V. E. 1987. Membrane conductance mechanisms in dissociated cells from the *Necturus* olfactory epithelium. Ann. N. Y. Acad. Sci. This volume.
17. FIRESTEIN, S. & F. WERBLIN. 1987. The interaction of generator current and voltage-gated currents in the olfactory receptor response. Ann. N. Y. Acad. Sci. This volume.
18. SUZUKI, N. 1987. Voltage-dependent currents in isolated olfactory receptor cells. Ann. N. Y. Acad. Sci. This volume.
19. ANDERSON, P. A. V. & B. W. ACHE. 1985. Voltage and current-clamp recordings of the receptor potential in olfactory receptor cells *in situ*. Brain Res. 338: 273-280.
20. GRAZIADEI, P. P. C. & G. A. MONTI GRAZIADEI. 1978. Continuous nerve cell renewal

in the olfactory system. *In* Handbook of Sensory Physiology. Vol. IX. Development. M. Jacobson, Ed.: 55-82. Springer. New York.

21. MARGOLIS, F. L., W. SYDOR, Z. TEITELBAUM, R. BLACHER, M. GRILLO, K. ROGERS, R. SUN & U. GUBLER. 1985. Molecular biological approaches to the olfactory system: Olfactory marker protein as a model. Chem. Senses **10:** 163-174.

22. MASUKAWA, L. M., B. HEDLUND & G. M. SHEPHERD. 1985. Electrophysiological properties of identified cells in the *in vitro* olfactory epithelium of the tiger salamander. J. Neurosci. **5:** 128-135.

Adaptation Processes in Insect Olfactory Receptors

Mechanisms and Behavioral Significance

K.-E. KAISSLING, C. ZACK STRAUSFELD,[a]
AND E. R. RUMBO[b]

Max-Planck-Institut für Verhaltensphysiologie
8131 Seewiesen
Federal Republic of Germany

INTRODUCTION

Insect olfactory cells adapt, that is, adequate stimulation alters their subsequent responses. Adaptation in the sensory organ is an early step in the processing of sensory information, which together with central nervous integration enables the organism to cope with complex natural stimuli. Our concern here is with air-borne odors, mainly pheromones, but which include food odors and also carbon dioxide or water vapor. We describe adaptation of extracellularly recorded receptor potentials and nerve impulses. These responses are most likely generated in different regions of the sensory structures, the so-called sensilla, which include, besides receptor cells, stimulus-conducting structures and auxiliary cells (FIG. 1). We also discuss the behavioral significance of adaptation especially for orientation in an odor plume. Our main subject will be the pheromone receptors of the silkworm moths *Bombyx mori* and *Antheraea polyphemus*.

ADAPTATION OF THE NERVE IMPULSE RESPONSE

Adaptation is evident in the so-called phasic-tonic response pattern of the nerve impulses to stimulation, known also from vertebrate olfactory cells as well as sensory cells of other stimulus modalities. Typically, an increase in odor stimulus intensity is followed by an initial burst of nerve impulses lasting 50-100 msec producing peak frequencies of up to 200 impulses per second; 500 impulses per second have been

[a] Present address: 2748 E. Drachman Str., Tucson, AZ 85716.
[b] Present address: CSIRO, Division of Entomology, P. O. Box 1700, Canberra City ACT 2601, Australia.

observed in CO_2 receptors of the honey bee.[1] The phasic peak is followed by a decrease in impulse frequency to a "tonic" plateau below 50 impulses per second. This fall, attributed to a decreased sensitivity of the nerve impulse generator, constitutes one example of adaptation. A subsequent stimulus may elicit a phasic peak of reduced height whereas the tonic plateau remains unadapted.[2] Several seconds or minutes are required before the cell recovers, that is, responds with the initial maximum frequency.

The tonic plateau of impulse frequency adapts in a different way. A weak stimulus will usually elicit a purely tonic response, capable of lasting many minutes without adaptation. The plateau is often not stable and declines slowly during stimuli lasting

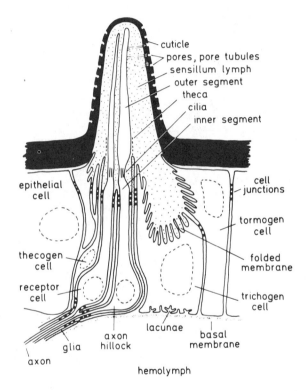

FIGURE 1. Schematic diagram of an olfactory sensillum with two receptor cells and three auxiliary cells, based on electronmicroscopic studies.[18,23,24]

several seconds or minutes as observed in pheromone receptors[3] and in humidity receptors.[4-6] A very strong stimulus may cause the tonic response to be reduced or abolished altogether, even when it is the first stimulus applied to a previously unadapted cell.[3] As a result, the ratio of phasic to tonic responses is dependent on stimulus intensity. The different adaptation behavior of the phasic and tonic responses suggests the existence of at least two separate adaptation mechanisms.

Upon removal of a stimulus, often after prolonged stimulation, a small transient reduction in impulse frequency may be observed.[3] This post-stimulus decrease is most obvious in water vapor receptor cells, which occur in pairs in the same sensillum and act antagonistically.[4–6] An increase in water vapor concentration elicits a positive peak of impulse frequency in one of the two cells, the "humidity" receptor, and a negative peak in the other, the "dryness" receptor. A decrease in humidity elicits the opposite effects in the respective cells.

RESPONSES OF PHASIC-TONIC RECEPTOR CELLS TO ODOR PULSES

The phasic-tonic behavior of olfactory cells is of functional significance. A well-defined peak of impulse frequency following increased stimulus intensity enables the animal to quickly recognize the onset of stimulation. At the same time, the end of a persistent stimulus, followed by the relatively small negative frequency peak, should be much more difficult to detect. However, constant stimuli do not represent the natural stimulus situation. Point-like odor sources such as a female moth give rise to a filamentous structure of odor plumes. A flying animal encounters frequent and sharp pulses of odor concentration on approaching such a source.[7] Single-cell recordings have shown that phasic-tonic cells are able to respond to single odor pulses of 20-msec duration at repetition rates of up to 10 odor pulses per second (FIG. 2).[3] At such frequencies each odor pulse may elicit only a single nerve impulse that is phase-locked to the stimulus pulse. The nerve impulse activity ceases immediately when the stimulating pulses are removed. Indeed, male moths of *Bombyx mori* stop walking upwind if a pause in stimulation lasts more than 0.7 sec.[8]

These investigations show that a peculiar combination of adaptation processes in phasic-tonic cells enables the animal to detect single odor pulses and also the end of a series of such pulses.

ADAPTATION OF NERVE IMPULSES VERSUS ADAPTATION OF THE RECEPTOR POTENTIAL

It is commonly held that nerve impulses in sensory cells are elicited by the receptor potential. It is, therefore, pertinent to ask whether adaptation of the nerve impulse response merely reflects adaptation of the receptor potential. In pheromone receptor cells on the antennae of male saturniid moths, adaptation of the nerve impulse response is greater than that of the receptor potential (FIG. 4).[9] This strongly suggests that the impulse generator and the receptor potential have separate adaptation mechanisms.

Furthermore, the receptor potential of olfactory cells usually exhibits a purely "tonic" time course; a phasic peak, if at all present, is much less pronounced than the peak in the impulse frequency.[10] Thus the phasic-tonic behavior of the impulse frequency response must be a property of the impulse generator.

Different adaptation mechanisms for receptor potential and nerve impulse frequency favor the idea of spatially separated generator regions for each type of response. These results, therefore, are in agreement with the view that the receptor potential is generated by conductance changes in the outer dendritic segment and that nerve impulses are generated near the cell soma.[10,11] In contrast to this, initiation of the nerve impulses in the outer dendrite has been proposed recently.[12] A truly consistent model of the receptor cell function still remains to be developed.

RESPONSE CHARACTERISTICS OF RECEPTOR CELL TYPES IN THE SAME SENSILLUM

While the nerve impulse generator determines the phasic-tonic behavior, the impulse response is also correlated with the time course of the receptor potential. This

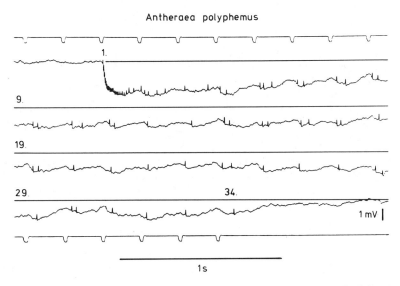

FIGURE 2. Receptor potential and nerve impulses from a single receptor cell elicited by a sequence of 34 stimuli of 20-msec duration each with the pheromone component E-6,Z-11-hexadecadienal. First and last trace: downward deflections show the transient increase of air pressure producing the odor pulses. The horizontal lines indicate the resting level of the receptor potential.

has been shown[14] by using certain derivatives of the pheromone that elicit receptor potentials with shorter rise and fall times, which are reflected in "faster" impulse responses, a sharper peak of impulse frequency and a more sudden drop in frequency after the end of stimulation (FIG. 3).

Typically, the time characteristics of receptor potential and nerve impulse response differ between cells in the same sensillum. In *Antheraea polyphemus* and in *Antheraea pernyi* the cell tuned to the pheromone component E-6,Z-11-hexadecadienal (AL) responds faster than the one tuned to E-6,Z-11-hexadecadienyl acetate (AC). This difference in response kinetics is independent of the general response amplitude. Thus, in *A. polyphemus* the amplitudes of receptor potentials and nerve impulses recorded

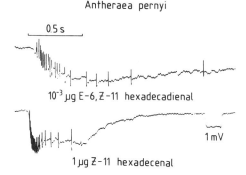

FIGURE 3. Receptor potential and nerve impulses from a single receptor cell elicited by a pheromone component (above) and by a derivative (below). The latter produces a characteristically different time course of the receptor potential and impulse frequency.[19]

from the acetate cell are larger than those of the aldehyde cell. In *A. pernyi* the situation is reversed, the aldehyde cell shows the larger responses.[14]

Different time characteristics of the cell response in the same sensillum are also evident from the resolving power of each cell type for pulsed odor stimuli. Thus the (slower) AC cell of *A. polyphemus* followed pulse frequencies up to about four pulses per second whereas the (faster) AL cell resolved frequencies up to 10 odor pulses per second (Kaissling and Rumbo, unpublished). The third receptor cell found in some sensilla[14] also belongs to the fast-response type. Its apparent key compound is the presumptive third pheromone component E-4,Z-9-tetradecadienyl acetate.[25] These results suggest that pheromone components perceived by cell types with a higher resolving power may play a special role for orientation in the close vicinity of the odor source.

CROSS-ADAPTATION OF RECEPTOR CELLS WITHIN A SENSILLUM

The presence of two (or three) receptor cells in the same sensillum (FIG. 1) allows a study of the extent to which adaptation is cell specific or involves changes in extracellular conditions. Each cell responds to a different key compound and can be stimulated separately. The characteristic size of nerve impulses indicates which cell was stimulated. Stimulation of one cell auto-adapts this cell but also cross-adapts the other cell.

Auto- and cross-adaptation have been studied for the two receptor cells responding to AC and AL in sensilla or *Antheraea polyphemus*.[9] These studies involved conditioning (adapting) stimuli of two-second duration and relatively high intensities, that is, producing more than half-maximal receptor potential. The degree of adaptation was determined by test stimuli applied 30-60 seconds after conditioning; thus, compared with the processes underlying phasic-tonic behavior discussed above, these experiments studied longer-term adaptation.

Cross-adaptation was much weaker than auto-adaptation. Therefore, cells can be selectively adapted by stimulation with their key compounds in order to block their responses and to facilitate measurements on other cells of the same sensillum that are sensitive to different key compounds.[14-17]

Cross-adaptation may be caused by alterations in stimulus-conducting structures of the sensillum common to both cells or by changes in the sensillum (receptor) lymph surrounding the outer dendrites of both cells. One possibility may be changes in the ionic composition of the sensillum lymph produced by strong stimulation of one cell. It remains to be shown whether cross adaptation can be produced during stimulation simply by depolarizing one cell, which lowers the transepithelial potential and, thereby, the membrane potential of all cells in the same hair. The cross-adaptation may also result from a direct inhibitory action of the conditioning stimulus on the cross-adapted cell. The "wrong" but structurally similar odor compound may possess some affinity to the cross-adapted cell and could block receptor sites for the key compound.

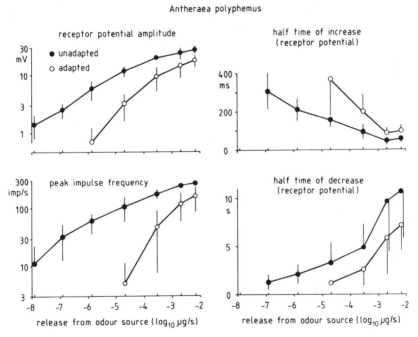

FIGURE 4. Unadapted and adapted dose-response curves for (**a**) the receptor potential amplitude, half-times of (**b**) rise and (**c**) fall of receptor potential, and (**d**) the peak impulse frequency obtained by averaging 10 consecutive inter-impulse intervals at peak frequency. Averages and standard deviations of the responses of 10 receptor cells.[22] The abscissa was calibrated using ^{3}H-labeled E-6,Z-11-hexadecadienyl acetate (Kaissling and Kanaujia, unpublished) supplied by H. Bestmann and G. Kasang. The air flow was 100 ml/sec; −8 corresponds to $2 \cdot 10^5$ molecules per cm[3] of air. The stimuli were given in order of increasing intensities because weaker stimuli had little or no effect on the response to succeeding stimuli of higher concentration. The first point of the adapted curve was measured 60 seconds after the strongest stimulus of the unadapted curve, serving as the conditioning stimulus. The intervals between stimuli were at least one minute for the unadapted curve and 30 seconds for the adapted curve. (Note: Peak impulse frequency adapts to a larger extent than receptor potential amplitude.)

It has not been demonstrated whether the observed cross-adaptation, brought about by relatively strong stimuli, has any biological function. Cross-adaptation effects at weaker stimulus intensities have, so far, not been systematically studied. There were no signs of cross-effects between the soma regions of receptor cells in the same sensillum, that is, no interrelations of nerve impulse responses between cells were observed. The receptor cell somata are morphologically well separated from each other by the thecogen cell and by a system of septate junctions (FIG. 1).[18]

LOCAL ADAPTATION

Stimulation of a small section of an olfactory hair locally adapts the receptor cell dendrite within the hair.[9,22] Such local adaptation has also been demonstrated in visual receptor cells (for references see Zack Strausfeld & Kaissling[9]). This effect shows that adaptation is at least partially due to a local alteration in the cell membrane, in the cytoplasm, or in both.

ADAPTATION EFFECTS OF THE RECEPTOR POTENTIAL

The characteristics of the receptor potential change in various ways with adaptation. The dose-response curves for the steady-state amplitude of the receptor potential and also its half-times for rise and fall are all shifted to higher stimulus concentrations. In addition the curve for the steady-state amplitude seems to approach a lower saturation value (FIG. 4). These changes could be understood in terms of a simple kinetic model involving the formation of a complex between stimulus molecules and hypothetical receptor molecules, activation of such complexes, and a conductance increase of the cell membrane.[20] The observed effects could then be explained by a reduced fraction of activated complexes or by a smaller increase of membrane conductance per activated receptor molecule. This interpretation is supported by the observation that, in parallel with the receptor potential, the stimulus-induced increase in preparation conductance also adapts.[20] However, further adaptation effects require additional assumptions. Most prominent of these is the increase in relative steepness of the dose-response curves (FIG. 4), which are comparatively flat in the unadapted state.[24] The slope (0.2 and below) may be partially explained by cable properties of the receptor cell dendrite.[21] Among other possibilities, there may exist a population of receptor molecules with distributed affinities to the pheromone molecule. A steeper slope would result if the receptor molecules with higher affinities would adapt, or be desensitized, to a larger degree than those with lower affinities.

Complete recovery from adaptation may require more than one hour, depending on the strength of the conditioning stimulus. Recovery takes longer in an isolated antenna than with the intact animal, indicating a dependence on the metabolic supply to the antenna.[22]

SUMMARY

Adaptation was studied in single olfactory receptor cells of male moths of *Bombyx mori* and *Antheraea polyphemus*. Receptor potential and nerve impulse generators have different and very likely, spatially separate adaptation mechanisms possibly located in the outer dendritic segment and the cell soma, respectively. Restricted portions of the receptor cell dendrite can be locally adapted. The impulse generator may exhibit at least two distinct adaptation processes with different kinetics, as deduced from a consideration of the phasic-tonic response and the different adaptation properties of each of these phases. The response characteristics of cells in the same sensillum are different. The "faster" responding cell types resolve odor pulses with frequencies up to 10 per second—a performance that is probably needed for orientation during flight toward a small odor source.

REFERENCES

1. LACHER, V. 1964. Elektrophysiologische Untersuchungen an einzelnen Receptoren für Geruch, Kohlendioxyd, Luftfeuchtigkeit und Temperatur auf den Antennen der Arbeitsbiene und der Drohne (*Apis mellifica L.*). Z. Vergl. Physiol. **48**: 587-623.
2. RUMBO, E. R. 1983. Differences between single cell responses to different components of the sex pheromone in males of the light brown apple moth (*Epiphyas postvittana*). Physiol. Entomol. **8**: 195-201.
3. KAISSLING, K.-E. 1986. Temporal characteristics of pheromone receptor cell responses in relation to orientation behaviour of moths. *In* Mechanisms in Insect Olfaction. T. L. Payne, M. C. Birch & C. E. J. Kennedy, Eds.: 193-194. Oxford University Press. Oxford, England.
4. WALDOW, U. 1970. Elektrophysiologische Untersuchungen an Feuchte-, Trocken- und Temperaturrezeptoren auf der Antenne der Wanderheuschrecke. Z. Vergl. Physiol. **69**: 259-283.
5. YOKOHARI, F. & H. TATEDA. 1976. Moist and dry hygroreceptors for relative humidity of the cockroach, *Periplaneta americana* L. Comp. Physiol. **106**: 137-152.
6. BECKER, D. 1978. Elektrophysiologische Untersuchungen zur Feuchterezeption durch die styloconischen Sensillen bei *Mamestra brassicae* L. (Lepidoptera, Noctuidae). Dissertation der Univ. Regensburg, FRG.: 1-80.
7. MURLISS, J. & C. D. JONES. 1981. Fine-scale structure of odour plumes in relation to insect orientation to distant pheromone and other attractant sources. Physiol. Entomol. **6**: 71-86.
8. KRAMER, E. 1986. Turbulent diffusion and pheromone triggered anemotaxis. *In* Mechanisms in Insect Olfaction, T. L. Payne, M. C. Birch & C. E. J. Kennedy, Eds.: 59-67. Oxford University Press. Oxford, England.
9. ZACK STRAUSFELD, C. & K.-E. KAISSLING. 1986. Localized adaptation processes in olfactory sensilla of saturniid moths. Chem. Senses **11**: 499-512.
10. BOECKH, J. 1962. Elektrophysiologische Untersuchungen an einzelnen Geruchsrezeptoren auf den Antennen des Totengräbers (*Necrophorus, Coleoptera*). Z. Vergl. Physiol. **46**: 212-248.
11. MORITA, H. 1959. Initiation of spike potentials in contact chemosensory hairs of insects. III. D. C. Stimulation and generator potential of labellar chemoreceptor of *Calliphora*. J. Cell. Comp. Physiol. **54**: 189-204.
12. WOLBARSHT, M. L. 1965. Receptor sites in insect chemoreceptors. Cold Spring Harbor Symp. Quant. Biol. **XXX**: 281-288.

13. DE KRAMER, J. J. 1985. The electrical circuitry of an olfactory sensillum in *Antheraea polyphemus*. J. Neurosci. **5:** 2484-2493.
14. KAISSLING, K.-E. 1979. Recognition of pheromones by moths, especially in saturniids and *Bombyx mori*. *In* Chemical Ecology: Odour Communications in Animals. F. J. Ritter, Ed.: 43-56. Elsevier/North-Holland Biomed. Press. Amsterdam, the Netherlands.
15. VARESCHI, E. 1971. Duftunterscheidung bei der Honigbiene—Einzelzell-Ableitungen und Verhaltensreaktionen. Z. Vergl. Physiol. **75:** 143-173.
16. Priesner, E. 1979. Progress in the analysis of pheromone receptor systems. Ann. Zool. Ecol. Anim. **11**(4): 533-546.
17. PAYNE, T. L. & W. E. FINN. 1977. Pheromone receptor system in the females of the greater wax moth *Galeria Mellonella*. J. Insect Physiol. **23:** 879-881.
18. KEIL, T. & R. A. STEINBRECHT. 1983. Beziehungen zwischen Sinnes-, Hüll- und Gliazellen in epidermalen Mechano- und Chemorezeptoren von Insekten. Verh. Dtsch. Zool. Ges. **76:** 294.
19. KAISSLING, K.-E. 1977. Structures of odour molecules and multiple activities of receptor cells. *In* International Symposium on Olfaction and Taste VI. J. Le Magnen & P. MacLeod, Eds.: 9-16. Information Retrieval. London, England.
20. KAISSLING, K.-E. & J. THORSON. 1980. Insect olfactory sensilla: Structural, chemical and electrical aspects of the functional organization. *In* Receptors for Neurotransmitters, Hormones and Pheromones in Insects, D. B. Sattelle, L. M. Hall, & J. G. Hildebrand, Eds.: 261-282. Elsevier/North-Holland Biomed. Press. Amsterdam, the Netherlands.
21. KAISSLING, K.-E. 1971. Insect olfaction. *In* Handbook of Sensory Physiology. L. M. Beidler, Ed. **4**(1): 351-431. Springer-Verlag. Berlin, Germany.
22. ZACK, C. 1979. Sensory adaptation in the sex pheromone receptor cells of saturniid moths. Dissertation Fak. Biol.: 1-99. Ludwig-Maximilians—Universität. München, Germany
23. KAISSLING, K.-E. 1986. Chemo-electrical transduction in insect olfactory receptors. Ann. Rev. Neurosci. **9:** 121-45.
24. KAISSLING, K. E. 1972. Kinetic Studies of Transduction in Olfactory Receptors of *Bombyx mori*. *In* Olfaction and Taste IV. D. Schneider, Ed.: 207-213. Wissenschaftl. Verlagsgesellschaft. Stuttgart, Germany.
25. BESTMANN, H., A. B. ATTYGALLE, T. BROSCHE, J. ERLER, H. PLATZ, J. SCHWARZ, O. VOSTROWSKY & C. H. WU. 1987. Identification of three sex pheromone components of the female saturniid moth *Antheraea Pernyi* (Lepidoptera: Saturniidae). Z. Naturforsch. **42c:** 631-636.

A Standardized Olfactometer in Japan

A Review Over Ten Years

SADAYUKI F. TAKAGI

Department of Physiology
School of Medicine
Gunma University
Maebashi 371, Japan

Since the end of 1960s, Japan has been very conscious of environmental pollution. Control of odor pollution became an important area for research. When the environmental bureaus embarked on the control of malodors, one of their first tasks was to select a panel of normosmic subjects who could properly inspect malodors; however, the government lacked solid criteria for selecting panel members.

An independent development was an increase in persons who complained of olfactory disorders. Otorhinolaryngologists, however, had not yet developed effective methods of treatment for such disorders.

To seek solutions for these two problems, I organized a research group, the Olfactory Test Committee, comprised of eleven professors of otorhinolaryngology, one professor of psychology, and two professors of physiology.

In the fall of 1971, these members and their staffs began working toward the common goals of manufacturing a standardized olfactometer and developing methods of remedying olfactory disorders. The details of some of this research were later published in a book entitled *Olfactory Disorders—Olfactometry and Therapy.*[1]

PRELIMINARY STUDIES FOR MANUFACTURE OF A STANDARDIZED OLFACTOMETER

The first task for the committee was to select appropriate odors for olfactory testing.

Selection of Test Odors

The conditions that the test odors must satisfy were:

(a) Each odor should be distinct from the others.
(b) They should be simple so that their qualities might be easily perceptible to many people.
(c) They should not alter in quality or strength over time.
(d) They should be artificially manufactured odorants of simple structures, because natural odorants in many cases are impure and their components are not constant.

Ten odors were then selected, as shown in TABLE 1. These ten odors were used as tentative test odors throughout our preliminary research activities. Hereafter, they will be referred to as "the ten odors" or "the tentative test odors."

Preparation of Odorous Solutions

The original substances or solutions of these ten odors were diluted successively by the 10-fold method (10^{-1}, 10^{-2}, 10^{-3}, and so forth, through 10^{-17}) so that a series of solutions were prepared. Solvents used were Nujol in most cases, but propylene glycol or diethyl phthalate was used in a few cases. These series of odorants were called the "10-odor series" for the sake of convenience.

A Method for Smelling Test Odors

After much discussion, the committee decided to use slips of nonodorous filter papers to test the olfactory ability of a subject. For olfactory testing, the tester dipped one end of the paper into an odorous solution and handed the other end to a subject, having him or her sniff the odorous end.

Determination of Order of Presentation

Sniffing bad odors in succession deprives subjects of their objectivity, so the order of smelling the ten odors was tentatively decided as shown in TABLE 1 (from 1 to 10) with the intention of alternating pleasant and unpleasant odors.

Determination of Detection Threshold

Detection thresholds were sought by beginning from the lowest concentrations at 10^{-17} and working stepwise upward. When a subject indicated that he or she could smell something, but could not identify the smell, that concentration was designated the detection threshold for that subject.

TABLE 1. Ten Test Odors[a]

Names of Odorants	Qualities of Odors	
β—Phenyl ethyl alcohol	Odor of rose Light sweet odor	A
Methyl cyclopentenolone (Cyclotene)	Burnt odor Caramel odor	B
Isovaleric acid	Putrid odor, odor of long- worn socks, sweaty odor, odor of fermented soybeans	C
γ—Undecalactone	Canned peach odor Heavy sweet odor	D
Skatole	Odors of vegetable garbage Oral odor or aversive odor	E
Cyclopentadecanolide (Exaltolide)	Musk odor Powder odor	F
Phenol	Odor of disinfectants Odor of hospitals	G
dl—Camphor	Camphor odor Pleasant, refreshing odor	H
Diallyl sulfide	Garlic odor	I
Acetic acid	Vinegar odor Pungent odor	J

[a] From *Olfactory Disorders—Olfactometry and Therapy,*[1] used with permission of Igaku-Shoin Ltd.

Determination of Recognition Threshold

After the detection threshold was determined, the subject in most cases could perceive the quality of an odor within one or two steps of higher concentration. Recognition thresholds were determined when a subject correctly named or described the quality of the odor he perceived. Many subjects had difficulty in naming an odor or describing the quality of an odor. For such subjects, appropriate expressive words were prepared for each of the ten odors (TABLE 1). Subjects could easily find one or two words from those on the list to express their impressions.

Olfactory Testing with the "10-Odor Series"

Using the above methods, members of the committee examined a large population of healthy young men and women between the ages of 18 and 25. Subjectively, the selected subjects offered no complaints of olfactory dysfunction, and objectively, they were found by rhinoscopy to be free from pathology. Olfactory detection and recognition thresholds were sought in these subjects. From these values, distribution curves of the detection and recognition thresholds were made for each odor, with the ordinate showing the relative distribution of subjects and the abscissa indicating the concentration expressed as x of 10^{-x} (x = 1, 2, ... ,12). The curves of these 10 odors seemed to be similar in shape to a normal distribution curve. As a trial, a χ^2 test was applied and null hypotheses that the distributions of these threshold values were normal were examined for each of the ten odors. The null hypotheses were rejected in most cases. These results might be attributed to the long tail that was found on the right side in most of the distribution curves. Strictly speaking, therefore, we could not say that all these curves represent normal distribution curves, but considering a quasi-temple bell type distribution, the averages ($\overline{X}$), variances (s^2), standard deviations (s), and coefficients of variation ($100s/X$) were calculated for the men and women combined, and also for men and women separately.

Classification of the 10 Odors

To classify the 10 odors, first a correlation matrix between pairs of odors among the 10 was computed from the data of detection threshold values of all subjects. Factor analysis by the principal factor method was then applied to the matrix, five factors were extracted, and the factor loadings of the 10 odors were sought. However, the ratio of factor sampling was insufficient and only about 65% accurate. According to the values of factor loadings, the 10 odors were classified into six groups. This classification later became useful when five odors were selected from the 10 odors to manufacture the T & T Olfactometer.

Manufacture of a Standardized Olfactometer

The Olfactory Test Committee then embarked upon a series of tests that would help determine a standardized olfactometer. From the practical viewpoint, clinicians complained that 10 odors were too many and that testing with all of them would be overly time-consuming. In response, the committee decided to select five odors.

Examination of the Suitability of the 10 Odors for Olfactory Testing

When the distribution curves of the detection and recognition thresholds for the 10 odors were examined, the curves of exaltolide, phenol, and dl-camphor were found to be shifted more than one unit further to the left than the others. This meant that these three odors were so weak that most people could perceive them only at nearly saturated concentrations. Hyposmic subjects who could not detect them at one (a one-tenth dilution of the saturated concentration) on the abscissa could not be tested further by this odor series. These three odors were thus judged to be less appropriate as test odors than the others.

A prerequisite condition for the olfactory test was that test odors should be as different from each other as possible. By calculating factor loadings, the 10 odors were classified into six groups. In this analysis, however, the factor sampling was only 65% accurate. With this in mind, the committee examined the appropriateness of the seven odors.

Although β-phenyl ethyl alcohol and Skatole fell into the same group, we could easily distinguish these two odors from each other. So we included both of them in the five test odors.

Iso-valeric acid, diallyl sulfide, and acetic acid all belonged to the same group in the analysis. Because sniffing bad odors deprived subjects of their objectivity in the olfactory test, the first two unpleasant odors seemed inappropriate as test odors. Still, iso-valeric acid was well known as an axillary odor and presumed to be a human pheromone, so this important odor was made one of the five odors. Acetic acid was well known as a stimulant of the trigeminal nerve, so it was considered inappropriate as a test odor. By removing acetic acid and diallyl sulfide, the five test odors were finally selected for the standardized olfactometer.

New Concentration Series for the Selected Five Odors

In the above preliminary tests, the averages ($\overline{X}$) of detection thresholds were calculated. Based on these x values, the committee determined a new concentration series having eight steps, from $\chi \times 10^{-2}$ to $\chi \times 10^{+4 \text{ or } +5}$ for each of the five odors.

Manufacture of the T & T Olfactometer

Using the five test odors and based on the averages of the threshold concentrations, the first standardized olfactometer in Japan was manufactured in 1975. The committee named this set the T & T Olfactometer ("T & T" stands for Toyota and Takagi). Dr. Toyota was the first chairman of the committee.

It was immediately approved as the sole standard olfactometer by the Otorhinolaryngology Society of Japan. Since then, the T & T Olfactometer has been widely used not only in many clinics and laboratories but also in many prefectures and cities. Olfactory testing with the T & T Olfactometer in clinics has been reimbursed by the government at a rate of 420 social welfare reimbursement points (one point represents 12 yen).

The Use of Olfactograms

An olfactogram was prepared for recording the results of testing with the T & T Olfactometer. In it, detection thresholds were marked by o—o and recognition thresholds by x—x.

To diagnose degrees of olfactory sensitivities, the committee agreed to categorize dysosmias into six classes according to the averages of recognition thresholds in the olfactogram. An average of less than −1 is hyperosmic, between −1 and +1 is normosmic, and between 1.1 and 2.5 is normosmic or mildly hyposmic. From 2.6 to 4.0, the averages indicate moderately hyposmic response, between 4.1 and 5.5, severely hyposmic. An average recognition threshold beyond 5.6 constitutes an anosmic response.

TEN YEARS' RESULTS OF OLFACTORY THERAPY IN JAPAN

A steroid hormone (betamethasone disodium) has been found most effective for treatment of olfactory disorders. It has been applied topically onto the olfactory mucosa with the patient's head fully inverted. L-Cysteine ethylester hydrochloride was also found to be effective. It has been used alone or in combination with the steroid hormone. Surgical treatment is often necessary to clear an obstructed airway in the nasal cavity.

During the years from 1972 to 1981, 3,498 patients visited the Olfactory Clinic of Showa University in Tokyo. Some 95-97% of the patients were found anosmic and the remaining 3-5% were hyposmic. After treatment, 29.2% of the patients recovered completely and 44.6% recovered partially. The remaining 26.2% did not show recovery at all (Asaka, unpublished data).

Among the 108 anosmic or severely hyposmic patients treated in the clinic of Kanazawa University Hospital, 29% of them recovered remarkably, 30% recovered partially, and the remaining 41% did not recover at all.[2] Among the 448 patients treated in the Olfactory Clinic of Gunma University from 1970 to 1983, 57.1% of them were anosmic, 20.8% severely hyposmic, 14.3% moderately hyposmic, 7.6% mildly hyposmic, and one had olfactory blindness. After treatment, 21.2% of the patients with intranasal causes recovered (Makino, unpublished data).

REFERENCES

1. TOYOTA, B., T. KITAMURA, & S. F. TAKAGI, EDS. 1978. Olfactory Disorders—Olfactometry and Therapy. Igaku-Shoin Ltd. Tokyo, Japan (in Japanese).
2. UMEDA, R. 1981. Diagnosis and Therapy of Olfactory Disorders Proceedings of the 82 General Assembly of Otorhinolaryngology Society of Japan (in Japanese). Department of Otorhinolaryngology, School of Medicine, Kanazawa University. Kanazawa, Japan.

Visual Approach to Fragrance Description

L. Givaudan & Cie S.A.
1214 Vernier-Geneva, Switzerland

WHAT IS A FRAGRANCE?

Fragrances are living works of art that change with the times. By its very essence, a fragrance evokes nature, a specific impression or ambiance. Like symphonies, fragrances are composed of notes and chords. Using these chords, these notes, the perfumer interprets a client's brief as closely as possible. Following his inspirations, a whim, an impulse, the perfumer creates what is for him the ultimate expression of dream and function. His inspiration springs from two basic concepts:

There is the classic concept: natural and synthetic fragrances are artistically used for their very intrinsic olfactive value, scents that evoke flowers, fruits, spices, precious woods, animals. The beauty and excellence of these fragrances are the perfumer's first source of inspiration. Ingeniously blended, for example, with amber, musk, mosses, and a thousand different nuances, they slowly transform into a masterpiece of harmony, perfumes rich in mystery, subtlety and warmth that underline and emphasize a woman's charm, her elegance, or natural freshness; or they become scents that stress a man's virility, sportsmanship, or feelings. This is how fine fragrances have played their classic role for centuries.

Using the same fragrances, the same notes, the perfumer creates compounds that evoke or create a specific ambiance, that underline a functional product's specific attributes and confer to it a unique personality. We call this the ambiance concept. Conceived to fire the imagination; to create intimate, cozy atmosphere; to transform bathrooms into environments of fresh, relaxing, and natural scents; to confer touches of softness and cleanliness to fabrics and household cleaners; to achieve stimulating effects on urges and passions.

Given the different ways fragrances are perceived by individuals, a perfume expresses in most cases both concepts. The balance between the two concepts depends on the final use of the fragrance. The more functional a product is, the more the ambiance concept prevails.

The importance of fragrances in creating certain ambiance concepts can be illustrated by the following example:

Rose:	Besides its olfactory note, well known by everybody, the rose odor suggests softness, femininity, and sensitivity.
Spicy notes:	These include clove, coriander, and cinnamon and express exoticism, light, warmth, and even arrogance.

119

Violet notes:	This fragrance lends elegance and distinction.
Lily of the valley:	With its flowery, fresh, green odor, it is reminiscent of spring, morning dew, youth, lightness, and delicacy.
Woody notes:	Sandalwood confers soft effects, vetiver expresses an impression of warmth, while patchouli brings embracing voluptuous sensations.
Balsamic and animal notes:	Depending on their nature, they can be calming, mysterious, sweet or sensual.
Fruity notes:	These evoke sparkle, mellowness but also have a voluptuous, glamorous shade.
Lemon odor:	Lemon expresses freshness, sun, beach, the South; associated with green notes, lemon may indicate sea, wind, and waves. Add some spices, and we can easily evoke the Caribbean Islands.

NECESSITY OF A COMMON LANGUAGE

To most people, the world of fragrance is a mysterious universe, where people with remarkable olfactive senses speak a language incomprehensible to anybody else.

In fact, perfumers have their own language to describe fragrances, their character and notes. The terms used relate to the olfactive world and are often rather confusing to the layman.

However, as fragrance specialists, we must interpret our client's wishes and the market needs and describe our creations through a language free from subjective attributes for perfumes such as "elegance," "richness," "sensuality," or "mystery"; definitions that fluctuate with the subjectivity of each person.

How much more difficult is the task for the layman to describe a compound, given its complex nature. Its character must be assessed, and its value and style appreciated. Almost impossible without a good knowledge of perfume bases, harmonies of scents, and of the classic main themes.

Music and painting are often equated with perfumery for their artistic value. But music can be written and therefore heard and appreciated by one's imagination. Painting can be seen, evaluated, and classified.

Not so fragrances: their classification is so subjective, with no recourse to a recognized, precise vocabulary, that communication becomes extremely difficult. This means that a marketer often knows what he wants the fragrance to do for his product, but he has difficulty in expressing his wishes to his fragrance supplier.

That is why our company, back in the early '70s, started by creating a simple language that sums up in one word the overall character of a given perfume. And this early approach is evolving all the time.

OUR PRESENT LANGUAGE

Having found that the creation and description of fragrances are being based on an overall olfactive note or on harmonies of scents, we have established a philosophy that permits us to group fragrances into these "overall olfactive notes" or "harmonies," and we call these groups analogies.

In order to group any fragrance into these "overall olfactive notes" or "harmonies," our company has adopted six analogies: basic floral/natural, aldehydic, Chypre, spicy, oriental, and Tabac.

Analogies

The Basic Floral/Naturals

Under this analogy falls all those fragrances with a dominant natural note originating from flowers, fruits, and green notes. These fragrances evoke an ambiance of freshness, refinement, and delicacy. Their overall olfactive accord is related to a natural theme.

Typical examples of this category are Joy (Patou), Eau Sauvage (Dior), the classic Eaux de Cologne, Paris (Y. Saint-Laurent), Anaís Anaís (Cacharel), Giorgio, les Jardins de Bagatelle (Guerlain).

The Aldehydic

The creation of the aldehydes in the 1920s led to an important trend in fine perfumery throughout the world. Aldehydic fragrances of this group are characterized by the powerful odor of these aroma chemicals. The best known and most classical of all aldehydic perfumes is, without a doubt, Chanel No. 5. Other typical perfumes of this category are Arpège (Lanvin), Calèche (Hermes), and Calandre (Paco Rabanne). In these perfumes a bouquet of aldehydes modifies and reinforces a harmony of wood, iris, and noble flower essences, with rose and jasmin often predominating. These are perfumes that recapture a rich, elegant, and distinguished atmosphere of high class.

The Chypre Mossy Woody

The Chypre or "mossy-woody" blend unveils sensual, heady perfumes where leather, precious woods, mosses, and animal scents unite intimately. The name "Chypre" was given to a substance originating from the distillation of labdanum, a gum resin secreted by cistus-type shrubs growing on the Chypre Island. Chypre-type perfumes occupy an important position among fragrances for both women and men.

Famous creations of the last 25 years include: Cabochard (Gres), Y (Saint-Laurent), Azzaro, Le Temps d'Aimer (A. Delon), Aramis for men (Estee Lauder), Diva (Ungaro), Ysatis (Givenchy), and Santos (Cartier).

Although these fragrances are very different, they all have in common the same underlying Chypre character. Their ambiance concept is described as sensual, opulent, and intimate.

The Spicy

The spicy group of perfumes finds its expression in a simple, floral yet spicy blend. The spicy character of these fragrances comes from the use of carnation flower and its synthetic derivatives such as Eugenol and Isoeugenol, basic elements of clove oil. For this reason, we do not consider fragrances like L'Air du Temps (Nina Ricci), Fidji (Guy Laroche), or Charlie (Revlon) as being floral, but rather spicy creations. Spicy perfumes create an ambiance of vigor, spirit, and warmth.

The Oriental

The outstanding perfumes of this century are very often representative of their era and time reflections of the epoch. The oriental types illustrate perfectly this statement.

The beginning of the 20th century was the era of the "new art" with its curved, sophisticated lines; of the oriental mode and exotic fabrics from China, Egypt, and Japan; and of perfumes like Jicky, L'Origan, and Shalimar. The oriental character of these fragrances is given by notes like amber, musk, vanilla, and heavy floral or fruity effects.

More recent orientals of great success are Opium (Saint-Laurent), Must (Cartier), Oscar de la Renta, Balahe (Leonard), Poison (Dior), Beautiful, and Obsession. Men's fragrances include the classical Old Spice, Brut, Lagerfeld, and JHL (Estee Lauder). Masculine variants are less sweet, less animal-like, with more emphasis on dry and virile notes, to suit the masculine image.

The Tabacs

Our sixth analogy comprises scents exclusively aimed at men: the Tabacs. Tobacco is a fancy reference that corresponds to a fern, lavender, woody, powdery, and often spicy accord. This blend gives a virile agrestic and exotic character to the perfumes of this group. Typical examples are Tabac Original, Paco Rabanne pour Homme, Alain Delon, Revillon pour Homme, and Halston 101.

The main components that distinguish the typical character of each analogy determine the classification of a fragrance. Obviously, no perfume will be made of just these raw materials, but will contain a variety of other products. For this reason, within each analogy, there exist secondary notes. Let us take the floral/natural analogy as an example. A perfume like Lagerfeld's Chloe, which is a basic floral theme, presents an oriental tendency. That is why we often describe a perfume with its basic and its secondary analogy, if necessary.

To conclude, we can say that before creating a fragrance and without taking into account crucial factors such as stability, suitability for the desired application and price constraints—the perfumer must first define the analogy and the balance between the two concepts. The combination of these two elements will lead as nearly as possible to the ideal profile.

Our six analogies have been conceived to simplify and improve the communication between our customers and us when talking about fragrances. The analogies allow us to talk about a very subjective theme, but with a high degree of mutual understanding.

Galaxies

Starting from these analogies, we have then set up our fragrance galaxies. We have two galaxies: one for the feminine perfumes, one for the masculine perfumes. We have chosen the term galaxy to express precisely this undefined space, in which we find odors floating, without sharp contours, just like nebulae. In this manner, we can easily place the existing perfumes and indicate the position of a future fragrance with regard to its note and to other fragrances already on the market.

Naturally, we can apply this method using the same analogies for all other market segments: cosmetics, soaps, shampoos, detergents, and so on. We must, however, take into account the ambiance concept of these products. In positioning the successful products of different markets, we can already gain a precise idea of the existing tastes and, as you are well aware, a clear knowledge of the fine fragrance trends is essential to us and to our customers because these trends are the starting points of the future fragrances for more functional products.

OUR FUTURE LANGUAGE

As is apparent, our present language is essentially based on the olfactive description. It seems to us increasingly necessary to make an easy and useful "functional" description available. We therefore intend to create "functional" analogies and "functional" galaxies. We would thus place the brief of a perfume in a functional context, in the same way that we characterize its olfactive description.

To be in a position to progress with this project, we have to wait for the results obtained from a variety of rather experimental consumer tests: tests of products in the market and tests on raw materials incorporated in functional products such as soaps, detergents, shampoos, and deodorants. We shall have to choose the adjectives (in this case, names for our functional analogies) that best describe feelings, images, and so on, that is, cleansing, softening, creamy, skin care, refreshing, and also the visual or perhaps audiovisual images that will help us to convey these functional analogies to the uninitiated person.

By combining our olfactive analogies and galaxies (continuously updated) and our functional ones, we hope, in the near future, to be able to communicate better with our customers as well as with our own perfumers.

Workshop Summaries

Trigeminal Chemoreception

DANIEL KURTZ, *Chair*

R. J. Reynolds Tobacco Company
Bowman Gray Technical Center
Winston-Salem, North Carolina 27102

DISCUSSANTS: YVES ALARIE (*University of Pittsburgh, Pittsburgh, PA*);
WILLIAM CAIN (*Yale University, New Haven, CT*);
HERBERT STONE (*Tragon Corporation*); RICHARD
COSTANZO (*Medical College of Virginia, Richmond,
VA*); JACK PEARL (*National Institutes of Health, Be-
thesda, MD*); MAXWELL MOZELL (*State University of
New York, Syracuse, NY*); and JOHN AMOORE (*Olfacto-
Labs*).

ALARIE: I have been interested in how one predicts how humans react to sensory
irritants from animal models. In industry we have what is called the "threshold limit
value" or TLV. TLVs are levels in the atmosphere (given in parts per million) that
a person can be exposed to for eight hours a day, five days a week for a lifetime with
no major health effects. We are interested in predicting these TLVs. There are about
600 substances that are listed for TLVs, of which 66% are sensory irritants. These
are trigeminal stimulants. We have used an animal model to predict the potency of
the sensory irritant, that is, to predict how humans will accept their chemical environment.

To do this, we measure respiration in mice. When there is a trigeminal stimulus
in the atmosphere, the mice will hold their breath in a very characteristic way. There
will be a pause during expiration. The length of this pause is proportional to the
concentration of the irritant. You can get a linear concentration–response curve simply
by measuring the respiratory rate and plotting the percent decrease versus the log of
the irritant concentration. From there you can get a 50% point. For 41 industrial
chemicals, the TLV is predicted by the mouse model with a correlation of 0.92. If
you know how potent the irritant is in the animal model, you can pretty much predict
the highest level that will be accepted by human workers in industry.

We found that trigeminal stimuli fall into three groups. The potency of nonelec-
trolytes or nonreactive chemicals can be predicted very accurately if their vapor
pressure is known. The potency of reactive chemicals (dienophiles or electrophiles)
can be predicted by knowing their reactivity toward an $-SH$ group. The last class
of trigeminal stimulants are more like pharmacological agents, that is, nicotine and
capsaicin. These must fit a receptor very nicely. We have tried to explain the relative
potency of these chemicals through interactions with a single transmembrane protein.
Those chemicals that act by physical activity alone simply change the free energy of
the membrane causing a conformational change in the protein. Those that react with

127

—SH groups attack the protein more directly. Several different pharmacological agents may fit in a particular pocket just like acetylcholine or nicotine fit in cholinergic receptors.

CAIN: I first got started looking at trigeminal stimulation by looking at some people who were unilaterally neurectomized and who therefore had no trigeminal sensitivity on one side. It was surprising to find how much the trigeminal nerve seemed to be contributing to the perceived magnitude of things that we would term odors and odor sensations. Why is it the case that trigeminal sensations take longer to be perceived than olfactory sensations? They do. If you inhale ammonia, you will get the odor first and then you will get the kick of the ammonia. For virtually all irritants, this will be true. As you get to very high levels of irritation, the irritation will move forward in time, but it will rarely, if ever, overtake odor latencies if odor is present. Another tendency is for perceived magnitude to march up with repeated exposures. This will happen breath by breath, or in the case of the eyes it will continue over time. The system seems to be integrating what it is getting. Whether this is a damage-induced integration, I don't know. But it does say that concentration is only one determinant of perceived magnitude.

In 1980, Claire Murphy and I published a paper in *Nature* having to do with interactions between odorants and irritants. If you ask somebody to judge the odor and the irritation of butanol (concentration versus perceived magnitude) you would find that irritation would go up, and that odor would go up less sharply and then would turn down. At first we thought that we had a weird subject, but that is always a bad instinct. Nonmonotonic things such as this are essentially always real. Claire and I made mixtures of carbon dioxide (a potent irritant at high concentrations) and the benign odorant amyl n-butyrate. We found there was indeed strong suppression of odor by irritation and, to some degree, suppression of irritation by odor. We could put the irritant in one nostril and the odorant in the other nostril and still get the suppression. There seemed to be some physiological interaction in the CNS and we used Herb Stone's electrophysiological data to back this up.

STONE: Most all consumer products (foods, personal care products, home care products, and beverages) have trigeminal components. Industry generally does not view these products in the context of trigeminal stimuli. In fact, if you were to talk about the trigeminal system in most companies, you would probably find that 99% or more of the people would not know exactly what the trigeminal system was. They would become concerned that they were selling painful products. However, you can relate the trigeminal characteristics of a product directly to changes in preference and to purchase behavior. In fact, the presence of a trigeminal sensation is often preferred and may indicate a certain efficacy.

COSTANZO: You mentioned that 66% of the chemicals with defined TLVs are considered sensory irritants. Do you presume that these sensations are mediated by the trigeminal nerve?

ALARIE: For the eyes and the nose, yes, since there is no other nerve there to mediate it. Certainly not olfaction. However, the term sensory irritant also covers throat irritation.

PEARL: When irritants are put to the mouth do you get the same type of breathing interruption as when you inhale it?

ALARIE: No, you can put an irritant in the eyes or the mouth and you will not get breathing interruption, which occurs only when you put it on the nasal mucosa.

MOZELL: If there were a stimulus that stimulates only the trigeminal nerve, could you adjust the concentration such that people would not call it irritating or adversive? Wayne Silver has a poster that says that you can't discriminate one irritant from another if you get rid of olfaction.

AMOORE: Something like pyridine is detectable at, say, 1 ppm by odor. Most people would say that they detect irritation at 20 ppm. People who have no olfactory sense due to something like a head injury cannot detect the irritation effect until 300 times the average odor irritation threshold for a normal person. There is a bit of a paradox here. It seems that the olfactory sense may be contributing to perceptual irritation.

CAIN: Henkin describes a class of hyposmia that most of the rest of us would describe as anosmia. There is no olfactory functioning, but people respond to vapors by giving responses that are apparently of a nonpainful sort and by making discrimination as well. I think that it needs to be interpreted in terms of time course, which may give you clues as to the type of trigeminal stimulus. But Henkin certainly sees it as a system that is working in a fashion other than just a pain system.

Temporal Aspects of Chemical Stimuli

Natural Stimulus Distributions, Receptor Cell Adaptation, and Behavioral Function

JELLE ATEMA, *Chair*

Boston University Marine Program
Marine Biological Laboratory
Woods Hole, Massachusetts 02543

DISCUSSANTS: PAOLA BORRONI (*Boston University Marine Program, Woods Hole, MA*); ROBERT FRANK (*University of Cincinnati, Cincinnati, OH*); ALAN GRANT (*The Worcester Foundation for Experimental Biology, Shrewsbury, MA*); STEVEN KELLING (*Cornell University, Ithaca, NY*); LINDA BARTOSHUK (*John B. Pierce Foundation, New Haven, CT*); BRUCE HALPERN (*Cornell University, Ithaca, NY*); KIM HOLLAND (*Hawaii Institute of Marine Biology, Kaneohe, Hawaii*); and REINHARD PREISS (*Max-Plank Institut für Verhaltenphysiologie, Germany*).

The discussion was introduced by Atema and centered on the behavioral importance of temporal patterns of chemical stimuli. In other sensory systems, temporal patterns carry important information and often elaborate processes exist to extract it. In chemoreception the emphasis of research has been and still is predominantly on determining the processes by which stimulus quality is coded. This includes primary transduction processes, and the sensory treatment of chemical mixtures. The study of temporal information processing in chemical senses has only just begun. Temporal coding depends initially on the adaptation and disadaptation of receptor cells.

An important point emerged in the discussion about the adaptive value of sensory adaptation. (Note: the term adaptation has an unfortunate dual meaning in biology as the previous sentence exemplifies purposefully. In an *evolutionary* sense *adaptive* means "enhancing reproductive success"; it signifies that a morphological, physiological, or behavioral character is contributing to the success of the species over evolutionary time. In *sensory* physiology *adaptation* refers to the reduction of response due to prior stimulation; one may choose—as is done here—to restrict the term adaptation to receptor cell processes and to use the term *habituation* for response reductions caused by higher order neural processes.) Two views about sensory adaptation were expressed. One view is that chemoreceptor cells suffer from an unfortunate but necessary constraint, presumably inherent in physiological-biochemical processes, that makes it impossible for cells to continue to respond unabatedly to a prolonged constant stimulus, that is, chemoreceptor cells adapt. Of course, nobody questions the phenomenon. The other view is that the adaptation characteristics of any receptor cell are as much part of its "design" as are its quality response characteristics; this view holds that sensory adaptation is evolutionarily adaptive and that both temporal tuning and spectral tuning are important features of receptor cells. The burden of proof now rests on those who hold the adaptive view of adaptation to show which temporal features of the stimulus are filtered and how they are used by the animal.

The example used to illustrate the two contrasting views was the behavior of insects in odor plumes. There was no disagreement about the phenomenon that insects orient better in pulsed plumes than in homogeneously distributed odor fields. The interpretation was different. Preiss felt that pulsed stimuli are more effective because pulsing prevents receptor cell adaptation, but that the animal does not use temporal information. Atema argued that, since a variety of chemoreceptor cells in insects and lobsters have different adaptation and disadaptation properties, the cells are specifically tuned to respond to different pulse features of odor plumes. He argued that all chemoreceptor organs (noses, tongues, antennae) experience different temporal features in their microenvironment and that perhaps each organ is tuned to its environment with a matching temporal filter. Kaissling argued strongly in favor of the importance of temporal information in chemoreception and presented data relating moth pheromone receptor responses to pulse patterns in pheromone plumes. In response to questions about whether frequency patterns of natural stimuli indeed correspond to physiological and behavioral responses, both Kaissling (for moths) and Atema (for lobsters) felt that the current evidence was insufficient but tantalizing. Moth pheromone receptor cells followed up to 10-Hz pulse frequencies and Kramer's work has shown that upwind flight behavior in moths was best at 3-Hz stimulus pulse rates.

The aquatic environment may contain lower dominant pulse frequencies than the aerial environment. In this context Halpern referred to McBurney's work on human taste perception, where as little as a 15% sinusoidal amplitude modulation with a very slow period significantly enhanced sensitivity. Here too, the initial interpretation was that amplitude changes serve to overcome the effects of (receptor cell) adaptation. This view may require re-examination in light of the new discoveries in arthropod chemoreception. Holland proposed that temporal stimulus patterns are important for distance orientation, but not to the tongue. Atema argued that one cannot automatically assume that the tongue does not also search, for instance for poisonous particles. He referred to Morita and Finger's work on the very accurate somatotopic maps of the palatal organ in goldfish, a morphological correlate of localization that seems adaptive for an animal that sorts its food inside the mouth and rejects unpalatable items.

Robert Frank addressed the question of temporal and spatial patterns of tongue stimulation that are created by chewing and tongue movement. Local temporal patterns of receptor stimulation would result from differential latencies by which different compounds have access to receptors. Differences in perceptual latencies are known. The discussion digressed from temporal to spatial patterns of stimulation of the (human) tongue and olfactory epithelium. Bartoshuk warned that one must distinguish between spatial separation of different molecular species on the tongue and spatial separation of odor (mixture) pulses in a plume. The physical/chemical properties of different compounds in a mixture may result in different time constants of tissue adsorption and desorption and thus cause temporal differences in the access of compounds to their receptors. Bartoshuk and Frank argued that this could have driven the evolution of different adaptation properties of receptor cells to match these stimulus time courses.

This led to a final point of discussion introduced by Halpern. It focused on the issue that psychologists have traditionally emphasized physiological performance limits while zoologists tend to concentrate on ecological relevance of the stimulus. Clearly, both are important and each depends on the ultimate goal of the research: mechanisms or consequences of the process under investigation. An awareness of the two approaches and their goals has often led to imaginative experimentation. The two approaches are mutually inspiring.

BIBLIOGRAPHY

1. ATEMA, J. 1985. Chemoreception in the sea: Adaptations of chemoreceptors and behavior to aquatic stimulus conditions. Soc. Exp. Biol. Symp. **39:** 387-423.
2. ATEMA, J. 1987. Aquatic and terrestrial chemoreceptor organs: Morphological and physiological designs for interfacing with chemical stimuli. *In* Terrestrial Versus Aquatic Life: Contrasts in Design and Function. P. Dejours, L. Bolis, C. R. Taylor, E. R. Weibel, Eds.: 303-316. Fidia Res. Ser. Liviana Press. Padova, Italy.
3. ATEMA, J. 1987. Distribution of chemical stimuli. *In* Sensory Biology of Aquatic Animals. J. Atema, R. R. Fay, A. N. Popper & W. N. Tavolga, Eds.: 29-56. Springer Verlag. New York.
4. BURKE, D., A. AKONTIDOU & R. A. FRANK. 1986. Time-intensity analysis of gustatory stimuli: Preliminary assessment of a new technique. Ann. N. Y. Acad. Sci. This volume.
5. ELKINTON, J. S. & R. T. CARDÉ. 1984. Odor dispersion. *In* Chemical Ecology of Insects. W. J. Bell & R. T. Cardé, Eds.: 73-91. Sinauer Associates. Sunderland, MA.
6. FINGER, T. 1987. Organization of chemosensory systems within the brains of bony fishes. *In* Sensory Biology of Aquatic Animals. J. Atema, R. R. Fay, A. N. Popper & W. N. Tavolga, Eds.: 339-363. Springer Verlag. New York.
7. KELLING, S. T. & B. P. HALPERN. 1983. Taste flashes: Reaction times, intensity and quality. Science **219:** 412-414.
8. KRAMER, E. 1986. Turbulent diffusion and pheromone triggered anemotaxis. *In* Mechanisms in Insect Olfaction. T. L. Payne, M. C. Birch & C. B. J. Kennedy, Eds.: 59-67. Oxford Univ. Press. Oxford, England.
9. MCBURNEY, D. H. 1976. Temporal properties of human taste system. Sensory Proc. **1:** 150-162.
10. MORITA, Y. & T. E. FINGER. 1985. Topographic and laminar organization of the vagal gustatory system in the goldfish, *Carassius auratus.* J. Comp. Neurol. **238:** 187-201.
11. MURLIS, J. & C. D. JONES. 1981. Fine-scale structure of odour plumes in relation to insect orientation to distant pheromone and other attractant sources. Physiol. Entomol. **6:** 71-86.
12. ZACK-STRAUSFELD, C. & K.-E. KAISSLING. 1986. Localized adaptation processes in olfactory sensilla of Saturniid moths. Chem. Senses **11:** 499-512.

How Is Peripheral Input Processed in the Central Nervous System?

JOHN SCOTT, *Chair*

Emory University
Atlanta, Georgia

PANELISTS: ERIC KNUDSEN (*Stanford University, Palo Alto, CA*); JUR-
GEN BOECKH (*University of Regensburg, Regensburg, Ger-
many*); and TAKASHI YAMAMOTO (*Osaka University,
Osaka, Japan*).

The workshop on central processing concentrated on issues related to the possible use of computational maps in the chemical senses. The discussion was focused on the model of barn owl sound localization presented in the symposium the previous morning. Vigorous discussion took place involving examples drawn from taste, vertebrate olfaction, and insect olfaction.

The two major characteristics of known computational maps are a systematic mapping of some properties of the stimulus along a spatial dimension in the nervous system and a transformation (other than a simple point-to-point representation) based on that mapping. In the sound localization example, a spatially systematic representation of the sound source is produced in the inferior colliculus by summation of inputs that code interaural intensity differences and interaural phase differences. Computational maps offer advantages over discontinuous categories in providing for rapid processing of fine discriminations; in providing for processing of unlimited combinations of sensory input; and in providing for simplicity in integrating mechanisms, such as spatially organized lateral inhibition.

In considering the analysis of chemical senses, it seems appropriate to ask whether a spatial continuum exists to form the basis of a map. There has been considerable debate over the possible continuous nature of the gustatory stimulus, but as yet no orderly change in a physical attribute of the stimulus has been shown to underlie these proposed orderings. The responses of the rat gustatory cortex described by Yamamoto show evidence of a map arrayed along an acceptance-rejection dimension. Studies of vertebrate olfaction have outlined at least one possibility for a physical stimulus continuum in the polar-nonpolar properties of odorant molecules that govern penetration into the mucosa and probably govern distribution of molecules across the receptor surface. On the other hand, those working in insect olfaction have not seen evidence for systematic continuities in chemical stimuli, although the data are consistent with recognition of a large set of odorants by different combinations of activity patterns across a limited number of receptors.

Computational maps share certain properties with across-fiber patterns, but there are important differences. Both concepts emphasize the importance of information distributed across many neural elements and provide means of analyzing inputs from broadly tuned receptors. This has the advantage that the system is sensitive to subtle variations and is not limited to a preset stimulus set. The concept of a computational map is more specific in its hypothesized mechanism of analysis, since the presence of a spatial gradient suggests that spatially organized factors, such as lateral inhibition, are involved in decoding. At the other extreme, an across-fiber pattern might be analyzed by having specific cells or groups of cells activated by specific patterns of input over the array of fibers. Some of the insect results described by Boeckh seemed to fit the latter model, there being no evidence that the CNS contains a map of any continuous function of the stimulus.

There was not agreement that all olfactory or gustatory stimuli can be arrayed along continuous dimensions without gaps. Even so, this does not preclude maps within categories. An example of continuous maps within discrete categories was cited in the presence of orientation columns within the occular dominance columns of the visual cortex. This arrangement produces small repeating micromaps that would not be noticed without fine-grain exploration of the cortical surface. Few examples of this type of fine-grain exploration of central nervous system regions can be found in the literature on the chemical senses. In this context it was suggested that analysis of chemical sensory data in terms of response types based on either stimulus properties or response properties may throw away information that is actually being used by the nervous system.

An important element of the studies of sound localization was the fact that the relevant stimulus parameters were already known in the sense that position of the sound source in space determined the direction of birds' flight during hunting. Discussion followed on the issue of whether it is necessary to use behaviorally relevant stimuli in the study of sensory systems and to look for mapping that corresponds to a feature of these stimuli. The successful study of auditory localization was an example of a top-down investigation where a precise behavioral correlate was available and research centered on finding cells that would correspond to that behavior. Some at the workshop disputed the general applicability of that approach. They pointed out that mammalian sensory systems have evolved to respond to a great variety of environments. Restriction to innately significant chemical stimuli such as sexual odorants sometimes leads to problems in identification and purity of the stimuli. On the other hand, one of the important characteristics of the mammalian system is its ability to discriminate an almost unlimited set of odorants.

Seeking the Mechanisms of Chemical Transduction

VINCENT E. DIONNE, *Chair*

Department of Medicine
University of California, San Diego
La Jolla, California 92093

PANELISTS: SUE C. KINNAMON (*Colorado State University, Fort Collins, CO*); and STEVEN J. KLEENE (*University of Cincinnati, Cincinnati, OH*).

Both olfactory and gustatory receptor cells respond to specific chemical stimuli in the environment by causing patterns of action potentials that are projected centrally, yet the transduction mechanisms that accomplish this remain unknown. This workshop brought together investigators studying ion channels in chemosensory receptor cells and investigators studying channels from those systems reconstituted into synthetic membranes to discuss the progress and prospects that relate channel function to chemical transduction. It provided a relaxed format in which to review work presented during the meeting and to air preliminary results of other recent work, and left an overall appreciation of the rapidly accelerating pace of single-channel studies in the chemical senses.

Presumably the appropriate combination of transduction processes, receptor types, and CNS connectivity allow the olfactory and gustatory systems to detect and distinguish among chemical substances. At this time single-channel studies seem to be focused mainly on the problem of stimulus detection, seeking to establish a relation between detection and channel function. The complexity of receptor cells with their variety of channel types, and the diversity of chemical stimuli appear to have combined to impede rapid progress. Ironically, the fundamental problems for single-cell studies of these two basic chemosensory systems are the converse of one another. For taste receptors the hurdle has been identification of the receptor cells after isolation; typically the cells lose their characteristic shape following dissociation from the gustatory epithelium and cannot be distinguished from other epithelial cells. In contrast, olfactory receptor cells retain their unique morphology when dissociated, but the number of potential stimuli is so vast that reliably inducing a chemical response is still an elusive goal.

After a brief introduction to single-channel techniques by the workshop organizers, the participants discussed a variety of specific problems. One issue was whether odorant-stimulated changes in channel activity are caused by a direct effect of the odorant on the channel, or are mediated indirectly, for example by changes in intracellular second messengers. Although biochemical studies have shown that cAMP production can be induced by odorants in preparations enriched with membrane from olfactory cilia, studies of reconstituted channels have found only direct effects of odorants on channel activity, that is, effects that are not mediated by cytosolic components. It was pointed out that odorants that act *in vitro* by modulating a second messenger may also act directly on channels in bilayers and cell-free membrane patches. Determining whether such effects are physiologically important or artefactual may be difficult. In addition, in the presence of some hydrophilic odorants, the activity of adenylate cyclase in membranes from olfactory cilia is unchanged. Perhaps compounds such as these act directly to modulate the membrane conductance or indirectly through noncyclase mechanisms such as the polyphosphoinositol pathway or via lipid-related effects. It seems quite probable that the diversity of odorants to which animals respond might necessitate many different mechanisms for transduction in order to accommodate substrate discrimination. In this regard the difficulty of distinguishing receptor and nonreceptor chemosensory mechanisms was lamented.

In further discussion it was noted that whole-cell membrane currents in olfactory neurons and taste receptor cells consist of several components carried by sodium, calcium, and an assortment of potassium channels. From cell to cell there is variability in the relative amounts of specific components. It is attractive to hypothesize that this difference relates to some level of chemical specificity, but it may reflect other factors such as age of the cell or deterioration of important membrane proteins brought on by the isolation procedures, especially those that involve enzymes.

The final problem discussed was the issue of species differences. The whole-cell currents from olfactory neurons isolated from related but different species do not appear identical. For example, in cells from the larval tiger salamander there is a prominent, fast-inactivating K^+ current; however, in the mudpuppy, which is also an aquatic salamander, none has been detected. What is not clear is whether such variations illustrate species-related differences or whether uncontrolled variables such as the dissociation procedures underlie them. If these turn out to be species-related, then making extrapolations from amphibia to mammals may be most unsatisfactory.

It was clear from the participants in this workshop that the interest in and application of single-channel recording techniques to studies of taste and olfaction are expanding rapidly. Just as for the visual and auditory systems, the promise is that this methodology will allow researchers to finally identify the intracellular elements and membrane channels that underlie the basic physiology. There is a renewed sense of excitement about these problems fueled by the hope that major advances in our understanding may be near at hand.

Salt Appetite

Intake and Preference for Salt in Man and Animals

Monell Chemical Senses Center
Philadelphia, Pennsylvania 19104

PANELISTS: SUSAN S. SCHIFFMAN (*Duke University, Durham, NC*);
DAVID HILL (*University of Toledo, Toledo, OH*); and
ALASTAIR R. MICHELL (*University of London, London,
England*).

S. Schiffman discussed the development of a substitute for NaCl and stressed that this is a difficult task due to the physiology of the taste system. Psychophysical and neurophysiological studies in her laboratory using pharmacologic probes indicate that there are at least three pathways that mediate the taste of salts on the tongue. Specifically, the taste effects of opening and closing ion channels on the human tongue with topical application of various chemicals were discussed. Sodium pathways are inhibited by amiloride[1] and potassium channels are selectively blocked by barium.[2] Trifluoperazine blunts the taste of calcium salts in man.[3] Multiple anion channels undoubtedly exist as well. She has shown, for example, that sodium salts with different anions vary considerably in taste quality[4]; suggesting that anions must make a differential contribution to the taste quality of sodium salts. She believes that as more is learned about cationic and anionic pathways in taste cell membranes, the design of salt substitutes will become more feasible.

It was emphasized that in humans certain conditions must be satisfied in order to demonstrate the amiloride effect of taste. Firstly, there are large individual differences in responses to amiloride with some subjects showing reductions in the perceived saltiness of NaCl while other subjects do not.[5] Secondly, one must use filter paper in applying both the amiloride and the taste stimulus. Whole-mouth rinses with amiloride solutions do not produce decreases in the scaled saltiness of NaCl solutions. Also, there is no reduction in perceived saltiness after a whole-mouth pretreatment with an amiloride solution, nor after pretreatment of the tongue with filter paper soaked in amiloride.[6]

D. Hill presented evidence for two components comprising the taste response to NaCl in developing rats, a sodium-sensitive component and a residual component. Previous work has shown that major changes occur during development in the neurophysiological (chorda tympani and parabrachial nucleus) responses of young rats.[7,8] Recent evidence supports the hypothesis that young rats first acquire the residual component and then later start adding sodium-sensitive channels. When the taste of NaCl is paired with illness, young postweaning rats form a conditioned aversion, which generalizes equally to NaCl, KCl, and NH$_4$Cl.[9] In adults, after the taste of NaCl has been paired with illness, an aversion generalizes to NaCl but not to KCl nor NH$_4$Cl. When adults are treated with amiloride, they respond like young rats and generalize a conditioned aversion to chloride but not sodium salts.[10] Thus it appears that there is a postnatal acquisition of amiloride-sensitive sodium channels on the tongue that may parallel the development of such channels in the pig colon. If young rats are maintained on a sodium-deficient diet during development, they seem to fail to form amiloride-sensitive channels. However, the amiloride channels open up within 24 hours after placement on a sodium-replete diet.[11] Thus the origin of these channels is related to the presence of a certain level of body sodium.

M. Bertino is studying rats' salt preferences in food. It is well known that sodium-replete rats prefer to drink salted water of suprathreshold isotonic and hypotonic concentrations to plain water. Preferences for salt in a variety of foods were examined using salt concentrations that were wt/wt equivalent to hypotonic, isotonic, and hypertonic salt water. When given a choice between salted and unsalted solid food, SD rats either prefer to consume the unsalted form or eat equal quantities of the salted and unsalted forms.[12] SHR rats who show heightened salt water preferences when compared with their normotensive controls, also do not prefer to consume salted solid food.[13] When salt preferences of SD rats were examined in liquid milk, some variation was seen. Given a choice between salted and unsalted heavy cream, they either preferred to consume the unsalted form or they ate equal amounts of the salted and unsalted forms similar to their responses to salted food. The avoidance was seen at the higher salt concentrations tested and was probably due to these concentrations being too salty.[14] However, the equal intake of salted and unsalted food observed at lower salt concentrations was not necessarily due to an inability of the rats to taste the salt. Conditioned aversion experiments demonstrated that when rats had ingestion of salted heavy cream paired with illness, they showed greater generalization of the resultant conditioned aversion to NaCl in water than control rats who were conditioned with unsalted heavy cream.[15] Unlike the results using heavy cream, rats prefer to consume salted skim milk. This difference may be related to the fat content of these products.

The lack of salt preference in food has been documented in baboons and is unaffected by maintenance on a high-salt diet during development.[16] Recent work from another laboratory suggests that salt in food may have an unpleasant taste for the rat.[17] Finally rats and humans may be similar in that salt preferences in food are dependent on the particular food tested.

A. Michell stressed that salt appetite is only one of a number of mechanisms regulating body sodium, and these are likely to converge on a concerted outcome rather than compete or correct for one another. Specifically, sodium appetite would be expected to go in the same direction as body mechanisms. Thus atrial natiuretic factor depresses salt appetite[18] reinforcing its renal effects rather than compensating for it.

Even though the physiological requirement for sodium is very low, rats, sheep, and humans will show a preference for salt when they have adequate dietary levels. The efficiency and speed of renal and enteric regulation of body sodium and the presence of extensive reserves in bone raises the question as to the physiological role of this salt appetite. Sodium deficiency is uncommon even among herbivores. In humans, the commonest clinical problem related to sodium intake is probably hypertension caused or exacerbated by excess intake.[19] The question was raised as to whether this need-free appetite for salt is an old appetite that evolved when environmental sodium was scarce and/or to save the animal from sodium depletion. A. Michell does not think that this evolutionary argument provides an adequate explanation for the need-free appetite because the sodium content of the diet that man was supposed to have evolved on was adequate. Another hypothesis is that the evolutionary diet was very high in potassium and extra sodium was needed for potassium excretion. Yet he has not observed an increased salt appetite after increasing dietary potassium content.

As far as the question of what the sodium is actually needed for, A. Michell believes there are essentially two answers: (1) to maintain the volume of extracellular fluid (its sodium concentration being primarily regulated through water balance); and (2) to be actively transported in exchange for K^+ and thus established gradients underlying both excitability and the movement of water and a variety of solutes. Since aldosterone stimulates the pump that is already a major consumer of ATP, problems faced by an animal with maximal salt conservation may not be sodium balance but excessive consumption of ATP in maintaining it. Thus mechanisms regulating body sodium including salt appetite should monitor not only extracellular volume or the amount of sodium in the body, but above all the rate of this crucial active transport.[20]

REFERENCES

1. SCHIFFMAN, S. S., E. LOCKHEAD & F. W. MAES. 1983. Proc. Natl. Acad. Sci. **80:** 6136-6140.
2. SCHIFFMAN, S. S. Barium selectively blocks potassium taste in rats. Submitted for publication.
3. SCHIFFMAN, S. S. Unpublished data.
4. SCHIFFMAN, S. S., A. E. MCELROY & R. P. ERICKSON. 1980. Physiol. Behav. **24:** 217-224.
5. SCHIFFMAN, S. S. Unpublished data.
6. DESOR, J. Unpublished data.
7. HILL, D. L., C. M. MISTRETTA & R. M. BRADLEY. 1982. J. Neurosci. **2:** 782-790.
8. HILL, D. L., 1983. Soc. Neurosci. Abstr. **9:** 378.
9. FORMAKER, B. K. & D. L. HILL. 1985. Abstr. Assoc. Chemorec. Sci. VII Meeting. No. 60.
10. FORMAKER, B. K., K. S. WHITE & D. L. HILL. 1985. Soc. Neurosci. Abstr. **11:** 1222.
11. HILL, D. L. 1985. Soc. Neurosci. Abstr. **11:** 448.
12. BEAUCHAMP, G. K. & M. BERTINO. 1985. J. Comp. Psychol. **99:** 240-247.
13. BERTINO, M. Unpublished data.
14. BERTINO, M. & G. K. BEAUCHAMP. 1987. Appetite **8:** 55-66.
15. BERTINO, M. Unpublished data.
16. BARNWELL, G. M., J. DOLLAHITE & D. S. MITCHELL. 1986. Physiol. Behav. **37:** 279-284.
17. K. CYBULSKI, D. F. JOHNSON & G. COLLIER. 1986. Abstr. Eastern Psychol. Assoc. 57th Meeting. p. 11.
18. FITTS, D. A., R. L. THUNHORST & J. B. SIMPSON. 1985. Brain Res. **348:** 118-124.
19. MICHELL, A. 1984. Perspect. Biol. Med. **27:** 221-233.
20. MICHELL, A. 1978. Perspect. Biol. Med. **21:** 335-347.

Recent Trends in Clinical
Measurement of Taste and Smell

JANNEANE F. GENT, *Chair*

University of Connecticut Health Center
Farmington, Connecticut 06032

DISCUSSANTS: HIROSHI TOMITA (*Nihon University School of Medicine, Tokyo, Japan*); R. GREGG SETTLE (*University of Pennsylvania, Philadelphia, PA*); R. COSTANZO (*Medical College of Virginia, Richmond, VA*); R. MATTES (*Monell Chemical Senses Center, Philadelphia, PA*); and S. HERNESS (*The Rockefeller University, New York, NY*).

GENT: Our speaker, Dr. Hiroshi Tomita, has developed procedures for localized testing of taste function using electrogustometry and filter paper discs. Up until three or so years ago, most clinical testing of taste function in the United States has been some version of a whole-mouth procedure, for example, the sip-and-spit method or Henkin's three-drop technique. Only recently have investigators in this country begun to explore the use of localized testing in the evaluation of taste disorders. Since 1976, Dr. Tomita has evaluated approximately 2,000 patients complaining of taste abnormalities.

[Dr. Tomita's 40-minute film was shown at this point. Many of the taste examination methods and results presented in the film are described in Tomita *et al.*[1]]

GENT: Dr. Tomita, could you please describe for us the types of complaints you see in your clinic?

TOMITA: Approximately 80% of the patients who come to the taste disorders clinic complain of hypogeusia or ageusia (reduced or absent sensitivity to all taste qualities). Of the remaining patients, some complain of pantogeusia (a constant bitter or other taste unrelated to eating) or dissociated taste disorder (reduced sensitivity to only one or two qualities[2]).

GENT: Do you ever use whole-mouth testing procedures other than to diagnose the ageusic patient?

TOMITA: No. In the clinic, only electrogustometry (EGM) and filter paper discs (FPD) are used. However, we use the whole-mouth testing procedure only for the patient who cannot experience a sensation with a concentration level five of FPD or lower.

SETTLE: Why do you use both electric and chemical taste thresholds? How do you resolve discrepancies between them?

TOMITA: EGM is useful since it is quick and exclusively quantitative. It is also possible to use EGM to detect taste disorders of which the patient is unaware. For example, in the facial palsy patients described in the film, it was possible to detect unknown taste disorders in 30% of the cases.[3] There is a good correspondence in the

140

results obtained using each method.[2] It is interesting that discrepancies in the two methods (good electric taste threshold, but poor discrimination of taste qualities) seem to occur in patients in the early stages of taste disorders due to zinc deficiency.[2]

SETTLE: As shown in your film, the glossopharyngeal nerve has a much wider area of innervation than previously thought based on animal studies. How long after chorda tympanectomy were these patients tested? Is it possible that the wider coverage by the glossopharyngeal nerve resulted from new growth of the nerve into regions previously innervated by the resected chorda tympani?

TOMITA: Patients who had undergone dissection of the chorda tympani during ear surgery were tested immediately after the operation, and also were tested three years after the operation. The longest observation was carried out through one year. The wide variation of ageusia appeared some months after the operation. A gradual absence of subjective taste abnormality was observed, but the area of ageusia still existed.

COSTANZO: Have you examined any autopsy material confirming your clinical findings of the area of innervation of the glossopharyngeal nerve?

TOMITA: Yes, we did. Please, see a report by M. Shonago.[4]

MATTES: With regard to your use of zinc therapy, have you done any studies during which you take the patients off of zinc, then retest function?

[For an answer, Dr. Tomita referred to two reprints in the handout material, Tomita,[5] and Sekimoto and Tomita,[6] that discussed the therapeutic use of oral zinc sulphate.]

HERNESS: In regard to patients in which the chorda tympani nerve has been dissected, have you tested patients for thermal or tactile sensitivity in the affected tongue area?

TOMITA: Patients were tested for pain sensation and that appeared to be normal.

REFERENCES

1. TOMITA, H., M. IKEDA & Y. OKUDA. 1986. Basis and practice of clinical taste examinations. Auris Nasus Larynx (Tokyo) 13: s1-s15.
2. TOMITA, H. & Y. HORIKAWA. 1986. Dissociated taste disorder. Auris Nasus Larynx (Tokyo) 13: s17-s27.
3. TOMITA, H. et al. 1972. Electrogustomery in facial palsy. Arch. Otolaryngol. 95: 383-390.
4. SHONAGO, M. 1971. Histological studies on the solid reconstruction of the human tongue's root. Jpn. J. Tonsil 10: 139-155.
5. TOMITA, H. 1981. Statistical analyses of 500 patients with taste disorder. Arch. Otorhinolaryngol. 231: 530-532.
6. SEKIMOTO, K. & H. TOMITA. 1986. Zinc chelation capacity of hypotensive agents causing taste disturbance. Nihon Univ. J. Med. 28: 233-252.

Poster Papers

Mechanism of Interaction between Odorants at Olfactory Receptor Cells[a]

BARRY W. ACHE,[b] RICHARD A. GLEESON, AND
HOLLY D. THOMPSON

C. V. Whitney Laboratory
University of Florida
St. Augustine, Florida 32086

Behaviorally relevant odors and tastes for most, if not all, organisms are mixtures of compounds, not single molecular species. Understanding chemosensory coding therefore requires understanding how multicomponent stimuli coactivate chemosensory pathways. Mixture interaction, in particular mixture suppression, is well known psychophysically, but the physiological basis of mixture suppression remains a mystery. We earlier reported that mixture suppression is the dominant type of interaction found in the antennular (olfactory) pathway of the spiny lobster,[1] that it is due in part to odorants interacting at the receptor cells,[2] and that mixture suppression in one type of taurine-sensitive receptor cell suggested that odorant molecules of different efficacies compete for common receptor sites.[3]

This study extends our initial observation[3] by characterizing the dose-response function of single chemoreceptor cells in the antennule of the spiny lobster to three stimulatory odorants in the presence and absence of single doses of one to three suppressive odorants. Odorants selected for testing were based on the results of an earlier study with this organism that identified the contribution of 31 components of a complex food odor, crab muscle tissue extract, to the stimulatory capacity of the extract.[1] Data for one cell (FIG. 1) show that the suppressive effect of both proline and arginine, evidenced by a right shift of the dose-response function of taurine when combined with these two suppressants, could be overcome by sufficiently high concentrations of taurine. Collectively, most of the cells followed this pattern (TABLE 1). Both proline and arginine antagonized the action of taurine in a manner consistent with competitive inhibition (parallel slope, right-shift, equal maximum response) (TABLE 1A). The maximum response of the taurine cells in this study ranged from 61 to 316 spikes/5 sec, indicating that suppression is not limited to low-response (< 20 spikes/10 sec)[3] taurine cells. Proline and arginine, as well cysteine, also suppressed responses to glycine, and did so in a manner similar to that found for taurine cells (TABLE 1B). In a single trail, proline suppressed the response to a third stimulatory component, the quaternary amine, betaine, and again, did so in a manner similar to that observed for taurine and glycine cells (TABLE 1C). In three (7.3%) of the 41

[a]This work was supported by National Science Foundation Award BNS 85-11256.

[b]Address for correspondence: Dr. Barry W. Ache, C. V. Whitney Laboratory, Rt. 1, Box 121, St. Augustine, FL 32086.

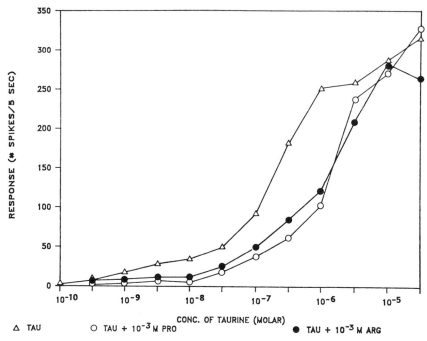

Δ TAU   O TAU + 10^{-3} M PRO   ● TAU + 10^{-3} M ARG

FIGURE 1. Dose-response relationship of an antennular (olfactory) chemoreceptor cell of the spiny lobster to taurine alone (Δ) and in the presence of a fixed dose of each of two suppressants, proline (O) and arginine (●).

TABLE 1. Effect of a Single, Fixed Dose of One to Three Suppressants on the Dose-Response Functions of Three Stimulatory Components of Crab Muscle Tissue Extract for Antennular Chemoreceptors of the Spiny Lobster

Odorant Stim.	Suppress.	No. of Cells Tested	Slope[a] ($\bar{X} \pm$ SEM)	EC_{50}[b] ($\bar{X} \pm$ SEM)	Max. Resp.[c] ($\bar{X} \pm$ SEM)
(A) Taurine	Proline	9	1.17 ± 0.34	0.74 ± 0.64	0.96 ± 0.07
	Arginine	8	1.09 ± 0.26	0.58 ± 0.41^d	0.96 ± 0.07
(B) Glycine	Proline	14	1.23 ± 0.39^d	0.78 ± 0.78^d	0.96 ± 0.20
	Arginine	3	1.47 ± 0.18^d	0.11 ± 0.84	0.85 ± 0.21
	Cysteine	3	1.34 ± 0.15^d	0.40 ± 0.42	1.11 ± 0.35
(C) Betaine	Proline	1	0.82	0.23	0.95

[a] Ratio of the slope of the linear part of the agonist + antagonist curve to that of the agonist-only curve.

[b] Ratio of the log-step shift in the EC_{50} of the agonist + antagonist curve to that of the agonist-only curve.

[c] Ratio of the maximum response (number spikes/first 5 sec) of the agonist + antagonist curve to that of the agonist-only curve.

[d] Significantly different at 95% confidence level as determined by two-tailed (one-tailed for EC_{50} column) t test performed on the values from which the ratios were determined.

agonist : antagonist pairings for which complete curves could be obtained (1 bet : pro, 1 gly : arg, 1 gly : pro), higher concentrations of agonist did not overcome the antagonist-induced suppression, allowing that other, perhaps noncompetitive, processes may contribute to peripheral mixture suppression.

Our results support our earlier observation,[3] as well as that of Johnson *et al.*,[4] that mixture suppression is in part a peripheral phenomenon in lobster olfaction. That five different components of food-related stimulus mixtures (three in this study, two in an earlier study[3]) suppress the action of three other components in a manner consistent with competition for common receptor sites, suggests that competitive inhibition may be an important underlying mechanism of peripheral mixture suppression in the lobster olfactory system.

REFERENCES

1. DERBY, C. D. & B. W. ACHE. 1984. Electrophysiological identification of the stimulatory and interactive components of a complex odorant. Chem. Senses **9:** 201-218.
2. DERBY, C. D., B. W. ACHE & E. W. KENNEL. 1985. Mixture suppression in olfaction: Electrophysiological evaluation of the contribution of peripheral and central neural components. Chem. Senses **10:** 301-316.
3. GLEESON, R. A. & B. W. ACHE. 1985. Amino-acid suppression of taurine-sensitive chemosensory neurons. Brain Res. **35:** 99-107.
4. JOHNSON, B. R., P. F. BORRONI & J. ATEMA. 1985. Mixture effects in primary olfactory and gustatory receptor cells from the lobster. Chem. Senses **10:** 367-373.

Reproductive Pheromones of the Landlocked Sea Lamprey (*Petromyzon marinus*)

Studies on Urinary Steroids

MICHAEL A. ADAMS, JOHN H. TEETER, YAIR
KATZ, AND PETER B. JOHNSEN

Monell Chemical Senses Center
University of Pennsylvania
Philadelphia, Pennsylvania 19104

Intraspecific chemical signals (pheromones) have been implicated in a variety of behavioral processes in fish, including schooling, parent-young interactions, homing, pair formation, and spawning.[1,2] Preferences of spawning-run landlocked sea lampreys (*Petromyzon marinus*) for substances released by sexually mature conspecifics of the opposite sex indicate that pheromones may play a role in the reproductive behavior of this species.[3] Pheromone release in sea lampreys coincides with the appearance of secondary sex characteristics. At this time, male sea lampreys release in their urine a pheromone that is attractive to female conspecifics (Teeter, unpublished observation). The possibility that this pheromone could be steroidal in nature led us to assay both pheromone-containing (behaviorally active) and pheromone-devoid (behaviorally inactive) male urine for its content of several metabolically important steroids; the values obtained from this analysis were then used to establish concentrations for examining the effects of these compounds on the preference behavior of female sea lampreys.

Assays of samples for pheromone activity were carried out using spawning-run female sea lampreys as test subjects in a two-choice preference tank (details of lamprey capture, urine collection, and bioassay techniques are presented in Lisowski *et al*.[4]). As an example of a representative test, females were given a choice between a compartment containing lake water and a compartment containing 12.8 μl/liter male urine. Data were analyzed using the Wilcoxon matched-pairs, signed-rank test. Presence of a pheromone was indicated if females spent a significantly longer time in the stimulus arm of the apparatus than in the lake water arm.

Samples of male sea lamprey urine were analyzed for the concentrations of nine steroids (dehydroepiandrosterone [DHEA], testosterone [T], dihydrotestosterone [DHT], progesterone [P], androstenedione [A], estrone [E_1], estradiol [E_2], corticosterone [B], and cortisol [F]) by radioimmunoassay (RIA). Samples analyzed included native urine and urine that had been enzymatically hydrolyzed with mixed β-glucuronidase/sulfatase. Values of the analyses were used to prepare solutions of the individual steroids for bioassay at concentrations that bracketed the urinary concentrations.

TABLE 1. Behavioral Preference Response Values for Several Male Sea Lamprey Urinary Steroids, Tested Individually, as Measured in a Two-Choice Test Tank[a]

Steroid	Stim. Soln. Concentration (ng/ml)	No. of Animals	Mean % Time ($\pm$ SE) on Stimulus Side	Attractive?[b]
DHEA	100[c]	24	56.9 $\pm$ (6.2)	N
	10[c]	23	53.4 $\pm$ (6.4)	N
	1[c]	24	56.7 $\pm$ (6.1)	N
	0.1[c]	24	52.9 $\pm$ (5.5)	N
DHT	0.2	24	42.1 $\pm$ (4.9)	N
P	0.4	24	57.1 $\pm$ 6.3	N
A	10	24	44.0 $\pm$ (6.5)	N
	1	24	51.7 $\pm$ (5.3)	N
	0.1	24	53.9 $\pm$ (5.3)	N
E_1	10	24	57.9 $\pm$ (5.0)	N
	1	24	52.4 $\pm$ (6.6)	N
	0.1	24	55.4 $\pm$ (5.8)	N
	0.01	24	53.8 $\pm$ (5.5)	N
B	100	18	57.7 $\pm$ (6.0)	N
	10	24	42.4 $\pm$ (5.4)	N
	1	30	52.7 $\pm$ (4.4)	N
F	100	24	53.0 $\pm$ (5.7)	N
	10	24	49.9 $\pm$ (6.0)	N
E_2	0.5	24	45.8 $\pm$ (5.8)	N
Urine control[d]		24	63.7 $\pm$ (5.0)	Y

[a] One milliliter of stimulus solution was used per test.
[b] $p < 0.5$, Wilcoxon matched-pairs, signed-rank test, two-tailed; Y = yes, N = no.
[c] pg/ml.
[d] One milliliter of bioactive male urine used as a control.

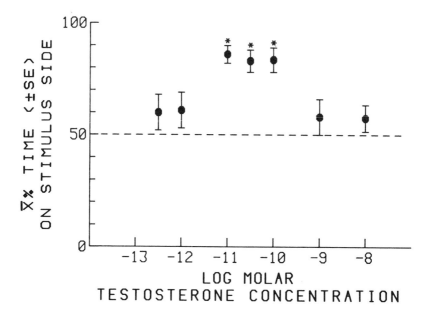

FIGURE 1. Plot of log molar testosterone concentration versus mean percent time that female spawning-run sea lampreys spent on the stimulus side of a two-choice behavioral preference tank. An asterisk denotes a significant preference for testosterone ($p < 0.5$, Wilcoxon matched-pairs, signed-rank test, two-tailed).

As little as 0.5 ml ($6.5 \, \mu1/1$ of water) of pooled urine from sexually mature male sea lampreys has been shown to elicit preference responses in females in our two-choice tanks.[3] We arbitrarily use 1 ml of urine per test ($13 \, \mu1/1$) when determining whether or not a particular urine sample contains pheromone. Consequently samples of the steroids were tested in the preference tank at the concentrations at which they were determined to be present in 1 ml of behaviorally active male urine by RIA (for details see Adams et al.[5]). Results of the behavioral tests are shown in TABLE 1. Only testosterone elicited a preference response in spawning-run female sea lampreys, and in concentrations three to four orders of magnitude greater than those found in behaviorally active unhydrolyzed male urine (FIG. 1).

It is possible that testosterone, or a closely related structural derivative, functions as a pheromone in sea lamprey when present at the appropriate concentration. This would occur when one or more males release urine in close proximity to a female. At such close range dilution would be minimal, and it is possible that the critical concentration might be reached. Even if this is the case and testosterone is functioning as a short-range attractant, our general bioassay results indicate that there is an additional substance that attracts females at much lower concentrations.

REFERENCES

1. SOLOMON, D. J. 1977. A review of chemical communication in freshwater fish. J. Fish. Biol.
   11: 363-376.

2. LILEY, N. R. 1982. Chemical communication in fish. Can. J. Fish. Aquat. Sci. **39:** 22-35.
3. TEETER, J. 1980. Pheromone communication in sea lampreys (*Petromyzon marinus*): Implications for population management. Can. J. Fish Aquat. Sci. **37:** 2123-2132.
4. LISOWSKI, J. J., P. G. BUSHNELL & J. H. TEETER. 1986. A two-choice water recirculation tank for assessing chemosensory preferences of landlocked sea lampreys. Prog. Fish-Cult. **48:** 64-68.
5. ADAMS, M. A., J. H. TEETER, Y. KATZ & P. B. JOHNSEN. 1986. Sex pheromones of the sea lamprey (*Petromyzon marinus*): Steroid studies. J. Chem. Ecol. **48:** 387-395.

Transduction Proteins of Olfactory Receptor Cells

Identification of Guanine Nucleotide Binding Proteins and Protein Kinase C

ROBERT R. H. ANHOLT,[a,b] SUSANNE M. MUMBY,[c]
DORIS A. STOFFERS,[a] PEGGY R. GIRARD,[d] J. F.
KUO,[d] A. G. GILMAN,[c] AND SOLOMON H. SNYDER[a]

[a]*Departments of Neuroscience, Pharmacology and Experimental
Therapeutics, and
Psychiatry and Behavioral Sciences
The Johns Hopkins University School of Medicine
Baltimore, Maryland 21205*

[c]*The Department of Pharmacology
The University of Texas Health Science Center at Dallas
Dallas, Texas 75235*

[d]*The Department of Pharmacology
Emory University School of Medicine
Atlanta, Georgia 30322*

We have analyzed guanine nucleotide binding proteins (G-proteins) in the olfactory epithelium of *Rana catesbeiana* using subunit-specific antisera. Immunoblotting with an antiserum reactive with a common sequence of the α subunits of G_s, G_i, G_o, and transducin reveals three immunoreactive polypeptides in isolated olfactory cilia and in deciliated olfactory epithelial membranes with apparent molecular weights of 40 kDa, 42 kDa, and 45 kDa corresponding to the relative mobilities of the α subunits of G_o, G_i, and G_s, respectively (FIG. 1A).[1-3] No immunoreactive species are detected in respiratory cilia. On a per milligram protein basis, the 45 kDa $G_{s\alpha}$ species is present in olfactory cilia in the same amount as in deciliated olfactory epithelial membranes and appears, in olfactory cilia, to be enriched relative to $G_{o\alpha}$ and $G_{i\alpha}$. This is in line with previous observations that have demonstrated a high activity of an odorant-sensitive, GTP-dependent adenylate cyclase in frog olfactory cilia.[4] In addition to $G_{s\alpha}$, a significant amount of the 40-kDa polypeptide, reactive with an antiserum specific for $G_{o\alpha}$, is detected in the olfactory cilia preparation (FIG. 1A and B). However, unlike $G_{s\alpha}$, the concentration of $G_{o\alpha}$ in olfactory cilia represents only a small fraction of that found in deciliated olfactory epithelial membranes (FIG. 1B). Immunoblotting

[b]Present address: Department of Physiology, Duke University Medical Center, Box 3709, Durham, North Carolina 27710.

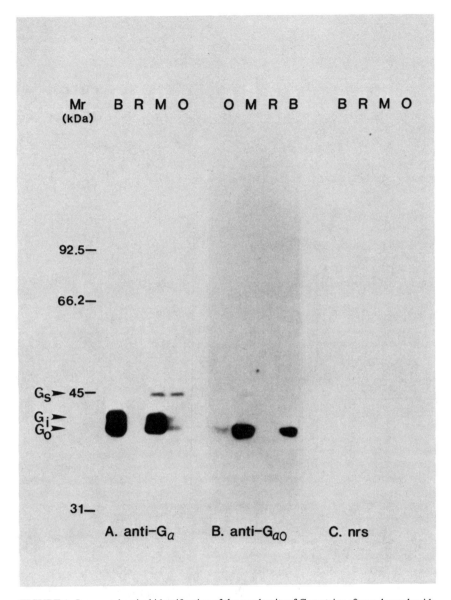

FIGURE 1. Immunochemical identification of the α subunits of G-proteins after polyacrylamide gel electrophoresis in SDS and electrophoretic transfer. The lanes were loaded with 100 μg protein of brain membranes (B), respiratory cilia (R), deciliated olfactory epithelial membranes (M), or olfactory cilia (O). Bound antibody was visualized with [^{125}I] protein A. (**A**) Immunoblot using a 2,500-fold dilution of an antiserum reactive with the α subunits of all G-proteins. (**B**) Immunoblot using a 500-fold dilution of an antiserum specific for the α subunit of G_o. (**C**) Immunoblot using a 500-fold dilution of normal rabbit serum (nrs).

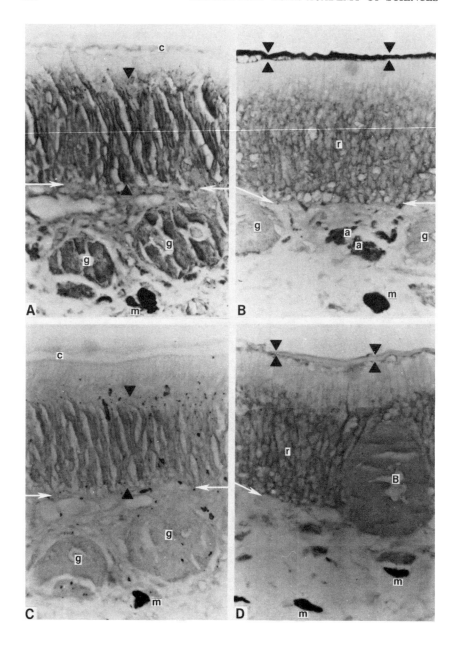

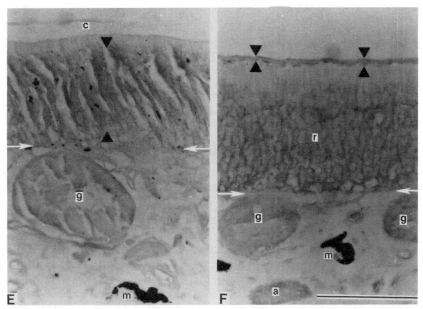

FIGURE 2. Immunohistochemical localization of G-proteins in the olfactory epithelium of *Rana catesbeiana.* (A) Immunohistochemical staining with the anti-G_α antiserum. The arrowheads indicate the layer of olfactory receptor cells; "g" indicates the submucosal glands; "c" indicates the ciliary surface; and "m" indicates a melanocyte. The position of the basement membrane is indicated by the horizontal arrows. (B) Immunohistochemical staining with the anti-G_β antiserum. "a" indicates the axon bundles in the lamina propria and "r" the layer of olfactory receptor cells. The ciliary surface is indicated by the arrowheads. (C) Immunohistochemical staining with the anti-G_α antiserum after preincubation with the synthetic peptide against which it was generated. (D) Immunohistochemical staining with the anti-G_β antiserum after prein-cubation with the synthetic peptide against which it was generated. "B" designates a Bowman's gland. (E) Immunohistochemical staining with preimmune serum obtained from the rabbit in which the anti-G_α antiserum was raised. (F) Immunohistochemical staining with preimmune serum obtained from the rabbit in which the anti-G_β antiserum was raised.
All sera were used at 1,000-fold dilution. The bar represents 100 μm.

with an antiserum reactive with the β subunits of G-proteins reveals in olfactory cilia a single immunoreactive band at 36 kDa (data not shown).

Immunohistochemical studies using the anti-G_α and anti-G_β antisera reveal unexpected differences in the apparent distribution of the α and β subunits. The anti-G_α serum stains primarily the membranes of the olfactory receptor cells and the acinar cells of the Bowman's glands and of the submucosal glands in the lamina propria (FIG. 2A). In contrast, the anti-G_β serum stains exclusively and intensely the ciliary surface of the epithelium and the axon bundles in the lamina propria (FIG. 2B). Staining by each of these antisera can be prevented by preincubation of the serum with the appropriate synthetic peptide used to generate the antiserum (FIG. 2C and D). The synthetic peptide used to generate the anti-G_β serum cannot prevent staining by the anti-G_α serum and *vice versa* (data not shown). In addition, no staining is apparent with preimmune sera obtained from the same rabbits (FIG. 2E and F). After extended incubation periods with the antisera, staining of the ciliary surface of the epithelium can also be observed with the anti-G_α antiserum and some staining of the glands and receptor cells becomes evident with the anti-G_β serum (data not shown). The cartilage, connective tissue, and a layer directly below the ciliary surface, primarily representing the sustentacular cells, are not stained. In addition, no staining is observed along the ciliary surface of the small patch of respiratory epithelium that separates the dorsal sheet from the ventral sheet of olfactory epithelium (data not shown).

This unexpected difference in staining patterns between the anti-G_α and anti-G_β antisera may reflect differences in association of the α and β subunits within different G-proteins. Alternatively, an excess of free β subunit may exist in the ciliary surface and axon bundles. One additional possibility is that a substantial fraction of the β subunit on the cilia is present in association with an as yet unidentified, novel G-protein that is recognized poorly by our anti-G_α antiserum.

In addition to G-proteins, we have identified protein kinase C in olfactory cilia by means of a protein kinase C specific antiserum and phorbol ester binding. However, in contrast to the G-proteins, protein kinase C occurs also in cilia isolated from respiratory epithelium (data not shown). Thus, we cannot exclude the possibility that protein kinase C in olfactory cilia may have a general ciliary function rather than a specific role in chemosensory transduction.

REFERENCES

1. GILMAN, A. G. 1984. Cell **36:** 577–579.
2. STRYER, L. & H. R. BOURNE. 1986. Ann. Rev. Cell Biol. **2:** 391–419.
3. MUMBY, S. M., R. A. KAHN, D. R. MANNING & A. G. GILMAN. 1986. Proc. Natl. Acad. Sci. USA **83:** 265–269.
4. PACE, U., E. HANSKI, Y. SALOMON & D. LANCET. 1985. Nature **316:** 255–258.

Granule Cell Development in the Ferret Olfactory Bulb under Normal Conditions and under Continuous Overexposure to a Single Odor

R. APFELBACH AND E. WEILER

Department of Biology
University of Tübingen
7400 Tübingen, Germany

In the ferret *(Mustela putorius f. furo L.)*, as in other mammals, granule cells (GC) are intrinsic neurons that modulate the output of the relay neurons (mitral and tufted cells). The proliferation in the number of GC occurs mainly postnatally, and therefore it is to be expected that they may exhibit clear changes in morphology during postnatal development. Using the Golgi staining method, the number of spines on the apical endings (50 μm in length) of GC dendrites in the external plexiform layer were therefore counted and taken as the criterion for the stage of GC maturation.

The number of dendritic spines increases from day 30 to day 60 by about 60%. The mean number of spines remains high between day 60 and 90 and then decreases again by about 22% reaching adult levels around day 150 (FIG. 2). Both the increase ($p < 0.001$; Kolmogoroff-Smirnoff test) and the decrease ($p < 0.001$) are highly significant.

Though difficult to quantify, a change in the shape, diameter, and length of the dendritic spines can be calculated quantitatively provided that large numbers of them are measured. As shown in FIGURE 1, the mean diameter of the dendritic spine stem increases after day 90, as does the mean diameter of the head of the dendritic spines after day 60. Concomitantly, the mean length of the spines shows a moderate reduction between day 30 and day 60, a somewhat greater reduction between day 60 and day 90, and a sharp reduction after day 120, that is, after the number of spines starts to decline.

In order to study the influence of early olfactory experience on the developmental changes in the number of GC spines, the effect of early olfactory deprivation on this parameter was investigated. Olfactory deprivation was induced by rearing litters of ferrets from birth in an artificial environment saturated with geraniol odor. The continuous overexposure to a single odor masks the ability of the animal to experience other odors in the environment bringing about a relative state of olfactory deprivation.

The results (FIG. 2) of this experimental condition reveal that the relative state of odor deprivation does not impart its effect on the initial phase of growth where this parameter is increasing; rather, on the later phase of growth where this parameter is either stable or decreasing. Geraniol odor overexposure significantly enhances (by an additional 25%; $p < 0.001$) the normal decline in spine number observed in the normal animals.

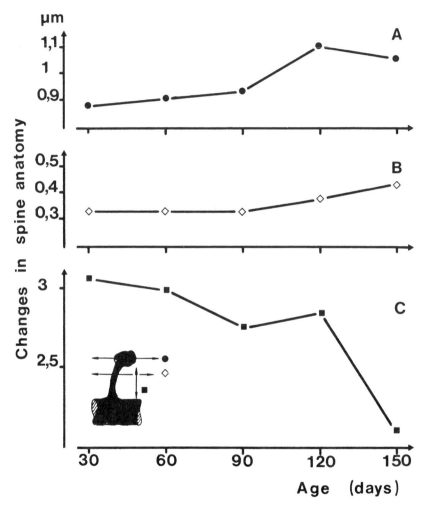

FIGURE 1. Age-dependent changes in the shape of dendritic spines. Each point represents the mean of 100 values.

According to Altman[1] it must be presumed that the maturation of GC is susceptible to environmental influences. We found that during the phase when dendritic spines show a normal overproduction and reduction, olfactory food imprinting is taking place in this carnivorous species.[2] Since the reduction process is enhanced by olfactory deprivation, one might suspect that the developing brain of the ferret undergoes a critical phase when it is especially susceptible to environmental influences.

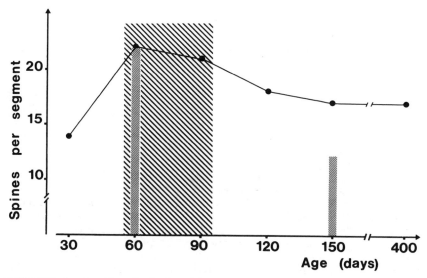

FIGURE 2. Age-dependent distribution of GC spines under normal conditions (●——●) and under geraniol overexposure (▨). Striped range between postnatal day 60 and 90 indicates the lifespan (sensitive phase) during which food odor imprinting is possible.

REFERENCES

1. ALTMAN, J. 1969. Autoradiographic and histological studies of postnatal neurogenesis IV. J. Comp. Neurol. **137:** 433-458.
2. APFELBACH, R. 1986. Imprinting on prey odours in ferrets *(Mustela putorius f. furo L.)* and its neural correlates. Behav. Proc. **12:** 363-381.

Morphological and Physiological Characterization of Interneurons in the Olfactory Midbrain of the Crayfish[a]

EDMUND A. ARBAS,[b] CAROL J. HUMPHREYS, AND
BARRY W. ACHE[c]

C. V. Whitney Laboratory
University of Florida
St. Augustine, Florida 32086

Olfactory integration in crustaceans has now been investigated at a number of different neuronal levels. At the receptor level, neurons have been identified that show both narrow and broad response spectra[1] suggesting that labeled line and across-neuron pattern discrimination may operate together to encode odor quality. We observe, for example, that while interneurons in general show broader response spectra,[2] some still retain the narrow response spectra characteristic of some receptors. These narrowly tuned interneurons either reflect nonselective convergence and subsequent selective processing or selective convergence of narrowly tuned afference. We are attempting to determine by intracellular recording and staining whether there is a functional relationship between the dendritic branching of olfactory interneurons in crustaceans, in particular those with projections in the olfactory lobes, where the afferent fibers converge onto interneurons,[3] and the response spectra of these cells. To date we have characterized several classes of interneurons:

1. Neurons whose arbors overlap with the projection of primary sensory axons in the olfactory lobe, and whose total projections are largely confined to the olfactory areas of the brain. These neurons are likely to be primary interneurons and therefore to subserve an intergrative function early in the pathway processing olfactory information. They can be divided into two or possibly three distinct subtypes based on features of their dendritic arborizations in the olfactory lobe, their projections to other regions of the brain and their physiology. Thirty-five percent of the total number of cells identified show complex, but orderly "tree-like" arborizations in the olfactory lobe (FIG. 1). Twenty-five percent showed more random arborizations that projected to both the olfactory and accessory lobes (FIG. 2). 10% of neurons identified had branching in the olfactory lobe but did not conform with the above two subtypes.

[a] This work was supported by National Science Foundation Award BNS 85-11256 to BWA and a NERC NATO research fellowship to CJH.

[b] Present address: Arizona Research Laboratories, Division of Neurobiology, University of Arizona, Tucson, AZ 85721.

[c] Address for correspondence: Barry W. Ache, C. V. Whitney Laboratory, Rt. 1, Box 121, St. Augustine, FL 32086.

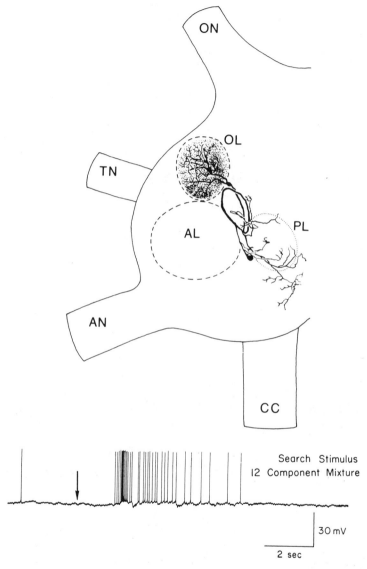

FIGURE 1. Camera lucida drawing of a typical neuron that fills the olfactory lobe with "tree-like" arborizations. The cell body lies within the dorsal medial cluster. This neuron shows the long, looped axon often present in this class and projections to the paraolfactory lobe. These neurons are spontaneously silent and give rise to action potentials when the olfactory organ is stimulated with chemicals. They show broad response spectra, give no evidence of mixture interaction and respond to ipsilateral stimulation.

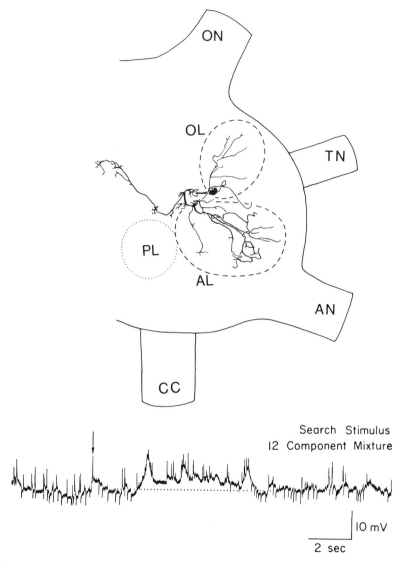

FIGURE 2. Camera lucida drawing of a neuron that branches randomly throughout the olfactory and accessory lobes. The cell body lies within the dorsal anterior cluster, and the neuron appears to project across the midline. These neurons typically show ongoing ipsps and epsps that are either excited or inhibited in response to odors. In the cell shown above, odors evoked an excitatory response in the neuron consisting of a depolarization and simultaneous cessation of a prominent inhibitory input (highlighted by dotted line). These interneurons also appear to be broad spectrum, but contrast to the other type (FIG. 1) in that they respond to contralateral as well as ipsilateral stimulation and show evidence for mixture interaction.

2. Neurons with all their projections intrinsic to the brain but that do not branch in the olfactory lobe. These cells account for only 10% of the identified cell population and are likely to participate at intermediate levels in the olfactory pathway receiving information from primary interneurons.

3. Projection neurons without arborizations in the olfactory neuropil areas and whose axons leave the brain via the circumesophageal connectives. These cells account for 20% of the identified cell population.

Cells in the different classes vary in their responses to different chemical stimuli, as well as in their ability to respond to stimuli of other modalities, suggesting that there is a functional relationship between the dendritic branching and the physiological response spectra of these cells. In particular the response spectra of interneurons in the first class, which are likely to be first order, were all broad and overlapping, but were not identical. The presence of broad-spectra interneurons in the olfactory lobe where the afferent neurons terminate, suggests that information is processed at an early synaptic level in the olfactory pathway by an across-fiber pattern. The presence of interneurons with nonidentical response spectra that have different morphologies suggests that patterns of dendritic branching may be related to the across-fiber pattern.

REFERENCES

1. ACHE, B. W. & C. D. DERBY. 1985. Functional organization of olfaction in crustaceans. Trends Neurosci. **8:** 356-360.
2. DERBY, C. D. & B. W. ACHE. 1984. Quality coding of an odorant in an invertebrate. J. Neurophys. **51:** 906-924.
3. SANDEMAN, D. C. 1982. Organization of the central nervous system. *In* The Biology of Crustacea. Vol. 3: Neurobiology: Structure and Function. H. L. Atwood & D. C. Sandeman, Eds.: 1-61. Academic Press. New York.

Dopamine and Substance P Are Contained in Different Populations of Tufted Cells in the Syrian and Chinese Hamster Main Olfactory Bulb

HARRIET BAKER

Department of Neurology
Cornell University Medical College
New York, New York 10021

Neurons synthesizing dopamine and substance P have been localized to a population of juxtaglomerular neurons in the main olfactory bulb of the hamster.[1-3] These neurons were classified, based on size and morphological characteristics, largely as tufted cells.[2-3] The congruence of both distribution and neuronal morphology suggested that the two transmitters might be colocalized, that is, contained within the same neurons. To investigate this hypothesis, a double-label immunohistochemical technique was used that employed different colors of chromogens.[4] Diaminobenzidine (a brown chromogen) was applied first, followed by 4-Chloronaphthol (blue). Neurons containing substance P were labeled with a specific antibody that reacted only with chemically authentic substance P.[5] Dopamine neurons were identified with a specific antibody to tyrosine hydroxylase (TH), the rate-limiting enzyme in the catecholamine biosynthetic pathway.[1] The antibodies were applied sequentially. The results were similar, independent of order.

Neurons containing both antigens were not observed in either species of hamster analyzed, Chinese or Syrian. However, perikarya labeled either with TH or substance P antibodies often were found adjacent to each other or even partially overlapping in the thick (30 μm) vibratome sections used in these studies. The labeled cells in the glomerular layer of both species were about the same size (12 μm), independent of antigen. However, in the external plexiform layer, larger (up to 20 μm) substance P-containing neurons were observed, especially in the Chinese hamster. The dendritic processes of the TH and substance P-containing processes also could be followed for long distances within the external plexiform layer (EPL). Interestingly, in the Syrian hamster, TH-, but not substance P-, containing neurons frequently were observed in the external plexiform layer while the reverse, substance P but not TH neurons, were observed in the EPL of the Chinese hamster. Differences between the two species also were observed in the distribution of substance P-containing centrifugal afferents. The internal plexiform layer of the Chinese, but not the Syrian, hamster contained a dense axonal arborization. Both species exhibited axonal terminals in the internal granule cell layer. It has been suggested that the terminals in these two layers are of centrifugal origin,[5] although the central nucleus that gives rise to them has not been identified.

These studies demonstrate that dopamine and substance P are not synthesized by the same neurons although these antigens are found in the same regions of the main olfactory bulb and in morphologically similar neurons. In addition, these data suggest that the distribution of neurons and the axonal arborizations of specific transmitter types may vary even in similar species.

REFERENCES

1. BAKER, H. 1986. Species differences in the distribution of substance P and tyrosine hydroxylase immunoreactivity in the olfactory bulb. J. Comp. Neurol. **252:** 206-226.
2. DAVIS, B. J., G. D. BURD & F. MACRIDES. 1982. Localization of methionine-enkephalin, substance P and somatostatin immunoreactivities in the main olfactory bulb of the hamster. J. Comp. Neurol. **204:** 377-383.
3. DAVIS, B. J. & F. MACRIDES. 1983. Tyrosine hydroxylase immunoreactive neurons and fibers in the olfactory system of the hamster. J. Comp. Neurol **214:** 427-440.
4. BAKER, H., D. A. RUGGIERO, S. ALDEN, M. ANWAR, & D. J. REIS. 1986. Anatomical evidence for interactions between catecholamine and ACTH containing neurons. Neuroscience **17:** 469-484.
5. KREAM, R. M., T. A. SCHOENFELD, R. MANCUSO, A. N. CLANCY, W. EL-BERMANI & F. MACRIDES. 1985. Precursor forms of substance P (SP) in nervous tissue: Detection with anti-sera to SP, SP-Gly and SP-Gly Lys. Proc. Natl. Acad. Sci. USA **82:** 4832-4836.

Tasting on Localized Areas[a]

LINDA BARTOSHUK, SALLI DESNOYERS,
COURTNEY HUDSON, LAURA MARKS, AND
MARGARET O'BRIEN

John B. Pierce Foundation
New Haven, Connecticut 06519

FRANK CATALANOTTO

University of Texas
San Antonio, Texas 78284

JANNEANE GENT, DORI WILLIAMS, AND
KAREN M. OSTRUM

University of Connecticut Health Center
Farmington, Connecticut 06032

The spatial properties of the tongue play an important role in taste phenomena observed in the clinic. Using a spatial screening test that compares six loci (the right and left sides of the front and rear edges of the tongue and the right and left sides of the palate), we have found localized losses of taste function in two etiological groups: head trauma and upper respiratory infection. Even patients with losses over extensive areas of the oral cavity can be unaware of the loss. They may also appear to be nearly normal on a conventional, whole mouth, sip and spit taste test. This occurs because relatively small areas of normal tissue can produce very intense sensations and the location of the taste sensations is not salient to the patient.

We have studied the spatial properties of the tongue in the laboratory (see FIG. 1) by scaling the taste intensities of solutions "painted" on the same loci as those used in the clinic test (front, rear edge, palate) but on the subject's left side only (Bartoshuk *et al.*[1]). The perceived intensities at each locus were compared to the intensities produced from the whole mouth (sip and spit procedure). Although the tastes from the three loci were less than from the whole mouth (see FIG. 2), they were much stronger than would be expected on the basis of the size of the area stimulated.

Q-tip scaling revealed losses that were specific to area and stimulus in two groups: the elderly and nontasters of PTC/PROP. For the elderly subjects, the sourness of citric acid was reduced on all three loci, the bitterness of quinine hydrochloride (QHC1) and the saltiness of NaCl were reduced in some areas but not others, and

[a] This work was supported by National Institutes of Health Grants NS 21600 and NS 16993.

166

the sweetness of sucrose was not reduced at all. For the nontasters, the overall trend was for slight reductions of all stimuli on the tongue but not the palate.

The anesthesia results (Ostrum *et al.*[2]) complemented those with Q-tip scaling. Anesthetization of small areas (unilateral chorda tympani block or bilateral palate block) did not cause any reduction of sip and spit tasting. In fact, some stimuli tasted stronger, suggesting a release of inhibition from the anesthetized area.

A test of Q-tips versus filter paper for the application of localized stimuli showed that the two were equivalent on the front but not on the rear of the tongue. On the rear edge, filter paper produced weaker sensations.

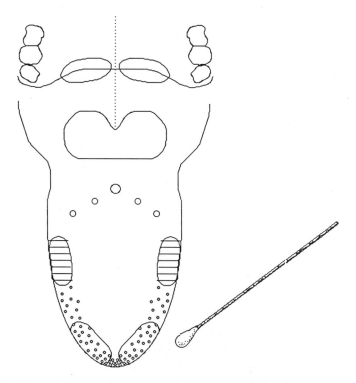

FIGURE 1. Schematic drawing of the tongue and palate showing the locations and approximate sizes of the areas tested in the clinical spatial test. Q-tip scaling was done on the same three loci (front, rear edge, and palate) but left side only.

In the course of conducting the spatial experiments, we discovered a taste illusion that underlines the importance of the tongue as a spatial organ. If taste solutions are painted from the side of the tongue to the tip, the taste sensation grows as the density of taste papillae increases toward the tip; however, if the solution is painted in the opposite direction, the taste intensity does not drop as rapidly as it rose. The taste system appears to borrow the tactile system for localization much as the thermal system does.[3]

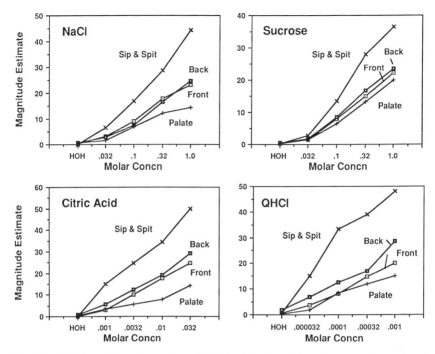

FIGURE 2. Q-tip scaling. Subjects ($n = 62$) judged the intensities of concentration series of NaCl, sucrose, citric acid, and quinine hydrochloride (QHCl) on the three loci shown in FIGURE 1.

REFERENCES

1. BARTOSHUK, L. M., S. DESNOYERS, M. O'BRIEN, J. F. GENT & F. C. CATALANOTTO. 1985. Taste stimulation of localized tongue areas: The Q-tip test. Chem. Senses **10**(3): 453.
2. OSTRUM, K. M., F. CATALANOTTO, J. GENT & L. BARTOSHUK. 1985. Effects of oral sensory field loss on taste scaling ability. Chem. Senses **10**(3): 459.
3. GREEN, B. G. 1977. Localization of thermal sensation: An illusion and synthetic heat. Percept. Psychophys. **22**(4): 331-337.

Effects of Vomeronasal Organ Removal in Lactating Female Mice

Dissociation of Maternal and Agonistic Behaviors[a]

N. JAY BEAN[b,c] AND CHARLES J. WYSOCKI[c]

[b]*Department of Psychology*
Vassar College
Poughkeepsie, New York 12601
[c]*The Monell Chemical Senses Center*
Philadelphia, Pennsylvania 19104

Female mice exhibit a number of characteristic behavioral patterns during lactation that are mediated in part by chemosensory cues. These studies were designed to elucidate the role of the vomeronasal organ in the mediation of several of these maternal behaviors. Vomeronasal organ removal (VNX) from the females before mating showed an unexpected sparing of some, but not all, behaviors. Fecundity was the same in the SHAM and the VNX females; litter size and survival rates were nearly identical. Additionally, no differences were observed in pup retrieval latencies, or the total time to retrieve both an alien and one of the mother's own pups. However, marked differences between the groups were noted in the behaviors of the females in the presence of a male introduced into the nest area. Twelve of the 17 SHAM females, but none of the VNX females were highly aggressive toward the intruder males (see TABLE 1). Indeed, none of the VNX females attacked, bit, or fought with the intruder male. These females were, however, highly interested in the intruder male and attempted to investigate and groom the intruder during the tests. A second study determined whether the lack of female aggression resulted from deprivation of important chemosensory cues during mating. In this study, the aggressive behavior of lactating females who had received VNX after one week of cohabitation with a male was analyzed. Results were consistent with the findings of the previous study; of the 20 SHAM females, 80% were highly aggressive toward the intruder males. The VNX females were once again found to be completely nonaggressive toward the intruders (see TABLE 1). A third study analyzed whether aggressive experience during a previous lactational period could ameliorate the effects of VNX. All of the SHAM females were highly aggressive during a second lactational period. Only two of the five VNX females fought with the intruder male, and one additional VNX female attacked but did not fight with the intruder (see TABLE 1). Although these three females attacked males, their aggression was markedly reduced from that exhibited by the SHAM group females.

[a]Supported in part by BNS83-16437 (CJW), National Science Foundation Research Opportunity Award to NJB and institutional support from Vassar College.

TABLE 1. Percentage of Females Fighting with Intruder Males

Subject Group	Exp. 1 Naive	Exp. 2 Social/Sexual Experience	Exp. 3 Aggressive Experience
VNX	0	0	40
SHAM	71	80	100

These results confirm and extend previous findings suggesting that the vomeronasal organ is extremely important for the mediation of aggressive behavior in mice. Additionally, these results show that different aspects of a lactating female mouse's behavior are differentially modulated by the vomeronasal system.

Preference for Extremely High Levels of Salt among Young Children[a]

GARY K. BEAUCHAMP AND BEVERLY J. COWART

Monell Chemical Senses Center
Philadelphia, Pennsylvania 19104

It has been stated without evidence that the human liking for salt (NaCl) is learned (e.g. Dahl[1]). Studies of infant and childhood responses to salt are required to evaluate this belief. We have been investigating the development of, acceptance of, and preference for salt in water and in soup in humans from birth on.

Previously, we reported that whereas infants less than four months of age ingested water and moderate concentrations of salt (0.10-0.20 M) in equal amounts, infants 4-24 months of age exhibited heightened acceptance of saline solution relative to water.[2] These shifts may be accounted for by the postnatal maturation of peripheral and/or central structures underlying NaCl perception. A postnatal change in saline acceptance, consistent with the change that we observed, has also been reported in rat pups.[3]

Children three to six years of age offered 0, 0.17, and 0.34 M salt in soup or in water, exhibited a striking context-specific reaction to the taste of salt. On both intake (acceptance) and paired-comparison (preference) measures, we found children responded positively to salt in soup but rejected salt in water.[4] The rejection of salt in water may reflect the novelty of salty-tasting water. Notably, the most preferred concentration of salt in soup was often 0.34 M, which is approximately twice the concentration in most commercial soups; many data indicate that peak preference for salt in soups among adults is about 0.20 M.

To further evaluate the level of salt in soup most preferred by children, a paired-comparison procedure was employed. Vegetable soup broth was prepared with several different concentrations of salt (0, 0.25 M, 0.50 M, 1.00 M). Subjects (aged 3-6, n = 23; 7-10, n = 18; 16-31, n = 15) were given every possible pair of stimuli twice and asked which one of each pair they liked best. All subjects were drawn from a healthy clinic population, were of low socioeconomic status and were black.

Generally, the three- to six-year-old and 7- to 10-year-old groups preferred substantially higher levels of salt in soup than did adults (TABLE 1). Specifically, 65% and 78% of the two younger age groups preferred one of the two highest concentrations. In contrast only 13% of the adults preferred one of these concentrations (X^2 = 15.2; $p < 0.001$).

These data demonstrate that children from this population exhibit a preference for remarkably high levels of salt in soup. Adults from the same population, however, prefer levels of salt in soup more consistent with the levels used in commercial

[a]Supported by National Institutes of Health Grant HL-31736. We thank the Campbell Soup Company for their assistance.

171

TABLE 1. Preferred Level of Salt in Soup in Three Groups of People

Age Group	No.	Number of Persons Preferring[a]			
		0.00 M	0.25 M	0.50 M	1.00 M
3-6 years	23	3	5	6	9
7-10 years	18	2	2	5	9
18-26 years	15	1	12	1	1

[a] Molar values refer to the amount of NaCl added to the low-sodium vegetable soup that is approximately 0.006 M NaCl as supplied.

preparations. The bases for, and the generality of, the very high level of salt preference in these children are under investigation. However, since these levels are considerably higher than they would ever be exposed to during home food consumption, it is unlikely that this preference is determined entirely by dietary exposure.

REFERENCES

1. DAHL, L. K. 1958. Salt intake and salt need. N. Engl. J. Med. **285:** 1152-1157.
2. BEAUCHAMP, G. K., B. J. COWART & M. MORAN. 1986. Developmental changes in salt acceptability in human infants. Dev. Psychobiol. **19:** 17-25.
3. MOE, K. 1986. The ontogeny of salt preference in rats. Dev. Psychobiol. **19:** 185-196.
4. COWART, B. J. & G. K. BEAUCHAMP. 1986. The importance of sensory context in young children's acceptance of salty tastes. Child Dev. **57:** 1034-1039.

Schematic Sweet and Bitter Receptors

HANS DIETER BELITZ, HARTMUT ROHSE,
WOLFGANG STEMPFL, AND HERBERT WIESER

Institut für Lebensmittelchemie
Technische Universität München and
Deutsche Forschungsanstalt für Lebensmittelchemie
D-8046 Garching, Germany

JOHANN GASTEIGER AND CHRISTIAN HILLER

Organisch-chemisches Institut
Technische Universität München
D-8046 Garching, Germany

For sweet-tasting molecules, two polar groups—the AH/B-system[1] or the electrophilic-nucleophilic (e_s/n_s) system[2]—with a definite steric arrangement are postulated as essential groups, which may be supplemented by a hydrophobic group.[3] In contrast, bitter compounds need only one polar group, which may be electrophilic or nucleophilic, and a hydrophobic group.[4]

These general models of sweet and bitter stimuli, based on taste thresholds for numerous compounds, suggest a hydrophobic pocket as the very simple schematic receptor[2,5] with a corresponding nucleophilic-electrophilic system (n_r/e_r), and two hydrophobic zones for "sweet" or "bitter" ($h_r$sw) and bitter ($h_r$bi), respectively.

By superposing suitable compounds out of different classes with their polar contact groups in equal positions, the areas that are allowed or forbidden for sweet or bitter taste quality can be estimated. In this way it is possible to get an impression of the shape of the schematic receptor's hydrophobic pocket. Sweet and bitter molecules can be superposed easily with the aid of the computer program MARILYN.

FIGURE 1 shows the superposition of D-leucine, different cyclic amino acids, and D-tryptophan as an example. The area occupied by the benzene ring of tryptophan seems to be favorable for sweet taste, because tryptophan has a significantly lower sweet threshold than the other D-amino acids. By superposing saccharin, acesulfam, and the cyclic amino acids, the favored sweet area occupied by the hydrophobic moieties of the former strong sweeteners clearly comes out.

FIGURE 2 summarizes some results obtained by superposing compounds in a schematic manner: It has been shown with sweet and bitter compounds that the arrangement of the polar groups and the hydrophobic moiety relative to each other determines the quality and intensity of taste. By matching of the bipolar systems of sweet compounds with different taste thresholds, it demonstrates that there are less and more favored areas for hydrophobic groups relative to the bipolar system. The areas occupied by the cyclohexane rings of cyclamate and 1-amino-cyclohexane-1-carboxylic acid are similar and lead to thresholds in the range of 2 mmol/l. The area

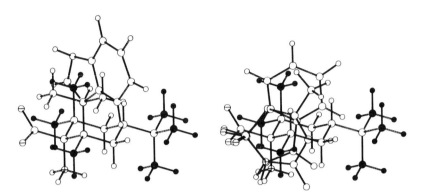

FIGURE 1. Superposition of 1-amino-cyclohexane-1-carboxylic acid (I), 2-methyl-I, 4-methyl-I, 4-*t*-butyl-I, D-leucine, with D-tryptophan (left), and saccharin/acesulfam (right) by the computer program MARILYN (large/small circles: carbon, nitrogen (+), oxygen (−)/hydrogen; open/filled circles: allowed/forbidden positions for sweet taste; plus/minus circles: electrophilic/nucleophilic groups).

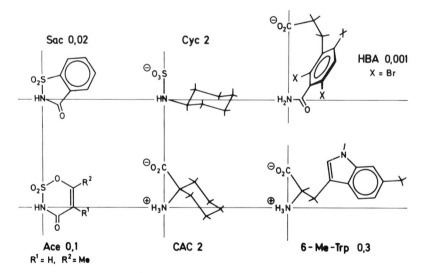

FIGURE 2. Schematic representation of some sweeteners with their bipolar systems in corresponding positions (sweet thresholds in mmol/1; Sac: saccharin; Cyc: cyclamate; HBA: 2,4,6-tribromo-3-carboxyethyl-benzamide; Ace: acesulfam, CAC: 1-amino-cyclohexane-1-carboxylic acid; 6-Me-Trp: 6-methyltryptophan).

occupied by the indole moiety of tryptophan seems to be more favored, because a threshold of 0.3 is reached. The hydrophobic moieties of acesulfame and saccharin are in a very suitable position for sweet taste: The thresholds are 0.1 and 0.02 mmol/l, depending on the different surfaces of the hydrophopic groups. Mostly favored is the perpendicular position of the benzene ring in the halogenated benzamides: A threshold of 0.001 mmol/l is reached with the tribromocarboxyethyl compound.

REFERENCES

1. SHALLENBERGER, R. S. & T. E. ACREE. 1971. Chemical Structure of Compounds and their Sweet and Bitter Taste. *In* Handbook of Sensory Physiology. L. M. Beidler, Ed. Vol. 4/2: 221-277. Springer-Verlag. Berlin, Germany.
2. BELITZ, H.-D., W. CHEN, H. JUGEL, R. TRELEANO, H. WIESER, J. GASTEIGER & M. MARSILI. 1979. Sweet and Bitter Compounds: Structure and Taste Relationship. *In* Food Taste Chemistry. J. C. Boudreau, Ed. ACS Symp. Ser. **115:** 93-131.
3. KIER, L. B. 1972. A molecular theory of sweet taste. J. Pharm. Sci. **61:** 1394.
4. BELITZ, H.-D., W. CHEN, H. JUGEL, W. STEMPFL, R. TRELEANO & H. WIESER. 1983. QSAR of bitter tasting compounds. Chem. Ind. **1:** 23-26.
5. BELITZ, H.-D., W. CHEN, H. JUGEL, W. STEMPFL, R. TRELEANO & H. WIESER. 1981. Structural Requirements for Sweet and Bitter Taste. *In* Flavour '81. P. Schreier, Ed.: 741-755. Walter de Gruyter. Berlin.

Early-Stage Processing of Odor Mixtures

GRAHAM A. BELL, DAVID G. LAING, AND
HELMUT PANHUBER

CSIRO Division of Food Research
North Ryde N.S.W., Australia 2113

Little is known about how the components of odor mixtures interact to produce the perceived quality and intensity of the mixture. We have shown recently that neural metabolic responses to odors, measured with the radioactive 2-deoxyglucose (2-DG) technique, can be used to test interrelationships between odor mixture components.[1]

Metabolic activity was suppressed in reliably identifiable patches of glomeruli in the rat main olfactory bulb, when the animals were exposed to a two-component mixture at concentrations at which human subjects had perceived masking of one of

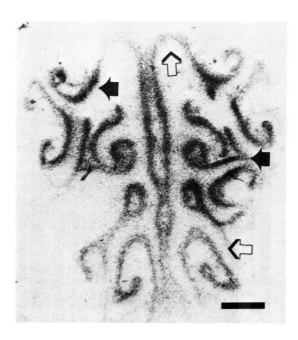

FIGURE 1. 2-DG autoradiograph from a coronal section taken midway through the olfactory epithelium of a rat stimulated with amyl acetate. Filled arrows show regions of high activity relative to those shown by open arrows. Bar = 1 mm.

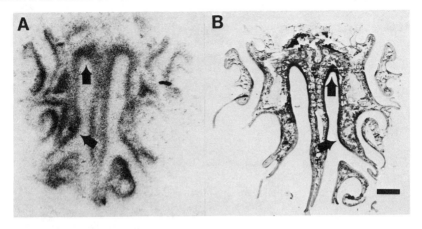

FIGURE 2. 2-DG image (**A**) and nearby section stained for cytochrome oxidase (**B**) from a rat stimulated with propionic acid. Arrows point to corresponding regions of high activity shown by both methods. Bar = 1 mm.

the components. In the same glomerular region, metabolic activity was not inhibited by a mixture in which masking was not perceived, but that showed activity consistent with stimulation by the two-odor components presented singly. Since the glomeruli contain the interfacing synapses between the nasal receptors and the brain, the processing of odor mixtures begins at an early stage in the system, either at the nasal receptor sheet or within the glomerular neuropil.

We now report on our current experiments using 2-DG and cytochrome oxidase aimed at detecting changes in regional metabolic activity at the olfactory receptor cell layer in the rat nose, which might reflect the mixture-induced suppression we have observed in the glomeruli.

Stimulation with single odorants, amyl acetate (FIG. 1) and propionic acid (FIG. 2), produced regional patterns of varying levels of metabolic activity throughout the whole olfactory epithelium.

The evidence presently available confirms earlier findings[2,3] of regional differences in metabolic activity in the olfactory epithelium after odor stimulation and suggests that odor-specific patterning of glomerular activity translates into patterns of activity across the receptor sheet. Hence, the processing of odor mixtures may begin at this early stage.

REFERENCES

1. BELL, G. A., D. G. LAING & H. PANHUBER. 1987. Odour mixture suppression: Evidence for a peripheral mechanism in human and rat. Brain Res. In press.
2. LANCET, D., J. S. KAUER, C. A. GREER & G. M. SHEPHERD. 1981. High resolution 2-deoxyglucose localization in olfactory epithelium. Chem. Senses **6(4):** 343-349.
3. NATHAN, M. H. & D. G. MOULTON. 1981. 2-Deoxyglucose analysis of odorant-related activity in the salamander olfactory epithelium. Chem. Senses **6(4):** 259-267.

Concentration-Dependent Responses of Sodium-Depleted Rats to NaCl in Food

MARY BERTINO, MICHAEL G. TORDOFF, AND
JOHN TLUCZEK

Monell Chemical Senses Center
Philadelphia, Pennsylvania 19104

Given a choice, sodium-replete rats of many strains ingest hypotonic or isotonic saline in preference to water. Rats allowed to choose between salted and unsalted forms of solid foods are either indifferent or avoid the salted alternative.[1]

Sodium-depletion by adrenalectomy[2] or by extended maintenance on a sodium-deficient diet[3] increases rats' preference for salt water and salted food. Combined administration of furosemide and a sodium-deficient diet produces a strong appetite for salt in water (e.g. Jalowiec[4]). This study investigated the relationship between salt concentration in food and the magnitude and duration of salt preference in furosemide-induced, sodium-depleted rats.

Each rat was tested under two conditions. In the experimental condition, rats were placed on a sodium-deficient diet for 48 hours and injected with furosemide (5 mg, s.c.) midway in this period. Twenty-four hours following the injection, salted Hartroft sodium-deficient diet (0.06, 0.12, 0.25, or 0.50% NaCl) and unsalted diet were placed in the rats' cages. Intakes were recorded at 1, 2, 4, 24, 48, 72, and 96 hours after salted food was first made available. In the control condition, rats were given the sodium-deficient diet for 48 hours and a mock injection. A second, similar experiment used different rats and different salt concentrations (0.50, 1.0, 2.0, 4.0, or 8.0% NaCl).

Salted food preference was observed in the first hour after concentrations of 0.25-8.0% NaCl in food had been made available; when 0.12% NaCl and 0.06% NaCl in food had been made available, the preference was significant only after four hours. Duration of the salt preference varied with concentration, lasting two hours in the group with 8.0% NaCl and up to three days in the group with 0.06% NaCl available.

The fact that it took rats four hours to show a salted food preference with 0.06% and 0.12% NaCl suggests that postingestional factors were primarily responsible for the preference. Although postingestional factors cannot be excluded, the latency of the response at other concentrations (0.25% NaCl and above) *suggests* that at these concentrations the rats could taste the salt.

REFERENCES

1. BEAUCHAMP, G. K. & M. BERTINO 1985. Rats (*Rattus norvegicus*) do not prefer salted solid food. J. Comp. Psychol. **99:** 240-247.

2. GRIMSLEY, D. L. 1973. Salt seeking by food selection in adrenalectomized rats. J. Comp. Physiol. Psychol. **82:** 261-267.
3. RODGERS, W. L. 1967. Specificity of specific hungers. J. Comp. Physiol. Psychol. **64:** 49-58.
4. JALOWIEC, J. E. 1974. Sodium appetite elicited by furosemide: Effects of differential dietary maintenance. Behav. Biol. **10:** 313-327.

The Structure of Chemosensory Centers in the Brain of Spiny Lobsters and Crayfish[a]

DAVID BLAUSTEIN, CHARLES D. DERBY, AND
ARTHUR C. BEALL

Department of Biology
Georgia State University
Atlanta, Georgia 30303

The crustacean olfactory system has been used in recent years to investigate basic questions about the functional organization of the chemical senses.[1] To provide a better understanding of the anatomical substrate upon which such chemosensory capabilities are based, we initiated this study of the morphology of the chemosensory pathways of two species of decapod crusteaceans often used in physiological studies—*Panulirus argus* (Florida spiny lobster) and *Procambarus clarkii* (freshwater crayfish). The results are based on several staining methods including silver, toluidine blue, horseradish peroxidase, and hexammine colbalt chloride stains.

The basic organization of the olfactory systems of *P. argus* and *P. clarkii* follows the general pattern described for other crustaceans.[1,2] Primary afferent chemoreceptor neurons in the antennules send axons to the paired olfactory lobes via the antennular nerves. Paired accessory lobes receive no direct input from antennular primary afferents, but are linked to the ipsilateral olfactory lobe via an interneuronal tract (olfactory-accessory tract). Output interneurons from the olfactory and accessory lobes project via the olfactory-globular (OG) tract to the medulla terminalis (MT), the most proximal of the four ganglia in the eyestalk.

The olfactory lobe of crustaceans, similar to the olfactory bulb of vertebrates[3] and the antennal lobe of insects,[4] possesses a distinct glomerular structure (FIG. 1A). The crustacean glomeruli are columnar compartments of chemical synapses that encircle a central core of interneuronal fibers and are surrounded by a marginal tract of axons from antennular primary afferents and interneurons. The glomeruli themselves have a distinct substructure. Each glomerulus of *P. clarkii* is composed of at least four regions (FIG. 1B): (1) an outer, darkly staining region, called the cap region,[5] containing chemical synapses between receptor cells and interneurons including interneurons of the lateral cluster whose fibers interconnect the cap regions of more than one glomerulus and appear to project to the medulla terminalis via the olfactory globular tract; (2) a lighter staining, subcap region, composed of horizontal fibers; (3) a darker staining layer of interglomerular horizontal fibers, perhaps functionally similar to periglomerular cells in the vertebrate olfactory bulb; and (4) a tapered base composed of interneuronal tracts and synapses. The structure of the glomeruli of the olfactory lobes of *P. argus* is roughly similar to that of *P. clarkii*.

[a]Supported by the National Institute of Neurological and Communicative Disorders and Stroke and a Whitehall Foundation Grant.

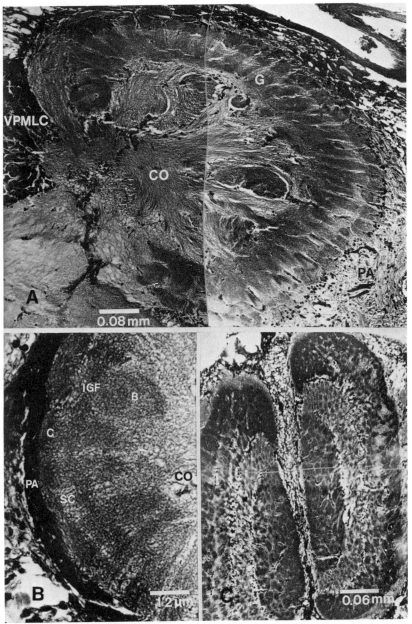

FIGURE 1. Olfactory and accessory lobes. (**A**) Olfactory lobe of *P. argus* demonstrating glomerular structure. G, glomerulus; CO, core of olfactory lobe, composed of interneuronal fibers; PA, antennular primary afferents; VPMLC, ventral paired mediolateral cluster of somata (toluidine blue, 8 μm). (**B**) Olfactory lobe glomeruli of *P. clarkii* showing regionalization. C, cap; SC, subcap; IGF, interglomerular fibers; B, base; CO, core of olfactory lobe, composed of interneuronal fibers; PA, antennular primary afferents (toluidine blue, 8 μm). (**C**) Accessory lobe of *P. argus* demonstrating substructure to lobe and spherical glomeruli (toluidine blue, 7 μm).

181

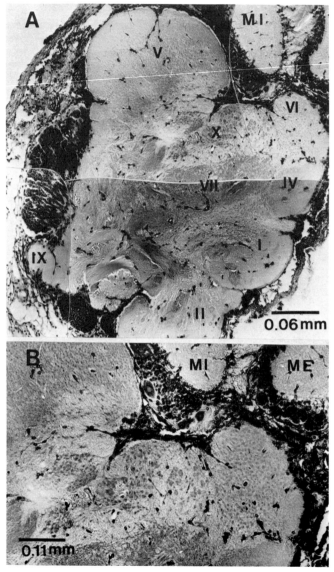

FIGURE 2. Medulla terminalis. (**A**) Horizontal section of medulla terminalis of *P. argus* demonstrating neuropile domains. Hemiellipsoid body consists of regions I and II; glomeruli centralis consists of regions III–X. MI, medulla interna. (**B**) Magnified view of regions V, VI, and X showing microglomerular neuropil of these regions. ME, medulla externa; MI, medulla interna (toluidine blue, 9 μm).

The accessory lobe (AL) is also organized into glomeruli (FIG. 1C). These glomeruli are spherical, smaller, and more numerous than the glomeruli of the olfactory lobes. In both species, these glomeruli are organized into discrete regions within the accessory lobe; however, this arrangement is much more pronounced in *P. argus*. The accessory lobe is connected by interneuronal tracts to regions in the protocerebrum and tritocerebrum as well as to the ipsilateral olfactory lobe. Interneurons with contralateral somata synapse within more than one of the glomeruli of the AL.

The medulla terminalis of decapod crustaceans consists of two major anatomical regions—the hemiellipsoid body (HE) and the glomeruli centrales (GC).[6] We have found at least 12 distinct regions within the MT, including two lobes of the HE, eight regions in the GC, and two accessory visual regions (FIG. 2). Interestingly, most of these areas of the HE and GC contain many tiny glomeruli, although the number and density of glomeruli in each region is variable. The OG tract, carrying chemosensory information from the olfactory and accessory lobes, ramifies in the HE and areas IV and VII of the GC. Tracts connect these regions to other regions of the MT. In general, projections from visual centers in the more distal eyestalk ganglia terminate in the lateral and ventrolateral aspects of the GC. The MT is also closely linked to a neuroendocrine organ, the X-organ and sinus gland.

REFERENCES

1. ACHE, B. W. & C. D. DERBY. 1985. Functional organization of olfaction in crustaceans. Trends Neurosci. **8:** 356-360.
2. SANDEMAN, D. C. 1982. Organization of the central nervous system. *In* The Biology of Crustacea. Vol. 3. Neurobiology: Structure and Function. H. L. Atwood and D. C. Sandeman, Eds.: 1-61. Academic Press. New York.
3. SCOTT, J. W. 1986. The olfactory bulb and central pathways. Experientia **42:** 223-232.
4. ERNST, K. D. & J. BOECKH. 1983. A neuroanatomical study on the organization of the central antennal pathways in insects. Cell Tissue Res. **229:** 1-22.
5. SANDEMAN, D. C. & S. E. LUFF. 1973. The structural organization of glomerular neuropile in the olfactory and accessory lobes of an Australian freshwater crayfish, *Cherax destructor.* Z. Zellforsch. Mikrosk. Anat. **142:** 37-61.
6. HANSTRÖM, B. 1931. Neue Untersuchungen über Sinnesorgane und Nervensystem der Crustaceen. I. Z. Morphol. Ökol. Tiere **23:** 80-236.

Self- and Cross-Adaptation of Single Chemoreceptor Cells in the Taste Organs of the Lobster *Homarus americanus*

PAOLA F. BORRONI AND JELLE ATEMA

Boston University Marine Program
Marine Biological Laboratory
Woods Hole, Massachusetts 02543

Chemosensory adaptation is well documented at the behavioral and central nervous system levels, or in whole-nerve recordings at the peripheral level; however, little is known about receptor cell processes, and the issue has only recently been addressed.[1,2] Because of their narrow tuning, lobster receptor cells are ideal for the study of this problem. Narrow tuning allows for precise manipulation of the receptor adaptation state in self-adaptation experiments, and testing of specific molecular models in cross-adaptation experiments.

Stimulus-response functions of narrowly tuned NH_4 receptor cells from lobster walking legs were recorded extracellularly with standard electrophysiological techniques. Stimuli were injected into constantly flowing backgrounds of either artificial seawater (ASW), or NH_4Cl solutions in ASW (for self-adaptation experiments), or Bet or Glu solutions in ASW (for cross-adaptation experiments). The background compounds in cross-adaptation experiments were chosen in the following way: though narrowly tuned, some NH_4 cells respond weakly to other nitrogenous compounds such as amino acids and amines; Bet and Glu (and to a lesser extent Hyp) are the most common and effective second-best compounds for NH_4 receptors. On average, these compounds elicit 10% or less of the response elicited by NH_4, at equimolar concentration. NH_4 receptors with Bet or Glu as second-best compounds were tested in a Bet or Glu background, respectively; cells with no second-best compounds were tested in either background, randomly.

Exposure to a constant self-adapting background eliminates the response of NH_4 receptors to stimuli of concentration lower than the background, and reduces the responses to all higher stimulus concentrations tested, by a nearly constant amount (~ 25 spikes/log unit of background concentration; TABLE 1). This results in a parallel shift of the stimulus-response function along the x-axis. Similarities in the slopes of the mean functions in each background suggest that the kinetics of stimulus-receptor interaction are not changed in the adapted cell.

Cross-adaptation has a different effect on NH_4 receptors, depending on their tuning breadth (FIG. 1): (1) Extremely narrowly tuned cells (that do not respond to any other compound than NH_4) are unaffected by exposure to a constant background of either Bet or Glu, at the concentration tested; (2) cells with Bet as second-best compound are still quite narrowly tuned, and are also not affected by a Bet background;

TABLE 1. Mean Responses of NH$_4$ Receptor Cells in Self-Adapting Backgrounds[a]

Stimulus NH$_4$Cl Concentrations	Background NH$_4$Cl Concentrations				
	1.6 μM	4.4 μM	38 μM	0.35 mM	3.5 mM
3 μM	14.0	7.6	NR[b]	NR	NR
30 μM	42.7	31.9	0.6	NR	NR
300 μM	71.4	56.2	27.7	2.4	NR
3 mM	100.0	80.4	54.8	20.8	1.4
30 mM	128.7	104.7	81.9	39.2	26.0

[a] Values obtained from regression equations; r > 0.97.
[b] NR = no response.

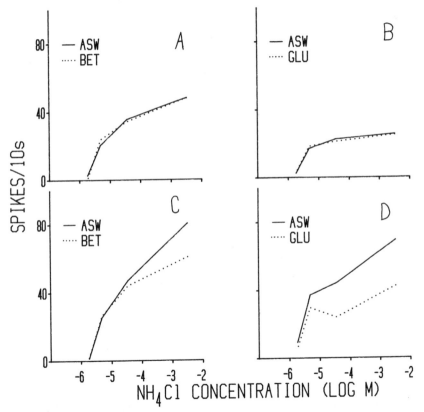

FIGURE 1. Mean stimulus-response curves of NH$_4$ receptor cells; responses of each cell were recorded in ASW background first and subsequently in a cross-adapting background. A and B = NH$_4$ receptor cells with no second-best compound, tested in ASW and 3 μM Bet or Glu background, respectively. C = NH$_4$ receptor cells with Bet as second-best compound, tested in ASW and 3 μM Bet background. D = NH$_4$ receptor cells with Glu as second-best compound, tested in ASW and 3 μM Glu background.

whereas (3) cells with Glu as second-best compound are less narrowly tuned giving approximately equal responses to equimolar NH_4 and Glu stimuli, and their responses to NH_4 are decreased by a Glu background.

From these data it is evident that self-adaptation plays a similar role in the chemical senses as in other sensory modalities: It allows the sensitivity of receptors to "re-set" in correspondence to changes in stimulus conditions. It therefore effectively expands the working range of the whole system beyond the dynamic range of each individual receptor. The preliminary cross-adaptation data suggest (1) a linear relationship between the stimulatory and adapting power of a compound; (2) a correlation between the tuning breadth of a receptor cell and its susceptibility to cross-adaptation.

REFERENCES

1. BORRONI, P. F. & J. ATEMA. 1985. Dynamic properties and self-adaptation of NH_4 cells in lobster taste organs. Chem. Senses **10**: 426.
2. ZACK-STRAUSFELD, C. & K. E. KAISSLING. 1986. Localized adaptation processes in olfactory sensilla of Saturniid moths. Chem. Senses **11**: 499-512.

Does the Trigeminal Nerve Control the Activity of the Olfactory Receptor Cells?[a]

JEAN-FRANCOIS BOUVET, JEAN-CLAUDE
DELALEU, AND ANDRE HOLLEY

Laboratoire de Physiologie Neurosensorielle associé au CNRS
Université Claude-Bernard
F-69622 Villeurbanne cedex, France

In a previous study, we have investigated the effects of acetylcholine and substance P (SP) on the electrical activity of the frog olfactory mucosa. Stimulation by these chemicals elicited low-threshold slow electrical potentials[1] and modified the spontaneous activity of olfactory receptor cells[2] (FIG. 1). These findings suggest that, in physiological conditions, acetylcholine and SP might be released from nerve terminals in the olfactory mucosa. As an argument for this hypothesis, nerve fibers displaying SP-like immunoreactivity were detected in the *lamina propria* and within the epithelium of the frog olfactory mucosa (unpublished observation). These fibers might be of trigeminal origin since the trigeminal nerve, which innervates the olfactory mucosa, contains SP-like immunoreactive neurons.[3]

In the frog, trigeminal neurons send fibers to the olfactory mucosa via the ophthalmic branch of the trigeminal nerve (NV-ob). We have investigated the effects induced by the antidromic electrical stimulation of NV-ob on the electrical activity of the frog olfactory mucosa. The stimulation of the peripheral stump of NV-ob was performed with 10-msec pulses at 30 V, 20 Hz.

Electrical volleys of one- to two-second duration delivered to NV-ob evoked slow electrical potentials in the olfactory mucosa, with the surface negative in relation to the indifferent electrode (FIG. 2). There was a delay of one to two seconds from the onset of the stimulation to the start of the response. In most preparations, the negative potential was a single wave that lasted for two to 15 seconds (FIG. 2A); the average amplitude was about 1 mV. In a few preparations, this wave was followed by a second wave that lasted for one to three minutes with a peak amplitude of 2-4 mV (FIG. 2B). During the single-wave potentials, the spontaneous firing rate of olfactory receptor cells was increased (FIG. 2A) or, more rarely, unchanged. During the last phases of the compound potentials, the firing rate of receptor cells was decreased (FIG. 2B); also, the peak amplitude of the electroolfactogram evoked by isoamylacetate was reduced; the unit responses to this odorant were reduced or suppressed.

Since capsaicin is known to deplete SP from primary sensory neurons, we examined its effects on antidromically generated potentials. A 10-minute exposure of the olfactory mucosa to capsaicin vapors suppressed these potentials. Moreover, SP ($10^{-7}M$) modified the responsiveness of the superfused mucosa to isoamylacetate. These findings

[a]This study was supported by the G.R.E.C.O.-C.N.R.S. "Sens chimiques."

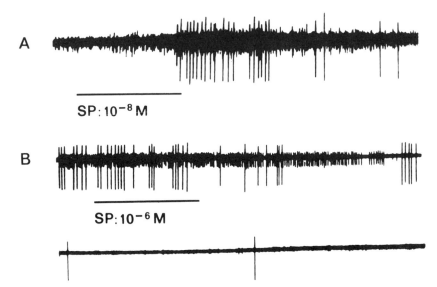

FIGURE 1. Effect of substance P (SP) on the spontaneous activity of olfactory receptor cells. Horizontal bars (10 sec) represent the passage of SP in a saline solution[1] over the superfused mucosa. Individual receptor cells displayed a selective sensibility to the peptide: (**A**) excitatory response at $10^{-8}M$; (**B**) inhibitory responses at $10^{-6}M$ recorded from two units in another preparation. The last two traces are continuous.

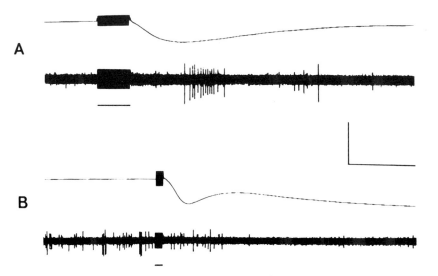

FIGURE 2. Modification of the electrical activity in the left olfactory mucosa by antidromic stimulation of the ophthalmic branch of the left trigeminal nerve. The stimulation (horizontal bars) elicited a single wave (**A**) or a compound wave (**B**) in the mucosa (upper traces); the spontaneous firing rate of single receptor cells (lower traces) was increased (**A**) or decreased (**B**). Horizontal calibration: 4 seconds in A, 10 seconds in B. Vertical calibration (slow potentials): 4 mV.

suggest that SP or a related peptide might be involved in the effects induced in the olfactory mucosa by electrical stimulation of NV-ob.

Experiments have been carried out that demonstrate that the olfactory and trigeminal systems interact at central levels.[4,5] Our study is the first to demonstrate that the functioning of the olfactory system might be controlled at the receptor cell level. It is suggested that, in physiological conditions, the trigeminal system could modulate the activity of the olfactory receptor cells via a local axon reflex triggered by odors and inducing the release of SP or a related tachykinin.

REFERENCES

1. BOUVET, J. F., J. C. DELALEU & A. HOLLEY. 1984. Réponses électriques de la muqueuse olfactive de Grenouille à l'application d'acétylcholine et de substance P. C.R. Acad. Sci. Paris 298 (III), n°6: 169-172.
2. BOUVET, J. F., J. C. DELALEU & A. HOLLEY. 1984. Electrical responses elicited by acetylcholine and substance P in the frog's olfactory mucosa. Communication to the European Chemoreception Research Organization. ECRO VI, Lyon-Ecully, France.
3. HÖKFELT, T., J. O. KELLERTH, G. NILSSON & B. PERNOW. 1975. Substance P: Localization in the central nervous system and in some primary sensory neurons. Science **190:** 889-890.
4. STONE, H., B. WILLIAMS & E. J. A. CARREGAL. 1968. The role of the trigeminal nerve in olfaction. Exp. Neurol. **21:** 11-19.
5. CAIN, W. S. & C. L. MURPHY. 1980. Interaction between chemoreceptive modalities of odour and irritation. Nature (London) **284:** 255-257.

Feeding Regulation and the Microstructure of Eating Tomato Leaf by Tobacco Hornworm Caterpillars[a]

ELIZABETH BOWDAN

Department of Zoology
Morrill Science Building
University of Massachusetts
Amherst, Massachusetts 01003

The tobacco hornworm caterpillar is an oligophagous feeder and one of its preferred foods is tomato leaf. The microstructure of feeding on tomato leaf was studied using an automated cafeteria.[1] In this device the leaf and the caterpillar are each a part of an electrical circuit that is closed when the caterpillar is touching the leaf. The activities of the feeding caterpillar generate characteristic, slow electrical changes (movement artifacts) that can be recorded (FIG. 1).

When the caterpillar touches the leaf, it explores for a while and then begins to bite. After some seconds (a chewing bout), it stops and may explore the leaf again or rest before resuming biting. This alternating pattern of chewing and not chewing continues for three minutes on average and constitutes a meal. The criterion for a minimum inter-meal interval (IMI) is a nonchewing period of 120 seconds derived, in part, from the log survivor curve of the durations of nonchewing intervals. Thus, measurements of chewing bout, exploration period, rest period, meal, and IMI durations as well as counts of the number of bites and bite frequency can be made.

Nine animals were tested, each on more than one day, to give a total of 30 experiments. Regulation of feeding could be demonstrated with respect to the weights of the animals between meals and also within meals.

The larger the animal the more it ate. There was a positive correlation ($p < 0.05$, or better) between weight and (a) bite frequency, (b) number of chewing bouts in meals, (c) meal durations, (d) normalized time spent chewing, and (e) normalized number of bites taken. Thus, larger animals bit faster and ate larger meals. There was no change in chewing-bout duration nor in meal frequency.

Between-meal regulation was demonstrated by several observations. In 11 of the 30 experiments, tails of the log survivor curves of not chewing (i.e. IMIs) were convex, indicating that in these experiments the probability that the IMI would end increased the longer the IMI became. In 6 of 29 experiments, there was a significant positive correlation between the time spent chewing within meals and the duration of the preceding IMI. And lastly, when data from all meals with up to six chewing bouts

[a] This work was supported by a grant from the Whitehall Foundation.

190

were pooled, there was a significant positive correlation between the number of bouts in meals and the duration of the preceding IMI. Thus length of time spent not eating (i.e. hunger) influences the size of the following meal but size of the meal (i.e. postingestive factors) has no effect on the following IMI.

Within-meal regulation was also demonstrated. Log survivor curves of both meal durations and number of bites within meals were convex for all experiments. That is, the longer the meal, the more bites that had been taken, the more likely the meal was

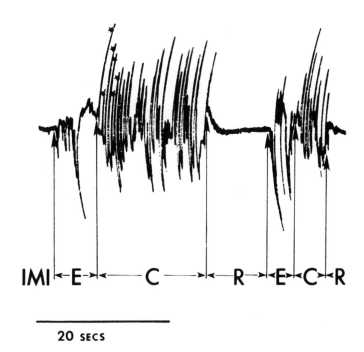

20 secs

FIGURE 1. Movement artifacts generated during feeding. At IMI (inter-meal interval) the caterpillar is not touching the leaf, the circuit is open, and the baseline is therefore broad and flat. At the first arrow the animal hits the leaf and begins to explore (E). At the second arrow it begins to bite. Each of the first six bites is indicated by an asterisk. Note the regularity of the biting artifacts compared with the exploring artifacts. Biting continues for a few seconds (C, a chewing bout), stops (3rd arrow), and the caterpillar then rests (R) creating a broad, flat, trace. The pattern is then repeated.

to end. When data for all meals were pooled, mean chewing-bout duration and mean number of bites per chewing bout were inversely correlated with the number of chewing bouts in the meal; that is, if the animal had eaten very little, it was more likely to begin eating again soon. Correlations between bite frequency and chewing-bout position within the meal were not significant but were negative in most cases. Not-eating durations correlated insignificantly but positively in most cases ($p < 0.02$, sign test); thus feeding slowed a little as the meal progressed.

One final observation of interest was that chewing bouts could last as long as four minutes. Since the caterpillar takes food into its mouth with each bite, this strongly suggests that biting and swallowing are independent of one another, totally unlike the situation in mammals.

REFERENCES

1. BOWDAN, E. 1984. An apparatus for the continuous monitoring of feeding by caterpillars in choice, or non-choice tests (automated cafeteria test). Entomol. Exp. Appl. 36: 13-16.

Enantiomeric Specificity of Alanine Taste Receptor Sites in Catfish[a]

JOSEPH G. BRAND,[b,c] BRUCE P. BRYANT,[b] ROBERT
H. CAGAN,[d] AND D. LYNN KALINOSKI[b]

[b]Monell Chemical Senses Center
Philadelphia, Pennsylvania 19104

[c]Veterans Administration Medical Center
University of Pennsylvania
Philadelphia, Pennsylvania 19104

[d]Colgate-Palmolive Company
Research and Development Division
Piscataway, New Jersey 08854

The mechanisms underlying discrimination of enantiomers by taste receptors are little understood. The cutaneous taste system of the catfish (*Ictalurus punctatus*) lends itself to analysis of enantiomeric discrimination because this system is differentially sensitive to enantiomers of amino acids.[1] Using this animal model, the molecular details of enantiomeric discrimination in taste can be studied both biochemically and electrophysiologically. Specific binding of stimulus amino acids has been demonstrated to receptor-containing membrane preparations from catfish taste epithelium.[2-4] Discrimination of the enantiomeric stimuli, L- and D-alanine, by the catfish cutaneous taste system has been investigated here using combined biochemical and electrophysiological approaches.

The electrophysiological assay showed that both L- and D-alanine stimulated the facial nerve when aqueous solutions of the stimuli flowed over the barbel, but that the relative stimulatory abilities of the enantiomers differed. Responses to L-alanine increased more sharply with concentration (10^{-9} to 10^{-3} M) than did those to D-alanine (10^{-9} to 10^{-3} M), even though the threshold values for both enantiomers appeared to be nearly equal (10^{-9} to 10^{-8} M). With most of the nerve bundle preparations studied, L- and D-alanine at equally stimulatory concentrations cross-adapted one another, but this cross-adaptation was not always complete. Additional experiments in which L- and D-alanine were present in a 1 : 1 mixture of equally stimulatory concentrations usually evoked a larger magnitude of response than did each single component alone. These results from cross-adaptation and mixture studies are consistent with the hypothesis that at least two populations of alanine-responsive transduction pathways are present.

This hypothesis was examined at the receptor level using a biochemical binding assay that measured the specific binding of the enantiomers to receptor tissue and that detected binding competition between the enantiomers. The enantiomers, L- and

[a]Supported in part by National Institutes of Health Research Grants NS-15740 and NS-22620, by NRSA Fellowship F32 NS07809-01-BNS-1 (to BPB) and by the Veterans' Administration.

D-alanine, each displayed specific binding to fraction P2 (prepared as previously described[2,3]) from catfish taste epithelium. Binding of each enantiomer was saturable, and each showed both high and lower affinity sites. The K_{Dapp} for L-alanine binding to its high affinity site was 1.5 μM, with a B_{max} of 60 pmol/mg protein; the respective values for D-alanine binding to its high-affinity site were 25 μM and 170 pmol/mg protein. L-Alanine and D-alanine were mutually competitive in the binding assay, with L-alanine showing a greater competitive activity. Saturation binding studies showed that D-alanine competed with L-alanine binding, although about five times more D-alanine than L-alanine was required to achieve the same level of inhibition of L-[³H]alanine binding (FIG. 1). The competition displayed kinetics indicative of simple competitive inhibition (FIG. 1). The competitive nature was also demonstrated through analysis of double-reciprocal plots of the data. Competition studies indicated that the lower affinity sites for D-alanine were less susceptible to L-alanine inhibition than was

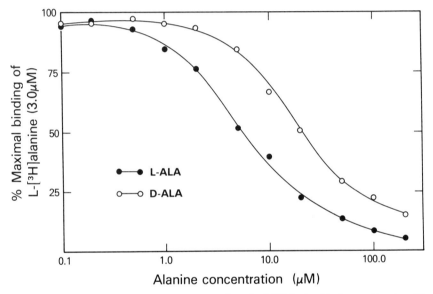

FIGURE 1. Percent saturation versus concentration of L-[³H]alanine binding to Fraction P2 in the presence of either unlabeled L-alanine (●) or unlabeled D-alanine (o). Concentration of L-[³H]alanine was 3.0 μM throughout. Unlabeled L-alanine or D-alanine varied in concentration from 0.1 μM to 200 μM. Amount of P2 protein per assay tube was 150 $\mu g/ml$. Each point represents the average value of triplicate samples.

the high-affinity site for D-alanine. Increasing concentrations of L-alanine led to inhibition of D-[¹⁴C]alanine binding that saturated near 95% when the D-[¹⁴C]alanine concentration was 10 μM. However, with D-[¹⁴C]alanine at 200 μM, increasing L-alanine concentrations inhibited this binding to a level of only around 70%.

Measured either electrophysiologically or biochemically, the taste system for L- and D-alanine is more sensitive to L-alanine. Yet, considerable overlap in recognition specificity exists for the two enantiomers. Experiments carried out at high concentrations of L- and D-alanine suggest the presence of populations of receptor sites and transduction pathways not common to both enantiomers. Taken together, these studies indicate that a portion of the responses to L- and D-alanine occur through a common receptor-transduction process, and also that more than one receptor-transduction process exist for each enantiomer.

REFERENCES

1. CAPRIO, J. 1975. High sensitivity of catfish taste receptors to amino acids. Comp. Biochem. Physiol. **52A:** 247-251.
2. KRUEGER, J. M. & R. H. CAGAN. 1976. Biochemical studies of taste sensation. IV. Binding of L-[^{3}H]alanine to a sedimentable fraction from catfish barbel epithelium. J. Biol. Chem. **251:** 88-97.
3. CAGAN, R. H. 1979. Biochemical studies of taste sensation. VII. Enhancement of taste stimulus binding to a catfish taste receptor preparation by prior exposure to the stimulus. J. Neurobiol. **10:** 207-220.
4. CAGAN, R. H. & A. G. BOYLE. 1984. Biochemical studies of taste sensation. XI. Isolation, characterization and taste ligand binding activity of plasma membranes from catfish taste tissue. Biochim. Biophys. Acta **799:** 230-237.

Behavioral Reactions to Taste Stimuli in Hatchling Chicks[a]

A. BRAUN-BARTANA, J. R. GANCHROW,
AND J. E. STEINER

Department of Oral Biology
The Hebrew University-Hadassah
Faculty of Dental Medicine
Jerusalem, Israel

Taste buds in chickens begin to develop about four days before hatching.[1] Differential behavioral responses to taste stimuli have been observed as early as one day before hatching, but only if the solvent is egg fluids and not water.[2] In contrast, adult chickens exhibit oral behaviors to chemical stimuli dissolved in water, but these give the impression of being mainly "aversive responses."[3] This research investigated spontaneous behavioral displays of 28 chicks within 24 hours after hatching.

Unrestrained, freely moving chicks were individually presented double-distilled water and aqueous taste solutions (0.001 and 0.02 M quinine, 0.01 and 0.1 N citric acid, 0.3 and 1.7 M fructose, and 0.005 and 0.02 M sodium saccharin) in a random sequence. Responses were videotaped (JVC-VHS) and scored under double-blind conditions. Options for slow-motion and single-frame analyses were available. Specific movement features were counted for one minute following the first detectable sampling. At the end of the minute, impressions of response intensity and hedonic tone were scored on a 100-mm visual analogue scale (total acceptance = 100 and total aversion = 0).

Percentage profiles of the main reactions to gustatory stimuli are presented in FIGURE 1. The top row of bar graphs compares the higher concentrations of each stimulus to water in the center, while the bottom row compares water (center) to the lower stimulus concentrations. Head-shaking and prolonged beak-clapping episodes dominated the response profiles to 0.02 M quinine and 0.1 N citric acid together with some beak wiping and gaping. These behaviors clearly indicated taste aversion to the observer whose average hedonic estimates for reactions to these stimuli were 12.5 and 11.5, respectively. Some aversion was also recognized when 0.001 M quinine was the stimulus (estimate 28.2) and estimates for these three stimuli were significantly different ($p < 0.025$) from those for water (estimate = 42.1). Reactions to all other stimuli produced estimates ranging from 38 to 48, and none of these was significantly different from the water estimate.

The dominant behavior for water, fructose, and saccharin exhibited in FIGURE 1 was repeated fluid contacts whether by drinking or by pecking (sampling?), with relatively lower incidence of the other behaviors. The percentage of animals expressing each behavior is arranged in a descending order using 0.02 M quinine as the standard. It may be seen that higher concentrations of quinine and citric acid produce similar

[a]This research was supported by Grant 3226/84 from the United States-Israel Binational Science Foundation (BSF), Jerusalem, Israel.

profiles (r_s = 0.96), which are different from all other stimulus compounds (r_s = 0.43). The response profiles to the two quinine concentrations were quite similar (r_s = 0.86) while the two citric acid concentrations were at variance (r_s = 0.29).

It was concluded that chickens, soon after hatching, are sensitive to gustatory stimuli dissolved in water, but, as in adults, the behavioral responses are predominantly interpreted as expressing aversion. Aversive reactions to quinine and citric acid in-

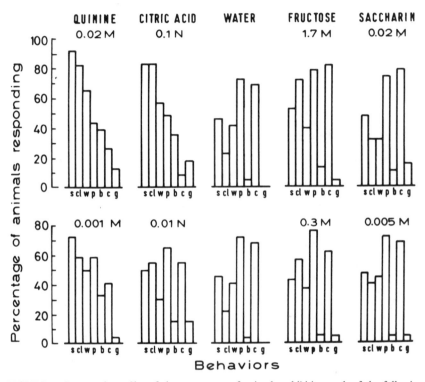

FIGURE 1. Bar graph profiles of the percentage of animals exhibiting each of the following behavioral reactions.

s (head shakes), repetitive, rapid, lateral-rotational movements of the head at a rate faster than 60 Hz;

cl (beak claps), repetitive (more than 20 per stimulus contact) opening and closing of the beak;

w (walking away), retreating from the stimulus container;

p (pecks), rapid pecking at the stimulus solution;

b (beak wipe), lateral head movements as beak makes contact with floor or foot;

c (drinking contacts), ≥ 3 beak contacts with fluid, each initiating drinking;

g (gape), beak held open for a sustained period of time.

cluded head shaking, prolonged beak clapping, beak wiping, gaping, and fewer fluid contacts in the one-minute interval. All of these behaviors are effective in removing fluid from the oral cavity. The head shaking associated with the other stimuli tended to be of very short duration (100-200 msec) and may function to remove fluid from the region of the nasopalatal opening, which could pick up fluid when the head tipped back to swallow. Very positive ingestive responses were rarely noted. In mammals, ingestive responses are often associated with stereotypic movements of mouth, lip, and tongue musculatures.[4-6] The rigid beak and tongue preclude such expressive motor features in avian species. Accordingly, only the number of drinking contacts might directly express taste acceptance and ingestion, but this variable could have been confounded with satiation during our half-hour testing sessions. Preliminary experiments with eight hatchling chicks tasting only distilled water and two concentrations each of sucrose and fructose did elicit significantly ($p < 0.05$) more fluid intake events to the "sweet" stimuli than to water, and this was the main feature that differentiated between them.

REFERENCES

1. GANCHROW, J. R. & D. GANCHROW. 1986. Development of taste buds in the chick embryo. Ann. N.Y. Acad. Sci. This volume.
2. VINCE, M. A. 1977. Taste sensitivity in the embryo of the domestic fowl. Anim. Behav. 25: 797-805.
3. GENTLE, M. J. 1982. Oral behaviour in response to oral stimulation in *Gallus domesticus.* *In* Determination of Behavior by Chemical Stimuli. J. E. Steiner & J. R. GANCHROW, Eds.: 127-136. IRL Press. London, England.
4. STEINER, J. E. 1979. Human facial expressions in response to taste and smell stimulation. *In* Advances in Child Development and Behavior. L. P. Lipsitt & H. W. Reese, Eds. Vol. 13: 257-295. Academic Press. New York.
5. GRILL, H. J. & R. NORGREN. 1978. The taste reactivity test. I. Mimetic responses to gustatory stimuli in neurologically normal rats. Brain Res. 143: 263-269.
6. GANCHROW, J. R., M. OPPENHEIMER & J. E. STEINER. 1979. Behavioral displays to gustatory stimuli in newborn rabbit pups. Chem. Senses Flav. 4: 49-61.

Taste Detection and Discrimination in Zinc-Deprived Rats

G. M. BROSVIC[a] AND B. M. SLOTNICK

Department of Psychology
The American University
Washington, D.C.

R. I. HENKIN

Georgetown University School of Medicine
Washington, D.C.

Both clinical and animal studies have implicated zinc as an important element in taste function. However, the results of clinical studies have been inconsistent (e.g. Henkin et al.[1]). The results of animal studies have demonstrated altered taste preferences for a number of tastants in zinc-deficient rats (e.g. McConnell and Henkin[2]).

It is difficult to interpret the results of clinical studies because of the difficulties of controlling for extraneous variables (e.g. heterogeneous disorders). While the results of the animal studies are more consistent, all have used the preference method and the outcomes do not necessarily reflect sensory (as opposed to hedonic) changes in any simple way, if at all.

METHODS

Phase 1: Baseline

Eighteen rats were maintained on 10 ml/day of zinc-supplemented water (100 ppm zinc ion/100 gr) and a zinc-free pelleted diet. Each rat was trained on a NaCl absolute threshold and on a two-tastant discrimination test. Animals were separated into control, moderately zinc-deprived, and severely zinc-deprived groups on the basis of serum zinc content.

[a] Current address: G. M. Brosvic, Psychology Department, Robinson Hall, Glassboro State College, Glassboro, NJ 08028.

Phase 2: Zinc Depletion

Zinc supplementation was continued for the control rats, reduced to 5 ppm zinc ion for the moderately zinc-depleted group, and withdrawn for the severely zinc-depleted group. Rats in the control and moderately zinc-depleted groups were individually yoked for paired feeding with the severely zinc-depleted group.

All animals were tested on the two-tastant discrimination and the NaCl threshold tasks on days 10 and 17 of Phase 2. Plasma samples were obtained on day 24.

Phase 3: Zinc Repletion

All experimental animals were again supplemented with 100 ppm zinc ion for 27 days and were tested on days 10 and 24. Plasma samples were taken on day 27.

RESULTS

Zinc depletion significantly reduced plasma zinc levels in the experimental groups. However, there was no appreciable increase in plasma zinc levels after 27 days of zinc repletion (see FIG. 1).

NaCl Threshold

In the baseline condition NaCl threshold was approximately 1 mM. There was no change in threshold for control rats over repeated testing during Phases 2 and 3. Both

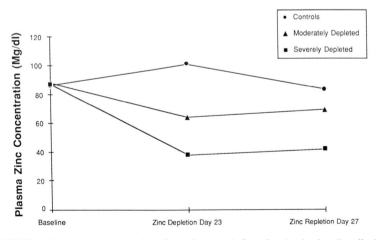

FIGURE 1. Mean plasma zinc values for each group before zinc deprivation (baseline), after 23 days of zinc deprivation, and after 27 days of zinc repletion.

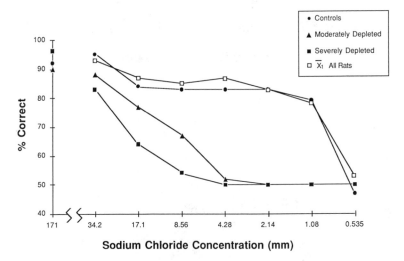

FIGURE 2. Mean percent correct responses as a function of NaCl concentration for the threshold test given after 17 days of zinc deprivation. The function connected by open squares represents the mean performance of all rats before zinc deprivation.

the moderately and severely zinc-depleted groups had an appreciable decrease in NaCl sensitivity relative to controls and their own baseline performance. A typical result of a threshold test is shown in FIGURE 2 (day 17 of zinc depletion). There was no increase in NaCl sensitivity during the zinc-repletion period (Phase 3).

Two-Tastant Discrimination

All zinc-depleted animals were able to discriminate between the 68 mM NaCl and the 88 mM sucrose stimuli. Their performance, however, was consistently more variable than controls, and the severely zinc-depleted group made significantly more errors than controls in each of these tests. Performance on these two-tastant tests did not improve during the zinc-repletion period (Phase 3).

REFERENCES

1. HENKIN, R. I., P. J. SCHECTER, W. T. FRIEDWALD, D. DEMETS & M. RAFF. 1976. A double-blind study of the effects of zinc sulfate on taste and smell dysfunction. Am. J. Med. Sci. **272**: 285-299.
2. McCONNELL, S. D. & R. I. HENKIN. 1974. Altered preferences for sodium chloride, anorexia, and changes in plasma and urinary zinc in rats fed a zinc-deficient diet. J. Nutr. **104**: 1108-1114.

Olfactory Recognition of Congenic Strains of Rats[a]

R. E. BROWN

Department of Psychology
Dalhousie University
Halifax, Nova Scotia, Canada

P. B. SINGH AND B. ROSER

Institute of Animal Physiology
Babraham, Cambridge, England

Individual mammals have an "olfactory signature" that can be recognized by conspecifics.[1,2] These individual odors may be controlled by genes of the major histocompatibility complex (MHC).[3] We provide the first evidence that two congenic recombinant strains of rats that differ only in some genes of the MHC can be discriminated by their urinary chemosignals.

Urine was collected from adult male PVG and PVG.R1 rats and presented to male PVG-RT1[u] rats in habituation-dishabituation tests.[4] Each test consisted of nine two-minute odor presentations. The time subjects spent rearing and sniffing at the urine presented on filter paper was recorded.

PVG-RT1[u] males ($n = 14$) could discriminate between urine from PVG and PVG.R1 strains (Overall, $F = 24.25$, df $= 8, 104$, $p < 0.001$; FIG. 1a). More time was spent investigating odors on tests four and seven than on any other test ($p < 0.01$; Newman-Keuls test). Individual PVG.R1, but not PVG, males were discriminated by their urine odor (overall, $F = 17.73$, df $= 8, 120$, $p < 0.001$; FIG. 1b). Both subgroups sniffed more on test four than on any other test ($p < 0.01$). The odor of PVG.R1 males was sniffed more on test seven than on test six ($p < 0.01$), but not the odor of PVG males. (Note: A retest with 12 males found no discrimination between odors of individual PVG.R1 males, so this result may be due to type I error.[5]) When two odors were taken from the same individual, there was no dishabituation on test seven (overall, $F = 21.05$, df $= 8,120$, $p < 0.001$; FIG. 1c). The time sniffing on test four was greater than on all other tests ($p < 0.01$).

Therefore, the PVG and PVG.R1 strains of rat, which differ only in the class I molecules of the MHC, produce urinary odors that can be used to discriminate between them. The critical chemical in the urine may be a fragment of the class I molecule or a small molecule associated with the class I molecule.[6] The MHC class I molecules that control these odor differences may provide the basis for individual and kin recognition and provide information for mate selection.

[a]This work was supported by grants from the Medical Research Council of Great Britain and the NSERC of Canada. P. B. Singh was supported by an AFRC studentship.

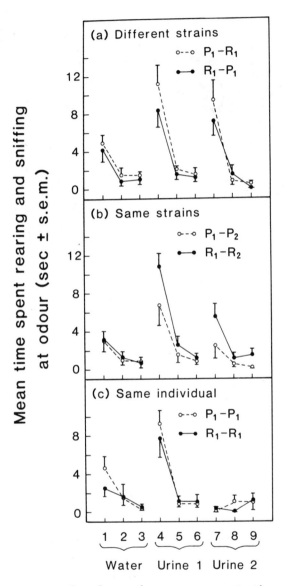

FIGURE 1. Mean time rearing and sniffing at odors by PVG-RT1^u males on each two-minute test.

REFERENCES

1. BROWN, R. E. 1979. Adv. Study Behav. **10:** 103-162.
2. BROWN, R. E. & D. W. MACDONALD, EDS. 1985. Social Odours in Mammals. Oxford University Press. Oxford, England.
3. YAMAZAKI, K., M. YAMAGUCHI, L. BARANOSKI, J. BARD, E. A. BOYSE & L. THOMAS. 1979. J. Exp. Med. **150:** 755-760.
4. SUNDBERG, H., K. DOVING, S. NOVIKOV & H. URSIN. 1982. Behav. Neur. Biol. **34:** 113-119.
5. BROWN, R. E., P. B. SINGH & B. ROSER. 1987. Physiol. Behav. **40:** 65-73.
6. SINGH, P. B., R. E. BROWN & B. ROSER. 1987. Nature **327:** 161-164.

Odorant- and Guanine Nucleotide-Stimulated Phosphoinositide Turnover in Olfactory Cilia[a]

RICHARD C. BRUCH AND TAUFIQUL HUQUE

Monell Chemical Senses Center
3500 Market Street
Philadelphia, Pennsylvania 19104

Since many cells respond to ligand occupation of appropriate receptors by undergoing increased phosphoinositide turnover,[1] we have investigated the possibility that olfactory neurons respond to odorants by this pathway. Olfactory receptors and their associated transduction elements are localized in the dendritic cilia of olfactory neurons.[2-4] A method was therefore developed for the isolation of olfactory cilia from the channel catfish (*Ictalurus punctatus*) with improved yield. The cilia were detached by calcium shock treatment of the olfactory epithelium in the presence of 300 mM sucrose. Following centrifugation to remove the deciliated epithelium, the cilia were centrifuged at 28,000 $\times$ g, resuspended, and layered on a 45% (wt/wt) sucrose cushion. The gradient was centrifuged at 50,000 $\times$ g and the cilia were collected from the top of the sucrose cushion, whereas pigmented material was pelleted at the bottom of the cushion. Under these conditions, the yield of cilia was 165 $\pm$ 9 μg protein/g tissue (n = 5 preparations). Using the same conditions, but with no sucrose in the deciliation medium, the yield of cilia was 92 $\pm$ 30 μg protein/g tissue (n = 3). The isolated preparations were characterized by transmission electron microscopy, SDS-polyacrylamide gel electrophoresis, and Western blotting to identify individual protein components such as tubulin.[5]

The isolated cilia exhibited phosphatidylinositol-4,5-bisphosphate phosphodiesterase (PIP$_2$-PDE, E.C.3.1.4.11) activity, which was enriched about fourfold over the activity of the deciliated epithelium.[6] This enzyme catalyzes the rapid hydrolysis of PIP$_2$ to form the potential second messengers diacylglycerol and inositol-1,4,5-trisphosphate.[1] Electrophysiological studies have demonstrated previously that amino acids are olfactory stimuli for the catfish.[7]. The isolated cilia preparations contained receptors for odorant amino acids as demonstrated by tritiated L-amino acid binding measurements.[2] PIP$_2$PDE activity was stimulated in the isolated preparations in the presence of odorant amino acids. For example, the enzyme activity was stimulated 57% over basal level in the presence of 5 μM L-alanine (TABLE 1). Furthermore, the PIP$_2$-PDE activity was also stimulated 39% over control levels in the presence of 0.1 μM GTP (TABLE 2). The nonhydrolyzable GTP analogues, Gpp(NH)p and GTPγS, were equally effective in their ability to stimulate PIP$_2$-PDE activity, but

[a]This work was supported in part by BRSG SO7-RR05825-06 from the Biomedical Support Grant Program, National Institutes of Health, and in part by a grant from the Veterans Administration to J. G. Brand.

were 78% as effective as GTP.[6] The latter observations, in combination with the identification of guanine nucleotide-binding proteins (G proteins) by immunoblotting,[6,8] strongly suggest the participation of a G protein in activation of phosphoinositide turnover. In combination, these results indicate that olfactory receptor occupancy stimulates phosphoinositide turnover by a mechanism that probably depends on the participation of a G protein.

TABLE 1. Odorant-Stimulated Activation of Olfactory PIP_2-PDE[a]

L-Alanine Added (μM)	Percent Stimulation
0	0
5	57
10	25
20	0
100	2

[a] Values are averages of duplicate determinations from a representative experiment using olfactory cilia from six fish.

TABLE 2. Nucleotide-Stimulated Activation of Olfactory PIP_2-PDE[a]

GTP Added (μM)	Percent Stimulation
0	0
0.1	39
1	28
10	5
100	0

[a] Olfactory cilia were incubated in the absence and presence of the indicated amounts of GTP at 37°C for five minutes before assay of the enzyme activity.

REFERENCES

1. HOKIN, L. E. 1985. Receptors and phosphoinositide-generated second messengers. Ann. Rev. Biochem. 54: 205-235.
2. RHEIN, L. D. & R. H. CAGAN. 1981. Role of cilia in olfactory recognition. In Biochemistry of Taste and Olfaction. R. H. Cagan and M. R. Kare, Eds.: 47-68. Academic Press. New York.
3. GETCHELL, T. V., F. L. MARGOLIS & M. L. GETCHELL. 1984. Perireceptor and receptor events in vertebrate olfaction. Prog. Neurobiol. 23: 317-345.
4. LANCET, D. 1986. Vertebrate olfactory reception. Ann. Rev. Neurosci. 9: 329-355.
5. BOYLE, A. G., Y. S. PARK, T. HUQUE & R. C. BRUCH. 1987. Properties of phospholipase C in isolated olfactory cilia from the channel catfish (Ictalurus punctatus). Comp. Biochem. Physiol. In press.
6. HUQUE, T. & R. C. BRUCH. 1986. Odorant- and guanine nucleotide-stimulated phosphoinositide turnover in olfactory cilia. Biochem. Biophys. Res. Commun. 137: 36-42.

7. CAPRIO, J. 1978. Olfaction and taste in the channel catfish: An electrophysiological study of the responses to amino acids and derivatives. J. Comp. Physiol. **123:** 357-371.

8. MUMBY, S. M., R. A. KAHN, D. R. MANNING & A. G. GILMAN. 1986. Antisera of designed specificity for subunits of guanine nucleotide-binding regulatory proteins. Proc. Natl. Acad. Sci. USA **83:** 265-269.

Use of Monoclonal Antibodies to Characterize Amino Acid Taste Receptors in Catfish

Effects on Binding and Neural Responses[a]

BRUCE P. BRYANT,[b] JOSEPH G. BRAND,[b,c] D. LYNN
KALINOSKI,[b] RICHARD C. BRUCH,[b] AND
ROBERT H. CAGAN[d]

[b]*Monell Chemical Senses Center*
Philadelphia, Pennsylvania 19104
[c]*Veterans' Administration Medical Center*
University of Pennsylvania
Philadelphia, Pennsylvania 19104
[d]*Colgate-Palmolive Company*
Piscataway, New Jersey 08854

One approach to investigating the specificity of taste receptor sites is through the use of specific, site-directed agents. Because antibodies are ideally suited for this, we previously developed monoclonal antibodies that interact with catfish taste epithelial plasma membranes and inhibit the *in vitro* binding of L-alanine.[1] The interaction of several of these antibodies with two putative amino acid taste receptors, the alanine and arginine receptors, has been further characterized with respect to the specificity of binding inhibition, their effect on neural responses, and the identity of the antigen(s). Monoclonal antibodies from two clones, termed G-7 and G-10, have been used in these studies following purification on Protein A-Sepharose. Both antibodies inhibited the binding of L-[^{3}H]alanine by the plasma membrane fraction (Fraction P2) of catfish taste epithelium at antibody protein concentrations ranging from 0.5 to 12 μg/ml. Controls such as bovine serum albumin (BSA) and nonimmune mouse IgG did not inhibit alanine binding. G-10 antibody also inhibited the binding of L-[^{3}H]arginine from 19.4-55.5%, depending on the concentration of ligand and antibody; inhibition decreased with higher concentration of arginine and lower concentration of antibody. This inhibition is similar in magnitude to the inhibition of L-alanine binding described earlier.[1] In a neurophysiological assay, antibodies G-7 and G-10 exhibited slight but sustained excitatory activity when applied to the catfish taste epithelium, while the controls, BSA and nonimmune mouse IgG, did not. Variable inhibition (5-20%) of

[a]Supported in part by National Institutes of Health Grants #NS-15740 and NS 22620 and NRSA Fellowship 1-F32-NSO7809-O1-BNS-1, and the Veterans Administration.

neural responses to both L-alanine and L-arginine was observed. Silver staining of taste plasma membrane proteins separated by SDS-PAGE revealed a major stained band of an approximate molecular mass of 110,000 daltons that was G-10 positive on an immunoblot. The band was also positive to concanavalin A and wheat germ agglutinin, indicating that it contained glycoprotein. Resolution of the taste plasma membrane proteins by two-dimensional electrophoresis (isoelectric focusing and SDS-PAGE) followed by immunoblotting using the G-10 antibody revealed a group of several acidic proteins with an approximate pI of 5.8 to 6.3. This group of antigens was not detected in immunoblots of SDS-PAGE separated proteins from catfish brain, gill, liver, intestinal epithelium, or olfactory mucosa. The taste tissue-specific distribution of antigen and excitation by antibody provide further support for the identification of G-10 antigens as taste receptors. The nonspecificity of inhibition of amino acid binding suggests either an antigenic site common to both receptors or that the antibody is acting on a component common to both alanine and arginine binding. Lack of strong inhibition of neural responses suggests that there may be differential access of antibody to antigenic sites in the *in vitro* and *in vivo* applications.

REFERENCES

1. GOLDSTEIN, N. I. & R. H. CAGAN. 1982. Biochemical studies of taste sensation: Monoclonal antibody against L-alanine binding activity in catfish taste epithelium. Proc. Natl. Acad. Sci. **79:** 7595-7597.

Time-Intensity Analysis of Gustatory Stimuli

Preliminary Assessment of a New Technique

DARLENE BURKE, ALIKI AKONTIDOU, AND
ROBERT A. FRANK

Department of Psychology
University of Cincinnati
Cincinnati, Ohio 45221

Much of the previous research in gustatory psychophysics has focused on the ratings of a stimulus made at a single point in time despite the fact that the stimulus may be changing continuously over the time between its detection and complete adaptation. A few experiments have investigated the temporal characteristics of various tastants, but most of this research has focused on adaptation time for sweeteners. This preliminary experiment explored the utility of a new approach for studying changes in taste intensity over time. Ten subjects made intensity judgments for two concentrations of sucrose, sodium chloride (NaCl), quinine sulfate (QSO_4), and citric acid. Whatman #5 filter papers measuring 13 by 20 mm were soaked in the solutions and placed on the dorsal anterior tongue. Placement of the filter paper on the tongue completed a lickometer circuit that triggered a stimulus marker on a polygraph. The subjects were instructed to move a small lever mounted on a response console from a rating position of 0 (no taste) to 10 (very strong taste) in proportion to the stimulus intensity and to rate the taste sensation continuously from the time they could first detect a taste until complete adaptation occurred. The lever was attached to a slide potentiometer whose output was amplified by the polygraph and translated into the movements of one of the chart pens.

Five measures were derived from the curves generated for each stimulus. These included onset or simple reaction time (the time between the placement of the filter paper on the tongue and the rating of the stimulus as greater than 0), stimulus rise time (the time between the first rating greater than 0 and maximum intensity), maximum intensity, decline time (the time between the decline from maximum intensity to complete adaptation), and total rating time (the time between stimulus onset and complete adaptation). These measures were sensitive to differences in both concentration and quality. Higher intensity ratings were associated with faster onset times, faster rise times, slower decline times and longer total response times (see TABLE 1). Quinine and citric acid were found to be less persistent than sucrose or sodium chloride.

Based on this and other preliminary work that is planned, this technique will be used to explore a number of problems in gustation including the psychophysics of taste mixtures, the relationship between hedonic and intensity ratings of gustatory

210

TABLE 1. Means of the Four Taste Qualities at Both Concentration Levels

Concentrations	Sucrose	NaCl	QSO$_4$	Citric Acid
Maximum intensity ratings (10-point scale):				
Lower conc.	6.21	6.80	6.51	5.69
Higher conc.	7.65	9.11	7.62	7.48
Overall mean	6.93	7.96	7.07	6.59
Onset (reaction) time (sec):				
Lower conc.	2.38	2.19	2.67	2.17
Higher conc.	2.33	1.71	2.67	2.07
Overall mean	2.36	1.95	2.67	2.12
Total response time (sec):				
Lower conc.	39.93	42.84	34.52	33.88
Higher conc.	62.27	66.55	40.06	43.69
Overall mean	51.10	54.70	37.29	38.79
Rise time (sec):				
Lower conc.	7.72	7.18	6.99	7.07
Higher conc.	6.46	5.28	6.27	5.33
Overall mean	7.09	6.23	6.63	6.20
Decline time (sec):				
Lower conc.	25.82	30.22	21.60	19.75
Higher conc.	42.45	49.42	26.38	30.25
Overall mean	34.14	39.82	23.99	25.00

stimuli over time, and the temporal characteristics of taste processing in phenylthiocarbamide (PTC) tasters and nontasters.

Influence of Aging on Recognition Memory for Odors and Graphic Stimuli[a]

WILLIAM S. CAIN

John B. Pierce Foundation Laboratory and
Yale University
New Haven, Connecticut 06519

CLAIRE L. MURPHY

Department of Psychology
San Diego State University
San Diego, California 92182

Aging impairs various aspects of olfactory functioning, most notably absolute sensitivity, suprathreshold perceived magnitude, quality discrimination, and odor identification. When asked to identify odors or to label them consistently, for example, the elderly perform worse than the young no matter what the precise demands of the task. Such tasks make demands on memory, however, and thereby leave open the question of whether the poorer performance of the elderly derives strictly from general memory losses in old age. We addressed the question in a comparison of recognition memory over time across modalities (olfaction, vision) and across levels of meaning and identifiability. Our results support the conclusion that aging takes a particular toll on odor memory.

METHOD

Sixteen young subjects (21.4 year) and 16 old (72.1 year), all healthy and active community dwellers, participated for pay in five 1.5-hour sessions over six to seven months. Tasks included: (1) rating of familiarity of memory stimuli (session 1) without knowledge that memory for these would ever be tested; (2) tests for recognition memory of the stimuli immediately after, two weeks after, and 6.5 months after first exposure (sessions 2, 3, and 4); (3) measurement of detection threshold for 1-butanol using a standard clinical procedure (session 3); and (4) attempt to identify the stimuli after all memory testing had concluded (session 5). The stimuli included common

[a] This work was supported by National Institutes of Health Grants AG04085 and NS21644.

odors (baby powder, leather, popcorn, cigarette butts, etc.), black-and-white pictures of American presidents and vice-presidents (denoted *faces* hereafter), and relatively obscure electrical engineering symbols. In the first session the subject rated the familiarity of 20 of each type of stimulus presented in intermixed fashion. In subsequent sessions, the subject received randomly drawn subsets of 10 of each for recognition, along with randomly drawn subsets of 10 distractor (or new) items of each type. The recognition judgment entailed a confidence rating with respect to whether a given stimulus was *old* or *new*. The ratings permitted erection of ROC curves. Area under a curve (A_z) represented strength of memory.

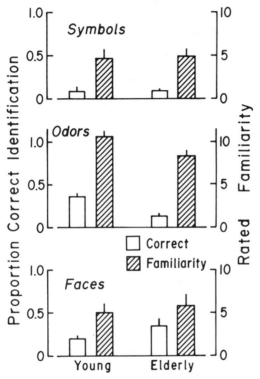

FIGURE 1. Showing how well young and old subjects identified (left ordinate) and how high they rated the familiarity of (right ordinate) sets of engineering symbols, common odors, and faces of American presidents and vice-presidents.

RESULTS

Among the three types of stimuli, odors received the highest familiarity ratings, well above the faces and symbols, which were essentially tied (FIG. 1). The ratings across types did not, however, predict memory performance. In the recognition task, where ANOVA uncovered significant effects of age ($p < 0.02$), stimulus type ($p < 0.001$), and stimulus type by age ($p < 0.001$), the relatively unfamiliar symbols proved easiest to remember (FIG. 2). Young and old found the task approximately equally easy. The faces proved on average next easier to remember, again roughly equally

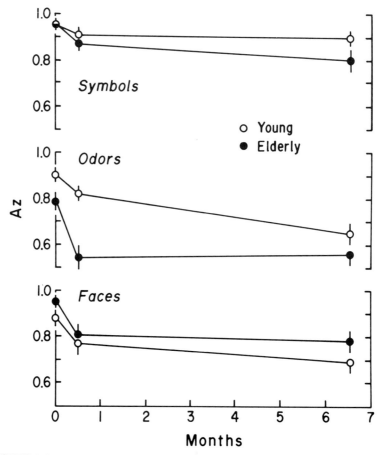

FIGURE 2. Recognition memory for the symbols, odors, and faces in young and old subjects. The index A_z, derived from ROC curves, indicates strength of memory between chance performance (0.5) and perfect performance (1.0).

easy for young and old. Odors yielded a different pattern. Whereas the young remembered the odors much like they remembered faces, the old forgot the odors completely after the first test. That is, for the test at two weeks and 6.5 months, the old performed no better than chance, an outcome that occurred nowhere else in the experiment.

DISCUSSION

Although their fivefold higher threshold ($p = 0.1$) and their weaker ability to identify odors ($p < 0.001$; FIG. 1) suggest, respectively, sensory and encoding disadvantages relative to the young, the old proved reasonably competitive to the young in the immediate test of odor memory. The serious deficiency came thereafter and would seem to require an explanation based on the durability of memory storage. Why odor memory proved so much more vulnerable than graphic memory is quite unclear.

Prediction of Olfactory Responses to Stimulus Mixtures by Cross-Adaptation Experiments[a]

JOHN CAPRIO, JOHN DUDEK, AND
JESSE J. ROBINSON II

Department of Zoology and Physiology
Louisiana State University
Baton Rouge, Louisiana 70803

In contrast to most experimental studies, olfactory receptors rarely encounter sequentially spaced single chemicals under natural circumstances. The normal chemical world is a mixture of substances varying in concentration and potency. Thus, it is critical to the basic understanding of the processes involved in olfaction to determine how this sense responds to the simultaneous presentation of compounds comprising, at least initially, simple mixtures. Numerous reports in the recent literature indicate that the response to a mixture cannot generally be predicted from knowledge of the responses to the individual components. This is attributed to the occurrence of both

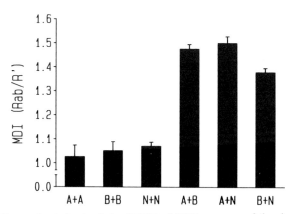

FIGURE 1. Mixture discrimination index (MDI) of EOG responses of the channel catfish to binary mixtures of amino acids. Since each component amino acid in the mixture is adjusted in concentration to provide approximately equal response magnitude, the MDI equals 1.0 when the response to the mixture (Rab; formed by equal aliquots of both components) equals the response to either component (R') at the concentration used to form the mixture. A, B, and N indicate acidic, basic, and neutral amino acids, respectively. MDI values are mean ± SE; $n = 195$ total observations.

[a]This work was supported by National Institutes of Health Grant NS14819.

mixture suppression and synergism among different components in the mixture. We report, however, that electroolfactogram (EOG) recordings in the channel catfish, *Ictalurus punctatus,* clearly show that responses to binary and trinary mixtures of amino acids are predictable with knowledge of the respective relative independence of the binding sites of the component stimuli obtained from cross-adaptation experiments.[1]

The present results indicate that a mixture of amino acids whose components interact with different binding sites (i.e., show minimal cross-adaptation) produce significantly larger responses than a mixture whose component amino acids stimulate the same site (i.e., show significant reciprocal cross-adaptation). For a mixture whose components show reciprocal cross-adaptation, the magnitude of the response to the mixture is equivalent to that produced by a higher concentration of any of the components and is predicted by the "stimulus substitution" model based on the

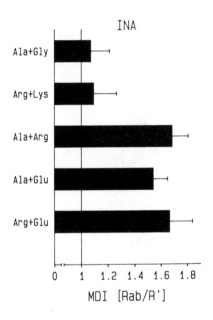

FIGURE 2. MDI of integrated olfactory receptor neural responses in the channel catfish to binary mixtures of amino acids. For definitions of Rab and R' see FIGURE 1 legend. MDI values are mean ± SE; $n = 70$ total observations. INA = integrated neural activity.

dose-response function of the stimuli (FIG. 1).[2–4] For a mixture of amino acids whose components show minimal cross-adaptation, the magnitude of the response is significantly larger than that predicted by the "stimulus substitution" model and may approach additivity (i.e., response summation model[2–4]) even though the power function exponent characterizing the dose-response functions of amino acid stimuli is 0.2 (FIG. 1). Since all stimuli were adjusted in concentration (and in pH 7.8-8.0 if necessary) to provide approximately equal response magnitude and no mixture suppression was observed, this suggests that some of the previous examples of mixture suppression may be explained by simple competitive binding among stimuli with differing potencies that share the same binding site. To ensure that the EOG was an accurate predictor of olfactory spike activity, integrated neural recordings from olfactory receptors to selected binary mixtures were performed, and these data confirmed the previous EOG results (FIG. 2). These studies also indicate that sensory stimulation of olfactory receptor cells by a finite number of stimulus molecules is enhanced more by the activation of different receptor site types than by stimulus interaction at a single site.

REFERENCES

1. CAPRIO, J. & R. B. BYRD, JR. 1984. Electrophysiological evidence for acidic, basic and neutral amino acid olfactory receptor sites in the catfish. J. Gen. Physiol. **84:** 403-422.
2. RIFKIN, B. & L. M. BARTOSHUK. 1980. Taste synergism between monosodium glutamate and disodium 5'-guanylate. Physiol. Behav. **24:** 1169-1172.
3. HYMAN, A. M. & M. E. FRANK. 1980. Effects of binary taste stimuli on the neural activity of the hamster chorda tympani. J. Gen. Physiol. **76:** 125-142.
4. CARR, W. E. S. & C. D. DERBY. 1986. Behavioral chemoattractants for the shrimp, *Palaemontes pugio:* Identification of active components in food extracts and evidence of synergistic interactions. Chem. Senses **11:** 49-64.

ATP-Sensitive Olfactory Receptors

Similarities to P$_2$-Type Purinoceptors[a]

WILLIAM E. S. CARR, RICHARD A. GLEESON,
BARRY W. ACHE, AND MARSHA L. MILSTEAD

C. V. Whitney Marine Laboratory and
Department of Zoology
University of Florida
St. Augustine, Florida 32086

Purinergic receptors, stimulated by ATP and other adenine nucleotides, are present in many internal tissues of vertebrate animals.[1] In an earlier physiological study of the spiny lobster, *P. argus*, antennular olfactory chemoreceptors with similarities to P$_1$-type purinoceptors were identified.[2] This report describes a second class of chemosensory purinoceptor present in the antennule of the spiny lobster. These ATP-sensitive cells have response spectra indicative of P$_2$-type purinoceptors.

METHODS

Neural activity of single units was recorded from ablated antennules as described elsewhere.[3] Chemical stimulation was achieved by injecting a stimulus solution into an artificial seawater flow that superfused the antennule. Evoked action potentials were recorded by suction electrodes applied *en passant* to fascicles of the antennular nerve.

RESULTS AND DISCUSSION

Six ATP-sensitive cells showed a potency sequence of ATP $\geqslant$ ADP > AMP or Ado over a concentration range of 1 to 1,000 μM. In 10 ATP-sensitive cells, tests of equimolar doses of 10 ATP analogues and derivatives provided the following additional insight into receptor specificity: The cells possessed a broad sensitivity to nucleotide triphosphates including those with modifications in the ribose moiety (2'-deoxyATP)

[a]This work was supported by National Science Foundation Grants BNS-84-11693 and BNS-86-07513.

219

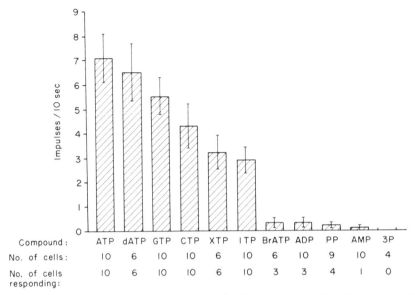

FIGURE 1. Relative activity of ATP and structurally related compounds. Bars show magnitude of mean response (± SEM) to 100 μM dose. Abbreviations: dATP = 2'deoxyATP; BrATP = 8-bromo-ATP; PP = pyrophosphate; 3P = tripolyphosphate.

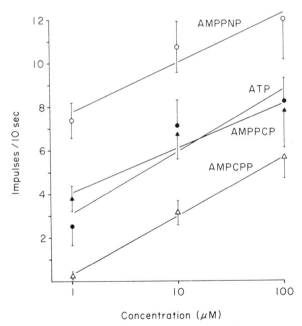

FIGURE 2. Dose-response relationships for ATP and slowly degradable analogues. AMPPNP = β, γ-imidoATP; AMPPCP = β, γ-methyleneATP; AMPCPP = α, β-methyleneATP. The relative potencies are as follows: ATP = 1; AMPPNP = 35; AMPPCP = 1.1; AMPCPP = 0.078.

and the purine moiety (e.g. GTP, CTP); deletion of a single phosphate group (e.g. ADP) resulted in a major decrease in activity; and the triphosphate chain alone (3P) was inactive (FIG. 1). Dose-response functions for ATP and slowly degradable analogues revealed that the imido isostere, AMPPNP, was about 35 times more potent than ATP, and the methylene isostere, AMPPCP, was equally potent to ATP (FIG. 2). All of the properties of the ATP-sensitive cells described above are similar or identical to the properties of P_2-type purinoceptors found in internal tissues of vertebrates.[4] See Carr *et al.*[5] for additional details.

Although the role of ATP as an olfactory signal in the lobster is uncertain, the role of this molecule as a gustatory stimulant is well known in blood-sucking insects.[6] The insect chemoreceptors have P_2-like properties slightly different from those found in the lobster.[5]

SUMMARY

The spiny lobster, *Panulirus argus,* has antennular chemoreceptors with the following response characteristics similar to those of P_2-type purinoceptors found in internal tissues of vertebrates: (1) potency sequence of ATP > ADP > AMP and adenosine; (2) excited by ATP analogues with changes in both the ribose and purine moieties; and (3) excited by imido and methylene isosteres of ATP.

REFERENCES

1. BURNSTOCK, G. & C. M. BROWN. 1981. An introduction to purinergic receptors. *In* Purinergic Receptors. G. Burnstock, Ed.: 1-45. Chapman and Hall. London. England.
2. DERBY, C. D., W. E. S. CARR & B. W. ACHE. 1984. Purinergic olfactory cells of crustaceans: Response characteristics and similarities to internal purinergic cells of vertebrates. J. Comp. Physiol. **155:** 341-349.
3. GLEESON, R. A. & B. W. ACHE. 1985. Amino acid suppression of taurine-sensitive chemosensory neurons. Brain Res. **335:** 99-107.
4. BURNSTOCK, G. & C. KENNEDY. 1985. Is there a basis for distinguishing two types of P_2-purinoceptors? Gen. Pharmacol. **16:** 433-440.
5. CARR, W. E. S., R. A. GLEESON, B. W. ACHE & M. L. MILSTEAD. 1986. Olfactory receptors of the spiny lobster: ATP-sensitive cells with similarities to P_2-type purinoceptors of vertebrates. J. Comp. Physiol. **158:** 331-338.
6. FRIEND, W. G. & J. J. B. SMITH. 1982. ATP analogs and other phosphate compounds as gorging stimulants for *Rhodnius prolixus.* J. Insect Physiol. **28:** 371-376.

Taste Perception of Salt in Young, Old, and Very Old Adults[a]

J. CHAUHAN, Z. J. HAWRYSH, C. KO, AND S. KO

Department of Foods and Nutrition
University of Alberta
Edmonton, Alberta T6G 2M8

This study investigated salt taste perception using aqueous and simple food systems in 180 young and elderly Albertans. Subjects, chosen randomly with extra volunteers, were divided into three age groups: young (20-29 years), old (70-79 years) and very old (80-99 years) with 30 women and 30 men per group. Magnitude estimations of the intensity and pleasantness of various concentrations of NaCl (20-640 mM) in deionized water (20 ± 3°C) and in a soup (55 ± 5°C) were obtained in triplicate, in separate sessions. Subjects were English speaking, relatively healthy, and noninstitutionalized. Subject attributes: education, salt use and so forth were obtained.

Initial analyses show significant age effects (water, $p = 0.025$; soup, $p = 0.001$) and age × concentration interactions ($p = 0.001$) on salt intensity estimates in water and soup. At low salt concentrations in both systems, the old gave larger intensity estimates than the young and very old, while at high salt concentrations, the old gave smaller intensity estimates than the young and the very old. In soup at high salt concentrations, the very old gave even larger ($p < 0.05$) intensity estimates than the young. Thus, very salty water and soups tasted less salty to the old than to the young and very old, while to the very old, salty soups tasted even saltier than to the young. Intensity estimates of the young and old for high salt concentration in water were significantly larger than those for soup, but for the very old a difference in system had no effect.

For salt in both systems, significant pleasantness estimates for age ($p = 0.001$) and age × concentration interactions ($p = 0.001$) were found. As salt concentrations in water increased, the young gave decreasing pleasantness estimates, while those of the old tended to plateau, rise slightly, and then drop. In water as salt concentration increased, the very old gave increasing pleasantness estimates up to the middle concentration, the breakpoint at which pleasantness estimates dropped markedly with increasing salt concentration. Pleasantness estimates for the soup showed no marked age differences. For soup, each age group gave increasing pleasantness estimates with increasing salt concentrations to a breakpoint concentration, which for the young and old was between the middle and the next highest concentration, and for the very old the breakpoint was the middle salt concentration.

For both systems, salt intensity estimates show significant sex effects (water, $p = 0.013$; soup, $p = 0.005$) and sex × concentration interactions ($p = 0.001$ for each system). At high salt concentration in both systems, the women gave larger intensity estimates than the men. Thus, very salty water and soups tasted saltier to the women

[a]Supported by Grant #69510, Alberta Heritage Foundation for Medical Research (AHFMR).

222

than the men. Salt intensity estimates in water also show a significant age × sex interaction ($p = 0.001$), with a trend toward significance for an age × sex × concentration interaction ($p = 0.078$). At high salt concentration in water the young and old women gave larger intensity estimates than the young and old men, whereas the very old showed no effect of sex.

At low salt concentrations in both systems, the frequency of zero estimates for the very old was very high compared to the young and old. These initial findings suggest deficits in salt perception in the elderly. Further analyses of the influence of age and subject attributes in relation to taste perception are required.

Pheromonal Information Coding by Projection Neurons in the Antennal Lobes of the Sphinx Moth *Manduca sexta*

THOMAS A. CHRISTENSEN AND
JOHN G. HILDEBRAND

Arizona Research Laboratories
Division of Neurobiology
University of Arizona
Tucson, Arizona 85721

The antennae of the male sphinx moth *Manduca sexta* bear numerous sexually dimorphic *sensilla trichodea,* each of which includes a pair of olfactory receptor cells

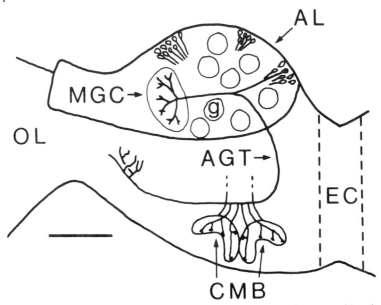

FIGURE 1. Diagram of the left hemi-brain of an adult male *Manduca sexta* (dorsal view, anterior at top) showing schematically an example of the most common type of sex pheromone specific PNs. This neuron has dendritic arborizations confined to the male-specific macroglomerular complex (MGC) in the antennal lobe (AL) and an axon that projects via the inner antenno-cerebral tract (or antenno-glomerular tract, AGT) to the calyces of the ipsilateral mushroom bodies (CMB) and the lateral protocerebrum. Other abbreviations: EC, esophageal canal; g, glomerulus in the AL; OL, optic lobe, Scale bar = 300 μm.

specialized to detect the female's sex pheromones.[1] One cell selectively responds to the principal female sex attractant, (E,Z)-10, 12-hexadecadienal (bombykal or Bal), and the other responds sensitively to (E,Z)-11,13-pentadecadienal ("C-15"), a potent mimic of a second, unidentified sex pheromone present in the blend elaborated by the female's lure gland (Hildebrand and Kaissling, unpublished). These receptors provide sex pheromone specific inputs to a male-specific olfactory subsystem that is responsible for processing sensory information about the pheromones and ultimately for their perception.[2] We study the first-order olfactory centers in the brain, the antennal lobes (ALs, FIG. 1), to reveal how the sensory information carried by these olfactory afferents is "coded" by central neurons in the pathway.

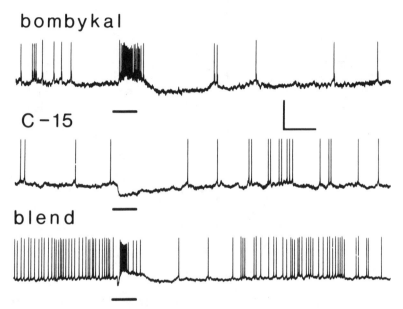

FIGURE 2. Physiological response profile of one of the types of pheromone-selective PNs represented by the morphological class portrayed in FIGURE 1. The cell's responses to stimulation of the ipsilateral antenna with the female's natural pheromone blend, the pheromone bombykal, and the pheromone-mimic (E,Z)-11,13-pentadecadienal (C-15) are shown. The bar beneath each record marks the time of odor stimulus onset and duration. Calibrations: 50 mV, 625 msec.

Using intracellular recording and staining techniques, we have discovered that projection or "output" neurons (PNs) of an important class in the male AL have dendritic arborizations confined to the male-specific macroglomerular complex (MGC) in the AL neuropil and send their axons in characteristic patterns to higher-order centers in the protocerebrum[3] (also Christensen and Hildebrand, in preparation). These projection neurons fall into at least three categories, each of which is characterized by different physiological representation of important features of the pheromonal stimulus.

PNs of the *first* type are purely excited by the female's natural pheromone blend and also by Bal or C-15 alone. About half of the cells of this type studied to date have responded equally to both Bal and C-15. Several of these PNs also have exhibited restricted arborizations confined to part of the MGC. The other half of these cells have been selective for Bal or C-15 and have given clearly dose-dependent responses. We speculate that one function of this first type of male-specific PNs may be to signal pheromone-triggered arousal to higher order centers in the brain. The axons of these cells project to the calyces of the mushroom bodies and further to the lateral protocerebrum (FIG. 1).

The *second* physiological type of male-specific PNs are characteristically excited by one pheromone and inhibited by the other. When the ipsilateral antenna is stimulated with the natural pheromone blend (or the corresponding blend of Bal and C-15), these cells give a complex response comprising inhibition followed by excitation (with rebound inhibition; FIG. 2). These neurons integrate inputs from the two pheromone-receptor pathways, and their mixed response suggests that these cells function as "blend detectors" that relay information about the pheromone mixture to higher centers in the brain.

The *third* type of male-specific PNs has so far been evident only through electrical stimulation of the antennal nerve. A single suprathreshold shock to the antennal nerve (to stimulate the olfactory afferent fibers) evokes tonic inhibition in these PNs.

Currently we are probing the synaptic interactions among cells in the AL by means of simultaneous intracellular recording from two AL neurons. Of particular interest is the role of inhibition mediated by local interneurons. Our goal is to understand how pheromonal information is integrated in the ALs in preparation for being relayed to the next levels in the CNS.

REFERENCES

1. SANES, J. R. & J. G. HILDEBRAND. 1976. Origin and morphogenesis of sensory neurons in an insect antenna. Dev. Biol. **51:** 300-319.
2. CHRISTENSEN, T. A. & J. G. HILDEBRAND. 1987. Functions, organization, and physiology of the olfactory pathways in the lepidopteran brain. *In* Arthropod Brain: Its Evolution, Development, Structure, and Functions. A. P Gupta, Ed. John Wiley. New York. In press.
3. MATSUMOTO, S. G. & J. G. HILDEBRAND. 1981. Olfactory mechanisms in the moth *Manduca sexta:* Response characteristics and morphology of central neurons in the antennal lobes. Proc. R. Soc. (London) **213B:** 249-277.

Experiential and Endocrine Dependence of Gonadotropin Responses in Male Mice to Conspecific Urine[a]

ANDREW N. CLANCY,[b,c] F. H. BRONSON,[d]
A. G. SINGER,[e] W. C. AGOSTA,[e] AND F. MACRIDES[c]

[c]Worcester Foundation for Experimental Biology
Shrewsbury, Massachusetts 01545

[d]University of Texas
Austin, Texas 78712

[e]The Rockefeller University
New York, New York 10021

Previous research has shown that a urinary pheromone of female mice acts via the vomeronasal organ to elicit rapid release of luteinizing hormone (LH) in conspecific males.[1-3] Several experiments were conducted to examine the importance of sexual experience for gonadotropin responses in male mice to female urine, male urine, saline, or mixtures of these stimuli, presented as an aerosol spray onto the snout. Both sexually naive and sexually experienced male mice had significantly higher plasma LH levels following presentations of female urine as compared to their plasma LH levels after presentations of male urine. However, only experienced males had significantly elevated plasma LH levels in response to female urine when compared to their LH levels after presentations of saline, and in naive males the LH levels tended to be lower following presentations of male urine as compared to their LH levels after saline presentations. Subsequent experiments with sexually experienced subjects demonstrated that male mouse urine produces a powerful suppression of LH release in other males, a novel chemosensory effect. Specifically, female mouse urine mixed with male urine failed to elicit LH responses in male subjects, whereas female urine mixed with saline was highly effective. Urine obtained from castrated male donors was as potent as urine from intact males in suppressing the gonadotropin response to female urine. The suppressive compound or compounds in male mouse urine thus do not appear to be critically dependent on gonadal hormones, as has also been found for the stimulatory pheromone in female mouse urine.[3] The pattern of results indicates that sexual experience is not necessary for the differential effects of female versus male urinary chemosignals on gonadotropin secretion in male mice, but that sexual expe-

[a]Supported by National Institutes of Health Grants NS12344 and HD19764.

[b]Address for correspondence: Andrew N. Clancy, Worcester Foundation for Experimental Biology, 222 Maple Avenue, Shrewsbury, MA 01545.

rience does enhance the potency of the female urinary pheromone relative to a sexually neutral component of urine such as sodium chloride.

REFERENCES

1. COQUELIN, A. & F. H. BRONSON. 1980. Secretion of luteinizing hormone in male mice: Factors that influence release during sexual encounters. Endocrinology **106:** 1224-1229.
2. COQUELIN, A., A. N. CLANCY, F. MACRIDES, E. P. NOBLE & R. A. GORSKI. 1984. Pheromonally induced release of luteinizing hormone in male mice: Involvement of the vomeronasal system. J. Neurosci. **4:** 2230-2236.
3. MARUNIAK, J. A. & F. H. BRONSON. 1976. Gonadotropic responses of male mice to female urine. Endocrinology **99:** 963-969.

Olfactory Discrimination of Plant Volatiles by the European Starling

LARRY CLARK AND J. RUSSELL MASON

Monell Chemical Senses Center
Philadelphia, Pennsylvania 19104

Birds that reuse old nest sites are likely to incur high parasite and pathogen loads. These same species also have a greater likelihood of including fresh, nonstructural vegetation into their nests during construction than species that nest at sites only once. One explanation for the utility of this behavior has become known as the nest fumigant hypothesis.[1] Chemical compounds contained within plants are known to be effective as biocides to ectoparasites, bacteria, and fungi. Furthermore, Clark and Mason[2] demonstrated that starlings chose a nonrandom subset of available vegetation for nest construction; and that plants preferred were more active as bacteriocides and pesticides than plants generally not used. More recently, we have evidence suggesting the presence of plants in nests improves the growth performance of chicks and depresses mite populations. Our objective has been to elucidate the proximal cues used by starlings in selecting appropriate plants. While visual identification is undoubtedly important, we focused our investigation on the chemosensory ability of birds because chemical cues are more directly related to the desired effect of fumigation, whereas visual cues, such as leaf shape, size, and color correlates less well with biocidal properties of plants.

Many, if not all, birds possess the requisite anatomical structures and neurophysiological capacity to perceive volatile cues.[3,4] In the laboratory and field, bird species have been able to behaviorally respond to odors. However, these studies focused largely on food gathering ability, concentrating species with large olfactory bulb to ipsilateral cerebellar ratios. Because of the small size of their olfactory bulb, passerines are commonly assumed to possess poor olfactory ability. Thus, a first step in testing whether starlings are able to use chemical information from plants to effect an adaptive behavior was to test the ability of these birds to respond to odorants.

We consistently were able to obtain multiunit recordings from twigs of the olfactory nerve of starlings in response to presentation of *n*-butanol and volatiles from six plant species. Birds responded equally well to plants preferred in nesting and to those not preferred. After pairings of plant volatiles with gastrointestinal malaise, birds exhibited conditioned avoidance in behavioral two-choice feeding experiments, making all possible pairwise discriminations between complex volatiles of the plant species. Bilateral olfactory nerve cuts before conditioning abolished the ability to acquire avoidance, suggesting olfactory cues mediated responding to plant species.

In cardiac conditioning experiments, we found evidence that sex and reproductive status may affect learning ability and olfactory perception of volatiles. Females responded to lower concentrations of ethyl buterate and cyclohexanone than males. Birds with high circulating levels of testosterone or estrogen (as indicated by beak color[5] were more sensitive to odorants than birds in a refractory breeding state. These results suggest that careful experimental control is needed in further testing of olfactory function of birds, especially where function is closely tied to a behavior pattern strongly influenced by genetic and hormonal effects.[6]

REFERENCES

1. SENGUPTA, S. 1981. Adaptive significance of the use of margosa leaves in nests of the house sparrow *Passer domesticus.* Emu **81:** 114-115.
2. CLARK, L. & J. R. MASON. 1985. Use of nest material as insecticidal and anti-pathogenic agents by the European starling. Oecologia **67:** 169-176.
3. BANG, B. G. 1971. Functional anatomy of the olfactory system in 23 orders of birds. Acta Anatomica. Suppl. **58:** 1-76.
4. TUCKER, D. 1965. Electrophysiological evidence for olfactory function in birds. Nature (London) **207:** 34-36.
5. BISSONNETTE, T. H. 1931. Studies on the sexual cycle in birds. IV. Experimental modification of the sexual cycle in males of the European starling (*Sturnus vulgaris*) by changes in the daily period of illumination and of muscular work. J. Exp. Zool. **58:** 281-318.
6. HINDE, R. A., E. STEEL & B. K. FOLLETT. 1974. Effect of photoperiod on oestrogen-induced nest-building in ovariectomized or refractory female canaries (*Serinus canarius*). J. Reprod. Fert. **40:** 383-399.

Olfactory Input to the Prefrontal Cortex in the Rat

MARIE-CHRISTINE CLUGNET
AND JOSEPH L. PRICE

Department of Anatomy and Neurobiology
Washington University School of Medicine
St. Louis, Missouri 63110

Previous studies have shown that there are several olfactory-related areas in the rat prefrontal cortex, especially in the orbital and agranular insular areas on the dorsal bank of the rhinal sulcus.[1,2] However, all parts of the prefrontal cortex, especially the medial and rostral areas, have not been explored, and the full extent of the olfactory-related region has been unclear. In this study, physiological responses to olfactory bulb stimulation were recorded throughout the orbital and mesial prefrontal cortex. The possible sources of olfactory inputs to the responsive areas were identified in other experiments using the retrograde axonal tracer WGA-HRP.

Positive responses to brief train stimulation of the olfactory bulb (3 shocks, 100 μsec in duration and 3 mA in amplitude, 3 msec apart) were found in the mesial prefrontal cortex as well as in the orbital and ventral agranular insular areas (FIG. 1). However, multiunit responses with short (< 10 msec to 20 msec) but variable latencies were typically recorded in the more lateral areas (ventrolateal orbital area, VLO; lateral orbital area, LO; ventral agranular insular area, AIv; FIG. 2A), while the responses in the more medial areas (prelimbic area, PL; infralimbic area, IL; medial orbital area) tended to be single units, with longer (20 msec to 50 msec), but invariant latencies (FIG. 2B). At low rates of stimulation, the latencies of these units are as constant as antidromic responses, but they do not follow high rates of stimulation.

Experiments with injections of WGA-HRP into the prefrontal cortex indicate that the more lateral areas (VLO, LO, AIv) receive inputs directly from the piriform cortex and other parts of the primary olfactory cortex. Both anterior and posterior parts of the piriform cortex (PCa and PCp), and the periamygdaloid cortex (PAC) send axons to AIv, while only PCa projects to VLO and LO. In contrast, the more medial areas (PL and IL) receive very few fibers from the primary olfactory cortex; only a few cells in PAC could be identified that project axons into these areas. However, these areas may also receive olfactory inputs from AIv and the medial part of the mediodorsal thalamic nucleus.

These experiments indicate that several areas on both the mesial and orbital surfaces of the rat prefrontal cortex are responsive to stimulation of the olfactory bulb. The strongest responses are found in the orbital and agranular insular areas in the dorsal bank of the piriform cortex, and are presumably evoked via the direct projections to these areas from the piriform cortex and other primary olfactory cortical areas. The much weaker responses in the medial prefrontal areas may be evoked through direct

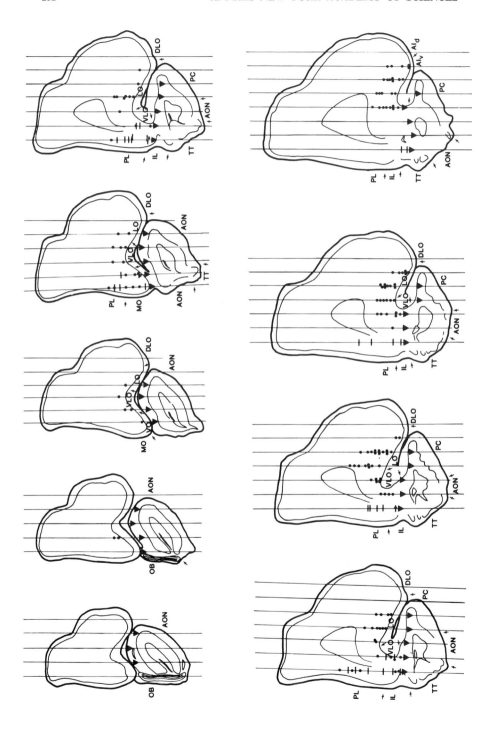

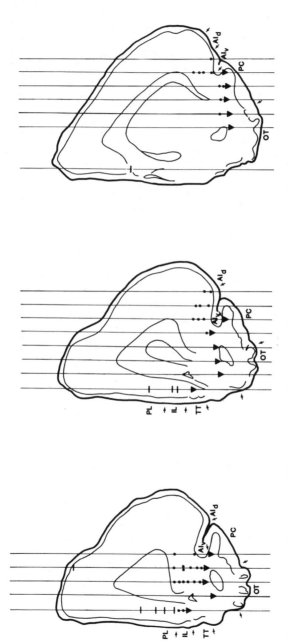

FIGURE 1. The distribution of positive responses to olfactory bulb stimulation in the orbital and mesial prefrontal cortex. The dots represent multiunit responses with short but variable latencies. The crossbars represent single-unit responses with typically longer but invariant latencies. At all points ventral to the inverted triangles, very short latency, intense responses typical of the primary olfactory cortex were recorded.

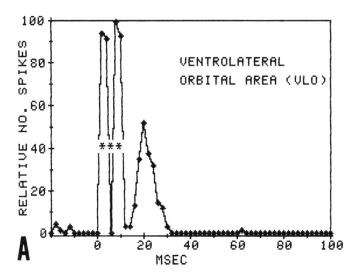

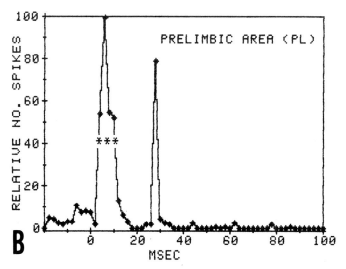

FIGURE 2. Typical peri-stimulus time histograms of unit activity evoked by olfactory bulb stimulation in (**A**) the lateral areas (e.g. VLO) and (**B**) the medial areas (e.g. PL) of the prefrontal cortex. The stimulus artefact is marked by ***.

projections from the periamygdaloid cortex, or through indirect pathways through the agranular insular cortex or the mediodorsal thalamic nucleus.

REFERENCES

1. KRETTEK, J. E. & J. L. PRICE. 1977. Projections from the amygdaloid complex to the cerebral cortex and thalamus in the rat and cat. J. Comp. Neurol. **171:** 687-722.
2. WIEGAND, S. J. & J. L. PRICE. 1980. Olfactory neocortical areas in the rat. Neurosci. Abstr. **6:** 307.

Ventilatory Response of the Tegu Lizard to Upper Airway Hypoxia and Hyperoxia during Normocapnia or Hypercapnia

E. L. COATES,[a] T. M. CATON, AND G. O. BALLAM

Department of Physiology
University of New Mexico School of Medicine and
Bioengineering Research Division
Lovelace Medical Foundation
Albuquerque, New Mexico 87108

It has been reported by Coates *et al.*[1] that the tegu lizard, *Tupinambis nigropunctatus,* depresses ventilatory frequency when CO_2 (0.4 to 4%) is delivered to the upper airways. The response originates from receptors innervated by the olfactory nerves. The experiment reported here was performed to test the ventilatory response to upper airway hypoxia and hyperoxia and to determine if O_2 concentration altered the ventilatory response to upper airway CO_2.

METHODS

Fresh air was inspired into the lungs throughout the experiment via an endotracheal T-tube inserted into the glottis. This isolated the upper airways (mouth and nasal cavities) and made possible the administration of gas mixtures to the upper airways independent of the fresh air delivered to the lungs. Gas mixtures were delivered directly into the external nares at 100 cc $\cdot$ min^{-1} by means of two tubes fixed in a silastic cap fitting over the nose. The hypoxic (2% O_2) and the hyperoxic (90% O_2) gases were delivered to the upper airways in square wave pulses, delivered at a frequency of 0.002 Hz during either normocapnia (0%) or hypercapnia (2%). To change gas concentrations, solenoids controlled by a function generator were opened or closed allowing preset gas mixtures to enter the upper airways. Care was taken to eliminate changes in flow. The ventilatory frequency (f) was measured using a pneumotachograph connected to the exit port of the endotracheal T-tube. Responses were analyzed by comparing the first minute following a change in O_2 (zone B), the last minute before a return to normoxia (zone C) and the first minute following the return to normoxia (zone D) to the minute of normoxia immediatley preceding the delivery of

[a] Address for correspondence: E. Lee Coates, Bioengineering Research Division, Lovelace Medical Foundation, 2425 Ridgecrest Dr., S.E., Albuquerque, New Mexico 87108.

TABLE 1. Ventilatory (f) Response to Upper Airway Hypoxia During Normocapnia and Hypercapnia[a]

Zone	Normocapnia	Hypercapnia
B	116.6 ± 10.4	118.6 ± 6.3[b]
C	108.4 ± 11.3	102.1 ± 5.9
D	113.4 ± 9.3	116.1 ± 9.6[b]

[a] Each zone was compared to the control Zone A, the resulting values are the mean f ± SEM; $n = 7$.
[b] Significant change ($p < 0.05$).

hypoxia or hyperoxia (zone A). Each lizard ($n = 7$) received five periods of both hypoxia and hyperoxia during normocapnia and hypercapnia.

RESULTS AND DISCUSSION

The lizards demonstrated the normal depression of f to 2% CO_2 in the upper airways, which has been previously reported.[2]

It was found that hypoxia delivered to the nasal passages during hypercapnia caused a significant ($p < 0.05$) increase in f in zones B and D. Although similar changes in f occurred during hypoxic normocapnia, they were not significantly different from hypoxic hypercapnia or normoxic normocapnia (TABLE 1). Neither hypoxic hypercapnia nor hypoxic normocapnia produced a significant change in f during zone C. Control studies have shown that the changes in f seen in this study were not due to changes in flow or stimuli such as the click of valves.

Hyperoxia delivered to the nasal passages caused a significant ($p < 0.05$) increase in f of zone B during both normocapnia and hypercapnia and in zone D during hypercapnia (TABLE 2). No significant ventilatory change occurred in zone C during hyperoxia or in zone D during hyperoxic normocapnia. In zone D, hyperoxic hypercapnia was significantly greater than hyperoxic normocapnia.

These results indicate an upper airway ventilatory response in the tegu lizard to hypoxic and hyperoxic hypercapnia. Changes in oxygen concentration during nor-

TABLE 2. Ventilatory (f) Response to Upper Airway Hyperoxia During Normocapnia and Hypercapnia[a]

Zone	Normocapnia	Hypercapnia
B	125.4 ± 10.7[b]	127.0 ± 15.6[b]
C	102.1 ± 5.7	106.5 ± 9.2
D	110.7 ± 8.0	186.8 ± 19.3[b]

[a] Each zone was compared to the control Zone A, the resulting values are the mean f ± SEM; $n = 7$.
[b] Significant change ($p < 0.05$).

mocapnia produced a statistically significant change in f only in zone B during hyperoxia. It is not clear from these results if there is an interaction between CO_2 and O_2. The results do indicate, however, that both increases and decreases in O_2 concentration produce a transient increase in f. It is not possible to determine from these results if the response is mediated by specific chemoreceptors or is a general response of nonspecific receptors to large changes in environmental O_2 concentration.

REFERENCES

1. COATES, E. L., R. M. PACHECO & G. O. BALLAM. 1985. Localization of upper airway CO_2 sensitive receptors in the tegu lizard *Tupinambis nigropunctatus.* Fed. Proc. **44:** 1348.
2. BALLAM, G. O. 1985. Breathing response of the tegu lizard to 1-4% CO_2 in the mouth and nose or inspired into the lungs. Respir. Physiol. **62:** 375-386.

Salt Deprivation and Amiloride-Induced Alterations in Neural Gustatory Responses Predict Salt Intake in Sham Drinking Rats

ROBERT J. CONTRERAS [a]

Department of Psychology
Yale University
New Haven, CT 06520-7447

An important goal of our research is to determine the role of the peripheral gustatory system in mediating changes in salt intake in the rat. Our strategy has been first to assess the plasticity and modifiability of the peripheral gustatory system at the neural level after dietary sodium deprivation[1] or after topical application of amiloride on the tongue.[2] Each manipulation alone reduces the chorda tympani responses to suprathreshold concentrations of sodium chloride. If the peripheral gustatory system is indeed important in controlling salt intake, then these changes in the NaCl neural response functions should predict corresponding changes in NaCl intake. Thus, we have begun to examine the separate effects of dietary sodium deprivation and amiloride on salt intake. As a way to isolate the contribution of orosensory cues, in the absence of visceral feedback, in determining salt intake, we have adopted the sham drinking preparation.[3]

Each rat ($n = 14$) was anesthetized with sodium pentobarbitol and surgically implanted with a stainless steel fistula in the rumen of the stomach along the greater curvature. The fistula was exteriorized outside the body wall and skin. A stainless steel screw closed the fistula, except when it was replaced with a collection tube. When the collection tube was attached to the fistula, the fluid ingested drained freely by gravity flow out of the stomach, down the collection tube and into a pan beneath the animal's cage. The rats' ($n = 5$) intakes of water and various NaCl concentrations were measured with fistula closed (control drinking) and open (sham drinking) in 30-minute drinking tests after overnight food and water deprivation. In addition, one set of rats were fed a sodium-deficient diet ($n = 4$) or a sodium-replete diet ($n = 5$) for 10 days before assessing their sham drinking of various NaCl concentrations. For all sham drinking tests, fluid intake, drainage volume, and Na^+ and K^+ electrolyte concentration by flame photometry were measured. Sham drinking was acceptable only when the volume of the gastric drainage was similar to that consumed.

The open fistula condition provided good sham drinking as the rats drank from 1.5-3.0 times more fluid depending on the concentration when the fistula was open than closed. Although sham drinking increased intake, the general shape of the function

[a] Address correspondence to: Robert J. Contreras, Department of Psychology, University of Alabama at Birmingham, University Station, Birmingham, Alabama 35294.

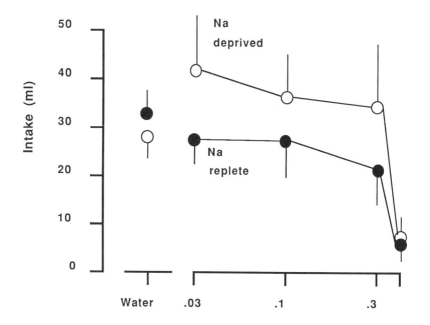

NaCl Concentration (M)

FIGURE 1. The average (± SEM) sham drinking of water and of various molar NaCl concentrations in four sodium-deprived (open circles) and five sodium-replete rats is shown. Their baseline water intakes are the same, yet sodium-deprived rats tend to consume more NaCl solution across concentration than sodium-replete rats.

was unchanged; intake was greatest for 0.03 M NaCl and then declined steadily with increasing concentration. The intakes of 0.3, 0.4, and 0.5 M NaCl were the same under both drinking conditions. Because the shape of the NaCl intake function was the same under normal and sham drinking conditions, the implication is that orosensory stimulation (presumably salt taste receptors) alone is critical in detemining NaCl intake. It should be noted that at the three highest NaCl concentrations, the Na^+ ion concentration consumed exceeded that collected. Thus, the sham drinking tests, although preferred over standard intake tests, were not perfect because some of the Na^+ ions consumed in the higher NaCl concentrations were presumably absorbed by the animal.

The effects of dietary sodium deprivation on the sham drinking functions for NaCl are shown in FIGURE 1. As shown in this figure, sodium-deprived and sodium-replete animals sham drank comparable amounts of water during the 30-minute intake test indicating that the two groups' baseline intakes were similar. Although there was a lot of variability, the data also show that sodium-deprived rats tend to drink more NaCl solution than sodium-replete animals. This is consistent with previous studies using standard one-bottle and two-bottle preference tests. Sham drinking may be a useful psychophysical procedure for obtaining NaCl intake functions from which to compare to corresponding neural functions from salt-sensitive neurons of the chorda tympani nerve. We plan to continue to use the sham drinking preparation to assess the effects of sodium deprivation and amiloride on the NaCl intake functions.

REFERENCES

1. CONTRERAS, R. J. & M. E. FRANK. 1979. Sodium deprivation alters neural responses to gustatory stimuli. J. Gen. Physiol. **73:** 569-594.
2. CONTRERAS, R. J., E. K. FARNUM & E. BIRD. 1985. Amiloride alters behavioral and neural gustatory responses to salt solutions. Chem. Senses **10:** 430.
3. GIBBS, J. & G. P. SMITH. 1984. The neuroendocrinology of postprandial satiety. *In* Frontiers in Neuroendocrinology. L. Martini & W. F. Ganong, Eds. Vol. **8:** 223-245. Raven Press. New York.

Neurosurgical Applications of Clinical Olfactory Assessment[a]

RICHARD M. COSTANZO,[b] PETER G. HEYWOOD,
JOHN D. WARD, AND HAROLD F. YOUNG

*Departments of Physiology and Biophysics and
Division of Neurosurgery
Medical College of Virginia
Richmond, Virginia 23298*

The neurosurgeon, through routine neurological examination, encounters olfactory deficits in those patients with skull fractures, contusions, lesions in the anterior fossa, and intracranial tumors. We randomly sampled 25 patients attending the head injury follow-up clinic and administered the CCCRC test of olfactory function[1] including a butanol detection threshold and odor identification subtest. Patients were of ages 9 to 61 and most had sustained a severe head injury as evidenced by a score of eight or less on the Glasgow coma scale. All patients were tested at least two months postinjury and were judged competent and coherent on the basis of an independent neuropsychological interview and assessment.

FIGURE 1 compares olfactory function in severe head injury patients to a group of 65 normal subjects.[2] Almost 60% of the head injury patients were found to be anosmic. About 15% showed some decrease in olfactory function and the remaining 25% appeared normal (fell within the normosmic or mildly hyposmic range). Further analysis revealed that of the 40% of patients who could detect olfactory levels of butanol (0.15% to 0.005%), half were unable to identify the standard number of odor stimuli. This suggests the possibility of a more central type of damage in these patients.

A separate group of "surgical anosmics," patients lacking olfactory connections as a result of surgical procedure, were also tested and their scores validate the olfactory function test's ability to define anosmia. All surgical anosmics fell to the left of the control distribution detecting only a 1.3% or stronger solution of butanol. Surgical anosmics were unable to identify any of the odor stimuli.

An example of a head injury patient illustrates the application of olfactory function testing in diagnosing and locating neural damage. The patient was a 26-year-old white male involved in a motorcycle accident who sustained a blow to the head. The olfactory function test was administered to the patient and results indicated a detection level within the normal range (0.016% butanol). However, the patient was unable to discriminate any of the odor identification stimuli. In FIGURE 2, a CAT scan of the patient's head revealed the presence of a large mass (hematoma) in a ventral region of the right frontal lobe (near olfactory pathways). In this case, the patient's inability

[a] Supported by National Institutes of Health Grants NS 16741 and NS 12587.

[b] Address correspondence to: Dr. Richard M. Costanzo, Department of Physiology and Biophysics, Box 551, MCV Station, Richmond, Virginia 23298.

to discriminate odors in the presence of normal detection thresholds indicated central damage.

In summary, following severe head injury, 60% of patients were found to be anosmic. Half of those detecting olfactory solutions of butanol were unable to correctly identify the normal number of stimuli suggesting central damage. Patients lacking olfactory nerve connections (surgical anosmics) validate test scores for the anosmic diagnostic category. Finally, olfactory function testing can assist the physician in locating and quantifying the extent of neural damage.

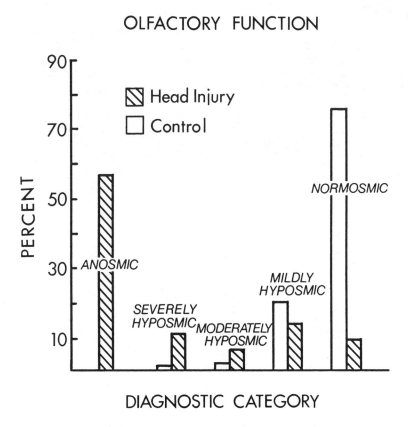

FIGURE 1. Distribution of olfactory function scores (diagnostic category) for severe head injury patients and control subjects. Categories correspond to composite scores obtained by summing the results of butanol threshold and odor identification subtests. Bar heights represent the percent of patients or control subjects in each category.

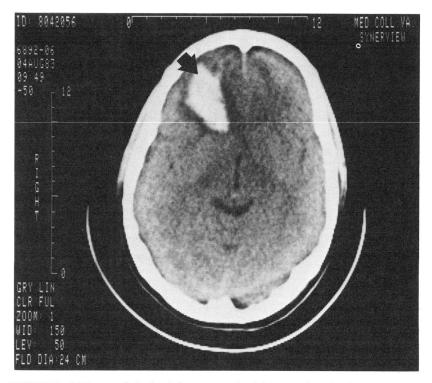

FIGURE 2. CAT scan of the head for a severe head injury patient showing a large mass (hematoma) in the right frontal lobe (see arrow). Confirmation of this lesion shows that the patient, capable of detecting olfactory stimuli but unable to identify odors, suffers from a more central (intercranial) damage.

REFERENCES

1. CAIN, W. S., J. GENT, F. A. CATALANOTTO *et al.* 1983. Clinical evaluation of olfaction. Am. J. Otolaryngol. **4:** 252-256.
2. HEYWOOD, P. H. & R. C. COSTANZO. 1986. Identifying normosmics: A comparison of two populations. Am. J. Otolaryngol. **7:** 194-199.

Perceived Burn and Taste Intensity of Physical Mixtures of Capsaicin and Taste Stimuli[a]

BEVERLY J. COWART

Monell Chemical Senses Center
Philadelphia, Pennsylvania 19104

Recent studies by Lawless and Stevens[1] and Lawless, Rozin, and Shenker[2] indicate there may be significant suppression of taste intensity when taste stimuli are interspersed with prerinses of a chemical irritant as compared to prerinses of solvent. In those studies, tastes and irritants were never presented in physical mixture, as they are typically consumed, and the periodic rinse format resulted in significant declines in perceived irritation during the time that taste sensations were being rated. Under those circumstances, any masking of taste would presumably be less than could be anticipated during the course of a normal spicy meal, when irritant stimuli may be present in every bite and perceived irritation would seem more likely to remain constant than to decline.

This phenomenon was examined further under conditions that seemed more ecologically valid and likely to produce a more consistent level of irritation. Both simple and complex taste stimuli were presented in physical mixture with capsaicin (CAP) to groups of consumers and nonconsumers of hot spices ($n = 12$/group). In two sessions, subjects used 13-point category scales to generate profiles of the sweetness, sourness, saltiness, bitterness, and burn of aqueous and chicken broth solutions containing either no added tastant, 0.34 M sodium chloride (NaCl), 0.01 M citric acid (CA), or both 0.34 M NaCl and 0.01 M CA. In one session, these stimuli were also rated with 1 ppm CAP added to each; in the other, the stimuli were also rated with 2 ppm CAP added.

Consumers and nonconsumers did not differ in their ratings of any perceptual attributes of the stimuli without CAP. Burn ratings of stimuli with CAP did differ significantly between groups with nonconsumers giving higher ratings at both 1 and 2 ppm CAP than did consumers. For both groups, and at each level of CAP, burn intensity at the time of taste ratings remained constant throughout the test session. Nonetheless, we observed no consistent changes in the taste intensity ratings of either group following the addition of CAP to stimuli.

In a series of pilot studies varying such parameters as rating procedure but retaining a mixture format, significant masking of salty or sour tastes was still never observed although there was often a tendency toward sour suppression. Finally, in a study directly contrasting the effects of rinse and mixture formats matched on all other parameters, the initial findings of Lawless and Stevens[1] were generally replicated in the rinse format in that significant suppression of both sour (0.0032 and 0.01 M CA)

[a] This work was supported by National Institutes of Health Grant #NS-20616.

and bitter (0.056 and 0.18 mM quinine hydrochloride), but not sweet or salty (0.18 and 0.56 M sucrose and NaCl), was observed. Given CAP-taste mixtures, the same subjects did show significant sour suppression, but they evidenced no bitter as well as no sweet or salty suppression. The major difference between bitter responses elicited in the two formats lay in the magnitude of ratings given under control conditions rather than under CAP-treatment conditions. Specifically, periodic presentation of water (which has a very weak bitter taste) as a control prerinse appeared to elevate ratings of the bitter stimuli, especially of the lower concentration. These findings indicate that irritant effects on taste may be neither as robust nor as general as the results of earlier studies suggested.

REFERENCES

1. LAWLESS, H. & D. A. STEVENS. 1984. Effects of oral chemical irritation on taste. Physiol. Behav. **32:** 995-998.
2. LAWLESS, H., P. ROZIN & J. SHENKER. 1985. Effects of oral capsaicin on gustatory, olfactory and irritant sensations and flavor identification in humans who regularly or rarely consume chili pepper. Chem. Senses **10:** 579-589.

Scanning and High-Voltage Electron Microscopy of Taste Cells in the Mudpuppy, *Necturus malculosus*[a]

T. A. CUMMINGS, R. J. DELAY, AND S. D. ROPER[a]

Department of Anatomy
Colorado State University
Fort Collins, Colorado 80523
and
Rocky Mountain Taste and Smell Center
University of Colorado Health Sciences Center
Denver, Colorado 80262

The initial interaction of tastants most likely occurs on the apical membrane of the taste cells since only that portion is exposed to the oral cavity. To gain better insight into this interaction, we examined the pore region of taste buds in *Necturus maculosus* using scanning electron microscopy (SEM), high-voltage electron microscopy (HVEM), and transmission electron microscopy (TEM).

In SEM, the taste papilla appeared as a raised mound of the lingual epithelium with a tufted cap. The tufted cap represents the taste pore, measuring 18-30 μm in diameter, where the apical ends of taste cells are exposed. The tufted appearance is due to microvilli present on taste cells that reach the surface. In thick sections viewed in HVEM, the pear-shaped taste buds, which contain up to 100 taste cells, measured 80-120 μm in diameter (FIG. 1). Individual taste cells extended 70-100 μm from the basal lamina to the taste pore. SEM of the pore revealed a patchwork distribution of three morphologically distinct types of apical specializations: long and branched (LB) microvilli, short and unbranched (SU) microvilli, and bundles of stereocilia (FIG. 2). SU microvilli were uniform in size and shape, whereas LB microvilli and the stereocilia were variable in size and shape.

As demonstrated in thin and thick sections, LB microvilli are specializations of dark cells. In the mudpuppy, dark cells contain packets of densely staining granular material in their apical region, which clearly distinguishes them from other taste cell types.[1] Further, we found that SU microvilli were the apical specializations of light cells. In the mudpuppy, light cells are characterized by their abundant smooth ER.[1] Stereocilia were present in most, but not all, taste pores. When present, they occurred in clusters, with 1-11 clusters found per taste pore. A cluster contained from 2-15 tall, slender, tapering stereocilia that extended far above the SU and LB microvilli.

[a] Supported by National Institutes of Health Grants #AG03340, #NS24107, and #NS20486. The Laboratory for High Voltage Electron Microscopy, University of Colorado, is supported in part by National Institutes of Health Grant PR 00592. The Scanning Electron Microscopy Center, University of Colorado, is supported in part by National Science Foundation Grant CPE 8406033.

The stereocilia arise from a cell that has the cytoplasmic markers characteristic of light cells. The underlying cytoskeleton of all types of apical specializations appeared to consist of densely packed parallel bundles of actin-like filaments. When no special efforts were made to remove the mucus from the tissue, the pore mucus completely covered the SU microvilli and partially covered the LB microvilli. Stereocilia projected above the mucus surface and thus are highly exposed to tastants in the saliva and oral cavity.

These three morphologically distinct types of apical specializations may reveal functional differences among taste cells. Murray[2] speculated that the differences in microvilli length, coupled with their differential submersion within the taste pore mucus, may impart chemoreceptive differences to light versus dark cells. This concept may be extended to include the highly exposed stereocilia. The morphology of the apical specialization may influence the initial interaction between taste stimulus and taste cell; thus, it is possible that the chemoreceptor ability and/or specificity of a taste cell may be imparted by its apical specialization.

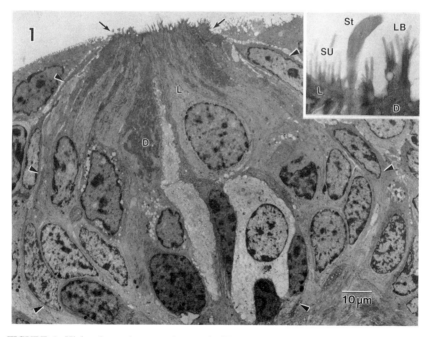

FIGURE 1. High-voltage electron micrograph (HVEM) of a longitudinal thick section of a taste bud (arrowheads) within the lingual epithelium. The apical ends of dark cells (D) and light cells (L) are directly exposed to the oral cavity at the taste pore (arrows). Insert shows a different thick section at higher magnification (magnification 6,240×). Short, unbranched microvilli (SU) are the apical specializations of light cells (L); long, branched microvilli (LB) are those of dark cells (D); and stereocilia are the apical specializations of cells that resemble light cells.

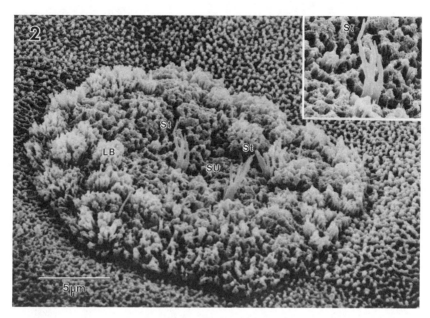

FIGURE 2. Scanning electron micrograph (SEM) of a taste pore showing three types of apical specializations. Bundles of stereocilia (St) project beyond the short, unbranched microvilli (SU) of light cells and the long, branched microvilli (LB) of dark cells. Insert shows high-magnification SEM of a cluster of stereocilia. The mucus partially covers the microvilli, whereas the stereocilia are highly exposed.

ACKNOWLEDGMENTS

Thanks are extended to Mr. George Wray and Mr. Douglas Wray, University of Colorado, for their excellent technical assistance.

REFERENCES

1. FARBMAN, A. I. & J. D. YONKERS. 1971. Fine structure of the taste bud in the mudpuppy, *Necturus maculosus.* Am. J. Anat. **131:** 353-369.
2. MURRAY, R. G. 1973. The ultrastructure of taste buds. *In* The Ultrastructure of Sensory Organs. I. Friedmann, Ed.: 1-81. North Holland. Amsterdam, the Netherlands.

Purinergic Receptors Occur Externally on the Olfactory Organs and Internally in the Brain of the Spiny Lobster[a]

CHARLES D. DERBY

Department of Biology
Georgia State University
Atlanta, Georgia 30303

WILLIAM E. S. CARR AND BARRY W. ACHE

C.V. Whitney Laboratory and
Department of Zoology
University of Florida
St. Augustine, Florida 32086

Purinergic receptors for adenosine and adenine nucleotides occur on diverse types of cells and tissues of vertebrate animals, including the CNS where they are thought to be involved in neurotransmission or neuromodulation.[1] Burnstock[2] recognized that purinergic receptors are not homogeneous and introduced the terms P_1 and P_2 to distinguish between two major types.

The presence of purinergic receptors on the primary chemoreceptor cells of marine crustaceans was indicated from behavioral experiments with the shrimp, *Palaemonetes pugio*, in which AMP was found to be one of the most potent chemoattractants in tissue extracts from several prey species.[3] An electrophysiological characterization of purinergic chemoreceptor neurons on the olfactory organ (antennule) of a marine crustacean was performed using a larger species, the spiny lobster *Panulirus argus*.[4] The responsiveness of the purinergic olfactory cells of lobsters correlated closely with the behavioral responses of shrimp. In both cases: (a) AMP was clearly the most effective purine tested; (b) there was a potency sequence of AMP > ADP > ATP > adenosine; (c) changes in the ribose phosphate moiety of the AMP molecule resulted in the greatest decrease in activity; and (d) the response to AMP was antagonized by theophylline. These results suggest that these marine crustaceans possess external purinergic receptors on olfactory receptor cells that are similar to the P_1-type purinoceptors found internally in vertebrates.

The hypothesis that some of the receptor types found in internal tissues of higher animals may have evolved from external chemoreceptors or primitive aquatic organ-

[a] Supported by National Science Foundation Grant No. BNS 84-11693, National Institute of Neurological and Communicative Disorders and Stroke Grant No. NS22225, and a Whitehall Foundation grant.

isms is supported by the fact that many neuroactive substances are known to activate receptors existing both in internal tissues and on external chemosensory surfaces.[5] Our finding of P_1-like purinoceptors in the olfactory organ of the spiny lobster and our ability to record from cells in the CNS of this animal compelled us to determine if receptors for purinergic substances also existed within this invertebrate.

An excised, perfused, anterior end preparation was used to compare the effect of different perfusion conditions on the spontaneous activity as well as olfactorily or

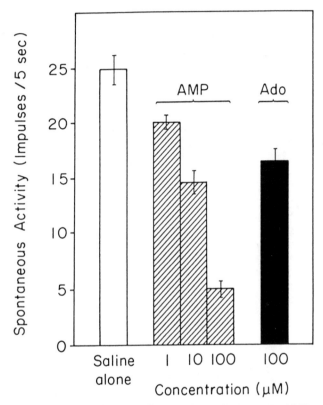

FIGURE 1. Modulation of spontaneous activity of a CNS neuron by AMP and adenosine. Spontaneous activity of brain neuron was measured during perfusion of CNS with lobster saline alone (open bars), lobster saline plus 1, 10, or 100 μM AMP (hatched bars), and 100 μM adenosine (solid bars). Spontaneous activity rates are mean values ($\pm$ 95% confidence limits) from 10-30 measurements. Values obtained during perfusion with saline alone were obtained both before and after perfusion with saline plus AMP or adenosine (from Derby *et al.*[6]).

electrically evoked activity of brain neurons having axons in the circumesophageal connectives.[6] Modulatory effects of AMP and adenosine were identified by comparing spontaneous activity and evoked activity under several different perfusion conditions: saline alone, saline plus 1, 10, or 100 μM AMP, and saline plus 100 μM adenosine. Perfusion with 100 μM AMP resulted in modulation of spontaneous activity in 71% of the spontaneously active brain neurons (FIG. 1) and modulation of the electrically or olfactorily evoked activity in 25% of the neurons (FIG. 2). Adenosine had modulatory effects similar to those of AMP. The effects occurred at concentrations as low as 1 μM, and were dose dependent, reversible, and usually depressive. This is the first report of purinergic receptors and their modulatory function in the nervous system of any invertebrate.

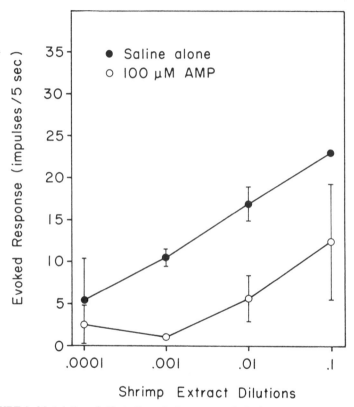

FIGURE 2. Modulation of olfactorily evoked responses of a brain neuron by AMP. Responses evoked by presentation of shrimp extract to the antennules were obtained during perfusion of CNS with lobster saline alone (solid symbols) and during perfusion with saline plus 100 μM AMP (open symbols). Values are means ($\pm$ 95% confidence limits) of one to four responses. Values for perfusion with saline alone were obtained both before and after perfusion with AMP. Values were corrected for spontaneous activity (from Derby et al.[6]).

The presence of purinergic receptors on the dendrites of primary olfactory neurons and on neurons within the brain of lobsters suggests that a receptor of the same general type may have been conserved through evolution and used for different functions. Any differences in specificity of external and internal purinoceptors, as well as the existence of other classes of external purinoceptors such as a P_2-type,[7] may reflect different selection pressures as the receptors evolved to monitor chemicals in the external versus the internal aquatic milieu.

REFERENCES

1. PHILLIS, J. W. & P. H. WU. 1981. The role of adenosine and its nucleotides in central synaptic transmission. Prog. Neurobiol. **16:** 187-239.
2. BURNSTOCK, G. & C. M. BROWN. 1985. The classification of receptors for adenosine and adenine nucleotides. *In* Methods in Pharmacology. D. M. Paton, Ed.: 193-212. Plenum Press. New York.
3. CARR, W. E. S. & H. W. THOMPSON. 1983. Adenosine 5'-monophosphate, an internal regulatory agent, is a potent chemoattractant for a marine shrimp. J. Comp. Physiol. **153:** 47-53.
4. DERBY, C. D., W. E. S. CARR & B. W. ACHE. 1984. Purinergic olfactory cells of crustaceans: Response characteristics and similarities to internal purinergic cells of vertebrates. J. Comp. Physiol. **155:** 341-349.
5. CARR, W. E. S., B. W. ACHE & R. A. GLEESON. 1987. Chemoreceptors of crustaceans: Similarities to receptors for neuroactive substances in internal tissues. Environ. Health Perspect. **71:** 31-46.
6. DERBY, C. D., B. W. ACHE & W. E. S. CARR. 1987. Purinergic modulation in the brain of the spiny lobster. Brain Res. In press.
7. CARR, W. E. S., R. A. GLEESON, B. W. ACHE & M. L. MILSTEAD. 1986. Olfactory receptors of the spiny lobster: ATP-sensitive cells with similarities to P_2-type purinoceptors of vertebrates. J. Comp. Physiol. **158:** 331-338.

Off Responses to Gustatory Stimuli in the Parabrachial Pons of Decerebrate Rats[a]

PATRICIA M. DI LORENZO

Department of Psychology
State University of New York at Binghamton
Binghamton, New York 13901

Although decerebrate rats have been shown to retain appropriate gustatory reactivity,[1] further studies suggest that they do not acquire conditioned taste aversions.[2] These results suggest the possibility that forebrain input to medullary and/or pontine gustatory relays may be important for the modification of gustatory reactivity as a result of experience. Little is known, however, about the precise influence of this input on these caudal structures. Recent investigations of taste responses in the nucleus of the tractus solitarius (the first central gustatory relay) in decerebrate rats have not reported changes in these responses as a result of this procedure.[3,4] The present experiment was designed to study the effects of decerebration on taste responses in the parabrachial nucleus of the pons (PbN), the second central gustatory relay.

Male albino rats (350-500 g) were decerebrated at the supracollicular level under ketamine (40 mg/kg i.p.) and ether anesthesia and prepared for electrical recording in the PbN. Thereafter animals were maintained under flaxedil and artificially respirated. All wound edges were frequently inundated with lidocaine HCl, and heart rate and core temperature were monitored throughout the experiment. Sapid solutions of NaCl ($0.1\ M$), HCl ($0.01\ M$), sucrose ($0.5\ M$), Na saccharin ($0.004\ M$) and quinine HCl ($0.01\ M$) were used as taste stimuli. Each trial consisted of a 10-second baseline, 10-second stimulus, 10-second wait and a 20-second rinse of distilled water. At least two minutes elapsed between trials. Gustatory responses from 32 parabrachial units in 13 decerebrate rats were recorded.

The most striking result of this study was the observation of OFF responses in a subset of taste-responsive parabrachial units. These responses took the form of a brief (1-2 sec) increase in the firing rate that occurred following the cessation of the stimulus presentation (see FIG. 1). This increase often exceeded the magnitude of the initial transient portion of the taste response and was superimposed on the steady-state firing rate that occurred during the stimulus presentation. Six out of 21 OFF responses occurred in the absence of a response to the stimulus. OFF responses were recorded in 14 (44%) units to at least one stimulus. All four taste qualities were capable of producing OFF responses in these units: NaCl produced OFF responses in five units, HCl in seven units, sucrose and Na saccharin in six units, and QHCl in three units (see FIG. 2). However, within a given unit, OFF responses were most often selective to a subset of taste qualities: Nine units showed OFF responses to only one taste

[a] Supported by National Institute of Neurological and Communicative Disorders and Stroke Grant 1 RO2 NS210578-01.

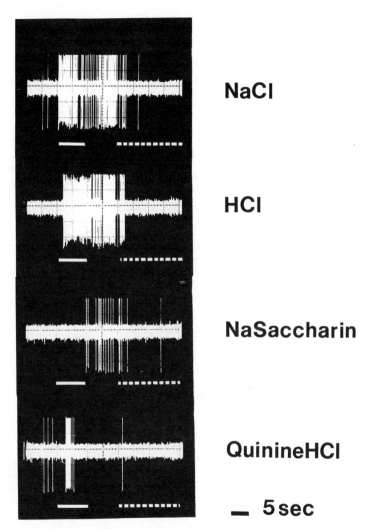

FIGURE 1. Responses of one unit in the PbN of a decerebrate rat to the four basic taste qualities. Solid white line indicates presentation of tastant. Dotted white line indicates distilled water rinse. OFF responses can be seen after NaCl and Na saccharin presentations.

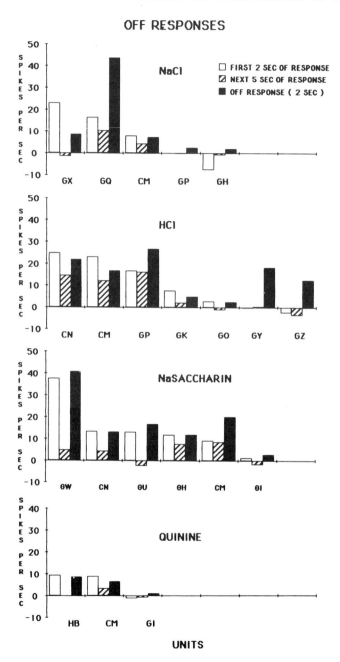

FIGURE 2. OFF responses to the four basic taste qualities in the PbN units in decerebrate rats. Average firing rate (spikes/sec) in first two seconds of response to the stimulus, the next five seconds of response, and the first two seconds of the OFF response are shown. Baseline firing rates were routinely subtracted from response measures.

quality, four units to two taste qualities, zero units to three taste qualities, and one unit to all four taste qualities. No relationship was found between the presence of OFF responses and the best-stimulus categorization of a particular unit.

In other respects, taste units in the PbN of decerebrate rats were similar to those in intact rats. Analysis of response profiles across stimuli showed that PbN units in the decerebrate rat were broadly responsive. NaCl and Na saccharin were the most effective stimuli: 23 of 32 (72%) responsive units each. HCl was slightly less effective: 19 of 32 (59%) response units. Sucrose, with 8 of 16 (50%) responsive units, and quinine HCl, with 13 of 32 (41%) responsive units, were the least effective stimuli. Comparison of PbN taste units in decerebrate versus intact rats revealed no differences in spontaneous rate of firing.

One possible implication of these results is that the suppression of OFF responses to taste stimuli in the intact animal by the forebrain may be important for the acquisition of conditioned taste aversions, particularly when the CS-US interval is long. In effect, the absence of a signal for the offset of a gustatory stimulus may enable the animal to form associations that bridge relatively long intervals of time. Conversely, the presence of OFF responses in the PbN of decerebrate rates may represent an explanation for the fact that decerebrate rats do not acquire conditioned taste aversions.

REFERENCES

1. GRILL, H. J. & R. E. NORGREN. 1978. The taste reactivity test: II. Mimetic responses to gustatory stimuli in chronic thalamic and decerebrate rats. Brain Res. **143:** 281-297.
2. GRILL, H. J. & R. E. NORGREN. 1978. Chronically decerebrate rats demonstrate satiation but not bait shyness. Science **201:** 267-269.
3. HAYAMA, T., S. ITO & H. OGAWA. 1985. Responses of solitary tract nucleus neurons to taste and mechanical stimulations of the oral cavity in decerebrate rats. Exp. Brain Res. **60:** 235-242.
4. MARK, G. P. & T. R. SCOTT. 1986. Gustatory activity in the NTS of decerebrate rats. Ann. N.Y. Acad. Sci. This volume.

Membrane Conductance Mechanisms of Dissociated Cells from the Olfactory Epithelium of the Mudpuppy, *Necturus maculosus*[a]

VINCENT E. DIONNE

Division of Pharmacology
Department of Medicine
University of California, San Diego
La Jolla, California 92093

Olfactory receptor neurons and other non-neuronal cells were dissociated from the olfactory epithelium of the adult mudpuppy, *Necturus maculosus,* and studied using whole-cell and patch-clamp methods. Single cells were isolated using collagenase and either trypsin or papain in Ca-free amphibian saline. Isolated cells retained their morphology, and on this basis six cell types were distinguished, only one of which was neuronal. Whole-cell membrane currents were recorded from two cell types: a non-neuronal, spherical type that was partially covered with asynchronous, motile cilia ("Medusa" cells) and olfactory neurons; columnar support cells did not survive the dissociation well and were not studied.

The whole-cell currents elicited from Medusa cells by depolarizing voltage steps consisted of a sustained outward membrane current, presumably carried by K^+. The currents had the kinetics and voltage dependence of a delayed rectifier. No transient, voltage-activated inward currents were observed in Medusa cells. Recording electrodes contained 120 mM KCl with 100 nM free Ca^{2+} buffered with BAPTA; pH was 7.2, HEPES buffer; in three cells GTP and ATP were also added.

Whole-cell currents from olfactory receptor neurons exhibited both inward and outward components in response to depolarizing voltage steps. The inward current was carried by Na^+ and Ca^2; its major component was a tetrodotoxin-sensitive Na^+ current that activated and inactivated rapidly. The outward current was carried by K^+; its time course resembled the delayed current seen in Medusa cells, but a component of the outward current was Ca^{2+}sensitive. The density of both the inward and outward currents declined with time when the internal solution was 120 mM KCl saline (above). The addition of 100 μM GTP plus 2.5 mM MgATP and lowering the free [Ca^{2+}] below 10 nM retarded this decline.

The input resistance of isolated olfactory neurons was measured and correlated with the zero-current (resting) membrane potential during whole-cell recording (FIG. 1). The mean input resistance was nearly 4 gigohms for cells with resting potentials ≤ -30 mV. The membrane capacitance in these neurons was 18.1 $\pm$ 5.4 pF, independent of voltage and [Ca^{2+}]. This compared well with the estimated surface area of the cell, assuming 1 μF/cm^2. In such small cells with this high input resistance, the activation of just one ion channel could significantly change the membrane potential to alter the cell's active properties.

[a]Supported by Grant #NS20962 from the National Institutes of Health.

Changes induced by putative odorants have been sought in the whole-cell membrane currents of olfactory neurons using a cocktail of twelve amino acids (Arg, Cys, His, Ile, Leu, Lys, Met, Phe, Thr, Trp, Tyr, Val). The cocktail was microperfused onto individual neurons. In six cells perfused repeatedly and briefly with the cocktail at pH 3.5, where many of these compounds are positively charged, the inward Na^+ and Ca^{2+} currents and the outward currents were reduced. However, the reduction

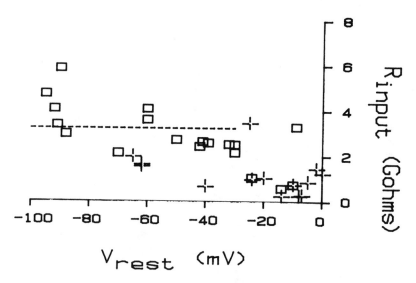

FIGURE 1. Input resistance versus resting membrane potential for dissociated olfactory neurons. The most negative resting potentials and the highest input resistances were obtained when the free $[Ca^{2+}]$ in the internal saline was below 10 nM ■. With 100 nM free Ca^{2+} (+) both parameters were low. The dashed line plots the mean input resistance for the low-Ca^{2+} cells with resting potentials ≤ -30 mV; it is 3.24 ± 1.07 gigohms.

was attributed to H^+ and not the amino acids since similar changes were induced by acid saline. Application of the amino acids at pH 7.3 had no effect on the membrane conductance mechanisms. Although it remains to be determined whether this pH-dependent modulation of the membrane conductance contributes materially to the olfactory response, it is certain that changes in the membrane conductance could account for chemical detection by olfactory neurons.

Olfactory Dysfunction in Alzheimer's Disease

A Summary of Recent Findings[a]

RICHARD L. DOTY[b]

Smell and Taste Center
Department of Otorhinolaryngology and Human Communication and
Department of Physiology
School of Medicine
University of Pennsylvania
Philadelphia, Pennsylvania 19104

PATRICIO REYES

Department of Neurology and Pathology
Jefferson Medical College
Thomas Jefferson University
Philadelphia, Pennsylvania 19107

Department of Neurology and Research
Coatesville Veterans Administration Medical Center
Coatesville, Pennsylvania 19320

Senile dementia of the Alzheimer's type (SDAT) is a chronic and debilitating neuropsychiatric disorder characterized by a progressive decline in intellectual faculties. This insidious disease accounts for at least half of demented patients over the age of 65 years,[1] and is characterized by the presence of large numbers of neuritic plaques and neurofibrillary tangles throughout cortical and subcortical brain structures, including the olfactory bulb, anterior olfactory nucleus, prepyriform cortex, and entorhinal cortex.[2-6] In light of such olfactory system involvement, we quantitatively examined the olfactory perception of a group of patients diagnosed as having mild to moderately severe SDAT. This article summarizes our findings, which are described in detail elsewhere.[7]

[a] Supported by National Institute of Neurological and Communicative Disorders and Stroke Grant NS 16365.

[b] Address for correspondence: Dr. Richard L. Doty, Smell and Taste Center, 5 Ravdin Building, Hospital of the University of Pennsylvania, 3400 Spruce Street, Philadelphia, PA 19104.

PROCEDURES AND RESULTS

Thirty-four patients who satisfied stringent criteria for the clinical diagnosis of Alzheimer's disease[8] were matched to 34 healthy noninstitutionalized control subjects on the basis of age, sex, and ethnic background. The subject details, including the criteria for inclusion in the study, are presented in a principal publication.[7] To assess odor identification ability, all 68 subjects were administered the University of Pennsylvania Smell Identification Test (UPSIT).[9-11] The SDAT patients were also administered the Picture Identification Test (PIT) to identify those too demented to comprehend nonolfactory aspects of the UPSIT.[12] In addition, 15 of the SDAT patients and an equivalent number of matched controls were administered a single-staircase, forced-choice, phenyl ethyl alcohol odor detection threshold test.[13]

Eight of the SDAT patients evidenced PIT scores < 35, and one was unable to complete the PIT. Thus, their data, along with those of their controls, were excluded from further consideration. The UPSIT scores of the 25 remaining Alzheimer's disease patients were significantly lower than those of their matched controls (respective median and interquartile range values equaled 17 [13-22] and 32 [26.5-38]; Wilcoxin matched-pairs signed-ranks test, $p < 0.001$). Only three evidenced scores above their individually matched controls, and only three had scores falling above the 25th percentile of published UPSIT norms.[14] Of those scores that fell below the 25th percentile, nine fell below the 10th percentile. No statistical difference was apparent between the test scores of the Alzheimer's patients in stage 1 ($n = 9$) and stage 2 ($n = 16$) of the disease (Mann Whitney U Test, $p > 0.20$).[8]

The olfactory deficit was not confined to odor identification ability, as indicated by significantly higher detection threshold values of the Alzheimer's patients relative to the controls (respective median vol/vol threshold and interquartile range values equaled $10^{-2.63}$ [$10^{-3.75}$- $10^{-1.76}$] and $10^{-5.13}$ [$10^{-7.50}$-$10^{-3.88}$]; Wilcoxin matched-pairs signed ranks test, $p < 0.001$). As in the case of odor identification, the decreased sensitivity was consistent, with only one patient evidencing a threshold value below its matched control. Similarly, no significant differences in the threshold values were apparent between the subjects in stage 1 and those in stage 2 of the disease.

DISCUSSION AND CONCLUSIONS

These results suggest that Alzheimer's disease is reliably accompanied, even in its earliest stages, by major alterations in the ability to detect and identify odors. These findings confirm recent brief communications noting odor identification problems in SDAT patients,[15,16] and demonstrate that olfactory sensitivity, per se, is altered by the disease process. Whether such alterations are the result of the destruction of neural elements within the olfactory system (e.g., by the action of environmental agents such as viruses or toxins) remains to be demonstrated, although it is noteworthy that the olfactory pathway is a major route for the penetration of neurotropic viruses into the central nervous system.[17-20]

Currently, clinical diagnosis of SDAT in living persons is based largely upon exclusion of other possible diseases, except in the few cases where a definitive diagnosis can be made from brain biopsy. These findings suggest that prudent use of quantitative olfactory testing may be of value in the early diagnosis of this debilitating and widespread disease.

REFERENCES

1. KATZMAN, R. 1976. The prevalence and malignancy of Alzheimer's disease: A major killer. Arch. Neurol. **33:** 217-218.
2. SCHNECK, M. K., B. REISBERG & S. H. FERRIS. 1982. An overview of current concepts of Alzheimer's disease. Am. J. Psychiatry **139:** 165-173.
3. ESIRI, M. M. & P. K. WILCOCK. 1984. The olfactory bulbs in Alzheimer's disease. J. Neurol. Neurosurg. Psychiatry **47:** 56-60.
4. REYES, P. F., P. L. FAGEL & G. T. GOLDEN. 1986. Olfactory cortex in Alzheimer's disease: Neuropathological studies with clinical correlation. J. Neuropathol. Exp. Neurol. **45:** 341.
5. REYES, P. F., G. T. GOLDEN, R. G. FARIELLO, L. FAGEL & M. ZALEWSKA. 1985. Olfactory pathways in Alzheimer's disease: Neuropathological studies. Soc. Neurosci. Abstr. **11:** 168.
6. HYMAN, B. T., G. W. VAN HOESEN, A. R. DAMASIO & C. L. BARNES. 1984. Alzheimer's disease: Cell-specific pathology isolates the hippocampal formation. Science **225:** 1168-70.
7. DOTY, R. L., P. F. REYES & T. GREGOR. 1987. Presence of both odor identification and detection deficits in Alzheimer's disease. Brain Res. Bull. **18:** 597-600.
8. CUMMINGS, J. L. 1983. Cortical dementias: Alzheimer's and Pick's diseases. In Dementia, A Clinical Approach. J. L. Cummings, Ed.: 35-72. Butterworth. Boston, MA.
9. DOTY, R. L., P. SHAMAN & M. DANN. 1984. Development of the University of Pennsylvania Smell Identification Test: A standardized microencapsulated test of olfactory function. Physiol. Behav. **32:** 489-502.
10. DOTY, R. L., P. SHAMAN, S. L. APPLEBAUM, R. GIBERSON, L. SIKORSKY & L. ROSENBERG. 1984. Smell identification ability: Changes with age. Science **226:** 1441-1443.
11. DOTY, R. L., M. G. NEWHOUSE & J. D. AZZALINA. 1985. Internal consistency and short-term test-retest reliability of the University of Pennsylvania Smell Identification Test. Chem. Senses **10:** 297-300.
12. VOLLMECKE, T. A. & R. L. DOTY. 1985. Development of the Picture Identification Test (PIT): A research companion to the University of Pennsylvania Smell Identification Test (UPSIT). Chem. Senses **10:** 413-414.
13. GHORBANIAN, S. N., J. L. PARADISE & R. L. DOTY. 1983. Odor perception in children in relation to nasal obstruction. Pediatrics **72:** 510-516.
14. DOTY, R. L. 1983. The Smell Identification Test Administration Manual.™ Sensonics. Philadelphia, PA.
15. CORWIN, J., M. SERBY, P. CONRAD & J. ROTROSEN. 1985. Olfactory recognition deficit in Alzheimer's and Parkinsonian dementias. IRCS Med. Sci. **13:** 260.
16. PEABODY, C. A. & J. R. TINKLENBERG. 1985. Olfactory deficits and primary degenerative dementia. Am. J. Psychiatry **142:** 524-525.
17. STROOP, W. G., D. L. ROCK & N. W. FRASER. 1984. Localization of herpes simplex virus in the trigeminal and olfactory systems of the mouse central nervous system during acute and latent infections by in situ hybridization. Lab. Invest. **51:** 27-38.
18. TOMLINSON, A. H. & M. M. ESIRI. 1983. Herpes simplex encephalitis. Immunohistological demonstration of spread of virus via olfactory pathways in mice. J. Neurol. Sci. **60:** 473-484.
19. MONATH, T. P., C. B. CROOP & A. K. HARRISON. 1983. Mode of entry of a neurotropic arbovirus into the central nervous system. Reinvestigation of an old controversy. Lab. Invest. **48:** 399-410.
20. GOTO, N., N. HIRANO, M. AIUCHI, T. HAYASHI & K. FIJIWARA. 1977. Nasoencephalopathy of mice infected intranasally with a mouse hepatitis virus, JHM strain. Jpn. J. Exp. Med. **47:** 59-70.

Rats Eating Together Prefer the Taste of Their Food

HEATHER J. DUNCAN,[a] AUDREY BUXBAUM, AND
MICHAEL G. TORDOFF

Monell Chemical Senses Center
Philadelphia, Pennsylvania 19104

One way of assessing the reward potential of a stimulus is to measure preferences for foods paired with that stimulus. The validity of this technique has been shown by shifts in preferences for neutral flavors paired with stimuli known to be rewarding in classical methodologies, such as sweet solutions or electrical brain stimulation.[1] Here, we show that this "conditioned taste preference" method can be used to examine the complex stimulus of the presence of another rat, a situation that would be difficult to study by operant techniques.

In Experiment 1, 16 male rats were fed regular chow in their home cages and, during the first 90 minutes of the dark period, fed flavored chow (0.8% chocolate or 0.8% chicken flavor) in training cages. Each rat always received one flavor in the company of another rat (the "together" flavor) and the other flavor when alone. This procedure was conducted for eight pairs of trials (16 days) according to a counterbalanced design. It was found that rats eating together consumed slightly but significantly less food in the 90-minute session than they did when eating alone (13% decrease), demonstrating a lack of social facilitation of feeding.[2] At the end of training, the rats were given a choice between the two flavors in the home cage. In a 90-minute test, 14 of 16 rats preferred the food they had previously eaten with a partner, and on average the group ate 232% more of this flavor than the other. Thus, food intake during training and food preference were dissociated.

The preference for food eaten with a partner was apparently a strong one. The rats were given free access to both flavors in their home cages for 11 days and throughout this period they maintained preferences for the "together" flavor, although by day 11 the overall preference was 61%, compared to 76% on day one. They then received four pairs (eight days) of reversal trials (pairing the other flavor with a partner rat). In a final two-choice test, 10 of 16 rats reversed their preferences and the difference in intake of the two flavors was abolished. TABLE 1 presents group preference data for the various phases.

In Experiment 2, using similar methods, 13 of 15 rats preferred the flavor eaten together after only one 90-minute exposure to the eating-together and eating-alone conditions.

These results suggest that (a) eating in the presence of another rat is a very rewarding and/or salient experience; (b) flavored food intake in the presence of a rewarding stimulus is not necessarily predictive of flavor preference in a choice test;

[a] Present address: University of Cincinnati Medical Center, Department of Anatomy and Cell Biology, ML521, Cincinnati, Ohio 45267.

TABLE 1. Percentage Intake of "Together Flavor"

Phase	Group 1		Group 2	
	Mean	SEM	Mean	SEM
	(chocolate together)		(chicken together)	
Experiment 1:				
Pretest	36.4	10.7	62.4	7.7
Training trial 8	53.3	1.6	48.6	4.6
Posttest	81.9	10.3	74.1	10.6
24-hour choice, day 11	65.4	12.0	56.1	10.2
	(chicken together)		(chocolate together)	
Reversal trial 4	48.3	2.8	37.0	1.6
Reversal test	47.0	14.8	51.0	12.2
	(chocolate together)		(chicken together)	
Experiment 2:				
Pretest	30.9	10.5	58.9	12.9
Trial 1	44.1	4.7	57.7	6.8
Posttest	58.9	12.3	82.9	8.7

and (c) flavor preference shifts can provide a sensitive measure of reward in situations that are not easily approached by other means.

REFERENCES

1. ETTENBERG, A. & N. WHITE. 1978. Conditioned taste preferences in the rat induced by self-stimulation. Physiol. Behav. **21**(3): 363-368.
2. TACHIBANA, T. 1974. Social facilitation of eating behavior in a novel situation by albino rats. Jpn. Psychol. Res. **16**(4): 157-161.

A Psychophysical-Decision Model for Sensory Difference Detection Methods

DANIEL M. ENNIS

Philip Morris Research Center
Commerce Road
Richmond, Virginia 23261

KENNETH MULLEN

Department of Mathematics and Statistics
University of Guelph
Guelph, Ontario, Canada N1G2W1

On reflecting about the perceived differences between chemosensory stimulants, there would appear to be significant value in attempting to arrive at a formal model in which stimulus and sensory parameters have been specified. Precise control of stimuli presented to the chemical senses is more difficult than in vision or audition where psychophysical models have been tested more extensively. For instance, temporal effects are large in experiments in olfaction and taste. For this reason, a model that specifies measurable stimulus variation would be helpful in interpreting results from discrimination tasks involving these senses and suggests that, in fact, the ideal model would be dynamic.

Chemosensory stimulation often leads to multivariate sensations. In order to understand the nature of these multivariate sensations, it will be necessary to build and test a theory that reliably models the effect of the perceptual parameters on performance. "Performance" is typically measured in method-specific terms, such as probability of a correct response, probability of confusing one stimulus with another, identification errors, and so on. Since the earlier work of Shepard,[1,2] Kruskal,[3,4] and their predecessors, scaling sensation magnitudes of stimuli with multivariate sensory attributes has been handled using a particular form of multidimensional scaling in which it is assumed that proximity measures and perceptual distances are monotonically related (maintain a rank-order relationship). Thus, the behavioral response (a measure of proximity) is assumed to depend on only a single parameter, the multidimensional distance, δ. Building on this theory, different assumptions have been made about the nature of the distance metric (for instance, city block or Euclidean), the relative weights given to different dimensions by subjects in determining δ, and even stronger assumptions about the relationship between δ and the proximity measure.

Recently it has been shown[5,6] that the monotonicity assumption is invalid under many multivariate conditions when methods are used that involve grouping stimuli on the basis of their similarity. A multivariate probabilistic model for grouping techniques has been derived[7] and discussed from the viewpoint of experimental work on the chemical senses[8] that does not invoke a monotonic relationship between proximity and distance. This model is Thurstonian in the sense that all of the variation is internal, and the stimuli are assumed to be physicochemically identical.

If stimulus variation is to be included in a model for discrimination, it becomes apparent that the model must specify the relationship between the stimulus and its corresponding sensory continuum. Although many models of this type are conceivable, there is considerable support for the power function.[9] FIGURE 1 shows a conceptual model, in the unidimensional case, in which the effect of stimulus and sensation variation on the behavioral response is specified by using Thurstonian and direct

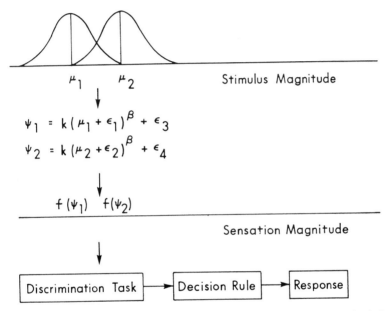

FIGURE 1. The conceptual framework for a psychophysical-decision model for discrimination methods. μ_1, μ_2: means of stimulus distributions; ψ_1, ψ_2: sensation values corresponding to stimuli randomly drawn from the stimulus sets; ϵ_1: random deviate from the first stimulus distribution; ϵ_2: random deviate from the second stimulus distribution; ϵ_3, ϵ_4: random deviates from sensation distributions (subject errors); β: power exponent; k: constant.

scaling concepts; the multivariate case is conceptually similar. Computer simulations of this model have shown that one can estimate β for different kinds of stimuli using a particular discrimination method. These results could then be compared with magnitude estimates or cross-modality matching experiments that assume that either magnitude estimates or matched intensities of a stimulus correspond exactly to the sensation magnitudes of the stimulus of interest. It would then be possible to test this assumption without using direct methods for estimating sensation magnitudes, and it would provide a new way of estimating psychophysical functions.

REFERENCES

1. SHEPARD, R. N. 1962. The analysis of proximities: Multidimensional scaling with an unknown distance function. I, II. Psychometrika 27: 125-140; 219-246.
2. SHEPARD, R. N. 1963. Analysis of proximities as a technique for the study of information processing in man. Hum. Factors. 5: 19-34.
3. KRUSKAL, J. B. 1964. Multidimensional scaling by optimizing goodness of fit to a nonmetric hypothesis. Psychometrika 29: 1-27.
4. KRUSKAL, J. B. 1964. Nonmetric multidimensional scaling: A numerical method. Psychometrika 29: 115-129.
5. ENNIS, D. M. & K. MULLEN. 1985. The effect of dimensionality on results from the triangular method. Chem. Senses 10: 605-608.
6. ENNIS, D. M. & K. MULLEN. 1986. A multivariate model for discrimination methods. J. Math. Psychol. 30(2):
7. MULLEN, K. & D. M. ENNIS. 1987. Mathematical formulation of multivariate Euclidean models for discrimination methods. Psychometrika 52: 235-249.
8. ENNIS, D. M. & K. MULLEN. 1986. Theoretical aspects of sensory discrimination. Chem. Senses 11: 513-522.
9. STEVENS, S. S. 1975. Psychophysics. John Wiley. New York.

Contributions of Smell and Taste to the Pleasantness of Flavor[a]

MELVIN P. ENNS [b] AND DAVID E. HORNUNG [c]

[b]*Department of Psychology and*
[c]*Department of Biology*
St. Lawrence University
Canton, New York 13617

The purpose of this study was to examine how the pleasantness of an odorant and the pleasantness of a tastant interacted to produce the overall pleasantness of an odorant-tastant combination. Twenty college undergraduates were asked to use the method of absolute magnitude estimation to estimate the pleasantness of the smell of ethyl butyrate (0.01, 0.04, 0.16% vol/vol), the taste of sucrose (5.0, 10.0, 20.0% wt/vol), and the overall pleasantness of each concentration of the odorant presented with each concentration of the tastant. That is, numerical estimates were used to rate the pleasantness of each stimulus, with the magnitude of the numerical estimate reflecting the degree of like or dislike. Stimuli that were liked were given a positive sign, and those disliked a negative sign. Stimuli that were neither liked nor disliked were given a zero. The stimuli were presented to the subjects via the Two-Module Delivery System, a device that allows the concentration of the stimulus delivered to the external nares to be varied independently from the concentration delivered to the mouth.[1]

Each of the subjects was first asked to rate the pleasantness of the odorant. Each solution was presented twice for a total of six trials. Next, each subject was asked to rate the pleasantness of the tastant (six trials). The order of stimulus presentation and the rinse procedure were the same as previously described.[1] Finally, each subject was asked to estimate the overall pleasantness of the odorant-tastant combinations. Each odorant/tastant pair was presented once (nine trials). The subjects were first instructed to smell then taste the solution, then answer the question, What is the overall pleasantness of the solution? Subjects rinsed both nose and mouth after each trial.

The arithmetical mean of the two estimates of the odor and taste were used for the data analyses. For ethyl butyrate, the number of subjects who reported "liking" the smell was inversely related to the concentration (0.01%, $n = 8$; 0.04%, $n = 8$; 0.16%, $n = 2$). The same relationship was found for the taste of sucrose (5.0%, $n = 10$; 10.0%, $n = 9$; 20.0%, $n = 6$).

When both the smell and the taste were given a positive hedonic rating, the overall hedonic (flavor) rating was most often positive. Likewise, when both the smell and taste were given a negative rating, the flavor rating was most often negative (see TABLE 1).

When the subjects gave a positive rating to the smell of ethyl butyrate and a negative rating to the taste of sucrose, or vice versa, the sign of the rating with the

[a]Supported by a grant from the General Foods Corporation.

TABLE 1. Positive, Negative, and Neutral Ratings of Flavor Stimuli as a Function of the Hedonic Ratings of the Odorant and Tastant

Odorant/ Tastant			Flavor					
			+	(%)	−	(%)	0	(%)
+ / +	Responses[a]	(24)	18	(75%)	4	(17%)	2	(8%)
	Subjects[b]	(7)	5	(72%)	1	(14%)	1	(14%)
− / −	Responses	(68)	4	(6%)	62	(91%)	2	(3%)
	Subjects	(12)	1	(8%)	10	(84%)	1	(8%)

[a] Total number of responses across subjects.
[b] Total number of subjects with at least one response to odorant-tastant as indicated.

TABLE 2. Hedonic Ratings of Flavor Stimuli When the Odorant and Tastant Have Opposite Hedonic Ratings

A. Absolute Value of Tastant > Absolute Value of Odorant

				Flavor			
				+	(%)	−	(%)
Tastant	(+)	Responses[a]	(19)	19	(100%)	0	(0%)
		Subjects[b]	(5)	5	(100%)	0	(0%)
Tastant	(−)	Responses	(15)	3	(20%)	12	(80%)
		Subjects	(5)	1	(20)	4	(80%)
Tastant	(−)	Subjects	(5)	1	(20%)	4	(80%)

B. Absolute Value of Odorant > Absolute Value of Tastant

				Flavor			
				+	(%)	−	(%)
Odorant	(+)	Responses	(9)	5	(56%)	4	(44%)
		Subjects	(3)	2	(67$)	1	(33%)
Odorant	(−)	Responses	(15)	3	(20%)	12	(80%)
		Subjects	(8)	2	(25)	(6)	(75%)

C. Absolute Value of Odorant = Absolute Value of Tastant

			Flavor			
			+	(%)	−	(%)
Tastant (+) / Odorant (−)	Responses	(15)	4	(27%)	11	(73%)
	Subjects	(8)	2	(25%)	6	(75%)
Tastant (−) / Odorant (+)	Responses	(3)	0	(0%)	3	(100%)
	Subjects	(2)	0	(0%)	2	(100%)

[a] Total number of responses across subjects.
[b] Total number of subjects with at least one response to odorant and/or tastant as indicated.

largest absolute hedonic value usually determined the sign of the overall rating (see TABLE 2A & B). In those instances where the hedonic ratings of the odorant and tastant were subjectively equal but opposite in sign, the flavor stimulus was usually given a negative rating (see TABLE 2C). This latter result suggests that in determining the overall hedonic rating a negative rating of one of the sensory components usually results in a negative rating of the odorant-tastant combination.

REFERENCE

1. HORNUNG, D. E. & M. P. ENNS. 1984. Chem. Senses **9:** 97-106.

Immunofluorescent Studies of the Development of Rat Olfactory Epithelium[a]

A. I. FARBMAN AND V. McM. CARR

Department of Neurobiology and Physiology
Northwestern University
Evanston, Illinois 60208

J. I. MORGAN AND J. L. HEMPSTEAD

Roche Institute of Molecular Biology
Nutley, New Jersey 07110

We have used monoclonal antibodies as specific probes to study the phenotypic expression of molecules during development of the olfactory epithelium. Our rationale was to monitor the first appearance of olfactory epithelial cell antigens during development and, ultimately, to identify those molecules that play significant roles. Antibodies were taken from three sources: a panel of monoclonals made from homogenates of adult rat olfactory mucosa;[1] monoclonals made to crude membrane fractions of adult and neonatal rat olfactory mucosa; and monoclonals made to total homogenates of E13-E14 fetal rat snouts. Screening was done by immunofluorescence on cryostat sections of adult, neonatal, and fetal rat nasal cavities. Various antibodies were found that are specifically immunoreactive to the luminal surface of olfactory epithelium, olfactory supporting cells, Bowman's glands, the luminal surface of respiratory epithelium, and some to both respiratory and olfactory luminal surfaces. This report will be limited to those antibodies that are immunoreactive with olfactory receptor cells.

In adult rat olfactory epithelium, Neu-5 antibody was bound primarily to olfactory nerve bundles and only weakly to perikarya. This antibody was first demonstrable in fetal olfactory epithelium at day E13, when it was immunoreactive with olfactory receptor cell surfaces and axons. Similarly, Neu-9 antibody, immunoreactive with both olfactory somata and axons in adults, was first demonstrable at E13. A third antibody, 1A-6, is first immunoreactive with E13 receptor cells and axons, but its most intense reactivity, particularly in E16-E18 animals, is seen at the luminal edge of the olfactory epithelium. These three antibodies apparently bind to newly expressed molecules present at E13 in rat embryos, but not at E12.

In adult rats, Neu-4 binds to both perikarya and axons. The antigen binding to Neu-4 is not clearly demonstrable at E13 but is at E14. The staining pattern is similar to that of Neu-9.

[a]Supported in part by National Institutes of Health Grants #NS18490, NS06181, and NS23348.

In the accompanying graph (FIG. 1), we have summarized the information available from this and other immunohistochemical studies, and, along the X-axis, have correlated these findings with data available on other developmental events.[2-6]

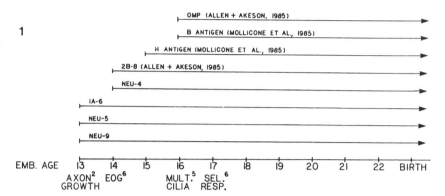

FIGURE 1. A summary diagram correlating the time in embryonic development when certain antigens make their appearance and the time when other significant cellular events occur. Cells are first immunoreactive to monoclonal antibodies 1A-6, Neu-5, and neu-9 on embryonic day 13, to Neu-4 and 2B-8 on E14; H antigen is first demonstrable on E15; OMP and B antigen appear on E16. Axonal growth from olfactory cells is first detectable on E13; the first demonstrable electro-olfactogram was shown on E14; multiple cilia first appear on receptor cells on E16; individual receptor cells first exhibit selective responsivity to odor stimuli on E17.

REFERENCES

1. HEMPSTEAD, J. L. & J. I. MORGAN. 1985. A panel of monoclonal antibodies to the rat olfactory epithelium. J. Neurosci. **5:** 438-449.
2. FARBMAN, A. I. & L. M. SQUINTO. 1985. Early development of olfactory cell axons. Dev. Brain Res. **19:** 205-213.
3. ALLEN, W. K. & R. AKESON. 1985. Identification of an olfactory receptor neuron subclass: Cellular and molecular analysis during development. Dev. Biol. **109:** 393-401.
4. MOLLICONE, R., J. TROJAN & R. ORIOL. 1985. Appearance of H and B antigens in primary sensory cells of the rat olfactory apparatus and inner ear. Dev. Brain Res. **17:** 275-279.
5. MENCO, B. PH. M. & A. I. FARBMAN. 1985. Genesis of cilia and microvilli of rat nasal epithelia during prenatal development. I. Olfactory epithelium, qualitative studies. J. Cell Sci. **78:** 283-310.
6. GESTELAND, R. C., R. A. YANCEY & A. I. FARBMAN. 1982. Development of olfactory receptor neuron selectivity in the rat fetus. Neuroscience **7:** 3127-3136.

Consistency of Preferences for Salt in Different Foods

RICHARD SHEPHERD AND CYNTHIA A. FARLEIGH

AFRC Institute of Food Research
Norwich, NR4 7UA, England

While there is evidence for preferences for salt concentrations in a particular food being predictive of normal salt intake of an individual,[1] the prediction is relatively weak. One reason for this may be that individuals do not have a consistent liking for high or low salt levels across different foods.

Thirty-two subjects (16 males and 16 females) took part in a study where they tasted four foods varying in salt content, and rated them on a 100-mm graphic relative-to-ideal rating scale. This was anchored at the left with the label "Not nearly salty enough," at the right with "Much too salty," and in the center with "Just right." The foods were tomato soup (34-1215 mg Na/100 g), bread (136-952 mg Na/100 g), boiled potato (8-1335 mg Na/100 g), and meat paté (73-1741 mg Na/100 g). Subjects took part in two sessions for each of the foods, with order of testing balanced.

Samples with ten concentrations of salt were available for each of the foods (six for the bread), but in order to minimize range bias the samples were presented in an order determined by the responses of the subjects.[2] The first sample presented was always the fifth concentration (third for the bread), and if the subject rated this below ideal then the next stimulus was of a higher concentration, whereas if it was rated above ideal then the next concentration was a lower one. The concentrations were subsequently presented to try to equalize the number of samples rated above and below ideal, and to equalize the average distance that these were rated from ideal. Ten samples were presented in all. From the plot of the ratings against log (concentration), the individual's most preferred concentration (ideal) of salt was calculated, for each of the foods, along with the slope, which gives a measure of how concerned he or she is about deviations from the ideal.

The mean ideal concentrations for the four foods were significantly different ($F(3,128)=135.7$, $p < 0.001$), as were the slopes of the functions $F(3,128)=11.7$, $p < 0.001$) (see FIG. 1). The subjects preferring a high concentration in one food also tended to prefer a higher concentration in the others; the correlations between the ideal concentrations for the foods are shown in TABLE 1. The correlations between the slopes of the function for the different foods were lower (see TABLE 1). Part of this may be accounted for by the lower reliability of this measure over the two sessions. The correlations between the determinations of the slope for each session varied between 0.31 (df = 30, NS) and 0.66 (df = 30, $p < 0.001$), whereas for the ideal concentration these correlations varied between 0.70 and 0.85 (df = 30, $p < 0.001$). This indicates a very reproducible estimate of the ideal concentration from only one test session.

These results demonstrate a consistent liking for high salt levels in some individuals across different types of foods. They also demonstrate the reproducibility of the estimate of ideal concentration from assessments in only one session.

TABLE 1. Correlations between the Estimates of Log (Ideal Concentration) and the Slope of the Function for the Four Foods ($n = 32$)

| | Correlations | |
Foods	Log (Ideal Concentration)	Slope of Function
Soup versus bread	0.66[a]	0.46[b]
Soup versus paté	0.78[a]	0.27
Soup versus potato	0.69[a]	0.41[c]
Bread versus paté	0.68[a]	0.43[c]
Bread versus potato	0.65[a]	0.20
Paté versus potato	0.60[a]	0.39[c]

[a] $p < 0.001$.
[b] $p < 0.01$.
[c] $p < 0.05$.

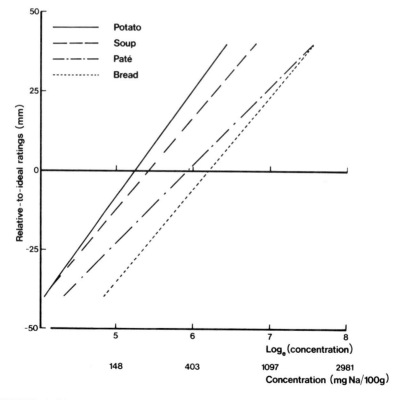

FIGURE 1. Mean regression lines fitted to relative-to-ideal ratings plotted against log(concentration) for the soup, bread, paté, and potato.

REFERENCES

1. SHEPHERD, R., C. A. FARLEIGH & D. G. LAND. 1984. The relationship between salt intake and preferences for different salt levels in soup. Appetite **5:** 281-290.
2. SHEPHERD, R., C. A. FARLEIGH & D. G. LAND. 1984. Effects of stimulus context on preference judgements for salt. Perception **13:** 739-742.

Human Taste Detection Thresholds

Each Subject's Threshold for a
Single Stimulus Is Unique[a]

A. FAURION, T. LARDIER, J. X. GUINARD,
AND B. NAUDIN

Laboratoire Neurobiologie Sensorielle, EPHE
91305 Massy, France

Taste thresholds were individually measured in human subjects. An up and down procedure[1] was applied to forced-choice pair presentations of stimuli. Subjects' responses were analyzed on line by a computer. Successive stimulus concentrations were automatically adjusted and delivered by the same microcomputer. Stable and reproducible thresholds were obtained under carefully controlled experimental conditions: Stimulus purity was checked by HPLC; retronasal olfaction was suppressed by a 200 l/min air stream flowing into the nostrils through silastic tubing inserted in a nonleaky plaster moulding of the subject's nose; paired solution temperature differences were lower than 0.1°C; all solutions were kept sterile; successive presentations of stimuli were made independent by intermingling at random several tests in the same session; subjects were trained for each stimulus until they produced stable thresholds. The mean of four to eight tests was used to assess, within half an hour, one subject's threshold for one stimulus. Each threshold was estimated at least twice after training. The reliability for each stimulus was tested across subjects by Pearson's r coefficients.

The individual detection thresholds of 58 subjects for 40 stimuli were collected in two matrices of, respectively, 29 subjects × 19 stimuli and 29 subjects × 21 stimuli. In both cases, and for each stimulus, the intra-individual variance was within a ratio of one to two, whereas the inter-individual variance was in a ratio of more than 1 to 10. All subjects' profiles were different as shown by hierarchical classifications (FIG. 1). Only two out of 40 intra-stimulus correlation coefficients were low, only 18 out of 381 inter-stimulus coefficients were equal to 0.6 or above (FIG. 2). Seventy-five percent of the data variance was represented in a seven-dimensional space and 90% in a 10-dimensional space. The partition between sweet and bitter stimuli was seen only on the first factorial plane (32% of information). The projection on the other dimensions showed that distances between stimuli of the same quality were nearly as large as distances between stimuli of different qualities. In conclusion, (1) each stimulus elicits a different and reproducible pattern; (2) a specific sensory image is recognized even at detection level since training for a given stimulus does not generalize to others; and (3) the bitter and sweet tastants constitute a continuum. These quantitative data fall in accordance with qualitative profile descriptions performed by the same subjects at supraliminar level. A few compounds were described as bitter by a few subjects,

[a]Supported by CVG-région Nord Picardie, ORSAN, and INSERM (CRE 856011).

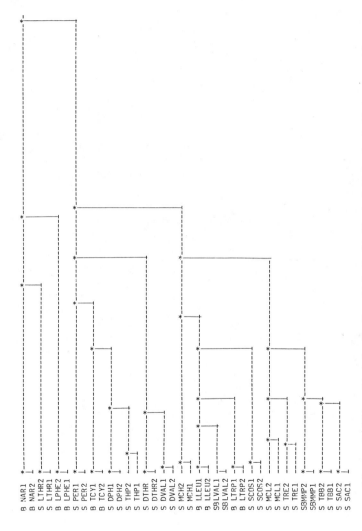

FIGURE 1. Dendrogram of the result of hierarchical classification obtained with Matrix I. Each stimulus is represented by two profiles (1 and 2) accounting for repeated tests. Notice the proximity of labels corresponding to the same replicated stimulus. Sweet and bitter stimuli are not partitioned in two groups; on the contrary, a continuum is observed. S: so-called sweet stimuli; B: bitter. LTHR, DTHR: L- and D-threonine; PERillartine; DPHEnylalanine; THP: trans hydroxyproline; D- and LVALine; MCH and MCL: two acesulfam derivatives; SCOS: Na cyclooctyl sulphamate; TREhalose; SACcharine; NARingin; D- and LPHEnylalanine; TCY: tetracycline; LLEUcine; LTRP: L-tryptophane; MMP: methyl mannopyranoside.

	SAC	DVAL	DPHE	DTHR	PER	LTHR	TBB	MCL	MCH	MMP	TRE	SCOS	THP	LLEU	TCY	NAR	LTRP	LVAL	LPHE
	.92	.90	.95	.95	.94	.90	.90	.80	.91	.85	.76	.90	.52	.86	.87	.83	.83	.86	.96
DVAL																			
DPHE	.51	.59																	
DTHR		.64	.59																
PER			.48	.48															
LTHR				.54															
TBB	.50		.65	.51															
MCL			.63	.48															
MCH							.68												
MMP			.57	.53	.61		.49	.52											
TRE				.71	.49		.46	.56	.48	.53									
SCOS					.47														
THP																			
LLEU	.43		.54					.46	.66	.46	.48	.61	.61						
TCY			.57	.60					.50	.46	.44	.46	.44						
NAR						.43	.43						.44		.46				
LTRP									.56	.53		.50							
LVAL				.60													.54		
LPHE										.58			.45				.56	.60	
	SAC	DVAL	DPHE	DTHR	PER	LTHR	TBB	MCL	MCH	MMP	TRE	SCOS	THP	LLEU	TCY	NAR	LTRP	LVAL	

FIGURE 2. Intra-stimulus and inter-stimulus correlation coefficients (Pearson's r) of matrix I. Coefficients below 0.4 do not appear.

as sweet by others, sweet and bitter by some, neither bitter nor sweet (although tasty) by others. The "sweet" and "bitter" semantic descriptors are not sufficient to describe the different qualities experienced by subjects. The licorice taste reported for glycyrrhizic acid was not qualitatively altered for any subject when *all* olfactory clues were suppressed. The interindividual qualitative description differences, the continuous space obtained with threshold data, and a previous work[2] allow us to ask a new question regarding whether some sweet and some bitter stimuli should not interact with sets of partially common receptor sites.

REFERENCES

1. DIXON, W. J. & MASSEY 1960. Introduction to Statistical Analysis: Sensitivity Experiments. MacGraw Hill. New York.
2. GENT, J. F. & L. M. BARTOSHUK. 1983. Sweetness of sucrose, neohesperidin dihydrochalcone, and saccharin is related to genetic ability to taste the bitter substance 6-*n*-propylthiouracil. Chem. Senses 7(3/4): 265-272.

Chemosensory Discrimination

Behavioral Abilities of the Spiny Lobster[a]

JACQUELINE B. FINE-LEVY, CHARLES D. DERBY,
AND PETER C. DANIEL

Department of Biology
Georgia State University
Atlanta, Georgia 30303

We are interested in the neural mechanisms by which organisms code the quality of chemical stimuli. We are using the Florida spiny lobster *(Panulirus argus)* to explore these mechanisms. This paper describes behavioral experiments that define the chemosensory discriminatory capabilities of spiny lobsters.

We have used an aversive conditioning paradigm that relies on the relative changes in responses to the conditioned chemical and to nonconditioned chemicals following a series of conditioning trials as an indicator of the similarity between the conditioned chemical and each of the nonconditioned chemicals.

All experiments consisted of three phases. During the preconditioning phase, each chemical at each concentration was presented to the animal during each test day with no reinforcement, thus providing baseline measurements of responsiveness to stimuli. For the conditioning phase, each chemical at each concentration was presented to the animal during each test day, with an aversive stimulus coupled to the designated conditioned stimulus (artificial crab mixture[1] for mixture experiments, taurine for single-chemical experiments). Postconditioning was identical to the preconditioning phase.

Two types of experiments were performed, one using chemical mixtures as stimuli, and the other using single chemicals. For the mixture experiment, two animals were conditioned to artificial crab mixture[1] at two concentrations, 0.5 mM and 0.05 mM. Three additional mixtures,[1] artificial mullet mixture, artificial oyster mixture, and artificial shrimp mixture, were also presented to the animals each test day at both concentrations, as well as artificial seawater, which served as a control. For the first single-chemical experiment, two animals were conditioned to taurine at a concentration of 1 mM. Glutamate at 1 mM and seawater were also presented to the animals each test day. For the second single-chemical experiment, two animals were conditioned to taurine at 1 mM and 0.1 mM, and taurine, AMP, glycine, betaine, and glutamate were presented each test day. For both types of experiments, the same aversive stimulus, a "pseudopredator," was employed. The pseudopredator consisted of a rapidly moving object that contrasted sharply with the substrate in each test tank, and thus provided visual as well as mechanical disruption of the animals' activities.

[a]Supported by National Institute of Neurological and Communicative Disorders and Stroke Grant No. NS22225 and a grant from the Whitehall Foundation.

Responses to chemical stimuli were quantified from the number of times certain behaviors were exhibited for three minutes after stimulus presentation. Only behaviors that showed a significant conditioning effect (i.e., a decrease in response due to aversive conditioning) were used in the statistical analysis. These behaviors included "search," "frantic search," and "grab at stimulator." Aversion to a chemical stimulus was calculated as the probit of $1 / (1 + \exp (-(\text{mean of preconditioning responses} - \text{postconditioning response})))$. Results of the chemical mixture experiment and the single-chemical experiments are presented in FIGURES 1 and 2, respectively.

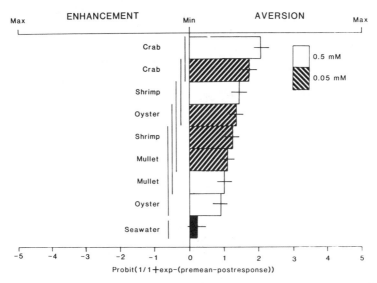

FIGURE 1. Behavioral discrimination among chemical mixtures by spiny lobsters. The ability of lobsters to discriminate between crab mixture (the conditioned mixture) and shrimp mixture, mullet mixture, or oyster mixture (nonconditioned mixtures) was determined by comparing the magnitudes of the conditioned aversions to these mixtures. The magnitude of each conditioned aversion was calculated as the probit of $1 / (1 + \exp (-(\text{mean of preconditioning responses} - \text{postconditioning response})))$. This measure gives a value between the limits of -5 and $+5$, with positive values representing conditioned aversion, negative values representing conditioned enhancement, and a value of zero representing no conditioning effect. Analysis of variance with GT2 or Tukey's multiple comparison tests ($\alpha = 0.05$) was used to compare the aversions, with vertical lines indicating aversions of similar magnitude. At 0.5 mM, the magnitudes of aversions are: crab > shrimp > mullet = oyster > seawater. At 0.05 mM, the magnitudes of the aversions are: crab > shrimp = mullet > seawater, but crab = oyster. Together, these results indicate that shrimp is perceived as being more similar to crab than is either mullet or oyster.

The results of these experiments indicate that lobsters are capable of associative learning, and that they can discriminate between certain mixtures as well as between certain single chemicals. The shrimp mixture is perceived as being more similar to crab than is either the mullet or oyster mixture (FIG. 1). These results are in concordance with the degree of similarity in the composition of the mixtures: Correlation analysis indicates that, compared to the composition of the crab mixture, shrimp is most similar, and mullet and oyster are significantly less similar to the same degree.

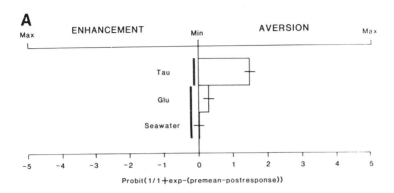

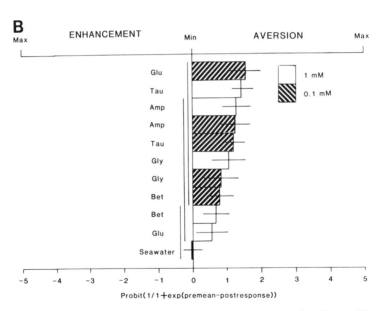

FIGURE 2. Behavioral discrimination among single compounds by spiny lobsters. The ability of lobsters to discriminate between taurine (the conditioned chemical) and glutamate (**A**), and between taurine (the conditioned chemical) and adenosine 5′-monophosphate, glycine, betaine, or glutamate (**B**), was determined using the same techniques as for the mixture experiments, as described in the legend to FIGURE 1. In (**A**), the magnitudes of the aversions, tested at 1 mM, are taurine > glutamate = seawater. Interestingly, the lobsters actively avoided taurine following conditioning. In (**B**), the magnitudes of the aversions at 1 mM are: taurine > betaine = glutamate = seawater, while taurine = AMP = glycine > seawater. This result confirms and extends the conclusion of (**A**). At 0.1 mM, there are no significant differences in the aversions to any of the chemicals, although all are different from that to seawater.

At 1 mM, glycine is perceived as being more similar to taurine than is either glutamate or betaine (FIG. 2). These results are in agreement with a neurophysiological study of the spiny lobster,[2] which showed that based on across-neuron correlations of responses of chemoreceptor cells and responses of chemosensory interneurons ascending the optic tract, glycine is more similar to taurine than is either glutamate or betaine.

REFERENCES

1. CARR, W. E. S. & C. D. DERBY. 1986. Behavioral chemoattractants for the shrimp, *Palaemonetes pugio:* Identification of active components in food extracts and evidence of synergistic mixture interactions. Chem. Senses **11:** 49-64.
2. DERBY, C. D. & B. W. ACHE. 1984. Quality coding of a complex odorant in an invertebrate. J. Neurophysiol. **51:** 906-924.

Immunoreactivity to Neuronal Growth-Dependent Membrane Glycoprotein Occurs in a Subset of Taste Receptor Cells in Rat Taste Buds[a]

THOMAS E. FINGER, HEMA SRIDHAR,
AND MARY WOMBLE

Department of Cellular & Structural Biology
University of Colorado Medical School
Denver, Colorado 80262

VAR L. ST. JEOR AND JOHN C. KINNAMON

Department of Molecular, Cellular, and Developmental Biology
University of Colorado
Boulder, Colorado 80309

The monoclonal antibody 5B4[1] recognizes a membrane glycoprotein similar to neuronal cell adhesion molecules (N-CAM), but which is expressed preferentially in growing neuronal systems. This monoclonal antibody reacts little with proliferative neuroblasts or with the mature nervous system except for the glomeruli of the olfactory bulb, which contain growing olfactory nerve elements. Since taste buds comprise neuron-like epithelial cells that undergo continuing histogenesis and alteration of neuronal connectivity, the 5B4 antibody was used to analyze this system. We hypothesized that the continuous rearrangement of nerve fibers at the base of the taste bud would result in immunoreactivity to the 5B4 antibody.

For immunocytochemistry, rats or mice were fixed by means of perfusion with a solution of 2% paraformaldehyde and 0.05% glutaraldehyde in 0.1 M phosphate buffer (pH 7.2) to which was added an additional 15% vol/vol of saturated picric acid solution. The tongue and olfactory bulbs were removed and placed in fresh fixative an additional three hours. The tissue then was washed in buffer and cryoprotected in sucrose-glycerin. After sectioning at 40-80 μm on a sliding freezing microtome, the tissue was treated for 30 minutes in a solution of 1% sodium borohydride in buffer and rinsed in buffer. The tissue then was placed in 1% normal goat serum

[a] Supported by National Institutes of Health Grants NS00772 and NS20486.

[b] Address for correspondence: Dr. Thomas E. Finger, Dept. Cellular & Structural Biology, B-111, University of Colorado Medical School, 4200 E. 9th Ave., Denver, CO 80262.

in phosphate buffer for one hour at room temperature followed by overnight exposure at 4°C to a 1 : 10 dilution of antibody 5B4 in phosphate buffer. Following several buffer rinses, the antibody was localized with standard peroxidase-antiperoxidase (PAP) or avidin-biotin complex (ABC) immunocytochemistry. Control sections were exposed to normal mouse serum or to other monoclonal antibodies (directed against enkephalin or substance P) in place of the 5B4 antibody. None of the control preparations exhibited the positive immunostaining described below.

Contrary to our initial hypothesis, no immunoreactive nerve fibers were seen. However, taste receptor cells within taste buds from all three lingual taste fields exhibit immunostaining with antibody 5B4. Virtually all taste buds contain some immunoreactive taste cells (see FIG. 1). Different cells within a single taste bud, however, exhibit different degrees of staining. The entire surface of each positive taste cell

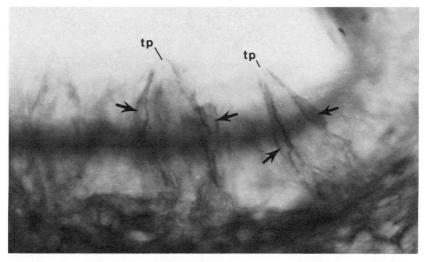

FIGURE 1. Light micrograph showing 5B4 immunoreactivity in the circumvallate papilla of a rat. Several immunoreactive taste cells are visible (arrows). The dark horizontal band is the out-of-focus edge of this thick section and does not represent immunostaining. *tp*, taste pore. Original magnification 720 ×; reduced by 25%.

appears to be immunoreactive, from near the apical pore to the basal region. With electron microscopy, the 5B4 immunoreactivity appears clearly membrane associated (FIG. 2), surrounding a relatively nonreactive cytoplasm. The majority of reactive taste cells are intermediate-type cells as identified by the heterochromatin content and slight invaginations of the nucleus.

In summary, our results demonstrate a subset of taste cells that are immunoreactive to an antibody that hitherto has exhibited reactivity only to neurons, and then principally to those in the process of adding neuronal membrane. These results may indicate that taste cells derive from a neuronal precursor cell line such as neural crest or placodal proliferative cells rather than from indifferent epithelium. Further, the results may indicate that some taste cells are actively adding or altering membrane, perhaps in response to taste stimulation or normal aging.

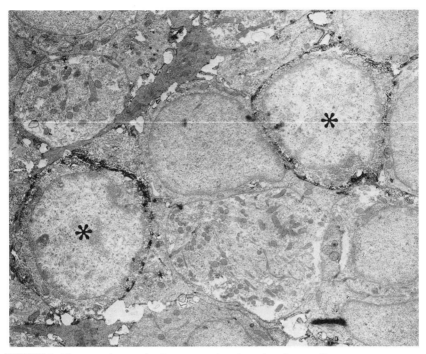

FIGURE 2. Electron micrograph of a cross section through a circumvallate taste bud in a rat. The 5B4 immunoreactivity shows as an electron density associated with the membrane of two cells (asterisks) in this micrograph. Both cells appear to be intermediate cells. Original magnification 7,500 ×; reduced by 25%.

REFERENCES

1. ELLIS, L., I. WALLIS, E. ABREU & K. H. PFENNINGER. 1985. Nerve growth cones isolated from fetal rat brain. IV. Preparation of a membrane subfraction and identification of a membrane glycoprotein expressed on sprouting neurons. J. Cell Biol. **101:** 1977-1989.
2. WALLIS, I., L. ELLIS, K. SUH & K. H. PFENNINGER. 1985. Immunolocalization of a neuronal growth-dependent membrane glycoprotein. J. Cell. Biol. **101:** 1990-1998.

Electrophysiological Basis of the Response of Olfactory Receptors to Odorant and Current Stimuli[a]

STUART FIRESTEIN AND FRANK WERBLIN

Graduate Group in Neurobiology
University of California
Berkeley, California 94720

The response properties of individual olfactory receptor cells have been analyzed using the whole-cell patch-clamp technique.[1] Cells were isolated from the nasal epithelium of the tiger salamander (*Ambystoma tigrinum*) by enzymatic treatment and gentle mechanical disruption.

In response to puffs of an odorant solution, cells responded as in FIGURE 1A. The membrane depolarized slowly to spike threshold near -50 mV when a single large action potential was generated. This was followed by one to six attenuated spikes that rode upon a sustained depolarization. Within 500 msec of the first spike, the membrane activity consisted only of oscillations about the depolarized plateau.

Responses to injected current, a more easily standardized stimulus, showed the same characteristics (FIG. 1B). Threshold could be attained with input currents as small as 3 pA. This is equivalent to a conductance of 90 pS or the opening of perhaps as few as three standard cation channels.

We were surprised to find responses that attenuated so rapidly and that were not graded with stimulus intensity. Nonetheless we believe that this response pattern can be adequately accounted for by an analysis of the gated currents underlying response production. We have previously described the gated currents of the cell soma.[2,3] A summary of these currents appears in FIG. 2. In addition to the three ionic currents shown, the outward $K+$ current can be further separated into three components: a Ca^{2+}-dependent [IK(Ca)], a voltage-dependent [IK(V)], and an inactivating (Ia) current.

As is commonly the case, the Na^+ and K^+ currents interact to create the action potential. The transient Na^+ current is responsible for the rapid depolarizing upswing of the action potential; inactivation of this current and activation of the K^+ currents serves to repolarize the cell. A substantial portion of the outward current (nearly one-third) is due to an inactivating conductance of the Ia type. This conductance inactivates rapidly (tau = 100 msec) at depolarized levels but recovers from inactivation much more slowly (tau = 5 sec) at hyperpolarized potentials.

This discrepancy in the time course of inactivation and recovery results in an accumulation of inactivation of the Ia current after each spike. Thus less and less outward current is available to repolarize the cell and less of the Na^+ is removed from inactivation; spike height is progressively reduced until no further action potentials can be generated.

[a] Supported in part by National Institutes of Health Grant #T32-GM07048.

Thus we are able to account for this surprising pattern of activity by the interaction of gated currents in the membrane that serve to rapidly attenuate action potential firing. This suggests that olfactory receptor neurons may be acting primarily as detectors and do not, in general, encode stimulus intensity information. Intensity discrimination may be left for a more central level of processing, perhaps at the first synapse where a single mitral cell integrates inputs from as many as 1,000 receptor cells.

Considered as detectors, these cells are exquisitely sensitive. A relatively positive resting potential near -65 mV, a high-input resistance (greater than 3 Gohms), and

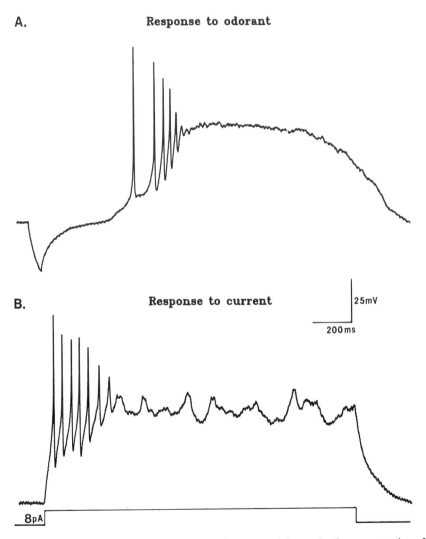

FIGURE 1. (A) Response to puff of odorant mixture containing equimolar concentration of amyl acetate, citral, and cinneole to a total concentration of 1 mM. Stimulus delivered at artifact. (B) Response to 8-pA step of depolarizing current.

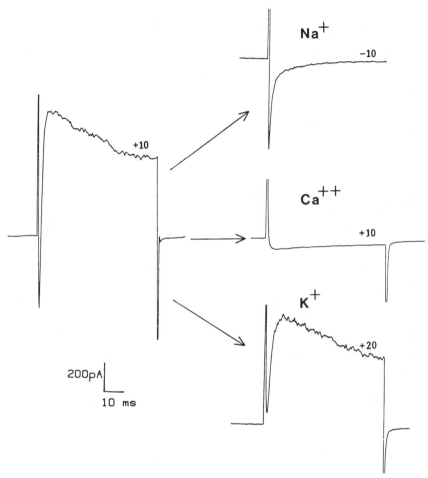

FIGURE 2. Summary of gated currents. Left-hand trace is composite response to a depolarizing voltage step to +10 mV. Capacitive artifacts mark beginning and end of record. On the right are the separate ionic conductances, as labeled. Number associated with each trace is the command potential.

a simple geometry all contribute to a low response threshold. We have shown that they are able to reach firing threshold with the opening of as few as three membrane channels.

REFERENCES

1. HAMILL, O. P., A. MARTY, E. NEHER, B. SAKMAAN & F. J. SIGWORTH. 1981. Pflugers Arch. Physiol **391:** 85-100.
2. FIRESTEIN, S. & F. S. WERBLIN. 1985. Soc. Neurosci. Abstr. **11,** 2: 970 (#286.3).
3. FIRESTEIN, S. & F. S. WERBLIN. 1987. Proc. Natl. Acad. Sci. USA, in press.

The Suppressed Response of NaCl following Amiloride

A Halogen-Specific Effect[a]

BRADLEY K. FORMAKER[b] AND DAVID L. HILL

Department of Psychology
The University of Toledo
Toledo, Ohio 43606

Previous investigations have shown that lingual application of the sodium transport blocker, amiloride, partially suppresses the whole-nerve response to NaCl.[1,2] It has been suggested that the chorda tympani response to NaCl is jointly composed of amiloride-sensitive and amiloride-insensitive sodium components.[3] However, a behavioral investigation from our laboratory suggests that the residual taste response to NaCl following amiloride treatment may be related to the chloride ion, rather than an amiloride-insensitive sodium component.[4]

To further study the residual response of NaCl after amiloride, we recorded multifiber chorda tympani responses to a concentration series (0.05 M, 0.1 M, 0.25 M & 0.5 M) of NaCl, NaBr, sodium acetate (NaAc), NaHCO$_3$, choline chloride (ChCl), NH$_4$Cl and ammonium acetate (NH$_4$Ac). Recordings were made before and after the lingual application of 500 μM amiloride hydrochloride. A response was defined as an increase in neural activity above baseline 20 seconds after stimulus onset. Since amiloride did not affect the whole-nerve response to NH$_4$Cl, all responses were expressed relative to 0.5 M NH$_4$Cl. In addition, 0.25 M ChCl and 0.25 M NaCl were used in a cross-adaptation procedure; both solutions were mixed in amiloride.

Responses to NaCl and NaBr were suppressed 65-90% by amiloride (FIG. 1). In contrast, responses to NaAc and NaHCO$_3$ were suppressed 95-100% (FIG. 2). Thus, responses to sodium salts containing a *halogen* anion were not completely suppressed by amiloride, while responses to sodium salts lacking a halogen anion were completely suppressed. Moreover, the responses of ChCl were equivalent to the residual responses of NaCl and NaBr following amiloride (FIG. 1) and ChCl cross-adapted with the residual NaCl response. Responses to NH$_4$Cl and NH$_4$Ac were unaffected by amiloride.

These results suggest that the residual response of NaCl following amiloride may be related to a halogen-sensitive transduction mechanism rather than an amiloride-insensitive sodium channel. The complete suppression of NaAc and NaHCO$_3$ by amiloride cannot be predicted by a model that postulates an amiloride-insensitive

[a]Supported by National Institutes of Health Grants NS20538 and RCDA Award NS00964 to DLH.

[b]Address for correspondence: Bradley K. Formaker, Department of Psychology, Gilmer Hall, University of Virginia, Charlottesville, VA 29903-2477.

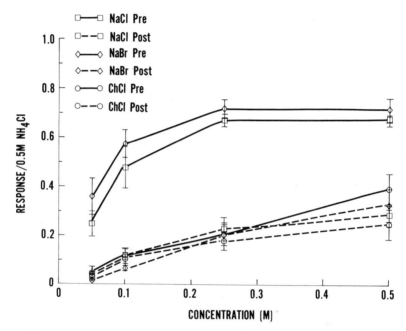

FIGURE 1. The average whole-nerve response to NaCl, NaBr, and ChCl before and after lingual application of 500 μM amiloride hydrochloride. Responses are expressed relative to 0.5 M NH$_4$Cl. Reliable differences ($p < 0.05$) between pre- and postamiloride treatments (solid and dashed lines, respectively) exist only for NaCl (X) and NaBr (diamond). ChCl (circle) responses before and after amiloride are similar to the sodium salt responses after amiloride. Standard errors are shown for each mean.

sodium transduction channel. Moreover, the similarity in neural response between ChCl and NaCl after amiloride, along with the cross-adaptation results, indicate that the residual response of NaCl following amiloride is related to the halogen anion. If this postulation is correct, then it can be predicted that mixing ChCl with either NaAc or $NaHCO_3$ will result in the same response pattern as that seen for NaCl. Preliminary results support this prediction.

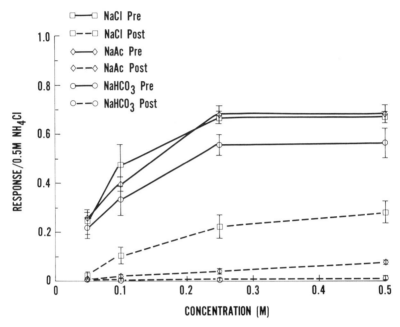

FIGURE 2. The average whole-nerve response to NaCl, NaAc, and $NaHCO_3$ before and after lingual application of 500 μM amiloride hydrochloride. Responses are expressed relative to 0.5 M NH_4Cl. All preamiloride responses are similar. All differences between pre- and postamiloride treatments are reliable ($p < 0.05$). Moreover, the NaAc and $NaHCO_3$ postamiloride responses are reliably less than the postamiloride NaCl responses at concentrations greater than 0.1 M. Standard errors are shown for each mean.

REFERENCES

1. BRAND, J. G., J. H. TEETER & W. L. SILVER. 1985. Brain Res. **334:** 207-214.
2. HILL, D. L. & T. C. BOUR. 1985. Dev. Brain Res. **20:** 310-313.
3. DESIMONE, J. A. & F. FERRELL. 1985. Am. J. Physiol. **249**(Regulatory Integrative Comp. Physiol. 18): R52-R61.
4. FORMAKER, B. K., K. S. WHITE & D. L. HILL. 1985. Soc. Neurosci. Abstr. **11:** 1222.

Analysis of Tastes in Mixtures by Hamsters[a]

MARION E. FRANK, SHILPA J. PATEL, AND
THOMAS P. HETTINGER

Department of Biostructure and Function
University of Connecticut Health Center
Farmington, Connecticut 06032

Chemical stimuli encountered by mammals in nature are often mixtures containing taste, olfactory, and other chemosensory features. Studies of the detection of chemosensory features within complex stimuli and the features' effects on ingestion suggest interactive processing of mixtures. Quantitative interactions, such as synergism and suppression[1] and qualitative interactions or syntheses,[2] have been seen in responses to mixtures of chemicals with distinct qualitative taste features.

Innate preference-aversions for mixtures of chemosensory stimuli, individually preferred or rejected over water by more than 20%, were measured for hamsters. In general, mixtures of preferred (0.5 M sucrose) and rejected (0.1 M NaCl, 0.001 M quinine hydrochloride, 0.01 M citric acid, 0.001 M dithiothreitol, half-saturated menthol, 0.005 M amyl acetate, 0.001% capsaicin) solutions were less palatable than the preferred but more palatable than the rejected solution alone. However, palatabilities did not add. Mixtures of rejected and preferred solutions were more palatable than expected. For example, volatile skunky dithiothreitol mixed with sucrose was as palatable as sucrose itself. The palatability of nonvolatile sulfurous 0.01 M Na mercaptoethane sulfonate (NaMES), strongly preferred itself (FIG. 1), was not increased by adding sucrose to it, nor were mixtures of two strongly rejected substances necessarily less palatable than the components, a result suggesting palatability is saturable.

Food aversions were established in hamsters to component 0.1 M sucrose, 0.1 M NaCl, 0.01 M HCl, and 0.001 M quinine solutions, each of which has distinct taste features, and solutions of their six binary mixtures. A learned aversion to a component generalized to mixtures containing the component but was reduced in strength, which suggested mixture suppression, since weaker aversions are shown to weaker stimuli (FIG. 2).[3] An aversion to a mixture generalized without reduction to either component; however, a reduced aversion was seen to mixtures containing only one of the two components, suggesting that presence of a novel component affected ingestion. An aversion learned to the innately preferred NaMES generalized (FIG. 2) to stimuli with three distinct chemosensory features: salty NaCl, skunky dithiothreitol (and sulfurous cysteine), and sweet sucrose. The significant generalizations of the NaMES aversion to sucrose and dithiothreitol indicate that NaMES has several chemosensory features and that the strong preference shown for NaMES, comparable to the strong preference shown for a sucrose-dithiothreitol mixture, may be an example of synergism. Since mixture components were identified, syntheses of unique qualitative prop-

[a]Supported by National Science Foundation Grant BNS 8519638.

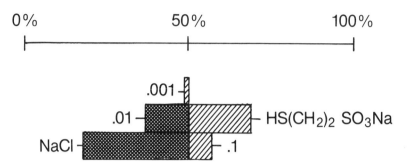

FIGURE 1. Percent preference-aversion (48 h, 2 bottle) for NaCl and Na mercaptoethane sulfonate (NaMES) solutions for hamsters. Means (n = 10-44) are shown by bars, standard errors by lines extending from bars. NaCl is increasingly rejected from 0.01 to 0.1 M and NaMES is more highly preferred at 0.01 than 0.1 M (t-tests, 2-tail, $p < 0.05$). Since NaMES is completely ionized, the reduced preference for 0.1 M NaMES may be due to the strong rejection of its Na taste component.

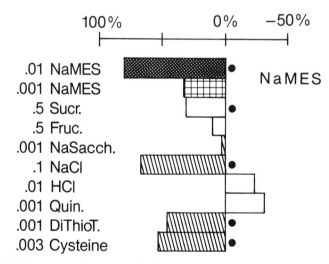

FIGURE 2. The pattern of cross-generalization of an aversion learned to 0.01 M Na mercaptoethane sulfonate (NaMES) for hamsters. The unconditional stimulus was an intraperitoneal injection of 0.15 M LiCl. The test stimuli, against which the generalizations of the aversion were measured, are listed to the left of the bars indicating percent suppression of drinking[3] due to conditioning. A dot to the right of a bar denotes statistical significance of the suppression (t-test, $p < 0.05$). The aversions to 0.01 M NaMES, 0.1 M NaCl, 0.001 M dithiothreitol, and 0.003 M cysteine are not significantly different from each other, but they are larger than the aversion to 0.5 M sucrose.

erties were not indicated; thus, perceptions of component stimuli may be quantitatively affected in mixtures, but qualitative features appear unchanged.

REFERENCES

1. BARTOSHUK, L. M. 1985. Taste mixtures: An analysis of synthesis. *In* Taste, Olfaction and the Central Nervous System. D. W. Pfaff, Ed.: 210-232. The Rockefeller University Press. New York.
2. ERICKSON, R. P. 1985. Definitions: A Matter of Taste. *In* Taste, Olfaction and the Central Nervous System. D. W. Pfaff, Ed.: 129-150. The Rockefeller University Press. New York.
3. NOWLIS, G. H., M. E. FRANK & C. PFAFFMANN. 1980. Specificity of acquired aversions to taste qualities in hamsters and rats. J. Comp. Physiol. Psychol. **94:** 932-942.

Mixture Integration in Heterogeneous Taste Quality Mixtures

An Assessment of Subadditivity for Total Mixture Intensity

ROBERT A. FRANK AND GARY ARCHAMBO

Department of Psychology
University of Cincinnati
Cincinnati, Ohio 45221

Using Anderson's[1] information integration approach, we recently reported that total intensity judgments for sucrose-sodium chloride mixtures showed increasing subadditivity as the solute concentrations increased.[2] (Subadditivity refers to the tendency for total mixture intensity to be rated as less than the sum of the unmixed component intensities.) In this series of experiments, the first experiment replicated and extended earlier work with sucrose-sodium chloride solutions to sucrose-citric acid and sucrose-quinine mixtures. Subjects rated the total intensity of the mixtures on a 21-point category scale using factorial combinations of the components. All stimuli were rated once per session using the sip-and-spit method (stimulus volume = 5.0 ml). A tap water rinse and 30-second intertrial interval followed the presentation of each stimulus.

The pattern of integration for the three types of solutions was essentially identical across mixture types, exhibiting increasing subadditivity as the concentrations of the solutes increased (see FIG. 1). The same pattern of integration was observed when magnitude estimates rather than category ratings were used, indicating that ceiling effects were not responsible for the extreme subadditivity noted at the higher solute concentration levels.

In a second set of experiments, subjects rated the individual sweet, salty, and sour components of the sucrose-sodium chloride and sucrose-citric acid mixtures. In this way, the contribution of mixture suppression[3] to the subadditivity of the solutions was assessed. As was expected, mixture suppression was observed for the component tastes of the solutions, but the magnitude of this effect accounted for only a fraction of the subadditivity observed for total intensity judgments.

In a final series of experiments, the role of psychophysical compression[4] in the subadditivity of the mixtures was assessed. This was accomplished by adding a constant amount of solute to increasing concentrations of the same substance and having subjects rate the intensity of the resulting solutions. This procedure was used for sucrose, sodium chloride, and citric acid solutions. It was found that the patterns of integration exhibited subadditivity as solute concentrations increased when either category scaling

296

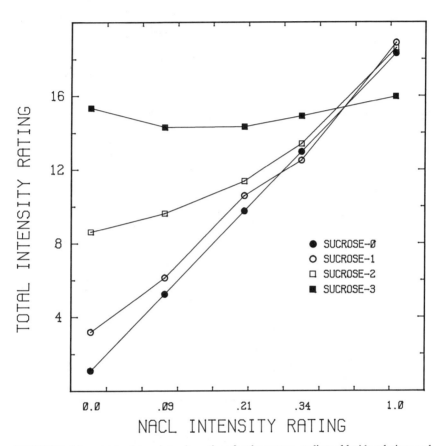

FIGURE 1. Mean total mixture intensity ratings for the sucrose-sodium chloride solutions and the unmixed components. The sucrose-0 curve represents the intensity ratings of the unmixed sodium chloride solutions plotted against themselves. The sucrose-1, -2, and -3 curves are for the sodium chloride components combined with 0.1, 0.3, and 1.0 M sucrose, respectively. The numbers along the abscissa refer to the molar concentrations of the sodium chloride stimuli.

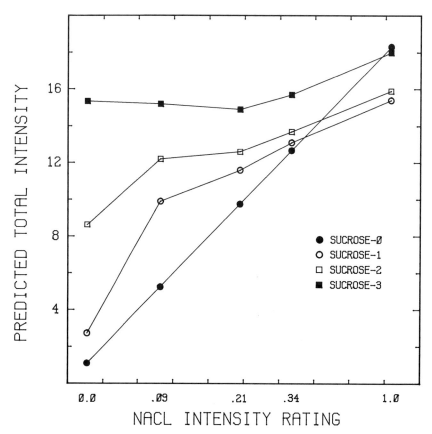

FIGURE 2. Suppression and compression corrected additive model of the sucrose-sodium chloride mixtures. The predicted values were generated by adjusting the ratings of the unmixed components for mixture suppression and psychophysical compression. The data are plotted as described in FIGURE 1.

or magnitude estimation was used. When a simple additive model of mixture intensity was corrected for both mixture suppression and psychophysical compression, reasonably accurate predictions of the observed ratings of the heterogeneous mixtures could be made (see FIG. 2).

Future studies will use the information integration approach to explore the role of suppression and compression in homogeneous taste quality mixtures, for example, mixtures of sweeteners.

REFERENCES

1. ANDERSON, N. H. 1981. Foundations of Information Integration Theory. Academic Press. New York.
2. FRANK, R. A., D. BURKE & J. ESTEP. 1985. Information integration in sucrose/sodium chloride mixtures: Global differences for intensity and hedonic judgments. Seventh Annual Meeting of the Association for Chemoreception Sciences. Sarasota, FL.
3. KAMEN, J. M., F. L. PILGRIM, N. J. GUTMAN & B. J. KROLL. 1961. Interaction of suprathreshold taste stimuli. J. Exp. Psychol. 62: 348-356.
4. BARTOSHUK, L. M. 1975. Taste mixtures: Is mixture suppression related to compression? Physiol. Behav. 14: 643-649.

Differential Effects of Cooling on the Intensity of Taste[a]

SANDRA P. FRANKMANN AND BARRY G. GREEN

Monell Chemical Senses Center
Philadelphia, Pennsylvania 19104

Although there have been numerous psychophysical studies of the effect of solution temperature on taste intensity, inconsistencies in the results obtained from different laboratories have yielded a confusing picture of temperature-taste interactions. Clear and convincing effects of temperature at suprathreshold concentrations have been reported only for sucrose[1,2] and quinine.[3] In both cases, intensity was found to decline with decreasing temperature, but only when solution temperature was quite cold, or when rinsing conditions would have favored significant cooling of the tongue. It was hypothesized, therefore, that the temperature of the tongue is the critical variable in temperature-taste effects, and that solution temperature itself is important only to the extent it changes tongue temperature.

The first experiment evaluated the effect of cooling both the tongue and the solution on the perceived intensity of five concentrations (quarter log units) of citric acid (0.001-0.01 M), sodium chloride (0.032-0.032 M), caffeine (0.0032-0.032 M), and sucrose (0.056-0.56 M) solutions. Using magnitude estimation, subjects ($n = 12$) were asked to evaluate taste intensity (one taste stimulus per session) at 20°, 28°, and 36°C. Mouth temperature was maintained at the same temperature as the stimulus by having the subject rinse repeatedly with cold (ca. 5°C) or warm (ca. 38°C) water as mouth temperature was monitored with a microthermocouple.

The results showed the perceived intensity of the sourness of citric acid and the saltiness of sodium chloride were unchanged across tongue temperatures of 20°, 28°, and 36°C (FIG. 1). However, the perceived intensity of the sweetness of sucrose was significantly lower at 20° than at 28° or 36°C ($F = 15.66$ (2,22), $p < .05$; FIG. 1). Although a similar trend appeared for caffeine, the reduction in perceived bitterness produced by cooling did not reach statistical significance. The data nevertheless implied the temperature sensitivity of the sensory mechanisms serving citric acid and sodium chloride are different from those for sucrose and perhaps for caffeine.

The second experiment evaluated the prediction that it is tongue temperature and not stimulus temperature that is more important for influencing taste intensity. To achieve this, the temperature of the subject's tongue was held at either 20° or 36°C while he or she was presented with a series of caffeine or sucrose solutions at 20° or at 36°C.

The results showed the bitterness of caffeine ($F = 21.26$ (3,33), $p < 0.01$) and the sweetness of sucrose ($F = 8.01$ (3,33), $p < 0.01$) were less intense when the mouth was cooled than when it was warm, regardless of solution temperature (FIG.

[a]Supported by the Dairy Research Foundation and a grant from the National Institutes of Health (NS20577).

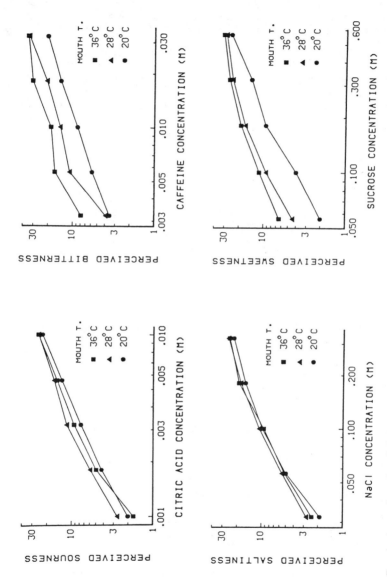

FIGURE 1. Perceived taste intensity of citric acid, NaCl, caffeine, and sucrose when both the mouth and the solution temperature are at 20°, 28°, or 36°C.

2). These data suggest tongue temperature is indeed the controlling variable in temperature-taste effects.

In summary, for caffeine and sucrose the intensity of bitterness or sweetness is reduced when tongue temperature is lowered to 20°C, and this effect is relatively independent of solution temperature. On the other hand, the intensity of the sourness of citric acid and the saltiness of sodium chloride are unaffected by the temperature of either the tongue or the solution. Differences in temperature sensitivity may reflect

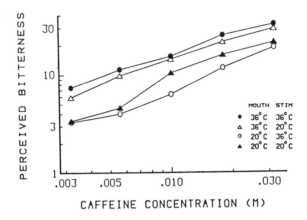

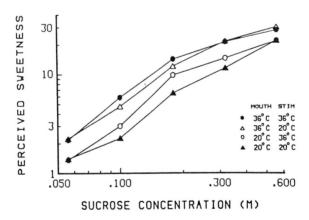

FIGURE 2. Perceived taste intensity of 20° or 36°C solutions of caffeine and sucrose when the mouth temperature is 20° or 36°C.

differences in the receptor mechanisms and/or the neural transduction processes for different taste stimuli. In addition, because for taste mixtures the contribution to flavor made by some of the taste components will vary with temperature, complex flavors will likely be perceived differently at different oral temperatures. Studies are planned to discover the extent to which such effects occur.

REFERENCES

1. BARTOSHUK, L. M. 1982. Effects of temperature on the perceived sweetness of sucrose. Physiol. Behav. **28:** 905-910.
2. CALVINO, A. M. 1986. Perception of sweetness: The effects of concentration and temperature. Physiol. Behav. **36:** 1021-1028.
3. MCBURNEY, D. H., V. B. COLLINGS & L. M. GLANZ. 1973. Temperature dependence of human taste responses. Physiol. Behav. **11:** 89-94.

How the Olfactory System Generates its Chaotic Background "Spontaneous" Electroencephalographic and Unit Activity[a]

WALTER J. FREEMAN

Department of Physiology-Anatomy
University of California, Berkeley
Berkeley, California 94720

The basal electroencephalograms (EEGs) of the olfactory bulb (OB), anterior olfactory nucleus (AON), and prepyriform cortex (PC) fluctuate continually and erratically. Their autocorrelation functions approach zero with time, showing their unpredictability. Their power spectra are broad with multiple low peaks. Their amplitude histograms are nearly Gaussian. Trains of units from single cells show interval histograms in the form of a Poisson distribution with a refractory period. Yet the activity is spatially coherent and has the low dimension of chaos.

A model in nonlinear differential equations of each structure (FIG. 1), that suffices to simulate averaged evoked potentials (AEPs), poststimulus time histograms, and odor-induced bursts fails to generate chaotic activity. A model consisting of OB, AON, and PC models coupled with both positive and negative feedback suffices. Two requirements are for distributed delays corresponding to the conduction velocities and distances of the lateral (LOT) and medial (MOT) olfactory tracts, and an excitatory bias of periglomerular neurons onto mitral-tufted cells that is subject to AON control. While each structure in isolation is stable, when coupled they drift into chaos. The chaotic activity is augmented by AON feedback to the periglomerular cells and quenched by AON and PC feedback to the bulbar granule cells. The instability is manifested in the steady chaotic unit and EEG activity of all three structures, subject to centrifugal modulation in respect to motivation (FIG. 2).

This conclusion is supported by the effects on the OB and PC EEGs and AEPs of partial or complete surgical section of the bulbar stalk, either sparing the LOT, the MOT, or neither.[1,2] Local anesthesia or cryogenic block of the stalk have essentially the same effects[3] of suppression of EEG activity and reduction of the oscillations of AEPs. The pars externa and pars lateralis of the AON have a relatively large influence on the PC AEPs.[2] This is consistent with the anatomical pattern of axonal projection from the AON to the glomerular layer of the OB.[4]

Chaotic basal activity in the central olfactory system has an obvious biological role of engaging all its neurons in continuous unstructured and unpatterned "noise"

[a] Supported by National Institute of Mental Health Grant MH06686.

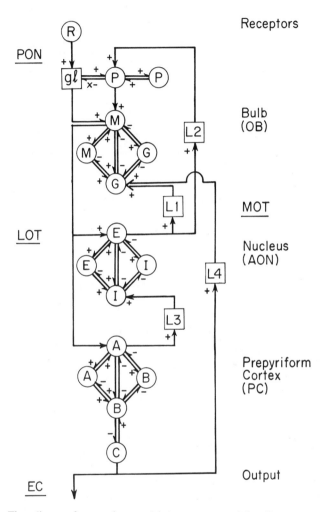

FIGURE 1. Flow diagram for equations modeled on anatomy of the olfactory system. L1-L4 are delays. Details are in Freeman.[6]

as a prevention of the atrophy of disuse. Its role in odor identification appears twofold.[5] It enables rapid and unbiased access to the basins of learned limit cycle attractors that constitute the memory bank of the olfactory system, and it enables the escape from the known repertoire into chaos in order to allow the formation by synaptic modification of a new attractor for a novel odor presented under reinforcement.

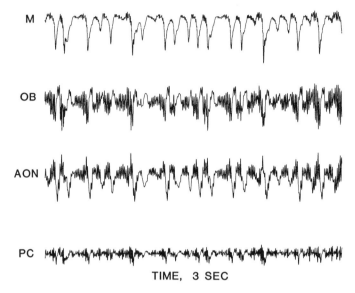

FIGURE 2. Outputs from four representative elements of a model of the olfactory system operating at a chaotic attractor. M: initial cell output (continuous pulse density); OB: granule cell (G) output (bulbar EEG); AON: nuclear EEG; PC: prepyriform EEG.

REFERENCES

1. FREEMAN, W. J. 1968. Effects of surgical isolation and tetanization of prepyriform cortex in cats. J. Neurophysiol. **31:** 349-357.
2. HIGHSTONE, H. H. 1968. Anterior olfactory nucleus and forebrain evoked potentials. Thesis in Physiology. University of California. Berkeley, CA. 145 pp.
3. GRAY, C. M., W. J. FREEMAN & J. E. SKINNER. 1986. Associative changes in the spatial amplitude patterns of rabbit olfactory EEG are norepinephrine dependent. Behav. Neurosci. **100:** 585-596.
4. LUSKIN, M. B. & J. L. PRICE. 1983. The topographic organization of associative fibers of the olfactory system in the rat, including centrifugal fibers to the olfactory bulb. J. Comp. Neurol. **216:** 264-291.
5. FREEMAN, W. J. & C. A. SKARDA. 1985. Spatial EEG patterns, non-linear dynamics and perception: The neo-Sherringtonian view. Brain Res. Rev. **10:** 147-185.
6. FREEMAN, W. J. 1986. Petit mal seizure spikes in olfactory bulb and cortex caused by runaway inhibition after exhaustion of excitation. Brain Res. Rev. **11:** 259-284.

Development of Taste Buds in the Chick Embryo (*Gallus gallus domesticus*)[a]

J. R. GANCHROW AND D. GANCHROW

Department of Oral Biology
The Hebrew University-Hadassah Faculty of Dental Medicine
Jerusalem, Israel

Department of Anatomy and Anthropology
Sackler Faculty of Medicine
Tel Aviv University
Tel Aviv, Israel

Chickens and rats share a 21-day period *in ovo* and *in utero,* respectively. Rats are altricial at birth and while the first immature taste bud appears at 20 days *in utero,* most bud development progresses postnatally[1] prior to weaning. In contrast, the present investigation demonstrates that in the precocial chick, dramatic taste bud development occurs during the last trimester *in ovo.*

Oral epithelium containing taste buds associated with the anterior mandibular glands was serially sectioned (10 μm) and stained with hematoxylin and eosin. Buds were unilaterally counted in 27 Anak (broiler breed) chickens at 16, 17, 18, 19, and 20 days of incubation (E16-E20), on the day of hatching (H1) and 50-60 days thereafter (H50-60).

While no buds were seen at E16, spheroidal collections of developing bud cells were observed in the basal epithelial regions in proximity to gland duct openings at E17. Fine tubules were observed intragemmally already at E17 and E18. By E19 these spheroidal clusters had begun to elongate and some displayed distinct taste pore openings. The ensuing two days were characterized by expanding epithelium as the cells therein continued to multiply such that by H1 the taste bud had a distinctly ovoidal shape and several cell types comprising the bud were easily distinguishable.

The number of taste buds continued to increase from E17, reaching a peak at E19 and an increasing amount of buds opened into the oral cavity at E20 and H1. FIGURE 1 demonstrates the dramatic increase in number of taste buds during E17-E19 as related to salivary duct openings. Since the absolute number of taste buds and gland openings varied from chick to chick, the more stable measure of number of buds per salivary gland duct opening was chosen. It may be seen in FIGURE 1 that after E19, the number of buds per anterior mandibular gland duct opening remained fairly constant at about 2.5.

It is concluded that although rats and chickens share the same gestational time span, the degree of taste receptor development just before birth (hatching) probably reflects demands that the immediate postnatal environment will place upon the chicks.

[a]This research was supported by Grant #3226/84 from the United States-Israel Binational Science Foundation (BSF), Jerusalem, Israel.

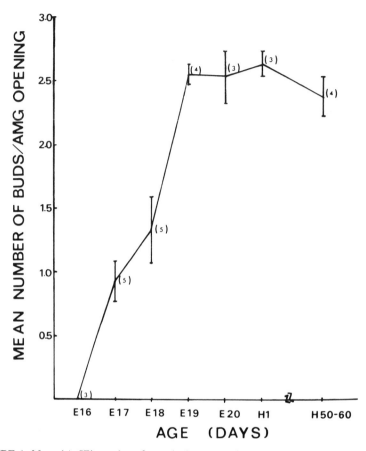

FIGURE 1. Mean (± SE) number of taste buds per anterior mandibular salivary gland duct opening as a function of age. The number of cases sampled at each age is in parentheses.

REFERENCES

1. MISTRETTA, C. M. 1972. Topographical and histological study of the developing rat tongue, palate, and taste buds. *In* Third Symposium on Oral Sensation and Perception: The Mouth of the Infant. J. F. Bosma, Ed.: 163-187. Charles C. Thomas. Springfield, IL.

Taste Signal Strengths of Salt and Water Stimuli Provide a Model for Sensory Difference Tests, Involving Adaptation, Learning, and Variation in Supra- and Subadapting Sensitivity

L. GOLDSTEIN, N. ODBERT, U. FARELLE, AND
M. O'MAHONY [a]

Department of Food Science and Technology
University of California
Davis, California 95616

Sensory difference tests like the triangle, duo-trio, and paired comparison tests are reported to vary in their sensitivity. O'Mahony and Odbert[1] developed a system called Sequential Sensitivity Analysis (SSA) as a predictive scheme to account for these differences in test discriminability. The scheme was first developed for a model system (3 mM NaCl versus purified water) and then generalized to food systems.

SSA considers the tests in terms of the sequences of tasting involved to perform the test. There are four possible orders in which salt and water stimuli can be tasted: salt tasted after water (W-S), water tasted after salt (S-W), water after water (W-W), and salt after salt (S-S). Considering a triangle test, there are six possible orders of tasting: (1) W-W-S, (2) W-S-W, (3) S-W-W, (4) S-S-W, (5) S-W-S, and (6) W-S-S.

The first three orders involve one salt stimulus and are comprised of two W-S sequences, two S-W, and two W-W. The second three involve two salt stimuli and comprise two W-S sequences, two S-W, and two S-S. It is evident that the sequences in the two triangular designs are different. The mean discriminability of the second stimulus in these pairs was measured using an R-index measure[1-3] for a sample of 13 subjects. It was found to be, in decreasing order of strength: W-S, S-W, W-W, and S-S. It is a simple matter to deduce that the first triangular design has a set of better discriminabilities and should thus provide a more sensitive test; this was confirmed for the same 13 subjects. The same result was noted for alternative forms of duo-trio test.

To investigate why the taste sequences should have this order of measured discriminability, the following experiments were performed. For 19 subjects, the same mean order was confirmed. The subjects were then trained, and it was found that the

---

[a] Address for correspondence: Professor Michael O'Mahony, Department of Food Science & Technology, University of California, Davis, California 95616.

order changed as they began to learn the range of sensations associated with a given stimulus. From the NaCl concentrations of the expectorated stimulus and saliva, it was apparent that the stimulus concentration was altered by secreted saliva and by residual stimulus from prior tastings. For example, the W-S sequence involved not a simple increase in NaCl concentration of $+3$ mM, but a range of increases in concentration from $+1.4$ to $+4.1$ mM. Changes in oral environment caused by secreted saliva and stimulus residuals would alter receptor sensitivity by adaption. It would seem that subjects initially had to learn a range of sensations, mediated by variations in physical signal strength and by adaptation, and associate them with the corresponding stimulus sequence. Inexperienced subjects would make more errors initially with the S-S sequence because they would expect it to have a salty taste. To a lesser extent, errors would be made in identifying the W-W sequence when it elicited supra- or subadapting tastes, if subjects expected water to be tasteless. Learning the sensations associated with these sequences would also be more difficult because they could be both supra- and subadapting, whereas W-S and S-W were either one or the other.

Further, using a stimulus flow delivery technique, it was found that subjects were more sensitive to supra- than subadapting stimuli; the former producing higher R-index values. This would explain why the W-S sequence was more discriminable than the S-W sequence.

Thus, the order of discriminability of the four possible sequences involved in tasting salt and water involves variation in physical signal strengths caused by secreted saliva and residual stimuli from prior tastings, stimulus learning, adaptation, and variations in supra- and subadapting signal detectability. These form the basis for explaining the differences in discriminability between the various sensory difference tests.

REFERENCES

1. O'MAHONY, M. & N. ODBERT. 1985. A comparison of sensory difference testing procedures: Sequential sensitivity analysis and aspects of taste adaptation. J. Food Sci. **50:** 1055-1058.
2. O'MAHONY, M. Short-cut signal detection measures for sensory analysis. 1979. J. Food Sci. **44:** 302-303.
3. O'MAHONY, M. 1986. Sensory difference and preference testing: The use of signal detection measures. *In* Applied Sensory Analysis of Foods. H. Moskowitz, Ed. C.R.C. Press. Boca Raton, FL.
4. O'MAHONY, M. 1979. Salt taste adaptation: The psychophysical effects of adapting solutions and residual stimuli from prior tastings on the taste of sodium chloride. Perception **8:** 441-476.

A Comparative Study of the Neurophysiological Response Characteristics of Olfactory Receptor Neurons in Two Species of Noctuid Moths[a]

ALAN J. GRANT AND ROBERT J. O'CONNELL

The Worcester Foundation for Experimental Biology
Shrewsbury, Massachusetts 01545

ABNER M. HAMMOND, JR.

Department of Entomology
Louisiana Agricultural Experiment Station
Louisiana State University Agricultural Center
Baton Rouge, Louisiana 70803

The maintenance of reproductive isolation among sympatric species of insects is thought to depend, in part, on the specific composition and release of pheromones from each insect. To investigate how such pheromones are differentially processed by the olfactory system, thereby providing the sensory input required for this isolation process, neurophysiological responses were recorded from single-antennal sensilla on the male soybean looper moth to stimulation with known amounts of the individual components of its pheromone blend and the individual components of the blend produced by a sympatric species, *Trichoplusia ni* (Hübner), the cabbage looper moth. These two Noctuid moths, both in the subfamily Plusiinae, have overlapping geographic and seasonal distributions. Additionally, they have similar temporal patterns of activity and share some of the same host plants.[1] Females of both insects release complex blends of 12–14 carbon esters that are used to attract males for mating. Each species releases (*Z*)-7,dodecen-1-ol acetate (Z-7,12:Ac) as the major component of their respective blend; however, the composition of the remaining components differ.

In common with the chemosensory system in the cabbage looper,[2] the soybean looper possesses two classes of morphologically distinct antennal sensilla, each containing two chemosensitive olfactory receptor neurons. In both species, one class of sensilla (HS) contains a receptor neuron sensitive to Z-7,12:Ac. The neurophysiological characteristics of the receptor neurons in this class of sensilla, including their unstimulated spontaneous activity, sensitivity to pheromone, and response spectra to other

[a]Supported by the Alden Trust and by Grant NS 14453 from the National Institute of Neurological and Communicative Disorders and Stroke.

compounds, appear identical in both species. The second class of sensilla (LS) in each species contains two receptor neurons, one of which is sensitive to one of the minor pheromone components. However, the compound found to be effective is not the same in the two species; the cabbage looper LS receptor neurons respond preferentially to Z-7,14:Ac (see Mayer and Mankin[3]) and the soybean looper to Z-5,12:Ac (FIG. 1). The existence of a receptor neuron sensitive to Z-5,12:Ac in the male soybean looper moth is interesting because this compound is thought to be absent from the pheromone bouquet released by conspecific females. To test the possibility that Z-5,12:Ac is actually present in the glands of female soybean loopers at concentrations below the detection limits of current analytical procedures, gland extracts were made from females of each species and tested against the LS sensilla in male soybean loopers (FIG. 2). A strong response was observed to stimulation with the female cabbage looper gland extract. This result was expected because of the presence of Z-5,12:Ac

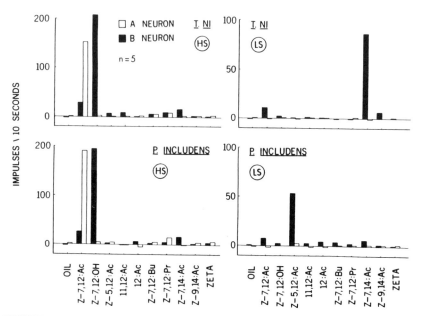

FIGURE 1. Histograms illustrating the responses from two classes of sensilla on the antennae of both male *T. ni* and *P. includens* to stimulations with fixed doses of ten behaviorally important compounds. Response indices, expressed as impulses/10 seconds, were calculated from 20 seconds of neural activity including 10 seconds of prestimulus activity, 2 seconds of stimulus activity, and 8 seconds of poststimulus activity. Response values represent the increment observed between the 10 seconds of prestimulus activity and the subsequent 10 seconds of activity. Stimulus cartridges were prepared by pipetting 1 μl of the desired pheromone, diluted in mineral oil (0.01 μg/μl), onto filter paper held in a glass cartridge (internal volume = 0.25 cc). Delivery of the stimulus was accomplished by passing filtered air (60 ml/min) through the cartridge for the two-second stimulus period. Each of the four sets of histograms represent the mean responses from five sensilla on five different antennae. A single set of 11 stimulus cartridges was used to establish all the responses illustrated for both species. (ZETA = (Z,E)-9,12,14:Ac.)

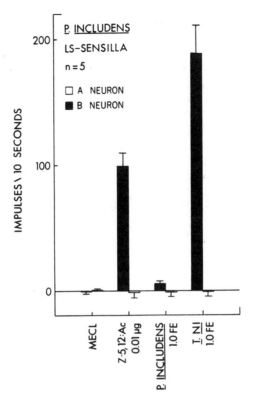

FIGURE 2. Histograms illustrating the responses from 5 LS sensilla on the antennae of male soybean looper moths to stimulation with methylene chloride control (MECL), Z-5,12Ac, and gland extracts from *T. ni* females and *P. includens* females. Female glands were removed and placed in chilled methylene chloride at a concentration of one female equivalent (FE)/3 μl. Cartridge preparation and stimulation were the same as described in FIGURE 1. Z-5,12:Ac (0.01 μg) was diluted in mineral oil and was presented to confirm the identity of the sensilla class. Bars indicate one standard error of the mean.

in its pheromone bouquet.[4] However, responses were not obtained after stimulation with female soybean looper gland extracts, even at intensity levels that are comparable to one female equivalent (FE), suggesting that this material (Z-5,12:Ac) is absent.

Mitchell[5] reported that trap catches of soybean looper males using either Z-7,12:Ac or calling female soybean loopers could be drastically reduced if calling female cabbage loopers were added to the lures. The presence of an olfactory receptor neuron on the antenna of male soybean loopers that is very sensitive to Z-5,12:Ac, a component absent in its own gland yet present in the effluvia of the female cabbage looper, suggests a mechanism whereby such discrimination between conspecific individuals and members of sympatric species might occur. Differences in the response characteristics of the receptor neurons in this second class of sensilla are thus thought to play a role in the reproductive isolation that exists between these two species of moths.

REFERENCES

1. LEPPLA, N. C. 1983. Environ. Entomol. **12:** 1760-1765.
2. O'CONNELL, R. J., A. J. GRANT, M. S. MAYER & R. W. MANKIN. 1983. Science **220:** 1408-1410.
3. MAYER, M. S. & R. W. MANKIN. 1987. Ann. N.Y. Acad. Sci. This volume.
4. BJOSTED, L. B., C. E. LINN, J.-W. DU & W. L. ROELOFS. 1984. J. Chem. Ecol. **10:** 1309-1323.
5. MITCHELL, E. R. 1972. Environ. Entomol. **1:** 444-446.

The Sensitivity of the Tongue to Ethanol

BARRY G. GREEN

Monell Chemical Senses Center
Philadelphia, Pennsylvania 19104

The purpose of this study was to quantify the sensitivity of the tongue to ethyl alcohol in terms of two psychophysical measures: perceived intensity of irritation and the latency to sensation onset. Two locations on the tongue (the dorsal tip and an area 3 cm posterior to the tip on the midline of the dorsal surface) were tested to learn if sensitivity to this common chemical irritant is greater at the tip than it is on the dorsum of the tongue. Such a spatial gradient was considered likely because it occurs for other somatosensory modalities (touch, temperature, and heat pain). Six concentrations of ethanol (35, 45, 55, 65, 75, and 85%) in solution with deionized water were presented to the tongue on saturated disks of filter paper (0.38 cm² in area). Subjects were instructed to respond with a button press when they felt the tactile sensation produced by the filter paper and again when they felt a sensation of irritation

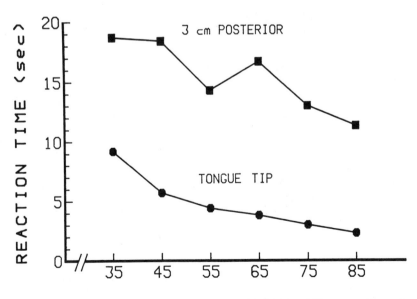

FIGURE 1

315

at the same site. The latency between the presses (timed via computer) was taken as the reaction time (RT) to ethanol-induced irritation. Magnitude estimates of perceived irritation were obtained 10 seconds later. Hence although the latency to sensation onset varied across concentrations and lingual locations, perceived intensity was always judged 10 seconds after irritation was detected.

The results are shown in FIGURES 1 and 2. Perceived irritation (FIG. 1) varied substantially between the tongue tip and the tongue's medial-dorsal surface. The 35% solution generated sensations four times more intense on the former area than on the latter area, and although the difference diminished with rising concentration, the 85% solution still induced sensations over two times stronger on the tip than on the dorsum. Consistent with the intensity data, latencies (FIG. 2) were much longer on the dorsum

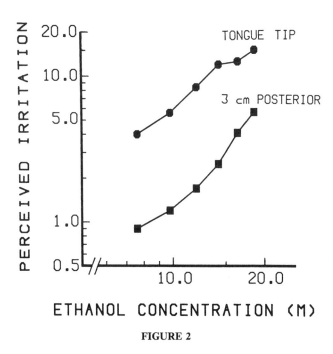

FIGURE 2

than on the tip, never falling below about 12 seconds even for the 85% solution. RTs on the tongue tip, though shorter than on the dorsum, were slow compared to typical latencies for taste sensations.[1,2] Ethanol RTs were never briefer than about two seconds. The latter finding implies gustatory information may precede common chemical information during ingestion of stimuli containing both ethanol and other sapid substances.

Experiments are continuing to determine the cause of the spatial variation in RT. The evidence accumulated so far indicates the more superficial location of free nerve endings in the epithelium of the anterior tongue[3] and the presence of somatosensory fibers in fungiform papillae[4] may combine to produce the superior sensitivity of the tongue tip to the irritation produced by ethanol.

REFERENCES

1. YAMAMOTO, T., T. KATO, R. MATSUO, M. ARAIE, S. AZUMA & Y. KAWAMURA. 1982. Gustatory reaction time under variable stimulus parameters in human adults. Physiol. Behav. **29:** 79-84.
2. KELLING, S. T. & B. P. HALPERN. 1983. Taste flashes: Reaction times, intensity, and quality. Science **219:** 412-414.
3. DIXON, A. D. 1962. The position, incidence and origin of sensory nerve terminations in oral mucous membrane. Arch. Oral Biol. **7:** 39-48.
4. FARBMAN, A. I. & G. HELLEKANT. 1978. Quantitative analyses of fiber population in rat chorda tympani nerves and fungiform papillae. Am. J. Anat. **153:** 509-521.

Conjugate Internalization of Apposed Dendritic Membranes During Synaptic Reorganization in the Olfactory Bulbs of Adult PCD Mice[a]

CHARLES A. GREER

Sections of Neurosurgery & Neuroanatomy
Yale University School of Medicine
New Haven, Connecticut 06510

During early development of the nervous system, the conjugate internalization (CI) of apposed neural plasmalemma membranes has been observed.[1,2] These appear as clathrin-coated pits that include a spinule of membrane and cytoplasm from an adjacent neuronal process.[2] The spinule of the evaginating process is pinched off along with the coated pit, forming double-walled coated vesicles. This event is distinct from the ubiquitous endocytosis of extracellular material. It is developmentally linked and includes the uptake of adjacent membranes and cytoplasm. Alternative hypotheses have been suggested regarding the function of CI including an inductive role in synaptogenesis,[1] or remodeling of immature membranes and the removal of proto-synaptic intercellular adhesion sites.[2] To investigate these hypotheses further, the morphology and incidence of CIs was examined in the olfactory bulb (OB) of the adult mutant mouse PCD. Between four to six months postnatally PCD mice lose a principal neuronal population from the OB, mitral cells.[3] As a consequence there is a period of reactive dendrodentritic synaptogenesis between tufted cells (TCs) and denervated granule cells (GCs), and extensive reorganization of the reciprocal dendrodentritic microcircuits in the external plexiform layer (EPL)[4] takes place.

Five-month postnatal PCD mice, homozygous recessive mutants and heterozygous littermate controls, and 6-12-day postnatal rats were perfused with aldehydes for conventional EM or prepared for Golgi-EM analyses of identified neurons. Micrographs of serial sections from the EPL were used to assess the morphological features and incidence of dendritic CIs.

The typical CI appeared as a coated pit composed of double plasmalemma membranes separated by 10-20 nm (FIG. 1). They were goblet shaped and extended a maximum of 250 nm into the endocytotic dendrite. They measured approximately 200 nm in diameter at the widest point and narrowed to 30-60 nm at the neck. It was not uncommon to recognize a vesicle within the evaginating spinule. Of particular interest, the typical CI was polarized from the GC gemmule into the TC dendrite and frequently occurred in close proximity to synaptic specializations. The comple-

[a]Supported in part by Grant NS19430 from the National Institute for Neurological and Communicative Disorders and Stroke and a C. Oshe Award from the Department of Surgery, Yale University School of Medicine.

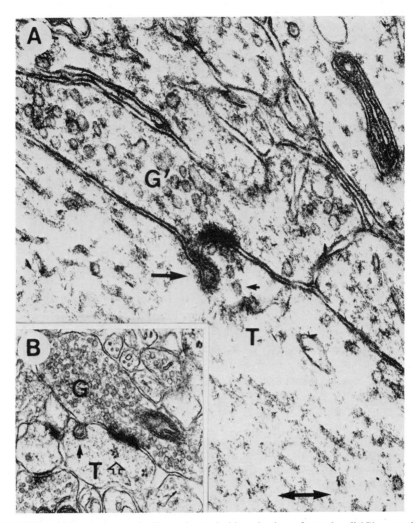

FIGURE 1. Electronmicrographs illustrating typical invaginations of granule cell (G) gemmule membranes into tufted cell dendrites (T) in mutant PCD mice. In (**A**) the large arrow denotes a clear double-walled invagination surrounded by the fuzzy clathrin coat. Vesicles (small arrow) associated with the immediately adjacent asymmetrical synapse from the tufted cell dendrite to the granule cell gemmule are seen clustered along the invagination. In (**B**) a second example of a double-walled invagination (solid arrow) is seen close to an apparent symmetrical synapse from the gemmule onto the tufted cell dendrite. Note the clusters of polyribosomes (open arrow) adjacent to the postsynaptic specialization. Such clusters have previously been reported to occur adjacent to newly developed synaptic specializations. Calibration bar = 100 nm (A); and 200 nm (B).

mentary transfer of material from the TC dendrite to the GC gemmule was not clearly documented in the PCD mice. The incidence of CIs increased significantly in the PCD mice relative to controls (FIG. 2). Moreover, the increase was proportionately greater in the deep sublamina of the external plexiform layer where synaptic reorganization is most evident following mitral cell loss.

These results demonstrate that normally quiescent developmental events can be reactivated in the adult. Also, the results suggest that CI plays a central role in

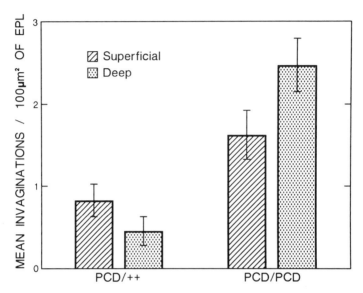

FIGURE 2. Mean (± SEM) conjugate dendritic membrane invaginations/100 μm^2 in the superficial and deep external plexiform layer (EPL) of mutant PCD mice (PCD/PCD) and control heterozygous littermates (PCD/ + +).

synaptogensis rather than simply in the remodeling of immature membranes. It is possible that CI contributes to the remodeling and *stabilization* of synaptic membrane specializations as previously discussed in regard to the early development of exuberant neuronal connections.[6]

REFERENCES

1. ALTMAN, J. 1971. Coated vesicles and synaptogenesis. A developmental study in the cerebellar cortex of the rat. Brain Res. **30:** 311-322.
2. ECKENHOFF, M. F. & J. J. PYSH. 1983. Conjugate internalization of apposed plasma membranes in mouse olfactory bulb during postnatal development. Dev. Brain Res. **6:** 201-207.
3. GREER, C. A. & G. M. SHEPHERD. 1982. Mitral cell degeneration and sensory function in the neurological mutant mouse PCD. Brain Res. **235:** 156-161.
4. GREER, C. A., N. HALASZ & G. M. SHEPHERD. 1983. Local circuit organization in the olfactory bulb following the loss of mitral cells. Int. Union Physiol. Sci. Abstr. **15:** 447.
5. CHANGEUX, J. P. & A. DANCHIN. 1976. The selective stabilization of developing synapses: A plausible mechanism for the specification of neuronal networks. Nature **264:** 705-712.

Taste Dysfunction in Burning Mouth Syndrome[a]

M. GRUSHKA, B. J. SESSLE, AND T. P. HOWLEY

Faculty of Dentistry
University of Toronto
Toronto, Ontario M5G 1G6

BMS is a poorly characterized pain disorder that appears to affect primarily post-menopausal women. In spite of its unclear cause, a neurological or psychogenic component is often inferred. Because a dysgeusic taste is a frequent symptom of BMS[1,2] studies were initiated to determine if any objective evidence could be found of taste dysfunction at threshold and suprathreshold levels of the four taste qualities in 47 BMS subjects (mean age $\pm$ SD., 55.2 $\pm$ 10.4 years); 27 age- and sex-matched normal subjects (52.8 $\pm$ 8.4 years) served as controls. The range of concentrations were, in molar units: 96.8×10^{-3} to 3.7×10^{-6} for sucrose and sodium chloride; 17.6×10^{-4} to 3.7×10^{-7} for citric acid, and 1.7×10^{-4} to 4.6×10^{-9} for quinine hydrochloride.

No statistically significant ($p > 0.05$) differences between BMS and control subjects were found in thresholds for salt, sour or bitter; however, thresholds for sweet were significantly higher for BMS than control subjects (TABLE 1, statistical analyses carried out on mean log molar units). When the BMS subjects were divided into those subjects who were experiencing a dysgeusic taste ("taste" group, approximately 75% of BMS subjects tested at threshold) and those without a dysgeusic taste ("no taste" group, approximately 25% of subjects) and compared with controls, significant differences ($p < 0.01$) were again found only for sweet. Post hoc comparisons, however, indicated no significant differences between any pair of mean values.

At suprathreshold concentrations, significant differences were not found between the BMS and control groups for salt and bitter. Perception intensity was, however, significantly higher for the BMS than control subjects for sweet and for sour (TABLE 2); post hoc analyses indicated these differences occurred at lower suprathreshold concentrations. After reclassification of the BMS subjects into the "taste" (60% of BMS subjects at suprathreshold level) and "no taste" groups (40% of subjects) significant differences were again found for sweet and sour, which post hoc analyses indicated originated at only lower suprathreshold concentrations and only between the "taste" and "no taste" groups and between the "taste" and control groups. For salt, estimations of taste intensity were also rated consistently (but not significantly) higher for the "taste" than the "no taste" and the control groups. No similar trend was noted for bitter.

These data provide evidence for taste dysfunction in BMS, especially in those individuals with self-reported dysgeusia, and support a neurological cause of BMS.

—

[a]Supported by a grant to M. Grushka from the Gerontology Research Council of Ontario.

TABLE 1. Taste Detection Thresholds (Mean ± Standard Deviation) for BMS and Control Subjects

Taste Stimulus[a]	BMS			Control		
	Arithmetic Means		Geometric Means Molar	Arithmetic Means		Geometric Means Molar
	Log Molar	Molar		Log Molar	Molar	
Sweet[b]	-2.224 ± 0.88	$15.52 \pm 14.72 \times 10^{-3}$	5.97×10^{-3}	-3.546 ± 1.50	$6.31 \pm 10.00 \times 10^{-3}$	2.84×10^{-4}
Salt	-3.164 ± 1.14	$3.74 \pm 5.32 \times 10^{-3}$	6.85×10^{-4}	-3.608 ± 1.26	$3.31 \pm 5.42 \times 10^{-3}$	2.47×10^{-4}
Sour	-4.648 ± 1.12	$1.58 \pm 3.00 \times 10^{-4}$	2.25×10^{-5}	-4.702 ± 1.03	$1.44 \pm 3.83 \times 10^{-4}$	1.98×10^{-5}
Bitter	-6.039 ± 1.15	$6.98 \pm 11.00 \times 10^{-6}$	9.14×10^{-7}	-6.451 ± 1.39	$1.10 \pm 3.00 \times 10^{-5}$	3.54×10^{-7}

[a] BMS: $n = 19$ for sweet and salt; $n = 18$ for sour and bitter. Controls: $n = 17$.
[b] $p < 0.01$ for log molar taste detection threshold between BMS and control groups.

TABLE 2. Repeated Measures Analysis of Variance with One Grouping Factor (BMS and Control Subjects) and One Repeated Measures Factor (Concentrations) of the Four Taste Modalities

Source	Sweet			Salt			Sour			Bitter		
	SS[a]	df	MS	F	SS	df	MS	F	SS	df	MS	F	SS	df	MS	F
Group	0.94	1	0.94	0.57	3.02	1	3.02	2.58	1.62	1	1.62	1.95	0.02	1	0.02	0.02
Concentration	58.26	7	8.32	75.74[b]	91.13	7	13.02	116.93[b]	152.31	9	16.92	166.10[b]	83.68	8	10.46	89.90[b]
Group × concentration	2.63	7	0.38	3.42[c]	1.26	7	0.18	1.61	1.94	9	0.22	2.12[d]	0.16	8	0.02	0.18
Error	116.45	71	1.64	—	82.08	70	1.17	—	53.18	64	0.83	—	73.38	61	1.20	—

[a] SS, sum of squares; df, degrees of freedom; MS, mean square; F, F value.
[b] $p < 0.001$.
[c] $p < 0.01$.
[d] $p < 0.05$.

REFERENCES

1. ZISKIN, D. E. & R. MOULTON. 1946. Glossodynia: A study of idiopathic orolingual pain. J. Am. Dent. Assoc. **33:** 1423-1432.
2. BASKER, R. M, D. W. STURDEE & J. C. DAVENPORT. 1978. Patients with burning mouth. Br. Dent. J. **145:** 9-16.

Beta-Adrenergic Control of von Ebner's Glands in the Rat[a]

SUAT GURKAN AND ROBERT M. BRADLEY

Department of Oral Biology
School of Dentistry
University of Michigan
Ann Arbor, Michigan 48109

Von Ebner's lingual salivary glands secrete through multiple ducts into the clefts of the circumvallate and foliate papillae. At the microscopic level, gland acini are filled with secretory granules.[1] Since 80% of the lingual taste buds are contained in the foliate and circumvallate papillae in the rat,[2] and since taste receptor sites on the microvilli are bathed in saliva,[3] these glands provide a microenvironment important in taste transduction in posterior tongue taste buds.

Little is known concerning the neural control of von Ebner's glands. Recently we have shown that the source of their parasympathetic control is via cells in the inferior salivatory nucleus in the medulla.[4] We now have applied the morphometric methodology used by Getchell and Getchell[5] in a study of Bowman's glands to characterize the extent of β-adrenergic control of the von Ebner's gland acini.

METHODS

Rats were starved overnight to cause accumulation of secretory granules in gland acini. At 8:30 A.M. five groups of 10 rats each were injected intraperitoneally with the β-adrenergic agonist isoproterenol (IPR) dissolved in saline at doses of 7.5, 15, 20, 30, and 60 mg/kg. Control rats were injected with saline. An additional group was injected with 30 mg/kg IPR and 42 mg/kg DL-propranolol hydrochloride (PROP), a β-adrenergic antagonist. Two hours later the rats were sacrificed, the tongue removed, and a standard portion of the von Ebner's glands dissected. The gland tissue was fixed by overnight immersion in cold, buffered, 4% glutaraldehyde and 1% paraformaldehyde, post-fixed with 1% osmium tetroxide, embedded in Epon, sectioned at 1 μm and stained with 1% toluidine blue. Five representative acini were randomly selected from each rat for photomicroscopy. Measurements of acinar and secretory granule area were made from 4" $\times$ 5" black and white prints using computerized planimetry and the percentage of the total acinar area occupied by the granules was calculated.

[a] Supported by National Institutes of Health Grant NS21764.

RESULTS

Secretory granules in control acini occupy the whole of the cell cytoplasm. Isoproterenol causes the percentage of the cell area occupied by the granules to be reduced. All doses of IPR produced a reduction in secretory granules that was significantly different from control values ($F(5,59) = 7.44$, $p < 0.001$). Even with the highest dose, however, reduction was never complete (TABLE 1). When both the agonist IPR and antagonist PROP were given together, there was no significant reduction.

TABLE 1. Effect of Increasing Concentration of Isoproterenol on Secretory Granule Depletion

Isoproterenol Dose mg/kg	% Acinar Area Occupied by Secretory Granules X ± SD
0	55 ± 8
7.5	45 ± 3
15	41 ± 3
20	43 ± 2
30	33 ± 3
60	24 ± 1

DISCUSSION

It is apparent from these results that von Ebner's glands are under the control of the β-adrenergic nervous system. Injection of the β-adrenergic agonist reduces the secretory granules in the acinar cells. The reduction is related to the concentration of the injected agonist, but never exceeds 50% of control values. The reduction produced by IPR is completely blocked by the β-adrenergic antagonist PROP.

These results are similar to the 41% reduction in secretory granules obtained by Getchell and Getchell[5] using β-adrenergic stimulation of Bowman's glands, and quite different from the total degranulation of the rat parotid gland.[6] One explanation to account for these differences is that for Bowman's and von Ebner's glands one group of secretory granules is under control of the β-adrenergic system, while a second group is under some other neural control. If this proves to be true and if the granules are of a different composition, it would mean that the characteristics of the saliva bathing the taste buds in the circumvallate and foliate papillae could be modulated under neural control. Since flow of saliva is initiated by gustatory stimulation, afferent neural activity from taste buds could alter salivary composition and thereby control their microenvironment.

REFERENCES

1. HAND, A. R. 1970. The fine structure of von Ebner's gland in the rat. J. Cell Biol. **44:** 340-353.

2. MILLER, I. J. 1977. Gustatory receptors of the palate. *In* Food Intake and the Chemical Senses. Y. Katsuki, M. Sato, S. F. Takagi & Y. Oomura, Eds.: 173-185. Japan Scientific Societies Press. Japan.
3. BEIDLER, L. M. 1961. Taste receptor stimulation. *In* Progress in Biophysics and Biophysical Chemistry. Vol. **12:** 107-151. Pergamon Press. London, England.
4. BRADLEY, R. M., C. M. MISTRETTA, C. A. BATES & H. P. KILLACKEY. 1985. Transganglionic transport of HRP from the circumvallate papilla of the rat. Brain Res. **361:** 154-161.
5. GETCHELL, M. L. & T. V. GETCHELL. 1984. β-Adrenergic regulation of the secretory granule contents of acinar cells in olfactory glands of the salamander. J. Comp. Physiol. A. **155:** 435-443.
6. LILLIE, J. H. & S. S. HAN. 1973. Secretory protein synthesis in the stimulated rat parotid gland. J. Cell Biol. **59:** 708-721.

Earthworm Alarm Pheromone Is a Garter Snake Chemoattractant[a]

JEFFREY HALPERN, NANCY SCHULMAN, AND
MIMI HALPERN

*Department of Anatomy and Cell Biology
SUNY Health Science Center at Brooklyn
Brooklyn, New York 11203*

Earthworms, when irritated, secrete a substance called earthworm alarm pheromone (EAP) that conspecifics avoid on contact.[1,2] To produce the EAP, 25 to 100 earthworms (*Lumbricus terrestris*) were placed in a plastic cone between two metal plates and received an instantaneous shock every six seconds for two minutes. The resulting cloudy, viscous, mucus-like secretions were collected in a beaker placed under the stimulation chamber.

Earthworms placed on glass plates (8.3 × 10.2 cm) spotted with 0.5 cc of EAP rapidly escaped from the plates, whereas earthworms placed on plates spotted with distilled water (dH$_2$0) did not escape. EAP may be diluted by one-third before it loses its alarm properties, and lyophilization does not alter its aversive properties.

Garter snakes (*Thamnophis sirtalis parietalis*) tested with EAP in a two-choice discrimination task[3] spent more time at and tongue flicked more frequently dishes coated with EAP than dishes coated with dH$_2$0. Earthworm AP lyophilized and chromatographed on an AcA 44 column yielded three peaks as measured by 230-nm absorbance (FIG. 1): a high molecular weight peak (F2 eluting between dextran blue and albumin), an intermediate peak (F4 eluting with myoglobin) and a lower molecular weight peak (F6 eluting just in front of DNPalanine). All of the chemoattractant for snakes was found in F2. Protein content (as measured by Lowry assay) and neutral carbohydrate content (as measured by phenol assay) were greatest in the F2 peak as compared to the other peaks and troughs. EAP retained its chemoattractant activity at pH2 and pH11 and following dialysis against 50K MW cut-off membranes. EAP boiled for one hour did not lose its chemoattractant properties but lost these properties when boiled for two hours. When EAP is precipitated with saturated ammonium sulfate, all chemoattractant activity is retained in the precipitate and none is found in the supernatant. Many of the properties described for EAP are similar to those described previously for earthworm was (EWW),[4] obtained by soaking earthworms in 60°C dH$_2$0. EWW is not, however, an aversive stimulus for earthworms, and, unlike EWW, EAP cannot be produced from dead worms. Both substances appear to be glycoproteins of molecular weight in excess of 50K and are resistant to heat denaturation.

[a]Supported by National Institutes of Health Grant NS11713.

328

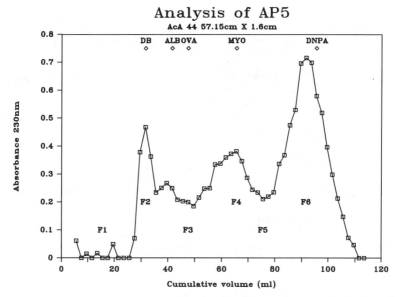

FIGURE 1. Gel chromatography of lyophilized and reconstituted alarm pheromone. Material was eluted from a 57.15 × 1.6 cm column of AcA 44 with 0.15 *M* saline. DB = blue dextran; ALB = bovine serum albumin; OVA = ovalbumin; MYO = myoglobin; DNPA = DNP alanine. F1–F6 are located in center of groups of fractions used to make up super fractions.

REFERENCES

1. RATNER, S. C. & R. BOICE. 1971. Psychol. Rec. **21:** 363-371.
2. RESSLER, R. H., R. B. GIALDINI, M. L. GHOCA & M. S. KLEIST. 1968. Science **161:** 597-599.
3. REFORMATO, L. S., D. M. KIRSCHENBAUM & M. HALPERN. 1983. Pharmacol. Biochem. Behav. **18:** 247-254.
4. HALPERN, M., N. SCHULMAN, L. SCRIBANI & D. M. KIRSCHENBAUM. 1984. Pharmacol. Biochem. Behav. **21:** 655-662.

Garter Snake Response to the Chemoattractant in Earthworm Alarm Pheromone Is Mediated by the Vomeronasal System[a]

NANCY SCHULMAN, EVELYN ERICHSEN, AND
MIMI HALPERN

Department of Anatomy and Cell Biology
SUNY Health Science Center at Brooklyn
Brooklyn, New York 11203

Garter snakes (*Thamnophis sirtalis parietalis*) were tested in a two-choice discrimination task[1] for differential responses to (1) earthworm alarm pheromone (EAP) produced by shocking earthworms with electric current, (2) earthworm wash (EWW) produced by bathing earthworms in a 60°C bath for two minutes; and (3) amyl acetate (AA) placed on a cotton plug in a perforated container.

Eighteen snakes were initially trained using EWW and subsequently tested with EAP and AA (nine snakes only). All snakes spent significantly more time and tongue-flicked significantly more frequently dishes coated with EWW or EAP as compared to distilled water (dH$_2$0) controls. No differential responses were observed to containers with AA as compared to containers with dH$_2$0. After preoperative testing nine snakes were tested for locomotor activity in an open field apparatus.

Snakes were subjected to bilateral vomeronasal ($n = 6$), olfactory ($n = 6$) or sham ($n = 6$) nerve lesions (FIG. 1). Snakes with olfactory or sham nerve lesions discriminated EWW and EAP from dH$_2$0 following surgery whereas snakes with vomeronasal nerve lesions discriminated neither EWW nor EAP from dH$_2$0. Locomotor activity was not differentially affected by surgery.

The heads of all snakes were decalcified, embedded in paraffin, and stained by the Bodian method. All lesions were verified microscopically by examining the site of the lesion as well as the corresponding sensory epithelium.

Studies by a number of investigators have previously demonstrated the importance of the garter snake vomeronasal system in the detection of and response to earthworm wash. This study adds another earthworm product, alarm pheromone, to the substances detected by the garter snake vomeronasal organ. The olfactory system does not appear to be necessary for differential responses to earthworm alarm pheromone.

[a] Supported by National Institutes of Health Grant NS11713.

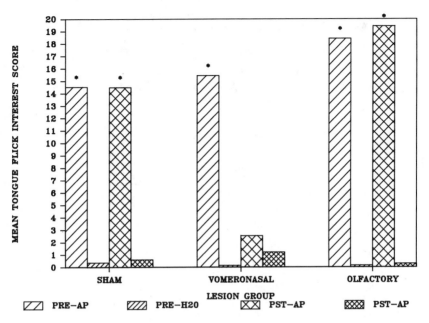

FIGURE 1. Mean tongue-flick interest scores of snakes to alarm pheromone (AP) and distilled water (H_2O) before (PRE) and following (PST) sham, vomeronasal, and olfactory nerve lesions. * above bar indicates that response to sample differed significantly ($p < 0.05$) from response to distilled water.

REFERENCES

1. REFORMATO, L. S., D. M. KIRSCHENBAUM & M. HALPERN. 1983. Pharm. Biochem. Behav. **18:** 247-254.

Latencies of Synaptic Potentials in Odor Responses of Salamander Mitral-Tufted Cells[a]

K. A. HAMILTON AND J. S. KAUER

Department of Neurosurgery and
Department of Anatomy and Cell Biology
Tufts University Medical School and
New England Medical Center
Boston, Massachusetts 02111

In the olfactory bulb of the tiger salamander, complex excitatory and inhibitory synaptic potentials underlie temporal patterns of spike activity that can be evoked in mitral and/or tufted relay neurons by odor stimulation.[1] In order to evaluate the timing of synaptic activity in odor-response patterns, we have measured the latencies of periods of depolarization and hyperpolarization and of initial spikes in intracellular recordings.

In a group of five mitral-tufted cells, stimulation of the olfactory mucosa with one-second pulses of cineole (CIN) elicited E-type responses, and isoamyl acetate (ISO) elicited S-type responses. The latencies of depolarizing and hyperpolarizing components of representative responses to CIN at 5×10^{-3}, 10^{-2}, and 10^{-1} times saturation and to ISO at 10^{-2} and 10^{-1} times saturation are plotted within the gray bars in FIGURE 1. The latencies observed in responses to electrical stimulation of the olfactory nerve and olfactory tract are plotted below for comparison.

Two important characteristics of E- and S-type odor responses are shown in FIGURE 1. The first is that periods of excitation tend to have shorter latencies in E-type responses than in S-type responses. For high odorant concentrations, depolarizing components of E-type responses can have latencies as short as 60 msec and usually precede the onset of hyperpolarization, as in responses to electrical stimulation. By contrast, depolarizing components of S-type responses generally have latencies longer than 200 msec and any spikes that are elicited follow the onset of hyperpolarization, even in responses to high concentrations.

The second characteristic is that, except in E-type responses to high odor concentrations, the initial event in both E- and S-type odor responses is hyperpolarization. The shorter latency of hyperpolarization compared to depolarization suggests that synapses that are inhibitory to mitral-tufted cells occur at a peripheral bulbar level, and can be activated without a preceding mitral-tufted cell impulse. In an accompanying report (Kauer and Hamilton, this volume), we describe results of 2-deoxyglucose autoradiographic studies that show that cells located in the periglomerular

[a]This work was supported by United States Public Health Service Grants NS-22035 and NS-20003.

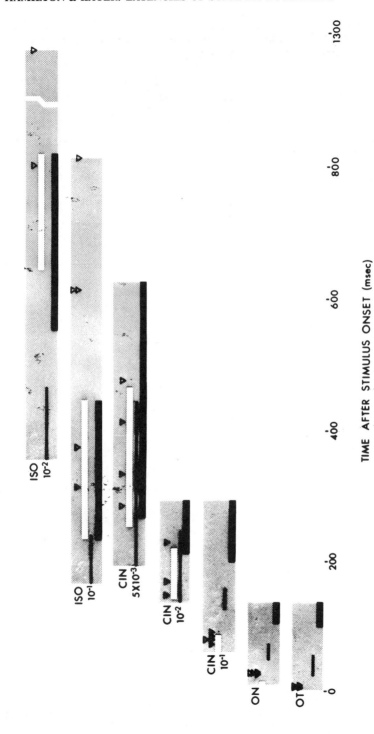

FIGURE 1. Latencies of components of mitral-tufted cell responses to odor stimulation with cineole (CIN) and isoamyl acetate (ISO), and to electrical stimulation of the olfactory nerve (ON) and olfactory tract (OT). White bars, onset of depolarization; thin black bars, onset of hyperpolarization; thick black bars, peak of hyperpolarization; triangles, initial spikes.

region of the bulb are activated by odor stimulation. We hypothesize that this activation contributes to the hyperpolarization of mitral-tufted cells and is important for the generation of temporal patterns of spike activity.

REFERENCES

1. HAMILTON, K. A. & J. S. KAUER. 1985. Intracellular potentials of salamander mitral/ tufted neurons in response to odor stimulation. Brain Res. **338:** 181-185.

Monoclonal Antibodies to Multiple Glutathione Receptors Mediating the Feeding Response of *Hydra*

KAZUMITSU HANAI,[a] MASAHIKO SAKAGUCHI,[a]
SACHIKO MATSUHASHI,[b] KATSUJI HORI,[b] AND
HIROMICHI MORITA[a]

[a]*Department of Biology*
Faculty of Science
Kyushu University
33, Fukuoka 812, Japan

[b]*Department of Biochemistry*
Saga Medical School
Saga 840-01, Japan

The feeding response of *Hydra japonica* can be quantitatively studied in terms of the duration of tentacle ball formation, a behavior associated with feeding in this species.[1] The effects of platelet proteins suggest that the response is mediated by at least five types of receptors (R1–R5), which are defined by the sensitivity to S-methylglutathione (GSM).[2]

Hybridomas were raised from a mouse immunized with galactose-binding proteins from tentacles, which were proposed as candidates for the receptor proteins on the basis of *in vivo*[3] and *in vitro* photolabeling studies.[4] We examined the effect of the culture supernatant of the hybridomas on the behavioral response to 0.1 μM GSM to pick up useful clones. We could detect specific antibodies to the receptors in the presence of the second antibody by the depression of the relevant response, because the effect of platelet proteins being contained in the culture supernatant was completely reversible for a brief incubation. The culture supernatant from hybridomas A3C3-4.8 and A3B5-8.6 (class IgM antibody) depressed the response significantly. In contrast, that of myeloma P3U1, which was used for fusion, or a hybridoma raised against unrelated antigens, did not. FIGURE 1 shows the response at varying concentrations of GSM in the presence of the purified IgM (final concentration 0.002 OD_{280}) from ascites fluid. More significant depression of the response below 0.1 μM indicates that the IgM is to the R1 receptor. FIGURE 2a shows numerous fluorescent spots obtained by an immunocytochemical method for the whole-mount preparation of a tentacle with A3B5-8.6, which is another IgM to the R1 receptor. These spots were not seen with nonimmunized mouse serum or IgM to unrelated antigens. FIGURE 2b shows a phase-contrast photomicrograph of the same field. (Note: A3C3-4.8 and A3B5-8.6 will be referred to as J1/0 and J1, respectively, in later publications.)

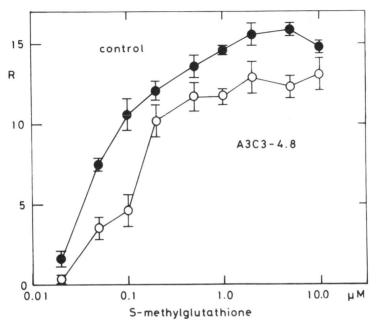

FIGURE 1. The response at varying concentrations of GSM in the presence of a monoclonal antibody A3C3-4.8.

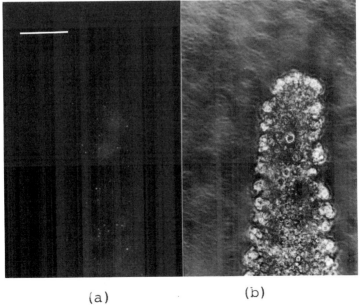

(a)   (b)

FIGURE 2. (a) Fluorescent immunocytochemistry for the whole-mount tentacle with a monoclonal antibody A3B5-8.6. (Culture supernatant was used.) (b) A phase-contrast photomicrograph for the same field. Bar is 50 μm.

REFERENCES

1. HANAI, K. 1981. J. Comp. Physiol. **144:** 503-508.
2. HANAI, K., H. KATO, S. MATSUHASHI, H. MORITA, E. W. RAINES & R. ROSS. 1987. J. Cell Biol. **104:** 1675-1681.
3. HANAI, K. & M. KITAJIMA. 1986. Comp. Biochem. Physiol. **83A:** 313-316.
4. KITAJIMA, M. & K. HANAI. 1986. Manuscript in preparation.

Taste Responsiveness of Hamster Glossopharyngeal Nerve Fibers[a]

TAKAMITSU HANAMORI,[b] INGLIS J. MILLER, JR.,[c]
AND DAVID V. SMITH[b]

[b]Department of Otolaryngology and Maxillofacial Surgery
University of Cincinnati Medical Center
Cincinnati, Ohio 45267

[c]Department of Anatomy
Bowman Gray School of Medicine
Winston-Salem, North Carolina 27103

Responses were recorded to five concentrations each of four stimuli from 83 taste-sensitive fibers in the hamster glossopharyngeal (IXth) nerve. Stimuli were delivered to either the vallate or foliate papillae via a syringe pump at 0.1 ml/sec at a temperature of 37°C. Stimuli, presented in ascending order, were: 0.01–1.0 M sucrose, 0.01–1.0 M NaCl, 0.0003–0.03 M HCl, and 0.0003–0.03 M quinine hydrochloride (QHCl), each in one-half log steps of concentration. Responses of each fiber were characterized as the number of impulses in the first 10 seconds after stimulus onset, corrected for the immediately preceding spontaneous rate.

Concentration-response functions were derived for each fiber for each of the four stimuli. The mean concentration-response functions for 56 fibers innervating the foliate papillae and for 27 fibers innervating the vallate papilla are shown in FIGURE 1A and 1B, respectively. The area under each concentration-response function (shaded) is proportional to the total number of impulses elicited across all five concentrations, which was used as a measure of responsiveness for each fiber. There was a significant difference in this measure between vallate and foliate papillae for QHCl ($p < 0.01$), which was a more effective stimulus when applied to the foliate papillae. Fibers of the IXth nerve are much more responsive to HCl and QHCl than to sucrose and NaCl ($p < 0.01$).

Cells were classified into best stimulus groups on the basis of their responses to the four stimuli. Depending on the concentrations chosen, the relative numbers of cells in each category and their breadth of responsiveness changes, as may be seen in FIGURE 2. At the same concentrations used to categorize chorda tympani (CT) fibers,[1] the mean response profiles of IXth nerve fibers are quite specifically tuned to their best stimulus, and there are relatively more sucrose- and NaCl-best fibers than at higher concentrations (FIG. 2A). As the concentrations are increased (FIG. 2B-2D), more cells are classified as HCl- and QHCl-best and each class of fibers becomes more broadly tuned across the four stimuli. Summing the responses across all concentrations (FIG. 2E) produces the most broadly tuned profiles. Using the total number of impulses as the measure of responsiveness, 8 (10%) of the 83 fibers were sucrose-

[a]Supported by Grant NS-23524 from the National Institute of Neurological and Communicative Disorders and Stroke (D.V.S.).

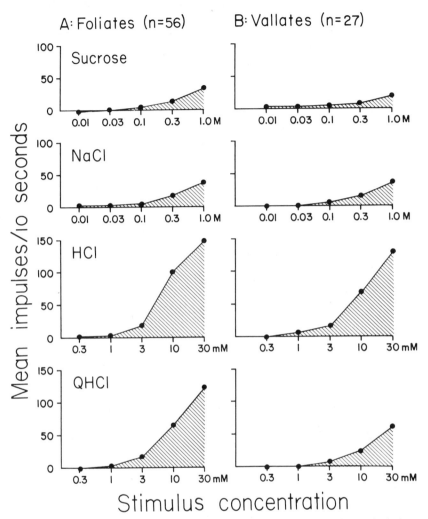

FIGURE 1. Mean concentration-response functions (impulses/10 sec) to each stimulus in fibers innervating foliate (**A**) and vallate (**B**) papillae.

best, 4 (5%) were NaCl-best, 52 (62%) were HCl-best, and 19 (23%) were QHCl-best. This distribution of sensitivities is quite different from that seen in the hamster CT nerve[1] where 25% of the fibers were sucrose-best, 53% were NaCl-best, 21% were HCl-best, and 1% were QHCl-best. At the same midrange concentrations used to classify CT fibers, IXth nerve fibers are much less responsive than CT fibers. The breadth of excitatory responsiveness (H) was calculated using the equation developed by Smith and Travers.[2] The mean breadth of tuning (H) of IXth nerve fibers was 0.470, compared to 0.559 in CT fibers stimulated with the same concentrations. However, when the responsiveness of the cells to the entire concentration range is used to measure their breadth of sensitivity, IXth nerve fibers are less specifically

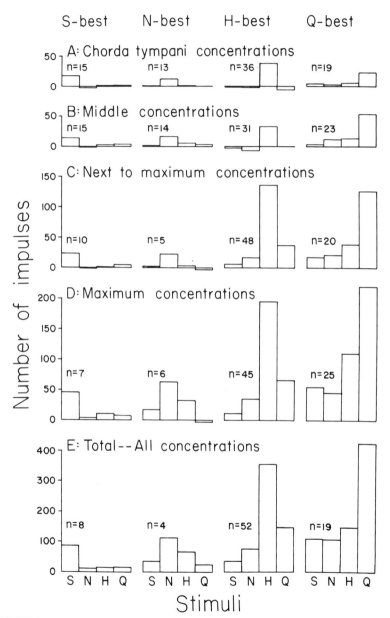

FIGURE 2. Mean response profiles for S-, N-, H-, and Q-best fibers in the hamster IXth nerve at different levels of stimulus concentration. Concentrations used in previous work on the chorda tympani nerve[1] are shown in **A**, with increasing concentrations in **B-D**. Total impulses across all concentrations are shown in **E**.

tuned ($H = 0.706$). The characterization of the response properties of gustatory fibers is enhanced by obtaining responses over a wide range of stimulus intensities.

REFERENCES

1. FRANK, M. 1973. An analysis of hamster afferent taste nerve response functions. J. Gen. Physiol. **61:** 588-618.
2. SMITH, D. V. & J. B. TRAVERS. 1979. A metric for the breadth of tuning of gustatory neurons. Chem. Senses Flavor **4:** 215-229.

Effects of Chemical Noise on the Detection of Chemical Stimuli[a]

LINDA S. HANDRICH AND JELLE ATEMA

Boston University Marine Program
Marine Biological Laboratory
Woods Hole, Massachusetts 02543

The differentiation of signals from background "noise" is a problem common to all animals and all sensory modalities. Chemoreception is no exception. The marine environment is chemically noisy in that it contains high levels of amino acids ($\sim 10^{-8}$ M) and ammonium ($\sim 10^{-6}$ M), compounds that are used as food signals by many marine species. We used the lobster, *Homarus americanus,* to investigate how its physiologically narrowly tuned chemoreceptors might aid the animal in recognizing food odors against chemical noise.

We recorded the searching and feeding behaviors of small lobsters during introduction of a food stimulus. Their 1-liter tanks had a constant flow of either raw

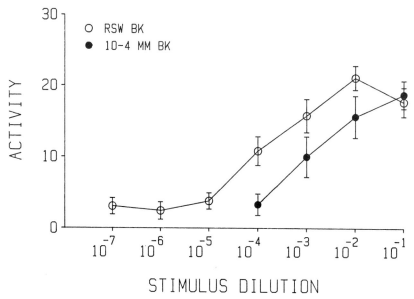

FIGURE 1. Activity (mean $\pm$ SEM) of lobsters ($n = 16$) in response to MM, measured in 10-second time intervals (30 intervals = 5 min) in RSW and in a 10^{-4} dilution of MM in RSW.

[a]This research was supported by grants from the Whitehall Foundation and National Science Foundation Grant BNS 85 12585.

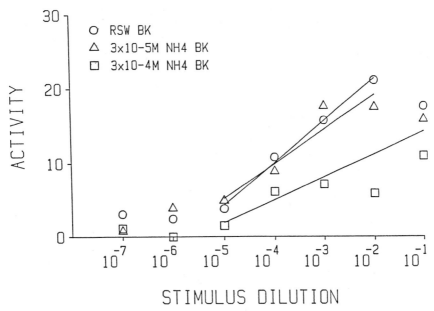

FIGURE 2. Activity (mean ± SEM) of lobsters in response to MM in RSW and in RSW containing $3 \times 10^{-5} M$ and $3 \times 10^{-4} M$ NH$_4$. Regression equations: RSW background, Y = 32.781 + 2.470 log X, r = 0.997, n = 16; $3 \times 10^{-5} M$ NH$_4$ background, Y = 28.538 + 2.025 log X, r = 0.943, n = 8; $3 \times 10^{-4} M$ NH$_4$ background, Y = 11.950 + 0.814 log X, r = 0.874, n = 8. Regression lines for responses in RSW and $3 \times 10^{-5} M$ NH$_4$ backgrounds are not significantly different from each other, but both differ significantly from the line for responses in $3 \times 10^{-4} M$ NH$_4$ background ($p = 0.05$; T test for the difference between slopes of regression lines).

seawater (RSW) or RSW with artificially elevated chemical backgrounds, that is, increased noise. The food stimulus (MM) was a solution of amino acids and ammonium based on mussel tissue composition.

In the first experiment, RSW backgrounds were alternated with elevated backgrounds, that is, RSW plus a 10^{-4} dilution of the stimulus mixture (FIG. 1). In the elevated background the stimulus-response curve to MM was shifted to the right along the concentration axis. Thus, it is the signal-to-noise ratio and not the absolute concentration of the stimulus mixture that determines the slope and height of the behavioral stimulus-response curve. Note that in RSW the response to a 10^{-1} dilution of MM has leveled off, perhaps indicating a behavioral maximum or "ceiling" effect. This same ceiling was reached in the elevated background, but here the highest signal-to-noise ratio was not tested.

Ammonium (NH$_4$) stands out when considering chemical noise in the lobsters' environment. NH$_4$ is present in high concentrations in the background ($5 \times 10^{-7} M$ in RSW), it is one of the more concentrated components of MM, and it is an excretory product of most marine organisms including lobsters' prey. NH$_4$ ($\sim 10^{-4}$-$10^{-2} M$) was the only one of six single compounds to elicit a good feeding response when applied to lobster legs.[1]

Our second experiment tested the effect of higher NH$_4$ concentration in the background seawater. A $3 \times 10^{-5} M$ NH$_4$ background had no effect but the $3 \times 10^{-4} M$ NH$_4$ background reduced the behavioral response to the mixture (FIG. 2). This reduction could be due to the lack of input from the ammonium-specific receptors that become self-adapted in higher ammonium backgrounds.[2] Ammonium may also have cross-adapting effects on other cell types, such as amino acid receptors, reducing their response to their best compounds.

In a third experiment responses to a mixture containing ten times less NH_4 than MM were the same as to MM, except at the highest concentration (10^{-1}). The response to MM declined at 10^{-1} but the response to low NH_4 MM continued to rise at that concentration. Perhaps the 10^{-1} stimulus contains sufficient NH_4 to suppress other cell types at the time of stimulation with the mixture. Reduced ammonium in the stimulus mixture could then lead to diminished suppression of amino acid cells and thus to a greater response.

Our results show that the lobster's behavioral response to a stimulus mixture adapts in a predictable way in a background containing higher concentrations of all compounds making up the stimulus mixture. Even the fluctuation of single compounds such as NH_4 can have effects on the perception of food mixtures. In principle, the results of these experiments can be explained on the basis of self- and cross-adaptation at the level of the chemoreceptors. However, central effects are, of course, not ruled out by these behavioral experiments.

REFERENCES

1. BORRONI, P. F. & J. ATEMA. 1985. Dynamic properties and self-adaptation of NH_4 cells in lobster taste organs. Chem. Senses **10**: 426.
2. BORRONI, P. F., L. S. HANDRICH & J. ATEMA. 1986. The role of narrowly tuned taste cell populations in lobster (*Homarus americanus*) feeding behavior. Behav. Neurosci. **100**(2): 206-212.

Neural and Behavioral Responses to Amino Acids in Mice and Rats

S. HARADA, T. MARUI, AND Y. KASAHARA

Department of Oral Physiology
Kagoshima University Dental School
Usuki-cho, Kagoshima 890, Japan

In order to clarify the gustatory effectiveness of amino acids, responses were examined using electrophysiological and behavioral methods in the mouse and rat. Electrophysiological experiments on the chorda tympani nerve have revealed that amino acids are classified into at least three groups according to basic (BA), neutral (NA), and acidic amino acids (AA).

The chorda tympani rsponses to both L- and D-BA hydrochloride salts (BA-HCl) in each animal were similar to those for NaCl, and adapted well with those for monovalent chloride salts but not with NA or sucrose in the cross-adaptation experiment[1] (see FIG. 1). Several structural analogues or derivatives of L-Arg and L-Lys were tested, to show that the α-amino group is essential for the strong stimulatory effectiveness of BA-HCl, although the BA of free-base form was less stimulatory than the corresponding HCl salts. Larger responses were elicitied when the pH of the solutions was lowered. A similar effect of lowering pH was also observed for NA and other substances that have more than one amino group. These results suggest that the charged amino group in the organic substances plays an important role in producing their strong effect. The relative stimulatory effectiveness of L-NA was larger in mice than in rats, and that for L-isomers was significantly larger than D-isomers in both species. Responses to these NA cross-adapted with sucrose, but not with NaCl or BA. The response characteristics of AA were similar to that for HCl.

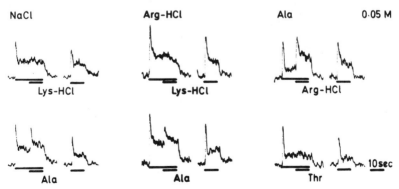

FIGURE 1. Adaptation effects of NaCl, L-Arg-HCl and L-Ala on the responses to amino acids. In each integrated recordings, an adapting solution was applied initially, followed by a test stimulus dissolved in the adapting solution. The resonse on the right in each recording is the test stimulus alone.

TABLE 1. Generalization of Taste Aversion for Seven Conditioning Stimuli in Mice

Conditioning Stimulus	NaCl	Q-HCl	Arg	Lys	T.S.[a] Suc	Ala	Ser	Gly	HCl
NaCl	+	−	−	−	−	−	− −	−	− −
Q-HCl	− −	+	+ +	+ +	− −		− −		− −
Arg	− −	+ −	+ +	+	+ −	−	− −	−	−
Suc	− −	− −	+ −		+	+ −	+ −	+	− −
Ala	−	−	−	− −	+	+	+ −	+	− −
Ser	− −	− −	+	− −	+ +	+ +	+ +	− −	− −
Gly	− −	− −	+	−	+ +	+ −	+	+	− −

[a] T.S. is the lick ratio, defined as the number of licks to the test stimulus divided by the number of licks to distilled water. The T.S. was classified into five categories (+ + 0-20, + 20-40, + − 40-60, − 60-80, − − 80-100). The concentration of each solution was 0.1 M except for 0.001 M Q-HCl, 0.001 M HCl, and 0.5 M Suc.

In order to clarify the taste sensation for amino acids in these species, the method of taste aversion conditioning by LiCl i.p. injection[2] was employed, and the behavioral ability to discriminate amino acids and four basic taste stimuli was examined in mice. The results summarized in TABLE 1 show that the taste of BA-HCl is generalized to other BA-HCl and guinine-HCl (Q-HCl) solutions, but not to NaCl. And the taste of Q-HCl is generalized to BA-HCl salts, but not to NaCl. Aversion conditioning to NaCl resulted in the rejection of only NaCl itself. These behavioral data suggest that, in mice, the taste of BA-HCl is mainly similar to that of Q-HCl, and quite different from that of NaCl. However, the electrophysiological data showed that the characteristics of the response to BA-HCl are quite similar to that for NaCl. One possible reason of this discrepancy is that NaCl adaptation depresses the receptive mechanisms for following application of bitter substances like Q-HCl or BA-HCl, although their receptive mechanisms are different. Another possibility is that the information from the bitter stimulus in the peripheral nerve is relatively important in the central nervous system and fewer impulses than for NaCl are necessary to produce avoidance behavior for harmful substances. On the other hand, both behavioral and electrophysiological data suggest that NA, which produces sweet taste in man, produces a sensation similar to sucrose in mice.

REFERENCES

1. SMITH, D. V. & M. FRANK. 1972. Cross-adaptation between salts in the chorda tympani nerve of the rat. Physiol. Behav. **8:** 213-220.
2. HALPERN, B. P. & D. N. TAPPER. 1972. Taste stimuli: Quality coding time. Science **171:** 1256-1258.

Quantitative Analysis of Feeding Behavior Stereotypes and of the Rejection of Aversive-Tasting Food in the Freshwater Prawn *Macrobrachium rosenbergii*

SHEENAN HARPAZ[a] AND JACOB E. STEINER[b]

[a]*Department of Zoology*
Life Sciences Institution

[b]*Department of Oral Biology*
Hadassah Faculty of Dental Medicine
The Hebrew University
Jerusalem, Israel

Food search and food intake behavior stereotypes of the freshwater prawn *Macrobrachium rosenbergii* were studied on a sample of 27 adult animals of both sexes in the intermolt phase of the molt cycle. Prawns starved for 96 hours and held individually in separate 10-liter aquaria, were stimulated to display their feeding behavior repertoire in one of the following manners: (a) offering an actual food pellet to which they were accustomed; (b) introducing small aliquots of betaine-HCl, which in previous studies[1] was found to act as a powerful chemoattractant for this prawn. The behavioral displays induced by these stimulants were videotaped and the recordings were evaluated by two independent viewers. The behavioral repertoire includes antennular flicking followed by sweeping scanning movements of the area immediately ahead of the animal performed by the first pair of walking legs, which precedes the locomotion of the entire body toward the source of chemotactic stimulation and finally grasping and lifting it to the mouth parts.[1] Frequency of antennular flicks, as well as that of searching and lifting movements of the first pair of walking legs, were used as measure units for the quantitative assessment of response intensity.

Results revealed that the intensity of food search and food intake behavioral sequences induced by actual food can be evoked in an identical manner, both quantitatively and qualitatively, when only betaine-HCl was used to simulate the stimulus of a nutrient (Harpaz & Steiner, unpublished). The betaine-induced features show a clear dosage-dependent response (FIG. 1).

The question that still remained open was, At what stage can the animal detect an aversive-tasting additive to the food, and how will this additive affect the feeding behavioral pattern? In order to explore this point, an experiment was carried out in which the identical food pellet used in the previously discussed results was made bitter by the addition of 0.07 M quinine-HCl. The bitter-tasting pellet was quantitatively equal to the unadulterated one in eliciting the entire behavioral pattern up to the point of ingestion. Only when the pellet reached the mouth parts, the chemoreceptors

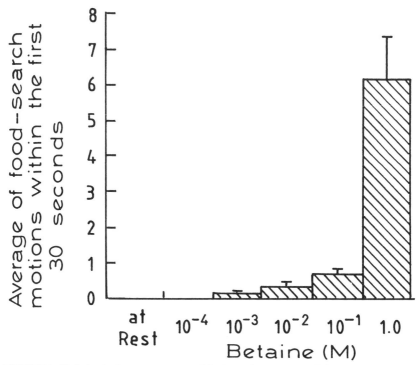

FIGURE 1. Variation in the average counts of food-search movements of the first pair of walking legs of *Macrobrachium rosenbergii,* within the first 30 seconds following betaine-HCl introduction, in relation to increasing molarity of the chemostimulant commencing from zero concentration (at rest).

located there brought about its rejection. It should, however, be added that in some instances the final rejection occurred after repeated "tasting" accompanied by mouth-cleaning movements that lasted up to 23 minutes.[2] An earlier indication that crustacean antennules are apparently indifferent to quinine could be found in the electrophysiological studies of Derby *et al.*[3] on the American lobster *Homarus americanus.* The practical implication of the above finding, from an aquacultural point of view, is that the mere onset of the feeding behavior pattern by any feed item is still no guarantee that the prawn will eventually swallow the chemotactically stimulating food.

REFERENCES

1. HARPAZ, S., D. KAHAN & R. GALUN. 1986. Variability in feeding behavior of the Malaysian prawn *Macrobrachium rosenbergii* during the molt cycle. Crustaceana, in press.
2. STEINER, J. E. & S. HARPAZ. 1987. Food intake behavior of the freshwater prawn, *Macrobrachium rosenbergii,* and the rejection of a bitter tasting food substance. Chem. Senses **12:** 89-97.
3. DERBY, C. D., P. M. REILLY & J. ATEMA. 1984. Chemosensitivity of lobster *Homarus americanus* to secondary plant compounds: Unused receptor capabilities. J. Chem. Ecol. **10:** 879-892.

A Diffusion Potential Model of Salt Taste Receptors[a]

HARRY WMS. HARPER

Eastern Research Center
Stauffer Chemical Company
Dobbs Ferry, New York 10522

Quantitative measurements were made of response magnitudes in the *chorda tympani* nerve of the hamster as the tongue was stimulated with solutions that varied in composition and concentration. These measurements eliminated a systematic error that has contaminated previous results based on whole-nerve responses: The mean population firing rate of the nerve fibers is proportional to the averaged *square* of the amplified nerve potential, not the average *rectified* value. This result, which follows directly from treating the whole-nerve potential as a random shot-noise (to which Campbell's theorem may be applied), has been confirmed by extensive analogue simulation studies.[1] Using this method, the form of sucrose concentration-response curves agrees closely with the relation expected on the basis of stimulus adsorption theory.[2] For salts, however, the data obtained are *not* consistent with a first-order stimulus-receptor adsorption process: At moderate to high concentrations, the response increases much more rapidly with concentration than theory predicts. The data *are* well accounted for by a model of receptor function based on diffusion potentials.[1] Two kinds of variable diffusion potential are involved: resting potentials of taste cell receptor membranes (between the variable taste pore contents and the constant interiors of taste cells), and a liquid junction potential (between taste pore contents and the constant interstitial fluid of taste buds). (Any model of taste reception that includes a current through the receptor membrane *must* include at least two variable potentials, because the current must traverse the changeable contents of the taste pore, and both upon entering and leaving the taste pore must therefore pass across a phase boundary, one phase of which is variable.) An excitatory current determined by these potentials passes through the taste cell membranes and nerve terminals deep within the taste bud (FIG. 1). Estimates of the conductances of the components of the proposed current path, using known taste bud anatomy and published data on nerve and epithelial tissues, are consistent with such a model. Nerve excitation may be directly due to the current, or may involve synaptic transmission. Two kinds of receptor cells are required to explain the data: one kind with receptor membranes selectively permeable to sodium-like ions, and another kind selective for potassium-like ions.

Receptor membrane resting potentials can be calculated using the Goldman-Hodgkin-Katz equation. Novel methods make possible accurate calculation of liquid junction potentials at experimental concentrations.[3] The diffusion potential

[a] The reported research was conducted while the author was Guest Investigator at The Rockefeller University.

DIFFUSION POTENTIAL MODEL

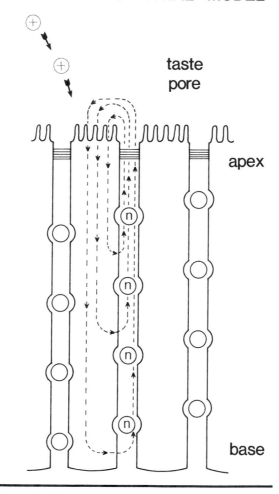

FIGURE 1. Single taste cells, each joined to its neighbors by a junctional complex near the apex. An excitatory current is shown passing into one cell through the receptor membrane; it spreads within the cell and passes out through the lateral membrane and returns to the receptor membrane by way of the extracellular space (containing afferent nerve terminals), the junctional complex (which is permeable to small ions), and the taste pore. Changes in the current are governed by two variable diffusion potentials: the resting potential of the selectively permeable receptor membrane and a liquid junction potential arising at the junctional complex. There are two types of cells like this, one permeable to sodium-like ions and the other permeable to potassium-like ions. The liquid junction potential is common to both, in series with the receptor membrane potentials.

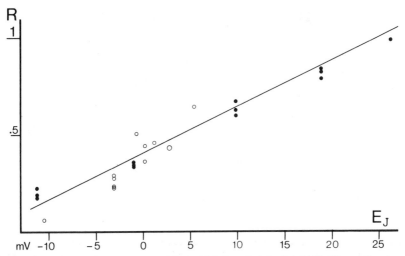

FIGURE 2. Normalized response magnitude "R" (equals 1 for 0.1 M NaCl) as a function of liquid junction potential "E_J", using stimulus ions not permeant in receptor membranes. Each point is a single response. The filled dots are data from three animals, using variable concentrations of $MgCl_2$ (0.005 M-0.5 M) as stimuli. The open dots are data from six animals, using fixed (0.1 M) concentrations of ions having different electrical mobilities. (Cations: methylammonium, tetramethylammonium, tetraethylammonium, tetrabutylammonium, choline. Anions: methylsulfonate, propionate, chloride.) The fitted line accounts for 94% of the variance; its slope gives the voltage sensitivity of the receptor mechanism.

model predicts, and neural data confirm, that responses should vary directly with the liquid junction potential (FIG. 2). In particular, it is found that the variation in responses to a given cation when paired with different anions is well accounted for by changes in E_J that the anions produce (due principally to differences in mobility). Observed responses are also consistent with the Nernst potentials expected at the receptor membranes of sodium (or potassium) cells when the tongue is exposed to varying concentrations of sodium (or potassium).

The existence of peripherally distinct sodium and potassium receptors, as required by this model, has been confirmed by the effects of amiloride on taste responses.[4] Available single-fiber data indicate that these two populations of receptor cells have independent afferent innervations.[5]

REFERENCES

1. HARPER, H. W. 1985. A quantitative study of taste responses in the *chorda tympani* of the hamster, with special reference to the mechanism of salt taste. Ph.D. Thesis. The Rockefeller University. New York.
2. BEIDLER, L. M. 1954. A theory of taste stimulation. J. Gen. Physiol. **38:** 133-139.
3. HARPER, H. W. 1985. Calculation of liquid junction potentials. J. Phys. Chem. **89:** 1659-1664.
4. HECK, G. L., S. MIERSON & J. A. DESIMONE. 1984. Salt taste transduction occurs through an amiloride-sensitive sodium transport pathway. Science **223:** 403-405.
5. FRANK, M. E., R. J. CONTRERAS & T. P. HETTINGER. 1983. Nerve fibers sensitive to ionic taste stimuli in *chorda tympani* of the rat. J. Neurophysiol. **50:** 941-959.

Analysis of Olfactory Neural Responses by a Method of Spike Train Matching

T. A. HARRISON, J. W. SCOTT, AND F. H. SCHMIDT

Department of Anatomy and Cell Biology
Emory University School of Medicine
Atlanta, Georgia 30322

Quantification of the relationship between neural activity and stimulus intensity is an important issue in the physiology of sensory coding. Studies of olfaction have had a particular problem in this respect. Recordings of olfactory responses by neurons in the olfactory bulb demonstrated that changes in both odor quality and odor intensity

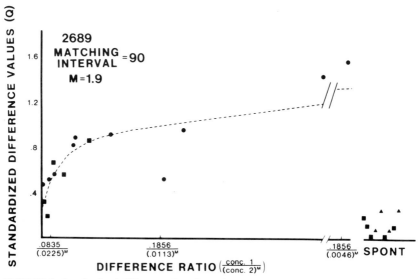

FIGURE 1. Standardized difference plot of a tufted cell's responses to six concentrations of amyl acetate. This series generated 15 pairwise response comparisons. Standardized Q scores are plotted against the disparity between each pair of stimulus concentrations represented as a transformed ratio of the concentrations (abscissa). Three of the 15 ratios are indicated along the axis. The statistical significance of each response comparison is shown by: circles, $p < 0.001$; triangles, $p < 0.05$; squares, $p < 0.05$ (randomization test for matched pairs). The points at the far right are for unstimulated activity. The dashed curve was fit by nonlinear regression. The optimal values of M (the exponent for the x-axis transformation) and of the matching interval are indicated.

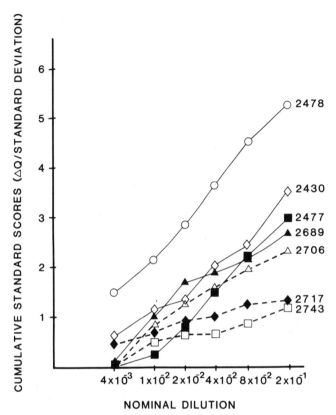

FIGURE 2. Cumulative differences between standardized Q scores for amyl acetate concentration series for seven olfactory bulb neurons. The points at the extreme left indicate the difference between the response at the lowest concentration and the spontaneous activity. The value plotted for each concentration is the cumulative sum of the standardized Q scores for the comparisons between each concentration and the adjacent lower concentration. The plot includes two mitral cells (2743 and 2706), three tufted cells (2689, 2430, and 2478), a superficial interneuron (2477), and a deep interneuron (2717).

can produce complex changes in the temporal firing patterns of these cells.[1] Quantification of these responses is important in characterizing the intensity-response function of a cell when attempting to determine which of several different odor qualities is the most effective stimulus.

We recently described a method for quantification of the change in response produced by a particular change in stimulus intensity.[1] We have applied this method to extracellular unit recordings from pentobarbital-anesthetized rat olfactory bulb. Two spike trains are compared by considering each spike in terms of its latency from the beginning of the trace. Spikes of similar latency in each trace are matched. The degree of disparity in these matches and the numbers of nonmatched spikes are summed to give a value we called the "Q index." The Q index can be used in two ways: (1) to measure response variability over repeated stimulus presentations and (2) to compare responses to different stimuli. Standardized Q scores are produced by dividing the mean Q for between-stimulus comparisons by the pooled standard deviation.

The validity of this approach can be demonstrated by plotting the standardized Q scores against a measure of the disparity between the odor concentrations being compared (standardized difference plot, FIG. 1). The particular measure chosen was the ratio of the higher concentration over the lower concentration raised to an arbitrary power. This transformation was necessary to plot all comparisons from a stimulus series on the same curve. When the data points lie close to a smooth function, as in FIGURE 1, this indicates that the Q index for between-concentration comparisons is systematically related to the magnitude of concentration differences.

The value of the Q index and the tightness of the standardized difference plots vary depending upon the difference in latency beyond which spikes are not allowed to match (matching interval). We have recently begun to choose the matching interval by performing a nonlinear regression to fit the data to the best of a series of curves by the single-process law.[2] Regression analysis was performed at each matching interval from 10 msec to 500 msec and the matching interval was chosen that produced the smallest sum of squared errors between observed and predicted values.

Computation of the standardized Q scores allows changes induced by a series of odor stimuli to be expressed independently of the particular response patterns observed. The stimulus response function can be expressed by cumulating these standardized Q's across the series from unstimulated conditions through the highest concentrations. FIGURE 2 illustrates cumulative response curves for seven cells showing that some have much steeper functions than others. This plot will allow comparisons of the response functions for different cell types and allow comparisons of the steepness of a particular cell's response functions for different odors.

REFERENCES

1. HARRISON, T. A. & J. W. SCOTT. 1986. Olfactory bulb responses to odor stimulation: Analysis of response pattern and intensity relationships. J. Neurophysiol. **56:** 1571-1589.
2. TURNER, M. E., R. J. MONROE & H. L. LUCAS. 1961. Generalized asymptotic regression and nonlinear path analysis. Biometrics **17:** 120-143.

Effects of Cadmium on the Rat Olfactory System[a]

LLOYD HASTINGS AND TRUE-JENN SUN

Department of Environmental Health
University of Cincinnati
Cincinnati, Ohio 45267

Clinical studies of workers exposed chronically to cadmium (Cd) suggest that prolonged Cd exposure may result in anosmia.[1] Chronic exposure of adult rats to Cd has been found to result in increased uptake of Cd by the olfactory bulb.[2] These two lines of evidence suggest that chronic Cd exposure might impair olfaction.

To investigate the effect of Cd on olfaction, 45 male adult rats were exposed to Cd0 via inhalation for five hours per day, five days a week for a total of 20 weeks. Exposure values were 250 $\mu g/m^3$ and 500 $\mu g/m^3$ (current threshold limit value for Cd0 is 50 $\mu g/m^3$). Before exposure, olfactory thresholds were obtained, using a conditioned suppression technique. In this paradigm, rats were water deprived and trained to lick a sipper tube for water reinforcement on a variable interval (VI) two-minute schedule. After licking behavior had stabilized, a conditioned stimulus (3 ppm isoamyl acetate) was presented for 20 seconds on a VI five-minute schedule. Each presentation of the conditioned stimulus was followed by a mild foot shock. Odor stimuli were generated by a flow-dilution olfactometer. After 7 to 14 days of conditioned suppression training, all rats had learned to suppress licking in the presence of the conditioned stimulus. By lowering the concentration of the isoamyl acetate, detection thresholds could be determined by the failure of the rat to suppress licking during its presence. Olfactory thresholds were obtained approximately every four weeks during the exposure period. At the end of the exposure, the rats were sacrificed and concentrations of Cd in various tissues determined.

After 20 weeks of Cd exposure, there was no evidence of anosmia in any of the rats nor were there any significant changes observed in olfactory thresholds. Cd content of both olfactory bulbs and nasal-olfactory epithelium was significantly elevated, compared to controls; however, histological examination revealed no alterations in the olfactory bulb. The lack of any demonstrable effect may be due to insufficient accumulation of Cd on the olfactory epithelium or in the bulb, due to the low solubility of the Cd compound used (Cd0).

REFERENCES

1. ADAMS, R. G. & N. CRABTREE. 1961. Anosmia in alkaline battery workers. Br. J. Ind. Med. **18**: 216-222.
2. CLARK, D. E., J. R. NATION, A. J. BOURGEOIS, M. F. HARE, D. M. BARKER & E. J. HINDERBERGER. 1985. The regional distribution of cadmium in the brains of orally exposed rats. Neurotoxicology **6**: 109-114.

[a]Supported by NIOSH Grant #2038.

Intralingual Stimulation with Sweet Proteins in Rhesus Monkey and Rat[a]

G. HELLEKANT, T. ROBERTS, C. HÅRD AF
SEGERSTAD, AND H. VAN DER WEL

*Department of Veterinary Science and
Wisconsin Regional Primate Center
University of Wisconsin
Madison, Wisconsin 53706*

It is thought that taste is the result of an interaction between a compound and a specialized set or sets of receptors on the microvilli (e.g., van der Heijden *et al.*[1,2]). This is also the case for thaumatin and monellin, which are extremely sweet to humans and give a nerve response in catarrhina primates,[3,4] although it does not taste sweet at all to the rat and some other noncatarrhine mammals.[5,6] Since it is known that compounds injected intralingually can elicit a taste[7] and a taste nerve response,[8,9] intralingual stimulation with monellin and thaumatin could give information on the localization of taste receptors on the taste cells.

Recordings from the chorda tympani proper nerve were obtained during intralingual stimulation through a catheter inserted in the external carotid artery, peripherally to its lingual branch to the tongue. The intralingual stimuli were dissolved in 0.9% NaCl and consisted of 0.04% monellin, thaumatin, and acetylated thaumatin (not sweet), and 0.01 M citric acid, 0.3 or 0.4 M NaCl, 0.3 M sucrose serving as control stimuli. A Taste-O-Matic system[10] delivered the oral stimuli.

FIGURE 1 shows that intralingual stimulation with thaumatin and monellin gave a significant response in the monkey but none in the rat. In the monkey, it should be noticed that the responses to oral stimulation with 0.3 M sucrose before and after the intralingual thaumatin or monellin stimulation have about the same heights. There was no cross-adaptation between the oral and intralingual stimuli.

This is further shown in FIGURE 2, in which responses from a monkey to oral sucrose were plotted after intralingual and oral thaumatin stimulation. The diagram shows cross-adaptation between oral thaumatin and sucrose but not between intralingual thaumatin and oral sucrose. Intralingual stimulation with nonsweet acetylated thaumatin (not sweet) did not elicit an increase of the nerve activity in the monkey or in the rat.

It may be suggested that the stimuli stimulated taste nerve fibers directly; however, the rat showed no nerve response to monellin and thaumatin while the monkey did. If the response in the monkey was caused by direct stimulation of nerve fibers, the capillary permeability to thaumatin and monellin should be different in the rat from

[a]This study was funded by National Institutes of Health Grant NS17021.

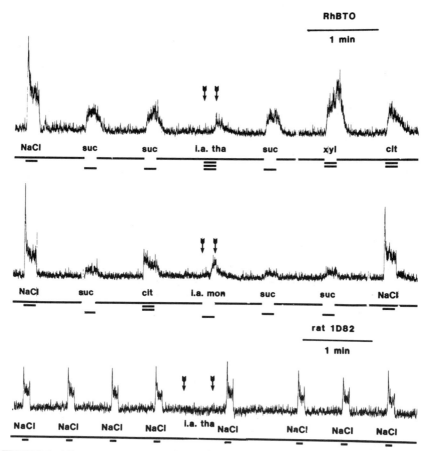

FIGURE 1. All responses, except the ones marked with arrows, were the results of oral taste stimulation. The top record was derived from a monkey while the bottom one is from a rat. The trace between the arrows shows the effect on the nerve activity of 7-10-second intralingual stimulation with 0.04% thaumatin and 0.04% monellin in a monkey (top) and a rat (bottom).

Abbreviations: tha, thaumatin; mon, monellin; suc, sucrose; cit, citric acid; asp, aspartame; D-try, D-tryptophane; asc, ascorbic acid; xyl, xylitol; and i.a., intra arterial.

the monkey, or the taste nerve fibers of the two species react differently to thaumatin and monellin. It seems more likely that the difference resided in the taste buds, which then must be concluded to be the site for stimulation. Since the tongue was constantly rinsed, rinsing of the tongue abolishes the response to orally applied stimuli, and there was no cross-adaptation between surface and intralingual responses, it is unlikely that the intralingual stimuli could have reached the taste buds via the taste pore. They must have reached the excitable structure via the tissue surrounding the taste buds. Assuming a stimulus-receptor interaction, this indicates that there are receptors to monellin and thaumatin on the intraepithelial parts of the taste cells in the rhesus monkey but none in the rat.

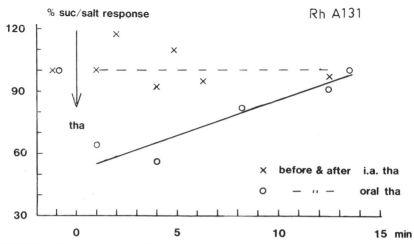

FIGURE 2. The diagram shows the response to oral stimulation with sucrose after intralingual thaumatin (x) and oral thaumatin (o). The sucrose responses were expressed in percent of nearest NaCl response. The response immediately before the thaumatin was then assigned the value 100% and the following responses calculated in percent of it.

REFERENCES

1. VAN DER HEIJDEN, A., H. VAN DER WEL & H. G. PEER. 1985. Structure-activity relationships in sweeteners. I. Nitroanilines, sulphamates, oximes, isocumarins and dipepetides. Chem. Senses **10:** 57-72.
2. VAN DER HEIJDEN, A., H. VAN DER WEL & H. G. PEER. 1985. Structure-activity relationships in sweeteners. II. Saccharins, acesulfames, chlorosugars, tryptophans and ureas. Chem. Senses **10:** 73-88.
3. GLASER, D., G. HELLEKANT, J. N. BROUWER & H. VAN DER WEL. 1978. The taste responses in primates to the proteins thaumatin and monellin and their phylogenetic implications. Folia Primatologica **29:** 56-63.
4. HELLEKANT, G., D. GLASER, J. BROUWER & H. VAN DER WEL. 1980. Study of behavioural and neurophysiological experiments on the sweet taste in five primates. In Olfaction and Taste VII. H. van der Starre, Ed.: 183-186. IRL Press Ltd.
5. HELLEKANT, G. 1975. Different types of sweet receptors in mammals. In Olfaction and Taste V. D. A. Denton & J. P. Coghlan, Eds.: 15-21. Academic Press. New York.
6. HELLEKANT, G. 1976. On the gustatory effects of monellin and thaumatin in dog, hamster, pig and rabbit. Chem. Senses Flavor. **2:** 97-105.
7. BROEBECK, J. R. 1979. Best & Taylor's Physiological Basis of Medical Practice. 10th edit.: 3-159. Williams & Wilkins. Baltimore, MD.
8. BRADLEY, R. M. 1970. Investigation of intralingual taste using perfused rat's tongue. Ph.D. Thesis. Florida State University. Tallahassee, FL. 122 pp.
9. BRADLEY, R. M. 1973. Electrophysiological investigations of intralingual taste using perfused rat tongue. Am. J. Physiol. **224:** 300-304.
10. BROUWER, J. N., D. GLASER, C. HARD AF SEGERSTAD, G. HELLEKANT, Y. NINOMIYA & H. VAN DER WEL. 1983. The sweetness-inducing effect of miraculin; behavioural and neurophysiological experiments in the rhesus monkey (Macaca mulatta). J. Physiol. (London) **337:** 221-240.

Ultrastructure of Mouse Fungiform Taste Buds[a]

DAVID M. HENZLER AND JOHN C. KINNAMON

Laboratory for High Voltage
Electron Microscopy
Department of Molecular, Cellular
and Developmental Biology
University of Colorado
Boulder, Colorado 80309

Like most mammals, the mouse normally has only one fungiform taste bud per papilla. The fungiform papillae are predominately distributed on the distal, posterior surface of the tongue, although the papillae are found on the anterior surface and the proximal, posterior surface. The fungiform taste buds are innervated by the chorda tympani branch of the facial nerve. Viewing semi-thin (0.25 μm) serial sections with the high-voltage electron microscope (HVEM) and using computer-generated, three-dimensional reconstructions has enabled us to examine morphological features of fungiform taste buds in an efficient manner.

MATERIALS AND METHODS

Young adult mice were anesthetized with sodium pentobarbital and perfused with 0.05 M sodium cacodylate buffer (pH 7.3) at 39°C. The animals were fixed with 39°C buffered 2.5% glutaraldehyde and 1% paraformaldehyde. The tongues were excised and papillae were removed and were processed normally for electron microscopy, stained *en bloc* with uranyl acetate and embedded in Epon 812/Araldite. After the taste buds were serially sectioned and post-stained, they were viewed with the HVEM. Contours from serial electron micrographs were entered into an IBM PCXT micro-computer and reconstructed using software developed by Dr. S. J. Young (University of California, San Diego).

RESULTS AND DISCUSSION

The shape of the fungiform taste bud is similar to a garlic bulb (FIG. 1), unlike the vallate or foliate taste buds, which are more ovoid shaped.[1] The taste pore opens

[a] This project was supported in part by Grants NS21688 and RR 00592 from the National Institutes of Health and by a grant from the Procter and Gamble Co.

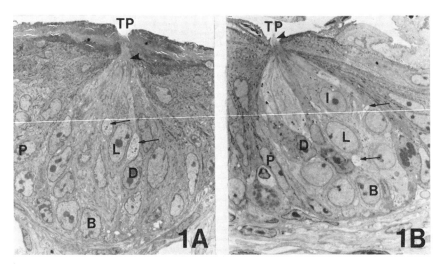

FIGURE 1. (A) High-voltage electron micrograph (HVEM) of a longitudinal section through a fungiform taste bud. Magnification 1,400×. (B) HVEM of a longitudinal section through a vallate taste bud. Note the different staining characteristics between the light and dark cells in the vallate bud as contrasted with the more homogeneous staining of the fungiform taste cells. Taste pore (TP); light (L), dark (D), intermediate (I), basal (B), and peripheral (P) taste cells; microvilli (arrowhead); nerve fibers (arrows). Magnification 1,300×.

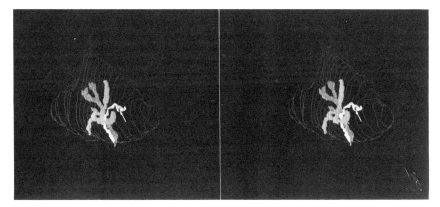

FIGURE 2. Three-dimensional reconstruction of three nerve fibers from the fungiform taste bud shown in FIGURE 1A. The taste pore is located at the top of the figure. The nerve fibers enter the base of the taste bud and course tortuously throughout the bud. A synapse (arrow) is shown on the white nerve fiber.

directly to the lingual surface and contains material with electron-dense granules. At the taste pore, microvilli extend from the taste cells to the exterior. The base of the taste bud is delineated by a basal lamina.

Each of the five types of taste cells is evident in fungiform taste buds: light, dark, intermediate, basal, and peripheral cells. The staining characteristics of the light and dark cells are similar, unlike in vallate and foliate taste buds in which the dark cells are more electron dense than the light cells.

In some of the taste cells, large, atypical mitochondria and subsurface cisternae can be found adjacent to nerve fibers. These features are also found in vallate and foliate taste buds.[2] The nerve fibers enter at the base of the taste bud, twisting and branching throughout the bud (FIG. 2). The diameters of nerve fibers in the fungiform taste bud are often larger than in the vallate or foliate taste buds. Although we have observed synapses from dark, intermediate, and light taste cells onto nerve fibers in fungiform taste buds (FIG. 2), the numbers of synapses per bud appears to be lower in fungiform taste buds compared with vallate taste buds.

REFERENCES

1. KINNAMON, J. C., B. J. TAYLOR, R. J. DELAY & S. D. ROPER. 1985. Ultrastructure of mouse vallate taste buds. I. Taste cells and their associated synapses. J. Comp. Neurol. **235:** 48-60.
2. ROYER, S. M. & J. C. KINNAMON. 1986. Interactions between taste cells and nerve fibers in murine foliate taste buds. Ann. N.Y. Acad. Sci. This volume.

Are Apical Membrane Ion Channels Involved in Frog Taste Transduction?

Neural and Intracellular Evidence[a]

M. SCOTT HERNESS

Laboratory of Neurobiology and Behavior
The Rockefeller University
New York, New York 10021

Recent studies with the diuretic amiloride in mammalian gustation[2,3] have implied that sodium channels may exist in the apical membrane of taste cells. This investigation tests if such apical ion channels might be involved in electrolyte transduction in frog gustation and tests if more than one type of channel may be present. Two channel inhibitors were used, amiloride, an inhibitor of amiloride-sensitive (AS) sodium channels, and cobalt chloride ($CoCl_2$), an inhibitor of calcium channels. Since both inhibitors operate from the outside channel surface, their effect on apical membrane channels would be demonstrable by topical application to the tongue surface. With topical application the solution comes in contact with only the external apical membrane surface of the taste cell.

Amiloride treatment of the frog tongue (a single four-minute treatment of 0.0001 M) showed little sodium specificity. NaCl or KCl responses were inhibited equally in glossopharyngeal nerve recordings (FIG. 1A). The neural response to 0.003 M $CaCl_2$ was also inhibited (32%) while responses to 0.01 M quinine and 0.01 M HCl were unaffected. Intracellular records indicate that amiloride treatment reduces the input resistance of the taste receptor cell (FIG. 1B). In cell one the input resistance was reduced by 48% and in cell two it was reduced by 67%. Resting potentials of both cells were depolarized by 5-10 mV posttreatment. Receptor potentials to 0.4 M NaCl and 0.4 M KCl were slightly smaller after treatment and still accompanied by a conductance increase. During the application of amiloride itself, the cell was hyperpolarized and its input resistance increased.

Cobalt chloride specifically inhibited the calcium response in the glossopharyngeal neural recordings when tested with stimulus-cobalt mixtures (FIG. 2A). Inhibition of the neural response to 0.003 M $CaCl_2$ was nearly complete and recovered immediately after removal of the cobalt. The same treatment enhanced responses to 0.5 M NaCl (135%) and 0.5 M KCl (25%). Such enhancements of NaCl and KCl by transition

[a] Supported in part by National Science Foundation Grant BNS 8111816 and in part by BRSG SO7 RR07065 awarded by the Biomedical Grant Program, Division of Research Resources, National Institutes of Health.

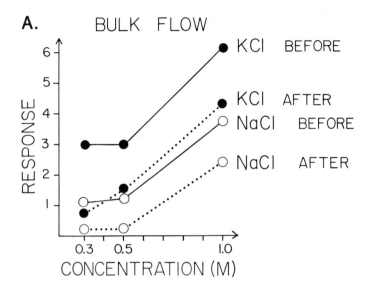

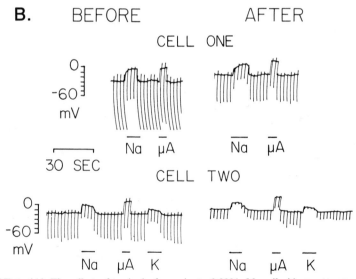

FIGURE 1. (A) The effect of a single four-minute 0.0001 M amiloride treatment on frog glossopharyngeal responses to KCl and NaCl of varying concentration. Note that both KCl and NaCl responses are inhibited. **(B)** Intracellular recordings from two different taste cells of the frog tongue before and after a brief exposure to 0.0001 **M** amiloride. While receptor potentials were not greatly affected, the resting potential was slightly depolarized and input resistance (as measured by the size of the conductance pulses) greatly reduced after amiloride exposure. Stimuli were 0.4 M NaCl (Na), $+30$ microamps (μA), and 0.4 M KCl (K).

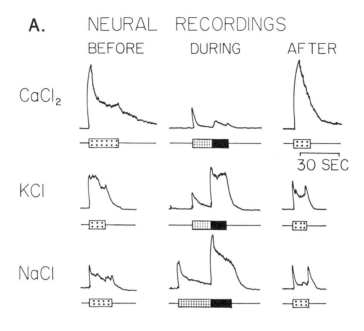

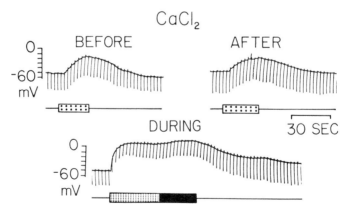

FIGURE 2. (A) The effect of cobalt chloride on the frog glossopharyngeal response to three electrolytes, 0.003 M CaCl$_2$, 0.5 M KCl, and 0.5 M NaCl. The dot-filled bar beneath the record indicates presentation of the stimulus alone, the checkered bar the presentation of 20 μM cobalt chloride, and the solid bar the presentation of the stimulus dissolved in 20 μM cobalt chloride; the line indicates the adapting fluid, 0.005 M NaCl. Cobalt chloride virtually eliminated the calcium response in a reversible manner while enhancing responses to KCl and NaCl. **(B)** An intracellular recording from a frog taste cell exposed to 0.003 M CaCl$_2$ (BEFORE and AFTER) and 0.003 M CaCl$_2$ mixed in, and preexposed to 20 μM cobalt chloride (DURING). Calcium had little effect on the receptor potential when the cell was preexposed to cobalt chloride.

metal ions such as $CoCl_2$ and $NiCl$ have been previously observed.[4] Intracellular recordings were made using the identical stimulus presentation paradigm used for neural recordings. These records indicate that cobalt depolarizes the cell with little conductance change, similar to receptor potentials to $CaCl_2$.[1] Application of the calcium-cobalt chloride mixture did not further affect the receptor potential. Receptor potentials to 0.003 M $CaCl_2$ alone presented before and after cobalt exposure were very similar.

The data are consistent with the notion that apical ion channels may be involved in taste transduction. However, both neural and intracellular evidence discount the role of an AS-sodium channel in frog apical membrane since amiloride treatment showed little sodium specificity. If an apical channel exists for sodium, it would have to be a general cation-type channel. In frog taste cells, ion permeability has been proposed for both the basolateral and apical membranes for sodium and potassium.[1,7] This hypothesis would require a mechanism for these cations to be transported across the membrane. Ion channels, specifically calcium and sodium channels, have been reported to exist in the basolateral membrane of frog taste cells,[5] but no specific mechanism has been proposed for ion transport across the apical membrane. A type of general cation channel would be feasible.

The lack of specificity, observed with amiloride was not observed with cobalt chloride giving more credence to the possibility of specific calcium channels. Different receptor sites on the apical membrane have been proposed for sodium and calcium in the frog based on pronase E treatment[6] suggesting that the receptor site for calcium may be composed of a protein. This observation is also consistent with the possibility of an ion channel as the receptor site for calcium. It should be emphasized, however, that as of yet there is no direct evidence for the existence of any of these channels and hence the possibility of other mechanisms has not been precluded.

REFERENCES

1. AKAIKE, N., A. NOMA & M. SATO. 1976. Electrical responses of frog taste cells to chemical stimuli. J. Physiol. (London) **254:** 87-107.
2. BRAND, J. G., J. H. TEETER & W. L. SILVER. 1985. Inhibition by amiloride of chorda tympani responses evoked by monovalent salts. Brain Res. **334:** 207-214.
3. HECK, G. L., S. MIERSON & J. A. DeSIMONE. 1984. Salt taste transduction occurs through an amiloride-sensitive sodium transport pathway. Science **223:** 403-405.
4. KASHIWAGURA, T., N. KAMO, K. KURIHARA & Y. KOBATAKE. 1978. Enhancement of frog gustatory response by transition metal ions. Brain Res. **142:** 570-575.
5. KASHIWAYANAGI, M., M. MIYAKE & K. KURIHARA. 1983. Voltage-dependent Ca^{++} channel and Na^+ channel in frog taste cells. Am. J. Physiol. **224**(Cell Physiol. 13): C82-C88.
6. KITADA, Y. 1984. Two different receptor sites for Ca^{++} and Na^+ in frog taste responses. Neurosci. Lett. **47:** 63-68.
7. OKADA, Y., T. MIYAMOTO & T. SATO. 1986. Contribution of the receptor and basolateral membranes to the resting potential of a frog taste cell. Jpn. J. Physiol. **36:** 139-150.

Amiloride Inhibition of Neural Taste Responses to Sodium Chloride[a]

THOMAS P. HETTINGER AND MARION E. FRANK

Department of BioStructure and Function
University of Connecticut Health Center
Farmington, Connecticut 06032

Amiloride is a rapid, reversible, and specific inhibitor of neural taste responses to sodium chloride.[1,2] The effect of amiloride on integrated responses of the hamster chorda tympani nerve is dependent on the concentrations of both NaCl and amiloride and on the order of application of the two substances to the tongue. FIGURE 1A shows the response to 0.1 M NaCl. Up to 90% suppression is observed for responses to amiloride-salt mixtures immediately following tonic responses to salt alone. For 0.1 M NaCl, inhibition by 1×10^{-5} M amiloride takes place within two seconds and complete recovery occurs within 10 seconds following amiloride removal (FIG. 1B). Recovery is slower at higher amiloride concentrations. The apparent dissociation constant for amiloride in the presence of 0.1 M NaCl has been estimated to be about 1×10^{-6} M and depends on NaCl concentration in a manner suggestive of competitive inhibition. Application of an amiloride-salt mixture to the tongue following water adaptation produces a phasic response nearly the same as for NaCl alone, but the tonic response a few seconds later is practically eliminated (FIG. 1C). When amiloride is applied one minute before the mixture, both the phasic and tonic levels are reduced compared to those in the absence of amiloride, but the suppression of the tonic level is not as great as in the other two cases (FIG. 1D). Since both the phasic and tonic responses to NaCl are inhibited by prior application of amiloride, the lack of effect of amiloride on the phasic response when applied simultaneously with NaCl probably results from slower diffusion of amiloride compared to sodium to the receptor sites.

Thus, amiloride causes both a rapid inhibitory response that is complete in about two seconds and a slower opposing sensitization that develops during one minute following application of amiloride. These results are wholly consistent with the idea that salt taste is mediated by receptor stimulation via amiloride-sensitive sodium channels.[1,2] The time course and specificity of inhibition, the reversibility, and the concentration parameters are within the ranges expected for such channels in epithelial tissue.[3] Inhibition is caused by nearly instantaneous blocking of sodium channels, resulting in decreased Na conductance, while sensitization can result from chronic decreased Na entry into receptor cells. The blocking of mucosal Na channels while maintaining normal serosal extrusion via the Na-K pump will slowly decrease intracellular sodium, so that the Na receptor potential becomes poised at a higher level of sensitivity to external sodium. The overall neural response to amiloride is usually one of suppression, but depends on opposing effects of conductance and voltage changes in the receptor cells.

[a] Supported by National Science Foundation Grant BNS 8519638.

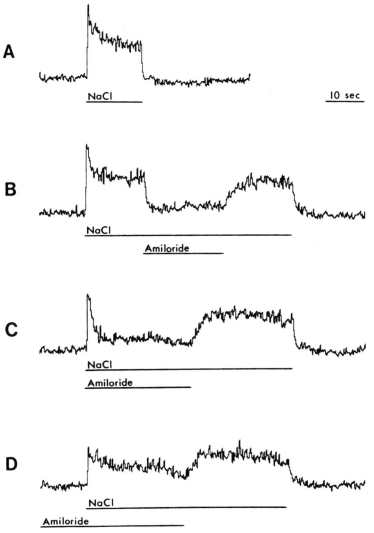

FIGURE 1. Effect of amiloride on hamster chorda tympani responses to sodium chloride. Amplified responses of the whole nerve were rectified and digitally integrated into 200-msec bins and plotted versus time. The tongue was placed in a flow chamber and thoroughly rinsed with water. Stimuli were applied where indicated by the bars under each record: NaCl, 0.1 M; amiloride, 1×10^{-5} M. Where both occur together, they were present in a single solution as a mixture. Water was on the tongue where no stimulus bars are shown. Flow rate = 3 ml/sec. (**A**) Control NaCl response with water rinse. (**B**) NaCl-amiloride mixture applied 15 seconds after NaCl, followed by NaCl, then water. (**C**) NaCl-amiloride mixture response after water adaptation, followed by NaCl, then water. (**D**) NaCl-amiloride mixture applied after one-minute treatment of the tongue with amiloride, followed by NaCl, then water.

REFERENCES

1. HECK, G. L., S. MIERSON & J. A. DESIMONE. 1984. Salt taste transduction occurs through an amiloride-sensitive sodium transport pathway. Science **223:** 403-405.
2. BRAND, J. G., J. H. TEETER & W. L. SILVER. 1985. Inhibition by amiloride of chorda tympani responses evoked by monovalent salts. Brain Res. **334:** 207-214.
3. BENOS, D. J. 1982. Amiloride: A molecular probe of sodium transport in tissues and cells. Am. J. Physiol. **242:** C131-C145.

Development of Amiloride Sensitivity in the Rat Peripheral Gustatory System

A Single-Fiber Analysis[a]

DAVID L. HILL

Department of Psychology
University of Toledo
Toledo, Ohio 43606

Response frequencies of rat peripheral and central taste neurons increase dramatically to many chemical stimuli during development.[1,2] The largest and most consistent developmental changes occur to NaCl and LiCl, with at least a twofold increase in response frequencies from preweaning to adult ages. However, response frequencies are similar throughout development to other salts such as NH_4Cl.[1,2] Recent research has shown that the epithelial sodium transport blocker, amiloride, suppresses multifiber chorda tympani responses to NaCl in adult rats by at least 60%.[3-5] In contrast, amiloride is ineffective in suppressing the NaCl response in rats aged 12-13 days.[5] Moreover, the NaCl response following amiloride in adults is the same as the response in young rats. Therefore, the developmental increase in sensitivity to NaCl appears to be related to a concomitant increase in amiloride sensitivity.

To explore the underlying peripheral events, neurophysiological taste responses were recorded from single chorda tympani fibers in rats aged 14-20 days ($n = 19$) and adults ($n = 37$). Responses were recorded to a concentration series (0.05 M, 0.1 M, 0.5 M, 1.0 M) of NaCl, sodium acetate (NaAc) and NH_4Cl before and after lingual application of 500 μM amiloride hydrochloride. Before application of amiloride, response frequencies to all concentrations of NaCl and NaAc were at least two times greater in adults compared to young rats; response frequencies to NH_4Cl were similar between age groups (FIG. 1). In contrast, the age-related differences in sensitivity to sodium salts were not apparent after lingual application of amiloride (FIG. 1a, b). That is, amiloride eliminated the developmental differences in sensitivity to sodium salts. Further analyses of the response characteristics of individual neurons revealed that amiloride did not selectively affect one group of neurons (e.g., neurons maximally responsive to NaCl or to NH_4Cl). In fact, amiloride appears to suppress sodium salt responses in a linear fashion, which is in contrast to preliminary results reported in the adult.[6] Response frequencies of neurons to NaCl and NaAc before amiloride were

[a] Supported by National Institutes of Health Grant NS0538 and RCDA NS00964.

[b] Address for correspondence: David L. Hill, Department of Psychology, Gilmer Hall, University of Virginia, Charlottesville, VA 22903-2477.

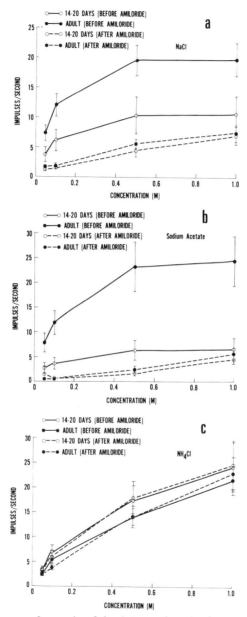

FIGURE 1. Mean response frequencies of chorda tympani neurons in rats aged 14-20 days and adults to NaCl (**A**), sodium acetate (**B**), and NH_4Cl (**C**) before and after lingual application of amiloride.

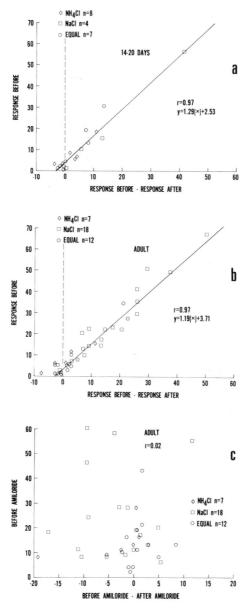

FIGURE 2. Single-fiber response frequencies to 0.5 M NaCl expressed relative to the change in NaCl response frequency due to amiloride in rats aged 14-20 days (**A**) and in adults (**B**). The same function is shown in (**C**) for responses to 0.5 M NH$_4$Cl in adult rats. Diamonds denote fibers maximally responsive to NH$_4$Cl, squares denote fibers maximally responsive to NaCl, and circles denote fibers equally responsive to NaCl and NH$_4$Cl.

highly correlated with the change in frequency (frequency before amiloride minus frequency after amiloride; $r = +0.93$ to $+0.99$), whereas no correlation existed for NH_4Cl (FIG. 2). That is, the greater the response frequency to NaCl and NaAc before amiloride, the greater the suppression by the sodium channel blocker. This function occurred for all concentrations of NaCl and NaAc and for both age groups. Moreover, the slopes and the y-intercepts were similar for both ages at each concentration of NaCl and NaAc (FIG. 2). Such similarities in the functions indicate that the non-amiloride-sensitive portion of the sodium salt response and the fundamental relationship between the sensitivity of a neuron to sodium and the magnitude of the amiloride effect is the same throughout development. Therefore, the developmental pattern of increased sensitivity to sodium salts seems primarily due to an increase in the number of functional amiloride-sensitive components. Such an increase in membrane components is reflected by increased average frequencies to sodium salts with age and alterations in the proportions of neurons maximally responsive to NH_4Cl or to NaCl as a function of age.

REFERENCES

1. HILL, D. L., R. M. BRADLEY & C. M. MISTRETTA. 1983. Development of taste responses in rat nucleus of solitary tract. J. Neurophysiol. **50:** 879-895.
2. HILL, D. L., C. M. MISTRETTA & R. M. BRADLEY. 1982. Developmental changes in taste response characteristics of rat single chorda tympani fibers. J. Neurosci. **2:** 782-790.
3. BRAND, J. G., J. H. TEETER & W. L. SILVER. 1985. Inhibition by amiloride of chorda tympani responses evoked by monovalent salts. Brain Res. **334:** 207-214.
4. HECK, G. L., S. MIERSON & J. A. DESIMONE. 1984. Salt taste transduction occurs through an amiloride-sensitive sodium transport pathway. Science **223:** 403-405.
5. HILL, D. L. & T. C. BOUR. 1985. Addition of functional amiloride-sensitive components to the receptor membrane: A possible mechanism for altered taste responses during development. Dev. Brain Res. **20:** 310-313.
6. NINOMIYA, Y., T. MIZUKOSHI, T. HIGASHI & M. FUNAKOSHI. 1984. Differential responsiveness of two groups of rat chorda tympani fibers to ionic chemical and electrical tongue stimulations. Proc. 18th Jpn. Symp. Taste Smell : 145-148.

Development of Olfactory and Vomeronasal Systems in the Red-Sided Garter Snake, *Thamnophis sirtalis parietalis*[a]

DAVID A. HOLTZMAN AND MIMI HALPERN

*Program in Neural and Behavioral Sciences and
Department of Anatomy and Cell Biology
SUNY Health Science Center at Brooklyn
Brooklyn, New York 11203*

This study describes the morphology of the olfactory and vomeronasal systems in the garter snake at three prenatal and two postnatal stages of development. Embryos from pregnant females, obtained from Dr. David Crews, were surgically removed and classified according to Zehr stages.[1] Embryos left within the female continued to develop and were born normally. Heads of embryos and neonates were fixed, by immersion, in 10% phosphate-buffered formalin, embedded in paraffin, sectioned, and stained using the Bodian method.

At Zehr stage 26/27, the vomeronasal epithelium (VNE) is composed of a layer 12-15 cells thick bordered dorsally and posteriorly by small, round clusters three to four cells in diameter (TABLE 1). The thicker layer appears to be the embryonic supporting cell (SC) layer of the adult, and the clusters appear to be undifferentiated cells and developing bipolar neurons (UD-BP). At Zehr stage 32, the SC layer (15 cells thick) is approximately equal in thickness to the UD-BP layer, which has begun to form columns. In addition, pigmentation of the connective tissue (CT) is sparse. At Zehr stage 36, the UD-BP layer is 17-22 cells thick and has formed tall columns. However, these columns appear separate from the SC layer, which has diminished to two to five cells in thickness. Pigmentation of the CT has greatly increased, but the VNE is not as pigmented as seen in the adult. Zero to two days after birth (0 days being the day of birth), the VNE is adultlike in many aspects. The SC is two to three cells thick, and the UD-BP layer is 24 cells thick. The UD-BP columns are no longer separate from the SC layer. The CT is pigmented as extensively as in the adult. At 12 days old, the SC layer is one to three cells thick and the UD-BP layer is approximately 30 cells thick.

Maturational changes are also observed in the olfactory epithelium (OE). At Zehr stage 26/27 and 32, the OE is 15 cells thick and homogeneous in appearance. At Zehr stage 36, the OE is six to eight cells thick and begins to show evidence of stratification. Developing Bowman's glands associated with the OE can also be seen. At zero to two days old and 12 days old, the OE is six to eight cells thick with adultlike organization and associated Bowman's glands.

[a]Supported by National Institutes of Health Grants NS11713 (to M. H.) and HD16687 (to Dr. D. Crews).

TABLE 1. Changes in Thickness of Vomeronasal and Olfactory Epithelia (Number of Cells in Depth) in Five Developmental Stages in *Thamnophis sirtalis parietalis*[a]

| System | Epithelial Thickness[b] | | | | |
	Stage 26/27	Stage 32	Stage 36	0-2 Days	12 Days
Vomeronasal					
SC	12-15	15	2-5	2-3	1-3
UD-BP	3-4	15	17-22	24	30
Olfactory	15	15	6-8	6-8	6-8

[a] Embryonic stages are those described by Zehr.[1] Day 0 represents the day of birth. SC: supporting cell layer; UD-BP: undifferentiated, bipolar cell layer.
[b] Expressed in number of cells.

At Zehr stage 26/27, neither the main olfactory bulbs (MOB) nor the accessory olfactory bulbs (AOB) have glomeruli, and both consist of undifferentiated cells concentrated around the olfactory ventricles. The MOB and AOB are connected with the OE and VNE, respectively. The bulbs can not be distinguished from each other except for the location of entering nerve fibers. The vomeronasal nerve enters the AOB from its dorsomedial aspect. The olfactory fila appear to enter the MOB ventrolaterally. At Zehr stage 32, glomeruli are still absent in both sets of bulbs. The MOB and AOB become distinguishable from each other, and most of the cells of both bulbs are concentrated on the medial side. At Zehr stage 36, glomeruli have started forming on the medial side of both sets of bulbs. The MOB and AOB are clearly distinguishable. The cells are still not as organized as in the adult, but hints of cell layer formation are apparent. At zero to two days and 12 days of age, both sets of bulbs have glomeruli and have distinct cell layers as in the adult.

The areas containing the tertiary neurons associated with these two systems first appear as distinct structures at Zehr stage 36.

REFERENCES

1. ZEHR, D. R. 1962. Stages in the normal development of the common garter snake, *Thamnophis sirtalis sirtalis.* Copeia **1962:** 322-329.

Possible Mechanisms for the Processes of Referred Taste and Retronasal Olfaction[a]

DAVID E. HORNUNG [b] AND MELVIN P. ENNS [c]

[b]Biology Department and
[c]Psychology Department
St. Lawrence University
Canton, New York 13617

This study is concerned with the possible mechanisms for the processes by which molecules that enter the nose via a sniff can produce the sensation of taste (referred taste) and liquids that are placed in the mouth can produce the sensation of smell (retronasal olfaction). College undergraduates used the method of absolute magnitude estimation to scale the intensity of the smell, taste, referred taste, and retronasal olfaction of at least three concentrations of one of the following stimuli: ethyl butyrate ($n = 11$), ethanol ($n = 14$), vanillin ($n = 14$), citral ($n = 13$), and almond extract ($n = 16$). For ethyl butyrate, almond, and citral, intensity estimates were first corrected for the taste and smell of distilled water. Then, for each stimulus, the psychophysical function (TABLE 1) describing the relationship between the arithmetic median of the estimate of intensity and the stimulus concentration was calculated (log-log coordinates).

For each of the five stimuli, the slope (TABLE 1) describing referred taste was closer to the slope describing smell than it was to the slope describing taste. In addition, for all concentrations of all stimuli, the median response assigned to referred taste was always less than the median response assigned to smell. This occurred even though the slope of the referred taste curve was sometimes greater then the slope of the smell curve. Further, in all cases where smell was more intense than taste (all concentrations of ethyl butyrate and the lowest concentrations of vanillin and almond) referred taste was always rated *more* intense than the actual taste (FIG. 1). Although these data must be considered very preliminary, they suggest that referred taste is a central (cognitive) confusion where a smell sensation is perceived as having a taste.

Although generally the slopes describing retronasal olfaction were somewhat closer to the slopes describing taste than they were to the slopes describing smell, the slopes describing retronasal olfaction were *always* less steep than the taste slopes (TABLE 1). With the exception of the lowest concentration of almond extract, the median response assigned to retronasal smell was always less than the median response assigned to taste. Further, in all cases where the taste was rated more intense than the smell (all concentrations of ethyl alcohol and the higher concentrations of almond), the retronasal smell was rated *more* intense than the actual smell intensity. These observations suggest that retronasal olfaction is also a central confusion. However, since

[a]Supported by a grant from General Foods Corporation.

375

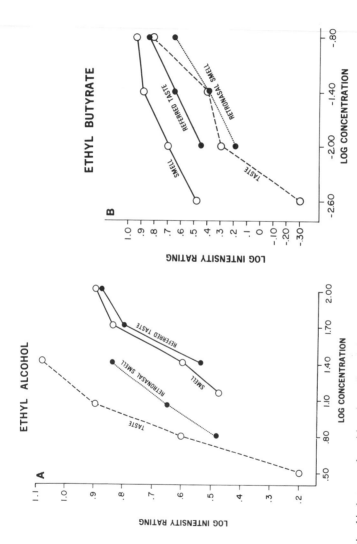

FIGURE 1. Relationship between the arithmetic median of the intensity estimate and stimulus concentration for ethyl alcohol (**A**) and ethyl butyrate (**B**). For all concentrations of ethyl alcohol, the median taste intensity (open circles, dashed line) was greater than the smell intensity (open circles, solid line). Retronasal smell (closed circles, dashed line) was rated less intense than the taste but more intense than the smell; referred taste (closed circles, solid lines) was rated less intense than the smell. For all concentrations of ethyl butyrate, the median smell intensity was greater than the taste intensity. Referred taste was rated less intense than smell but more intense than taste; retronasal smell was rated less intense than the taste.

TABLE 1. Slopes of the Psychophysical Functions

Stimulus	Smell (r)	Referred Taste (r)	Taste (r)	Retronasal Smell (r)
Ethyl butyrate	0.26 (0.97)	0.34 (1.00)	0.57 (0.97)	0.40 (1.00)
Almond	0.28 (1.00)	0.16 (1.00)	0.83 (0.94)	0.60 (1.00)
Ethyl alcohol	0.50 (0.98)	0.55 (0.94)	1.00 (0.98)	0.61 (0.99)
Citral	0.38 (0.99)	0.21 (1.00)	N.S.[a]	N.S.
Vanillin	0.24 (1.00)	0.31 (0.94)	0.74 (0.99)	0.50 (0.91)

[a] N.S. = r value less than 0.80.

all the retronasal smell slopes were less than the taste slopes, it is possible that part of the mechanism of retronasal smell may also be peripheral, as molecules diffuse from the mouth to the headspace above the olfactory receptors.

For retronasal smell, a sniff of distilled water vapor usually removed the relationship between the response and concentration. However, with a tastant in the mouth, the smell of distilled water was still scaled as more intense than the smell of distilled water alone. Distilled water in the mouth had a similar effect on referred taste. These observations would suggest that the mechanisms (cognitive and otherwise) for the processes of referred taste and retronasal smell are complex.

Gustatory, Thermal, and Mechanical Responses of Cells in the Nucleus Tractus Solitarius of the Frog

NOBUSADA ISHIKO,[a] TAKAMITSU HANAMORI,[a,b]
AND DAVID V. SMITH [b]

[a]Department of Physiology
Miyazaki Medical College
Miyazaki, 889-16, Japan

[b]Department of Otolaryngology and Maxillofacial Surgery
University of Cincinnati Medical Center
Cincinnati, Ohio 45267

Although responses of single fibers in the glossopharyngeal nerve of frogs have been studied,[1] the response characteristics of medullary cells have never been investigated. The responses of 216 neurons in the nucleus tractus solitarius (NTS) of American bullfrogs (*Rana catesbeiana*) were recorded following taste, temperature, and tactile stimulation of the tongue. Taste stimuli were: 0.5 M NaCl, 0.5 mM quinine hydrochloride, 0.01 M acetic acid, 0.5 M sucrose, each dissolved in 0.01 M NaCl and deionized water. These were delivered to the tongue via gravity flow, as was 30°C 0.01 M NaCl for temperature stimulation (warming). Tactile stimuli were delivered to the tongue and body surface using a small brush. Cells were classified on the basis of their responses to these stimuli.

Neurons showing excitatory responses to one, two, three, or four of the five kinds of taste stimuli were named Type I, II, III, or IV, respectively.[2] Cells whose spontaneous rate was inhibited by taste and/or tactile stimulation of the tongue were termed Type V. Type VI neurons were excited by tactile stimulation alone. Of the 216 cells, 115 were excited or inhibited by taste stimuli (Types I-V), with 35 being Type I, 34 Type II, 40 Type III, 2 Type IV, and 4 Type V. The remaining 101 cells were responsive only to tactile stimulation (Type VI).

Of those 111 cells excited by taste stimulation (Types I-IV), 106 (95%) responded to NaCl, 66 (59%) to acetic acid, 44 (40%) to quinine, 10 (9%) to water and 9 (8%) to warming. No cells responded to 0.5 M sucrose. The responses of these cells to NaCl, quinine, acetic acid, water, touch, and warming are shown in FIGURE 1. The cells are divided into types (I-IV) and ranked within each type according to the magnitude of their response to NaCl. Most Type I cells were sensitive only to NaCl or to NaCl and touch. Type II cells were primarily responsive to NaCl and acetic acid and touch, whereas Type III cells showed additional sensitivities to quinine and warming, which were not prevalent in Type I or Type II cells. Of these 111 cells of Types I-IV, 76 (68%) were sensitive to mechanical stimulation of the tongue. There was a significant correlation ($p < 0.001$) across cells among the responses to quinine, acetic acid, and warming and between the responses to NaCl and acetic acid. The

FIGURE 1. Responses (impulses/2.5 sec) of 111 NTS cells (Types I–IV) to gustatory, mechanical, and thermal stimulation of the tongue. Cells are divided into types and ranked within each type according to their response to NaCl.

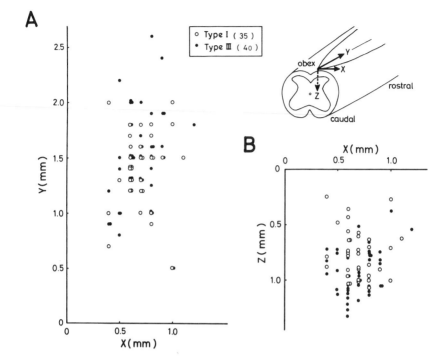

FIGURE 2. Distribution of NTS cells (Types I and III) in the brainstem shown in three-dimensional coordinates. X and Y are medial-lateral and rostral-caudal dimensions, respectively. Z is the dorsal-ventral dimension. Measurements are in mm from the midline, obex, and dorsal surface of the brainstem.

responses to NaCl and water, NaCl and touch, and acetic acid and touch were also significantly correlated ($p < 0.01$).

There was some differential distribution of these neuron types within the NTS, with more narrowly tuned cells (Type I) being located more dorsally in the nucleus than the more broadly tuned (Type III) neurons ($p < 0.01$). The distribution of these cell types is shown schematically in FIGURE 2, where Type I cells are depicted by open circles and Type III by filled circles. Cells responding exclusively to touch (Type VI) were also significantly ($p < 0.001$) more dorsally situated than those responding to two or more taste stimuli (Types II and III). The convergence of tactile and chemical sensitivities on many of these medullary cells suggests a close interaction between gustatory and mechanosensitive afferent inputs in the control of feeding behavior.

REFERENCES

1. KUSANO, K. 1960. Analysis of the single unit activity of gustatory receptors in the frog tongue. Jpn. J. Physiol. **10:** 620-633.
2. MARUI, T. 1977. Taste responses in the facial lobe of the carp, *Cyprinus carpio L.* Brain Res. **130:** 287-298.

Steroid-Dependent Anosmia[a]

B. W. JAFEK,[b] D. T. MORAN,[c] P. M. ELLER,[c]
AND J. C. ROWLEY, III[c]

[b]Department of Otolaryngology/Head and Neck Surgery
[c]Department of Anatomy
University of Colorado School of Medicine
Denver, Colorado 80262

Steroid-dependent anosmia (SDA) has been newly named, but long recognized. It is a syndrome characterized by inhalent allergy, nasal polyps, and anosmia. The anosmia is temporarily reversible by high doses or oral steroids.

Two patients with SDA were studied in detail. The first patient, a 49-year-old man, had anosmia and polyps. Previous evaluation at two other centers showed "anosmia." When tested in our clinic, his UPSIT score was 10/40 (anosmia). Taste testing was normal. Electron microscopy of a biopsy of olfactory epithelium revealed a normal number of olfactory receptors. He underwent intranasal ethmoidectomy, followed by low-dose (5 mg prednisone q.d.) oral steroids. His sense of smell one year later is "normal" and his UPSIT score 31/40.

The second patient, a 47-year-old man, also had anosmia and polyps. His UPSIT score was 9/40, indicative of anosmia. He underwent an olfactory biopsy, which showed olfactory receptors to be normal in number and fine structure, followed by bilateral intranasal sphenoethmoidectomy. His postoperative UPSIT score was 38/40 as he remained on low-dose (triamcinolone 4 mg q.d.) oral steroids.

From these two patients it is concluded that:

1. The anosmia of SDA is not due to neuronal or epithelial change, but blockage of access of the odorant molecules to the receptor sites.

2. The olfactory receptors appear normal in number and fine structure.

3. SDA can be corrected by appropriate surgery, followed by long-term low-dose oral steroids.

4. SDA is the first total anosmia in which active restoration of function can be achieved using appropriate medical and surgical means; this is documented by state-of-the-art methods.

[a]Supported in part by National Institutes of Health Grant #1-P01-NS 2048 20486-01 (Rocky Mountain Taste and Smell Center).

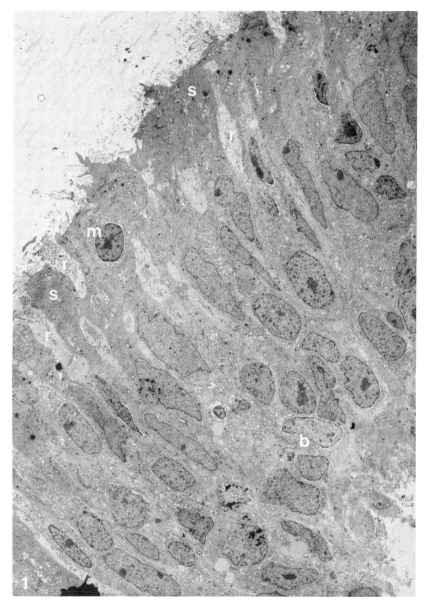

FIGURE 1. Low-magnification view of olfactory epithelium from patient. Receptors (r), support cells (s), a microvillar cell (m), and basal cells (b) are seen in appropriate numbers and orientation.

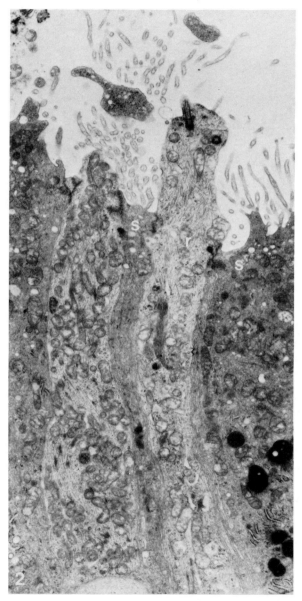

FIGURE 2. High-magnification view of apical region of receptor (r). Adjacent support cells (s) are bound by junctional complexes.

Investigations of Receptive Fields of Olfactory Bulb Neurons in the Frog

TAO JIANG AND ANDRÉ HOLLEY

Laboratoire de Physiologie Neurosensorielle associé au CNRS
Université Claude-Bernard
69622 Villeurbanne cedex, France

Anatomical[1] and electrophysiological[2,3] studies have shown that a restricted area of the olfactory bulb (OB) receives projections from extensive areas of the olfactory epithelium (OE). We are interested in the functional aspect of this convergence and in the spatial integration properties of OB neurons.

METHODS

In curarized frogs, an array of nine micropipettes (diam. 100 μm) filled with saline was placed on the ventral aspect of the OE (FIG. 1A), covering 2.5-4 mm^2. These electrodes were connected to an electrical generator operating under constant current or constant voltage. It delivered long-lasting pulses of positive polarity with a rising slope, a plateau, and a progressive termination (FIG. 1B). The stimulation induced phasic-tonic discharges from receptor cells located close to the electrodes[4] (FIG. 1B). Extracellular unitary responses of OB neurons evoked by focal increase in receptor cell activity were recorded from six regions of the ipsilateral OB.

RESULTS AND DISCUSSION

Fifty-two percent of the 222 recorded cells responded to at least one of the nine stimulated sites of the OE. Eleven types of response patterns were distinguished depending on the occurrence of variations in spontaneous activity associated with the different phases of the stimulation cycle (FIG. 1B). They resembled odor-evoked responses.[5,6] The most frequently observed types were phasic-tonic excitation (IV), suppression (IX), and poststimulus activity rebound (XI). The 11 response types could be grouped into three classes according to their prominent features: early excitation (E), delayed excitation (D), and suppression and/or rebound activity (S).

The proportion of neurons affected by stimulation did not markedly differ from one bulbar region to another. Important differences appeared when considering the mean number of electrodes that could influence unit bulbar activity (FIG. 1C). Twenty-six percent of the responding neurons were sensitive to all nine electrodes while the

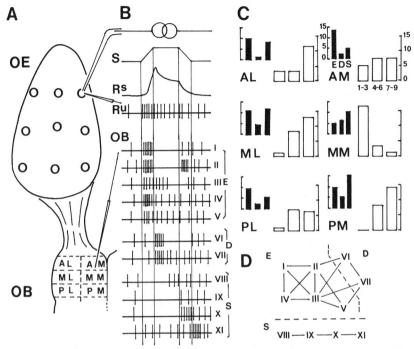

FIGURE 1. (A) Scheme showing the location of stimulating electrodes on the ventral olfactory epithelium (OE) and the six regions of the olfactory bulb (OB) where neurons were recorded.

(B) Examples of unitary (Ru) and summated (Rs) responses of receptor cells evoked by long-lasting (5-sec) stimulation (S) of an OE site. (OB) 11 types of response patterns recorded from bulbar neurons and grouped in three classes: excitation (E), delayed excitation (D), and suppression (S).

(C) Regional distribution of OB neurons as a function of the number of OE electrodes to which they responded (empty bars) and according to the category (classes E, D, S) of their response patterns (filled bars).

(D) Significant associations among different types of response patterns.

mean number of efficient electrodes per neuron was six. A particularly low value was found in the MM region of the OB (2.9).

Regional differences also appeared in the distribution of the response types (FIG. 1C). Class E was relatively dominant in AL, AM, and PL regions, and class S was better represented elsewhere.

Since a response type could be found associated with several others for a single neuron, we calculated an association index for all pairs of response types (FIG. 1D). Some types (e.g., IV and IX) appeared to be mutually exclusive although they were often encountered in different neurons. Significant associations were present among response types within each class and between class E and class D, but not between the latter classes and class S. These results suggest that the receptive fields of OB neurons are not organized in adjacent excitatory-inhibitory zones with clear-cut separation.

The spatial integrative properties of OB neurons were further investigated with stimulation of several OE sites in various combinations. Complex results were obtained.

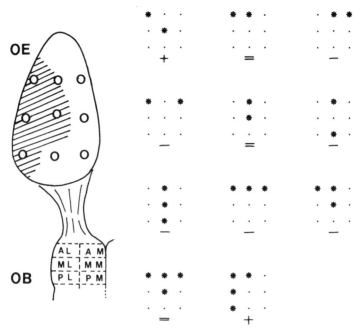

FIGURE 2. Interactions between different sites of the olfactory epithelium (OE) in the response of an olfactory bulb (OB) neuron. The hatched area in OE includes four sites evoking excitatory responses. Different combinations of stimulated sites (*) resulted in partial summation (+), partial suppression (−) of individual effects, or did not change the response (=).

In the example illustrated in FIGURE 2, combined stimulation of individually excitatory sites (hatched area) resulted in a stronger excitatory response (+), indicating partial summation of individual effects. In contrast, combined stimulation of these efficient sites with others that evoked no change in activity when individually stimulated, resulted in a partial suppression (−). In other examples (not shown), even some individually excitatory sites could loose a part of their efficiency when activated together.

CONCLUSION

Our results confirm that frog OB neurons have extensive receptive fields and suggest that the excitatory and inhibitory components of these fields are poorly segregated.

REFERENCES

1. ASTIC, L. & D. SAUCIER. 1986. Brain Res. Bull. **16:** 445-454.
2. COSTANZO, R. M. & M. M. MOZELL. 1976. J. Gen. Physiol. **68:** 297-312.
3. KAUER, J. S. & D. G. MOULTON. 1974. J. Physiol. (London) **243:** 717-737.
4. JUGE, A., A. HOLLEY & J. C. DELALEU. 1979. J. Physiol. (Paris) **75:** 919-927.
5. DÖVING, K. B. 1964. Acta Physiol. Scand. **60:** 150-163.
6. KAUER, J. S. 1974. J. Physiol. (London) **243:** 695-715.
7. DUCHAMP, A. 1982. Chem. Senses **2:** 191-210.

Tuning of Olfactory Neurons Sensitive to Hydroxy-L-Proline in the American Lobster[a]

BRUCE R. JOHNSON, CARL L. MERRILL,
ROY C. OGLE, AND JELLE ATEMA

Boston University Marine Program
Marine Biological Laboratory
Woods Hole, Massachusetts 02543

The lateral branch of the lobster's biramous antennule functions behaviorally as an olfactory organ, triggering search for and orientation to food odor sources.[1] A prominent population of olfactory receptor cells from the lobster *Homarus americanus* is narrowly tuned to trans-4-hydrox-L-proline (Hyp)[2] suggesting that Hyp is an important olfactory signal. However, Hyp is found free in lobster food extracts at concentrations of at most 10^{-4} M; it is normally found bound as one of the major amino acids in collagens. Thus, free Hyp may not be available in sufficient quantities to act as a feeding signal. In this study we examined the tuning properties of Hyp-sensitive olfactory receptors from *H. americanus* to determine other stimuli for these cells.

Responses of single olfactory receptors ($n = 77$) sensitive to Hyp were identified in ablated lateral antennular branches using extracellular hook electrode recordings from small nerve bundles. Prepared stimulus concentrations, reported below, were diluted approximately 30 times after injection into the test chamber. The tuning of these cells was tested in three series of experiments. First, Hyp, L-proline (Pro), *cis*-4-hydroxy-L-proline (cl-Hyp), *cis*-4-hydroxy-D-proline (cd-Hyp), hydroxy-L-lysine (Hyl), kainate (Kai), serotonin (5-HT), DL-octopamine (Oct), all at 10^{-4} M, α-ecdysone (A-Ecd), β-ecdysone (B-Ecd), at 2×10^{-6} M, and a saturated Sigma gelatin solution (SG1) and its one-tenth dilution (SG1/10) were tested on 29 cells. Only SG1 and SG1/10 consistently stimulated Hyp-best cells. Gel electrophoresis showed that SG is composed of gelatin fragments ranging from single amino acids to pieces of 150 kDa. Second, since SG is denatured collagen and contains 14% Hyp, we then tested the following purified macromolecules on 27 Hyp-sensitive cells: Sigma gelatin (SG2), SG2 12-kDa retentate (SG2-12), human placental collagen and gelatin (HPC and HPG), rabbit rib cage collagen and gelatin (RRC and RRG), chicken skin collagen hydrolysate (greater than 12-kDa retentate; CSH), and its 12-kDa filtrate (CSHF), all at 2 mg/ml; rat tail collagen and gelatin (RTC and RTG) at 1 mg/ml, earthworm collagen-like protein and gelatin (EWC and EWG) at 0.1 mg/ml, and *Nereis* cuticle gelatin (NCG) at 0.5 mg/ml. Again, only the SG solutions were consistently stimulatory for Hyp-best cells. Third, we further examined the stimulatory ability of

[a] Supported by grants from the Whitehall Foundation and the National Science Foundation (BNS-8512585) to J. A. and a Grant-in-Aid-of-Research from Sigma XI to B. R. J.

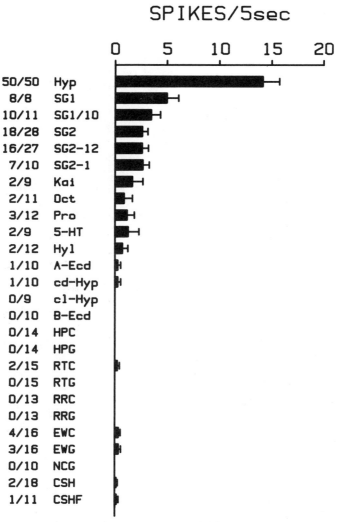

FIGURE 1. Mean response spectrum of 50 Hyp-best cells from the olfactory organ of *H. americanus.* Numbers to the left are the number of times a substance elicited a response over the number of cells on which a substance was tested. Bars: mean response ± SEM. See text for abbreviations.

different molecular weight fractions of SG (SG2, SG2-12 and an SG2 1-kDa retentate, SG2-1) on 21 Hyp-sensitive cells. Ten of these cells responded best to Hyp; the other 11 cells responded best to either SG2, SG2-12, or SG2-1. The response spectrum of cells ($n = 50$) responding best to Hyp from all experiments is shown in FIGURE 1; the remaining cells ($n = 27$) responded best to one of the gelatin solutions (24 cells) or one of the other substances. SG-best cells responded stronger to an SG solution

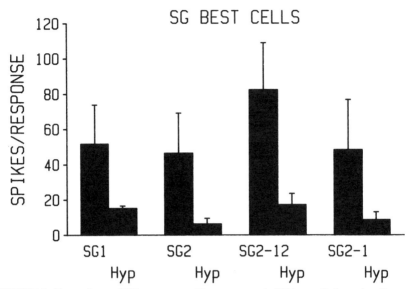

FIGURE 2. Comparisons of SG responses to Hyp responses in SG-best cells from the olfactory organ of *H. americanus.* Bars: mean response ± SEM. See text for abbreviations.

than to Hyp (FIG. 2) and SG solutions stimulated SG-best cells more effectively than Hyp stimulated Hyp-best cells (FIGS. 1 and 2).

We demonstrate here two common olfactory receptor types both responding to Hyp: 65% Hyp-best cells and 31% gelatin-best cells. Hyp-best cell stimulation by SG solutions may be due to Hyp residues in gelatin fragments. The stimulation of Hyp cells by SG solutions and the presence of cells highly sensitive to SG fractions suggests that degradation products (peptides) of connective tissue proteins may be important signal compounds for lobster feeding behavior.

REFERENCES

1. ATEMA, J. 1985. Chemoreception in the sea. Soc. Exp. Biol. Symp. **39:** 387-423.
2. JOHNSON, B. R. & J. ATEMA. 1983. Narrow-spectrum chemoreceptor cells in the antennules of the American lobster, *Homarus americanus.* Neurosci. Lett. **42:** 145-150.

Peripheral Coding by Graded Overlapping Reaction Spectra?

Sensitivity and Selectivity of Olfactory Receptor Cells in *Antheraea polyphemus* L.

WOLF A. KAFKA

Max-Planck-Institut für Verhaltensphysiologie
D-8131 Seewiesen
Federal Republic of Germany

In the general context of primary olfactory coding, quantitative electrophysiological measurements were made of the responses of 50 olfactory receptor cells to a total of 54 substances. The cells are distributed among three morphologically distinguishable sensilla on the antennae of the male silkmoth *Antheraea polyphemus:* sensillum basiconicum (b), and two types of sensillum trichodeum (ts, tl). The (up to 3) receptor cells in a given sensillum are distinguished by the amplitude of the recorded nerve impulses. On the basis of these two criteria, each receptor is assigned to one of nine morphological-electrophysiological categories (bs, bm, bl; tss, tsm, tsl; tls, tlm, tll; the latter indices, s, m, l, indicate the recorded amplitudes small, medium, and large, (Fig. 1).

The responses of all the receptor cells are compared with regard to reaction spectrum and to a substance-efficacy index defined, on the basis of dose-response curves, as the stimulus intensity (molecules/cm^3) required to elicit 25 impulses/sec.

By treating these efficacy indices as vector components and applying a vector-correlation procedure,[1] it has been possible to obtain a single quantitative measure of the similarity, one to one, of the reaction spectra of the receptor cells in a 50 × 50 matrix. The resulting degrees of similarity are grouped in a manner consistent with the morphological-electrophysiological categorization. There is no appreciable specialization to substances of particular chemical classes (alcohols, esters, acids, aldehydes, amines, carbohydrates); nevertheless, and despite considerable overlap among the various reaction spectra, the cells are highly selective (TABLE 1).

In view of the grouping of the reaction spectra found here, the traditional concept of dichotomy between so-called generalists and specialists[1,2] requires modification.

It has to be assumed that hypothetical entities called "acceptors," each capable of interacting only with a specific substance or set of substances, are distributed over the dendrite membrane in well-organized sets of compositions. The sets might be composed of a larger number of different acceptor types in generalists than in specialists.

Considering a limited capacity for the uptake of acceptors on the receptor cell membranes, the width of a reaction spectrum could very likely be correlated with the sensitivity of the cells: the broader their spectra the less their sensitivity.

TABLE 1. Patterns of Similarity and Specialization of the Reaction Groups[a]

c	st		(bs)	(bm)	(tsm)	(tss)	(tsl)	(tlm)	(tls)	(tll)	(bl)
36	.6	Cinnamaldehyde	+++++	+++++	+++++	++	0	+	+	+	++
41	1.4	Citral	++++	++	++++	++	+	+	−	−	+
48	3.1	Decanal	++++	++	0+	++	+	++++	−	++++	0
21	2.2	3-phenyl-1-propenal	+	0	0	−	−	−	−	−	0
18	8.9	Benzaldehyde	0	0	0	0	0	−	−	−	++
22	.71	Anisaldehyde	−	0	−	−	−	+	−	−	0
24	2.6	Menthol	+++	+	++	0	++	++++	+++	+	0
28	2.6	Isopulegol	+++	++	−	−	−	−	++	−	++
39	10	2-octanol	++	++	++	++	+	++	+	++	+
40	.57	Eugenol	+	+	0	0	+	0	0	0	+++
37	5	Linalool	+	+	−	−	−	−	−	0	+++
1	9.6	P-cymol	+	+	++	+++	+	+	+	+	++
16	2.2	Terpineol	+	+	++	++	+	+	++	+	+++
33	1.4	Geraniol	+	+	++	+	+	+	+	+++	+
29	.95	Isoeugenol	0	0	+++	−	−	0	0	0	0
19	3.5	Alpha-phenyl ethanol	+	+	+++	+	0	0	0	−	+
17	3.3	Benzyl alcohol	+	+++	0	−	−	−	0	−	0
20	2	Beta-phenyl ethanol	+	+++	−	−	−	−	0	−	0
26	19	Cyclohexanol	+	++	−	−	−	−	0	0	+++
32	1.8	Citronellol	0	0	−	−	−	0	0	0	+++
35	1.2	Thymol	0	+	−	−	−	0	0	0	+++
4	36	Cyclopentanol	0	0	−	−	−	0	0	+	+
25	3.3	Guajacol	0	0	−	−	−	0	0	−	0
9	67	Benzoic acid	++	0	+	++++	0	0	0	0	0
8	15	Methyl acrylic acid	++	+	−	−	−	−	−	−	0
7	36	Acrylic acid	+	0	−	−	−	0	0	0	0
13	4.3	Tiglic acid	+	0	−	−	−	−	−	−	0
14	4	2-methyl valeric acid	+	++++	−	−	−	−	−	−	0
12	3	z3-hexenoic acid	+	0	−	−	−	−	−	−	+

52	1.2	Phenyl ethyl acetate
51	50	Amyl acetate
43	2.1	Phenyl methyl acetate
44	4.3	Methyl benzoate
31	1.1	Anethole
49	2.8	Nitrobenzene
27	2.6	Ethyl benzoate
53	3.1	Menthone
10	160	3-methyl-3-buten-2-one
34	1.3	Carvone
30	1.7	Pulegone
15	20	Cyclohexanone
23	2.2	1-phenylpropanone
11	58	Hexan-2-one
54	4	Alpha-thujone
2	9.6	Limonene
6	20	Alpha-pinene
5	14	Beta-pinene
42	17	Camphene
38	.1	Cumarin
3	290	Cyclohexane
47	90	Tributyl amine
46	10	Octyl amine
45	44	Cyclohexyl amine

[a] The values for each group represent the mean reaction spectra within each morphological-electrophysiological category (see text). c = substance number, st = molecules per cm^3 of stimulus air $\times$ 10^{12}. Beginning with the broadest spectrum, that of reaction group bs, the symbols in the row following each substance name represent factors of st required to elicit a response of 25 imp/sec: $+++++ = 0 - 30$; $++++ = 30 \leq 100$; $+++ = 100 \leq 300$; $++ = 300 - 1,000$; $+ = 1000 \leq 3000$; $0 =$ ineffective at all steps of stimulus concentrations; $-- =$ no evaluation. Clear-cut similarities (outlined by solid and broken lines for the individual cell groups) refer only to a small number of compounds.

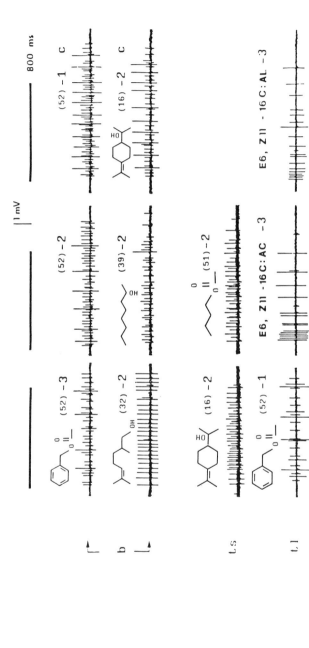

FIGURE 1. Selectivity and overlap of the reaction spectra of receptor cells in the sensillum basiconicum (b), s. trichodeum (ts and tl); sections of longer recordings. Black bars at top: stimulus duration. Numbers above the recordings: the first number represents the name of substance and the second the dilution in powers of st*10 (−1, −2, −3) see TABLE 1. Each line represents a record of one and the same sensillum. First line: dose–response relations to phenylethyl acetate (52) to three decadic steps of concentrations. Citronellol (32) mainly excites the "large" cell and 2-octanol (39) the "small" cell; terpineol (16) excites both cells. Terpineol also excites both the "medium" and the "large" cell of a short s. trichodeum (ts). The medium and large cells are also excited by amyl acetate (51). Phenylethyl acetate is also slightly effective on the pheromone-specific (E6, Z11-16:AC or E6,Z11-16:AL) receptor cells.

The findings thus might enhance the assumption that the mode of the primary encoding of olfactory signals is not only overcome by a mere specialization of different types of receptor cells, but also by different degrees of specialization.

REFERENCES

1. KAFKA, W. A. 1986. Similarity of reaction spectra and odor discrimination: Single receptor cell recordings in *Antheraea polyphemus* (Saturniidae). J. Comp. Physiol. **600:** In press.
2. SCHNEIDER, D., V. LACHER & K. E. KAISSLING. 1964. Die Reaktionsweise und das Reaktionsspektrum von Riechzellen bei *Antheraea pernyi* (Lepidoptera Saturniidae). Z. Vgl. Physiol. **48:** 632-662.
3. KAFKA, W. A. 1974. Physicochemical aspects of odor reception in insects. Ann. N.Y. Acad. Sci. **237:** 115-128.
4. SELZER, R. 1981. The processing of a complex food odor by antennal olfactory receptors of *Periplaneta americana*. J. Comp. Physiol. **144:** 509-519.

Specific L-Arginine Receptor Sites

Biochemical and Neurophysiological Studies

D. LYNN KALINOSKI,[a] BRUCE P. BRYANT,[a] GAD
SHAULSKY,[a] AND JOSEPH G. BRAND [a,b]

[a]Monell Chemical Senses Center
Philadelphia, Pennsylvania 19104

[b]Veterans Administration Medical Center
University of Pennsylvania
Philadelphia, Pennsylvania 19104

Amino acids are effective taste stimuli for the channel catfish, *Ictalurus punctatus*. From the results of neurophysiological[1] and behavioral studies,[2] at least two different classes of taste receptors have been inferred: one responding to alanine and some other neutral amino acids, the other responding to arginine. While the alanine binding site has been extensively characterized,[3] less is known of the site(s) for L-arginine binding. We report here the initial characterization of the arginine binding site(s). Binding of L-[^{3}H]arginine to the sedimentable Fraction P2 from the taste epithelium was measured by a modification of the method of Krueger and Cagan,[4] in which an isotope dilution step was included to reduce nonspecific binding. Time and pH for measuring maximal binding activity were established. At pH 7.8, the rate constant for association at 4°C was $1.1 \times 10^5 \, M^{-1} \, \text{min}^{-1}$. Dissociation was more complex, yielding rate constants of $1.4 \, \text{min}^{-1}$ and $4.1 \times 10^{-2} \, \text{min}^{-1}$. These data suggest the presence of at least two binding sites for arginine with K_Ds of $1.3 \times 10^{-5} \, M$ and $3.7 \times 10^{-7} \, M$. Homologous inhibition studies of L-arginine binding yielded a complex curve (which did not fit a simple mass action relationship [Hill coefficient 0.79]). These data are consistent with the presence of high and low affinity sites for L-arginine. The $K_{D\text{app}}$ as calculated from IC_{50} values are $4.2 \times 10^{-7} \, M$ for the high-affinity state and $8.4 \times 10^{-4} \, M$ for the low-affinity site. The ability of D-arginine, L-lysine, L-α-amino-β-guanidino-propionic acid (L-AGPA), and L-arginine to displace L-[^{3}H] arginine binding was tested under equilibrium conditions. The complex patterns of inhibition by L-arginine and L-lysine also support the existence of at least two different affinity classes of the receptor. L-AGPA and D-arginine inhibited binding of only the lower affinity L-arginine site. Significant inhibition of L-glutamate, glycine, and L-alanine occurred only above 10^{-3} *M,* indicating that the binding site(s) for L-arginine are selective. Using multiunit recordings, neurophysiological studies examined the stimulatory effectiveness of a number of guanidinium-containing compounds. Only L-arginine, L-arginine methyl ester, and L-AGPA were effective stimuli. Cross-adaptation experiments using the same preparation examined the ability of 10^{-4} *M* glycine, L-AGPA, L-lysine, D-arginine, L-glutamate, and L-alanine to inhibit responses to 10^{-6} *M* L-arginine. Only L-AGPA was an effective cross-adapting stimulus. These results indicate that effective agonists of L-arginine receptors must contain a guanidinium group and an unblocked

L-α-amino group. Taken together, the biochemical and neurophysiological results indicate that site occupation alone does not necessarily lead to receptor activation and the generation of neural events.

REFERENCES

1. CAPRIO, J. 1982. High sensitivity and specificity of olfactory and gustatory receptors of catfish to amino acids. *In* Chemoreception in Fishes. T. J. Hara, Ed.: 109-134. Elsevier Scientific Publishing Co. Amsterdam, the Netherlands.
2. STEWART, A., B. BRYANT & J. ATEMA. 1979. Biol. Bull. **157:** 396-407.
3. CAGAN, R. H. 1981. Recognition of taste stimuli at the initial binding interaction. *In* Biochemistry of Taste and Olfaction. R. H. Cagan and M. R. Kare, Eds.: 175-203. Academic Press. New York.
4. KRUEGER, J. M. & R. H. CAGAN. 1976. J. Biol. Chem. **251:** 88-97.

Cell Suspension from Porcine Olfactory Mucosa

Changes in Membrane Potential and Membrane Fluidity in Response to Various Odorants

MAKOTO KASHIWAYANAGI, KIMIE SAI, AND
KENZO KURIHARA

Faculty of Pharmaceutical Sciences
Hokkaido University
Sapporo, Japan 060

Several investigators have attempted to isolate olfactory cells but none of them have reported that the cells obtained retain the physiological functions necessary to respond to odorants. We have succeeded in obtaining a cell suspension that has the ability to respond to various odorants from porcine olfactory mucosa.

The porcine olfactory mucosae were gently removed with a spatula and dispersed in the culture medium by gentle pipetting. The cell suspensions thus obtained (cell suspension I) contained cells that have the characteristic structure of the olfactory cell. The content of cells having such morphology was 10% at least. Some olfactory cells were deformed and hence the actual content of the olfactory cells may well be more than 10% because the deformed cells were not counted as olfactory cells. Cell suspension I was fractionated further by means of a bovine serum albumin density gradient. The fraction of 4% BSA (cell suspension II) contained about 30% olfactory cells. The responses of cell suspension II to odorants were essentially similar to those of cell suspension I and hence cell suspension I was used for most experiments. The results obtained are as follows:

1. The membrane potential of the cell suspension, which was monitored by measuring the fluorescence changes of rhodamine 6G, was depolarized by an increase in K^+ concentration in external medium, but the magnitude of the depolarization was rather small. The reduction of concentrations of Na^+, Ca^{2+} and Cl^- in the external medium did not affect the resting potential, suggesting that the number of selective channels open in the resting state is rather small. These results are consistent with the recent results that the membrane resistance of the olfactory cells is extremely high.

2. Various odorants such as muscone, cyclooctane, β-ionone, nonanol, menthone, and amyl acetate depolarized the cell suspension dose dependently. The cell suspension that contained no viable cells did not respond to odorants.

3. The magnitude of depolarization by odorants was unchanged or a little increased by reduction of concentrations of Na^+, Ca^{2+}, and Cl^- in external medium.

In a previous paper[1] we showed that the elimination of Na^+, Ca^{2+}, and Cl^- from a solution perfusing the olfactory epithelium did not significantly affect the olfactory responses, suggesting that the changes in the permeabilities of such ions across the apical membranes of olfactory cells are not concerned with the generation of the receptor potentials. These results suggest that changes in the permeabilities of specific ions at the basal membrane as well as the apical membrane are not concerned with the generation of the receptor potential.

4. Application of various odorants to the cell suspension induced changes in the membrane fluidity at different sites of the membranes, which were monitored with various fluorescence dyes (ANS, 7-AS, 12-AO, and DPH), suggesting that odorants having different odors are adsorbed on different sites in the membranes. In previous papers,[2,3] we showed that the mouse neuroblastoma cell (N-18 cell) was depolarized by various odorants, and the membrane fluidity changed in response to various odorants. The profiles of the membrane fluidity changes monitored with various fluorescence dyes in the porcine olfactory cell suspensions were similar to those in the N-18 cell. These results suggest that, like the N-18 cell, specific receptor proteins unique to olfactory cells are not concerned with odorant reception in the porcine olfactory system.

REFERENCES

1. KURIHARA, K. & K. YOSHII. 1983. Role of cations in olfactory reception. Brain Res. **274:** 239-248.
2. KASHIWAYANAGI, M. & K. KURIHARA. 1983. Neuroblastoma cell as model for olfactory cell: Mechanism of depolarization in response to various odorants. Brain Res. **293:** 251-258.
3. KASHIWAYANAGI, M. & K. KURIHARA. 1984. Evidence for non-receptor odor discrimination using neuroblastoma cells as a model for olfactory cells. Brain Res. **359:** 97-103.

Odor Information Processing in the Olfactory Bulb

Evidence from Extracellular and Intracellular Recordings and from 2-Deoxyglucose Activity Mapping

J. S. KAUER AND K. A. HAMILTON

Department of Neurosurgery and
Department of Anatomy and Cell Biology
Tufts University Medical School and
New England Medical Center
Boston, Massachusetts 02111

Extracellular recordings from single mitral-tufted cells (MTs) in the salamander olfactory bulb have shown several stereotyped categories of response to controlled odor pulses. Excitatory responses (E-type) are categorized by increases in firing frequency at threshold that are curtailed by suppression at higher concentrations. These responses are said to be concentration tuned. Suppressive responses (S-type) are categorized by cessation of firing at odor onset with excitation sometimes elicited later in the pulse. Pooled extracellular data from many animals show that each response type is elicited by any of the odors tested (albeit in different cells), suggesting that a single pulse of any one odor triggers activity within many bulbar elements, each responding in a stereotyped way.

We have recently obtained intracellular recordings from salamander MT cells using the same stimulation methods (see Hamilton and Kauer[1]). These studies show that the periods of excitation and suppression in extracellular records correlate with periods of depolarization and hyperpolarization, indicating that these response patterns are generated by bulbar circuits and are not a simple reflection of activity patterns in the receptor cells. These intracellular data have also shown that the first event after the odor pulse onset is hyperpolarization in both E- and S-type responses at moderate concentrations. The presence of this initial hyperpolarization in both E- and S-type responses without a preceding spike suggests that this inhibition may be generated by peripheral bulbar elements such as periglomerular cells (PGs).

Our anatomical data[2] indicate that there are convergent projections connecting widespread epithelial loci to restricted glomerular regions. We suggest that this connectivity pattern may underlie the broad receptive fields seen in S-type responses and the restricted receptive fields seen in E-type responses (see Kauer and Moulton[3]). Thus receptor cells distributed widely across the olfactory mucosa may connect directly to PGs associated with a particular MT, thereby giving rise to the early hyperpolar-

400

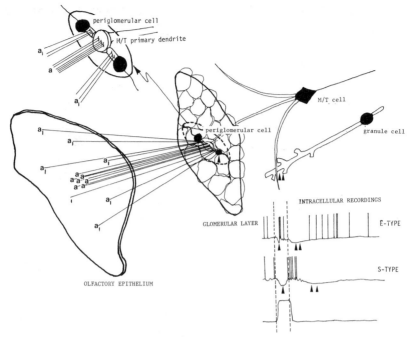

FIGURE 1. Hypothetical connections of olfactory receptor cell axons onto MT and PG cell dendrites in the glomerular layer of the salamander olfactory bulb. Left: dispersed (a_1) and clustered (a) receptor cells responding to odor "a" are shown on schematic view of ventral olfactory mucosa. Right (and inset): schematic view of flattened glomerular layer showing distribution of afferents from "dispersed" receptor axons onto PG dendrites (a_1's) and clustered reception cell axons onto MT dendrites (a's). Lower left: intracellular recording from one MT cell showing E and S responses to two different odors. Initial hyperpolarization is shown by a single arrowhead; secondary hyperpolarization is shown by double arrowheads. We suggest that the initial hyperpolarization is generated by projections of dispersed receptors to PGs, that depolarization is mediated by direct projections onto MTs, and that secondary hyperpolarization is generated via granule cells.

izization seen in the intracellular recordings, and clustered receptor cells may connect directly to MTs thereby giving rise to E-type excitation (see FIG. 1).

Since the activity of all bulbar cells cannot be observed simultaneously using electrophysiological methods, we have used enhanced resolution 2DG mapping to describe odor-elicited activity throughout the bulb. By superimposing 2DG autoradiographs onto their matched histological sections, we have shown that the highest 2DG levels in the glomerular layer occur over PG cell somata. We hypothesize that this activity may be a manifestation of widespread inhibition laid down by projections of receptors onto PG cell dendrites.

REFERENCES

1. HAMILTON, K. A. & J. S. KAUER. 1985. Intracellular potentials of salamander mitral/tufted neurons in response to odor stimulation. Br. Res. **338:** 181-185.

2. KAUER, J. S. 1980. Some spatial characteristics of central information processing in the vertebrate olfactory pathway. *In* Olfaction and Taste VII. van der Starre, Ed.: 227-236. IRL Press. London, England.
3. KAUER, J. S. & D. G. MOULTON. 1974. Responses of olfactory bulb neurones to odour stimulation of small nasal areas in the salamander. J. Physiol. (London) **243:** 717-737.

Lectin-Binding Sites in Olfactory Sensilla of the Silkmoth, *Antheraea polyphemus*

THOMAS A. KEIL

Max-Planck-Institut für Verhaltensphysiologie
Gruppe Kaissling
D-8131 Seewiesen
Federal Republic of Germany

In single-walled olfactory sensilla of insects, the extracellular "pore tubules" are thought to be the pathways by which odor molecules traverse the cuticular hair wall and reach the membranes of the olfactory dendrites.[1] For the formation of contacts between tubules and membranes, their polyanionic surface coats, which have been demonstrated by application of the cationic markers La^{3+}, ruthenium red, and cationized ferritin,[2] seem to be responsible.

The cellular surface coat consists mainly of oligosaccharide side chains of membrane-integrated glycoproteins, but also of glycolipids. Its responsibility for cell-cell and cell-substratum recognition and adhesion is generally accepted. The different sugar molecules that make up the surface coat can be marked by application of sugar-specific lectins.[3]

In order to investigate the composition of the surface coats present in silkmoth olfactory sensilla, different lectins (e-y laboratories, San Mateo, CA; conjugated with ferritin for visualization in the electron microscope) have been applied to apically opened hairs that were then processed for the EM (incubation time: 1 h; fixation in glutaraldehyde and OsO_4). TABLE 1 lists the lectins that showed recognizable binding

TABLE 1. Lectins That Showed Binding to Hair Structures[a]

Lectin *(source)* (specific sugar)	Dendrites	Pore Tubules	Cuticle
Con A *(Canavalia ensiformis)* (mainly mannose, glucose)	+ +	−	+ +
LFA *(Limax flavus)* (sialic acid)	+ +	+ +	−
BPA *(Bauhinia purpurea)* (galactose, galactosamine)	+ +	−	−
SBA *(Glycine max)* (galactose, galactosamine)	+	−	−
GS II (Griffonia simplicifolia) (glucosamine)	+	−	−
UEA I *(Ulex europaeus)* (fucose)	+	−	−

[a] (+ +) = strong reaction; (+) = positive reaction; (−) = negative reaction.

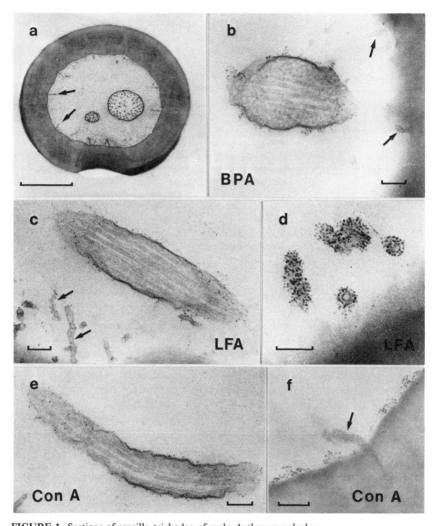

FIGURE 1. Sections of sensilla trichodea of male *Antheraea polyphemus.*

(a)   Cross section, not incubated, freeze-substituted. A thick and a thin dendrite are present in the hair lumen, containing cytoskeletal microtubules. Numerous extracellular pore tubules are protruding from the cuticular hair wall (arrows).

(b)   Incubation in BPA, which binds only to the dendritic membrane, but not to the pore tubules (arrows) and hair cuticle.

(c)   Incubation in LFA, which binds to the dendritic membrane, but also to the pore tubules (arrows).

(d)   Heavy binding of LFA to pore tubules, some of which are shown side-on and two end-on.

(e)   Incubation in Con A, which binds to the dendritic membrane.

(f)   Binding of Con A to the hair cuticle, but not to the pore tubules (arrow).

Magnification scale bar for a, 1 μm; for b-f, 0.1 μm.

to the different structures in the hairs, as well as their binding specificity. FIGURE 1 shows the morphological appearance of these structures after application of the three most effective lectins.

The strongest affinity for the dendritic membranes have Con A (which is said to have a relatively low specificity), LFA, and BPA. Whereas the former two lectins are distributed quite evenly over the dendritic membranes, BPA seems to bind more in the form of clusters. A relatively weak affinity for the dendritic membranes is shown by SBA, GS II, and UEA I. The inner surface of the hair cuticle binds only Con A; the pore tubules bind only LFA very strongly.

Thus it seems that the surface coats of the dendritic membranes are composed of the sugars that are typical for the nervous system.[4] The surfaces of the pore tubules are characterized by a high amount of sialic acid. This is the first direct hint about the chemical composition of these structures and might indicate the presence of glycolipids. The experiments have as yet yielded a way for discriminating pore tubules and dendritic membranes. They might finally help to identify receptor molecules in the membranes, which at least in frog olfactory cilia seem to be glycoproteins.[5]

REFERENCES

1. KEIL, T. A. & R. A. STEINBRECHT. 1984. Mechanosensitive and olfactory sensilla of insects. *In* Insect Ultrastructure. R. C. King & H. Akai, Eds. Vol. **2**: 477-516. Plenum. New York.
2. KEIL, T. A. 1984. Surface coats of pore tubules and olfactory sensory dendrites of a silkmoth revealed by cationic markers. Tissue Cell **16**(5): 705-717.
3. SHARON, N. & H. LIS. 1972. Lectins: Cell-agglutinating and sugar-specific proteins. Science **177**: 949-959.
4. KELLY, P. T. 1984. Nervous system glycoproteins. *In* The Biology of Glycoproteins. R. J. Ivatt, Ed.: 323-369. Plenum. New York.
5. CHEN, Z. & D. LANCET. 1984. Membrane proteins unique to vertebrate olfactory cilia: Candidates for sensory receptor molecules. Proc. Natl. Acad. Sci. USA **81**: 1859-1863.

Comparisons between Human Gustatory Judgments of Square Wave Trains and Continuous Pulse Presentations of Aqueous Stimuli

Reaction Times and Magnitude Estimates[a]

STEVEN T. KELLING [b] AND BRUCE P. HALPERN

Department of Psychology
Field of Physiology
Section of Neurobiology and Behavior
Cornell University
Ithaca, New York 14853-7601

Many studies on gustatory intensity of aqueous solutions have been reported. In most cases, stimuli were either a single pulse to the anterodorsal tongue,[1,2] or whole mouth stimulation.[3] Intermittent stimuli have been used, but the open dorsal flow delivery systems permitted neither rectangular waveforms nor trains with periods of less than a few seconds.[4,5] In these experiments, square wave gustatory trains with periods of 100 or 200 msec were compared with continuous stimuli. A closed-flow liquid delivery apparatus was used.[6]

Magnitude estimates of 2 mM sodium saccharin (NaSac) total taste intensity and reaction times, were obtained in proportion to a 2,000-msec continuous-flow standard.[7] The stimuli were presented either as a single, continuous pulse or as a 5- or 10-Hz square wave train of several different pulse durations or train lengths (FIG. 1).

Judged intensity increased with greater stimulus pulse duration or train length (FIG. 2). No significant differences in intensity judgments or reaction times were found between square wave trains at either frequency ($p > 0.05$). When the total solution presentation durations of square wave trains were compared with single pulses of the same duration, intensity judgments and reaction times were significantly different ($p < 0.05$; FIG. 2). Trains were more intense than comparable single pulses, and their reaction times were faster.

Gustatory intensity processing is an accumulation over time. At and above 5 Hz, stimuli presented as square wave trains are more effective than continuous-flow presentations. The efficacy of long-period gustatory trains in preventing adaptation over 60 seconds or more has been reported.[4] This data may indicate that gustatory receptor adaptation occurs rapidly and can be reversed by brief flows of carrier liquid. Alternatively, the enhanced effectiveness of gustatory trains may be related to the repeated on, or on-and-off, transients.

[a] Supported by National Science Foundation Grant BNS-8213476.

[b] Address for correspondence: Steven T. Kelling, Department of Psychology, Uris Hall, Cornell University, Ithaca, NY 14853-7601.

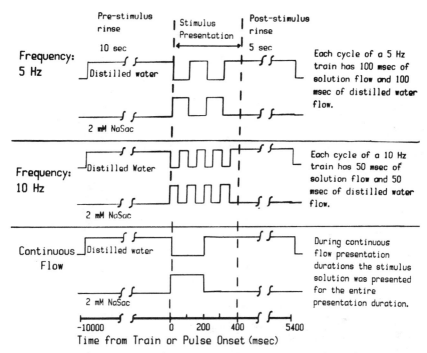

FIGURE 1. Schematic diagrams of gustatory stimulus square wave trains at frequencies of 5 Hz and 10 Hz, and a single continuous pulse presentation. The total stimulus solution presentation duration of a 400-msec square wave train at either frequency is equal to the presentation duration of a 200-msec duration continuous pulse.

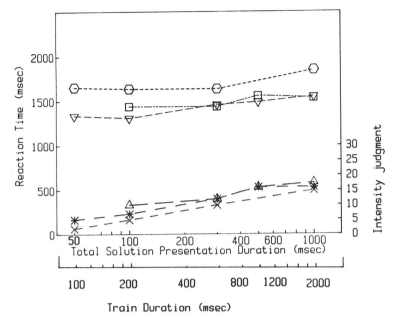

FIGURE 2. Median magnitude estimations, and their reaction times, to gustatory square wave train presentations, and to gustatory single, continuous-pulse presentations. All standard errors of the median were less than one for intensity judgments; less than 100 msec for reaction times. A throat microphone connected to a digital timer recorded the onset of a judgment. Five subjects were tested. They received two stimulus solution presentations, and two control stimulus (distilled water) presentations, at each train length or pulse duration, for the five sessions of an experiment. In this graph, hexagons and the dashed line represent magnitude estimation reaction times for single, continuous-pulse presentations; squares and the dash-double-dot line represent magnitude estimation reaction times for 5-Hz trains; while inverted triangles represent magnitude estimation reaction times for 10-Hz trains. Asterisks, triangles, and Xs represent the median magnitude estimations for 10-Hz trains, 5-Hz trains, and single, continuous pulses, respectively.

REFERENCES

1. BUJAS, Z. & A. OSTOJCIC. 1939. L'evolution de la sensation gustative en fonction du temps d'excitation. Acta. Inst. Psychol. Zagreb **3**: 1-24.
2. KELLING, S. T. & B. P. HALPERN. 1983. Taste flashes: Reaction times, intensity, and quality. Science **219**: 412-414.
3. MCBURNEY, D. H. 1984. Taste and olfaction: Sensory discrimination. *In* Handbook of Physiology—The Nervous System III, Sensory Processes. J. Brookhart and V. B. Mountcastle, Eds.: 1067-1086. Williams and Wilkins. Baltimore, MD.
4. HALPERN, B. P., S. T. KELLING & H. L. MEISELMAN. 1986. An analysis of the role of stimulus removal in taste adaptation by means of simulated drinking. Physiol. Behav. **36**: 925-928.
5. MCBURNEY, D. H. 1976. Temporal properties of the human taste system. Sensory Proc. **1**: 150-162.
6. KELLING, S. T. & B. P. HALPERN. 1986. The physical characteristics of open flow and closed flow taste delivery apparatus. Chem. Senses **11**: 89-104.
7. KELLING, S. T., E. SCHWARTZCHILD & B. P. HALPERN. Human gustatory judgments of aqueous square wave trains: reaction times and magnitude estimations. Chem. Senses **11**: 621.

Temporal Analyses of the Actions of Normal Alcohols on Taste Receptor Cell Responses to Sucrose[a]

LINDA M. KENNEDY

Department of Biology
Clark University
Worcester, Massachusetts 01610

Normal alcohols suppress fly behavioral and taste receptor cell responses to sucrose.[1,2] Preliminary data indicate that the receptor cell effects are biphasic over time—firing is first suppressed and then increased and irregular—as are effects of the gymnemic acids and ziziphins (GAs and Zs).[3] These similar biphasic effects suggest that mechanisms by which n-alcohols act on receptor cells could be similar to mechanisms of the GAs and Zs. I studied the temporal effects of a series of n-alcohols to (1) confirm similarities with effects of GAs and Zs and (2) test a prediction of the biphasic membrane penetration model[4] for the actions of GAs and Zs. The prediction is that the firing increase will occur sooner for less polar than for more polar modifiers, and it holds for the GAs and Zs according to their relative polarities.[3] If n-alcohols act according to the model and by mechanisms similar to those of the GAs and Zs, then the onset of firing increase should vary consistently with n-alcohol chain length (polarity).

Methanol, propanol, hexanol, and octanol were tested at the behavioral threshold concentrations[1] in two-trial sequences on single *Phormia regina* taste hairs in isolated proboscises. In the first trial, hairs were treated with Tris buffer (15 mM, pH 7) for two minutes, and then action potential responses to sucrose (50 mM in NaCl 50 mM) were tip recorded for 10 minutes. In the second trial, hairs were treated with n-alcohols (in Tris) or with a Tris control for two minutes and responses to sucrose recorded as before. For analysis, ratios were calculated as numbers of spikes in 0.1- and 1.0-second samples taken at 15- and 30-second intervals from second-trial responses per numbers of spikes in 0.1- and 1.0-second samples taken at the same times from first-trial responses.

Biphasic effects occurred: Firing was initially suppressed (ratios < 1), then increased and became irregular (ratios > 1; FIG. 1). Initial suppressions and increases over time were significant for each alcohol (Kramer, $p \leq 0.05$). Rates of onset of increased firing differed significantly after each alcohol except octanol in comparison with those after the Tris control (Hollander, $p \leq 0.01$). A plot of least-squares slopes for rates of firing increase against alcohol polarity (number of carbons) showed an inverse relationship (slope $= 0.12$, $r^2 = 0.69$), which was not significant ($p > 0.05$).

Propanol was tested at the behavioral threshold concentrations for methanol and hexanol. The initial suppression was of significantly greater magnitude and longer

[a] This work was supported by Clark University Faculty Development and Biomedical Research Grants to the author.

duration after 9 *M* propanol than after 9 *M* methanol (Mann Whitney, $p < 0.03$; FIG. 2). In contrast to 12 m*M* hexanol, 12 m*M* propanol did not suppress responses initially. However, each of three flies tested in extended trials (10-second stimulations every two minutes) showed a suppression and subsequent increase 10-30 minutes after 12 m*M* propanol. In the single extended Tris control trial, no such time-varying effects occurred (FIG. 2).

Thus the prediction was not met. Whereas the onset of increased firing may or may not be related to alcohol chain length, the initial suppression is. The different results for GAs and Zs may reflect effects of mixtures of saponins with various polarities in plant extracts. These data remain consistent with a membrane penetration process and suggest that physicochemical properties play a role in the initial suppression of responses to sucrose.

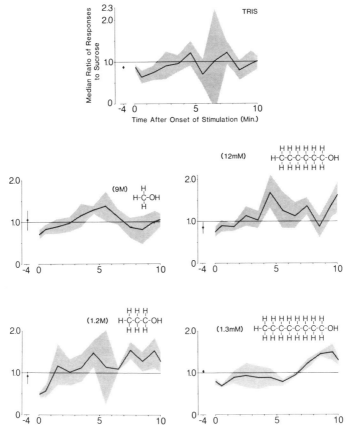

FIGURE 1. Ratios of responses to sucrose before and after treatment with Tris or with behavioral threshold concentrations of normal alcohols. Response ratios for methanol and propanol treatments are on the left; those for hexanol and octanol treatments are on the right. The solid lines are drawn from median ratios for the first 0.1 second (0 min) through median ratios averaged over subsequent one-minute intervals of responses. Shaded areas represent semi-interquartile ranges. Data points at −4 minutes give median ratios for the first 0.1 second of responses to five-second sucrose stimulations given one minute before treatments. For each alcohol, $n = 6$; $n = 7$ for the Tris treatment. See text for ratio formation procedure.

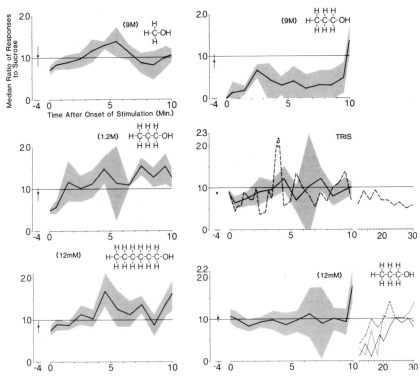

FIGURE 2. Ratios of responses to sucrose before and after treatment with equimolar concentrations of normal alcohols or with Tris. Graphic representations and *n*s are as in FIGURE 1. On the left are response ratios before and after behavioral threshold concentrations of methanol, propanol, and hexanol; on the right are response ratios before and after propanol concentrations equimolar to the methanol and hexanol threshold concentrations and before and after Tris. The fine, dotted, and dashed lines in the 12 m*M* propanol condition give individual ratios for three flies in extended trials. The dashed line in the Tris condition gives ratios for one fly in an extended trial.

ACKNOWLEDGMENTS

I thank M. Whitney and D. Kolodny for assistance, L. Wall for graphics, and T. Livdahl for comments on the statistics and manuscript.

REFERENCES

1. DETHIER, V. G. & L. E. CHADWICK. 1947. Rejection thresholds of the blowfly for a series of aliphatic alcohols. J. Gen. Physiol. **30:** 247-253.
2. STEINHARDT, R. A., H. MORITA & E. S. HODGSON. 1966. Mode of action of straight chain hydrocarbons on primary chemoreceptors of the blowfly, *Phormia regina.* J. Cell. Physiol. **67:** 53-62.

3. KENNEDY, L. M. & B. P. HALPERN. 1981. Action of gymnemic acids and ziziphins: Dose-effect and time-course relationships. Assoc. Chemorecept. Sci. **III:** 23.
4. KENNEDY, L. M. & B. P. HALPERN. 1980. A biphasic model for action of the gymnemic acids and ziziphins on taste receptor cell membranes. Chem. Senses **5:** 149-158.

Voltage-Dependent Ionic Currents in Dissociated Mudpuppy Taste Cells

SUE C. KINNAMON [a] AND STEPHEN D. ROPER

Department of Anatomy and Neurobiology
Colorado State University
Fort Collins, Colorado 80523

Rocky Mountain Taste and Smell Center
University of Colorado Health Sciences Center
Denver, Colorado 80262

Although taste receptor cells from a variety of species have been studied with intracellular microelectrodes, the mechanisms involved in the production of the taste receptor potential are unknown. Recent studies on the large taste cells of amphibians have shown that taste cells are electrically excitable and can generate action potentials in response to current injection and chemosensory stimulation.[1,2] In this study, we have applied the whole-cell configuration of the patch-clamp recording technique to enzymatically dissociated *Necturus* taste cells so that the voltage-dependent currents underlying the action potential could be identified and characterized in isolation. We report here that mudpuppy taste cells have voltage-dependent K^+, Na^+, and Ca^{2+} currents.

Taste cells were isolated from the surrounding nongustatory lingual epithelium by incubating the stripped lingual epithelium in collagenase (Type 3; 1.5 mg/ml) for 30 minutes or until the nongustatory epithelium could be gently peeled from the underlying connective tissue, isolating the more adhesive taste buds on prominent papillae. The isolated taste buds were then dissociated in Ca^{2+}-free amphibian physiological saline containing trypsin (0.5 mg/ml) for 10 minutes. Taste cells were drawn into fire-polished glass pipettes and plated onto concanavalin A-coated glass coverslips.

Whole-cell currents were recorded from isolated taste cells and nongustatory surface epithelial cells. *Surface epithelial cells* had no voltage-dependent currents in response to depolarizing voltage steps to +80 mV from a holding potential of −100 mV. In sharp contrast, *taste cells* had transient inward currents, followed by sustained outward currents in response to depolarizing voltage steps from a holding potential of −100 mV (FIG. 1A). The sustained outward current usually activated at −60 to −20 mV, reached a peak between 10 and 20 msec after onset of the voltage pulse, and did not inactivate for at least 10 seconds. When depolarizing voltage steps did not exceed 0 mV, the outward current was completely blocked by 8 mM TEA, by substituting KCl in the pipette with CsCl, or by replacing Ca^{2+} in the bath with Ba^{2+}. The pharmacology, kinetics, and voltage dependence of this current suggest that it is carried by K^+ ions and that it is a major component of the resting conductance of taste cells.

---

[a] Address for correspondence: Dr. Sue C. Kinnamon, Department of Anatomy and Neurobiology, Colorado State University, Fort Collins, CO 80523.

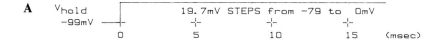

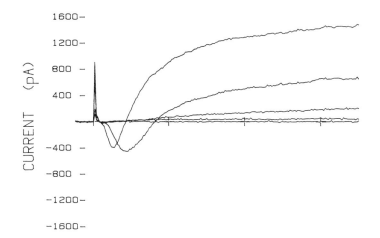

A

Leak Subtracted

805

2105.86

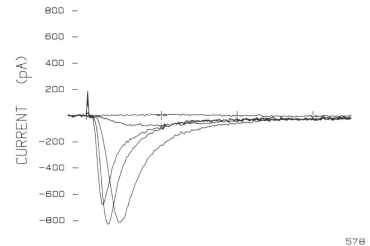

B

Leak Subtracted

578

1605.86

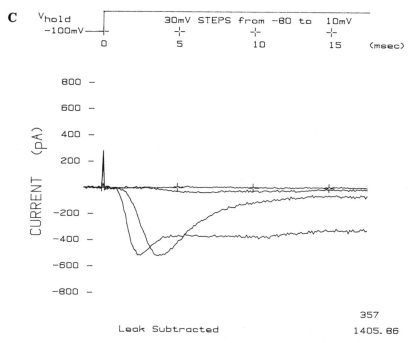

FIGURE 1. Whole-cell currents in isolated taste cells. (**A**) Patch pipette contained 140 mM KCl, 2 mM MgCl$_2$, 1 mM CaCl$_2$, 10 mM HEPES buffer (pH 7.2), 2 mM BAPTA, and 5 mM ATP. Currents were recorded in response to depolarizing voltage steps to -80, -60, -40, -20, and 0 mV from a holding potential of -100 mV. (**B**) Transient inward currents, revealed by replacing KCl in patch pipette with CsCl to block outward current. Currents were recorded in response to depolarizing voltage steps to -60, -40, -20, 0, and $+20$ mV from a holding potential of -100 mV. (**C**) Sustained and transient inward currents from a different taste cell, revealed by replacing KCl in patch pipette with CsCl. Currents were recorded in response to voltage steps to -80, -50, -20, and $+10$ mV from a holding potential of -100 mV. Leak and linear capacitative currents were subtracted from all records by computer.

Inward currents were studied by replacing KCl in the patch pipette with CsCl to block outward currents. Two pharmacologically and kinetically distinct inward currents were observed in taste cells. One current activated at approximately -40 mV, reached a peak in a few milliseconds, and completely inactivated (FIG. 1B). This transient inward current was completely and reversibly blocked by tetrodotoxin (TTX), suggesting that it is a Na$^+$ current. The other inward current activated at approximately 0 mV and usually did not inactivate over the duration of the voltage pulse (FIG. 1C). This sustained current was blocked by CdCl$_2$ or replacement of Ca^{2+} in the bath with Mg^{2+}. These data suggest that the sustained inward current is a Ca^{2+} current.

Outward currents were observed in all taste cells examined. In contrast, inward currents were not observed in all taste cells, and the type of inward current observed varied in different cells. Some cells possessed only Na$^+$ currents, whereas other cells possessed both Na$^+$ and Ca^{2+} currents. The magnitude of the Ca^{2+} current was highly variable, ranging from a few pA to almost 1 nA. It is possible that these differences in whole-cell currents can be attributed to differences in taste cell type, but this remains to be determined.

These data corroborate the results from intracellular recordings in pieces of lingual tissue[1,2] and indicate that voltage-dependent currents can be isolated and characterized in taste cells. In future studies the effects of taste stimulation on these identified currents can be determined.

REFERENCES

1. ROPER, S. D. 1983. Regenerative impulses in taste cells. Science **220:** 1311-1312.
2. KINNAMON, S. C. & S. D. ROPER. 1987. Passive and active membrane properties of mudpuppy taste receptor cells. J. Physiol. **383:** 601-614.

Inhibitory Effects of Ca^{2+} on the Mg^{2+} Response of Water Fibers in the Frog Glossopharyngeal Nerve

YASUYUKI KITADA

Department of Physiology
Okayama University Dental School
Okayama 700, Japan

In the frog glossopharyngeal nerve, the water-sensitive fiber (the water fiber) responds to $CaCl_2$ and $MgCl_2$. However, application of a mixture solution of $CaCl_2$ + $MgCl_2$ to the tongue led to only a small response. This indicates that the Ca^{2+} response is inhibited by the presence of Mg^{2+}, and the Mg^{2+} response by the presence of Ca^{2+}. The mutual antagonism between Ca^{2+} and Mg^{2+} suggests that there exist at least two different receptor sites in the water fiber, a calcium receptor site (X_{Ca}) for the Ca^{2+} response and a magnesium receptor site (X_{Mg}) for the Mg^{2+} response. In this study, properties of X_{Mg} were quantitatively studied.

Unitary discharges were recorded from single water fibers of the frog glossopharyngeal nerve during stimulation of the tongue with salt solutions. To explain the inhibition of the Mg^{2+} response by Ca^{2+}, it is assumed that Ca^{2+} competes with Mg^{2+} for X_{Mg} by forming an inactive Ca-X_{Mg} complex. The assumption of competitive inhibition by Ca^{2+} is supported by a double-reciprocal plot analysis of the data. The apparent dissociation constants for Mg-X_{Mg} and Ca-X_{Mg} were 7.0×10^{-2} M and 6.3×10^{-4} M, respectively.

In contrast to its excitatory effect, Mg^{2+} inhibits the Ca^{2+} response by competing with Ca^{2+} for X_{Ca};[1] thus, a comparison of apparent dissociation constants for cation-receptor complexes was made between X_{Mg} and X_{Ca} (TABLE 1). The value of the apparent dissociation constant of cation-X_{Mg} complex was quite different from that of cation-X_{Ca} complex for each divalent cation. Our findings indicate that X_{Mg} differs from X_{Ca}.

Each divalent cation has a dual action, excitation and inhibition, as described above. Furthermore, it was found that an increase of ionic strength in the stimulating

TABLE 1. Comparison of Apparent Dissociation Constants for the Receptor-Agonist and Receptor-Inhibitor Complexes

Cation	Calcium-Receptor Site	Magnesium-Receptor Site
Ca^{2+}	Agonist 5.9×10^{-5} M^a	Inhibitor 6.3×10^{-4} M
Mg^{2+}	Inhibitor 5.2×10^{-5} M^a	Agonist 7.0×10^{-2} M

[a] Kitada and Shimada.[1]

417

solution did not affect the Mg^{2+} response. These results imply that the salt response cannot be explained in terms of the surface potential on the outer membrane. It is concluded that there exist multiple receptor sites for cations in salt taste reception of the frog.

REFERENCE

1. KITADA, Y. & K. SHIMADA. 1980. A quantitative study of the inhibitory effect of Na$^+$ and Mg^{2+} on the Ca^{2+} response of water fibers in the frog tongue. Jpn. J. Physiol. **30:** 219-230.

Central Projections of Major Branches of the Facial Taste Nerve in the Japanese Sea Catfish

S. KIYOHARA [a] AND S. YAMASHITA

Biological Institute
College of Liberal Arts
Kagoshima University
Kagoshima 890, Japan

T. MARUI

Department of Oral Physiology
Kagoshima University Dental School
Kagoshima 890, Japan

J. CAPRIO

Department of Zoology and Physiology
Louisiana State University
Baton Rouge, Louisiana 70808

The Japanese sea catfish, *Plotosus anguillaris,* has four pairs of barbels around the mouth: nasal, maxillary, medial mandibular, and lateral mandibular. These barbels are almost equal in length and similar in structure. They have densely concentrated taste buds that are innervated by facial nerve fibers. The external taste buds are also distributed on the entire skin from the lips to the caudal fin. The sea catfish is a bottom feeder. It is shown that this fish does not move its barbels actively in localizing food but detects the food after its barbels or lips have come into contact with the food.[1] In conjunction with the high degree of development of the facial taste system, the primary facial taste center (facial lobe, FL) is extraordinarily developed in the medulla oblongata. The FL in the sea catfish is more differentiated than that of the North American ictalurid catfish, which shows three distinct lobules.[2] Examination of the FL of *Plotosus* in normal sections reveals that the FL is subdivided by fascicles of nerve fibers into five distinct lobules constituting five longitudinal columns through the anterior two-thirds of the FL (FIG. 1). These are termed (proceeding from medial to lateral): medial, intermediate 1, intermediate 2, lateral, and dorsal lobules. In the posterior one-third of the FL, these lobules become less distinct. To reveal the top-

[a] Address for correspondence: Dr. Sadao Kiyohara, Biological Institute, College of Liberal Arts, Kagoshima University, Kohrimoto 1-21-30, Kagoshima 890, Japan.

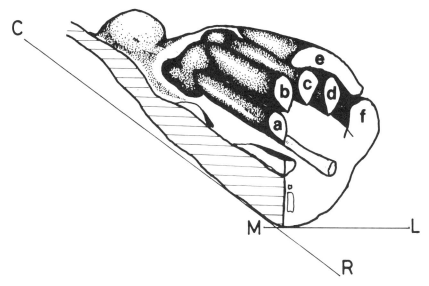

FIGURE 1. Schematic representation of the medulla oblongata of *Plotosus anguillaris* showing five lobules of the facial lobe: (**a**) medial lobule, (**b**) intermediate 1 lobule, (**c**) intermediate 2 lobule, (**d**) lateral lobule, (**e**) dorsal lobule, and (**f**) lateral line lobe.

ographical projections of peripheral facial fibers to these lobules of the FL, the central projections of seven branches (nasal barbel, palatine, maxillary, maxillary barbel, mandibular, mandibular barbel, recurrent) of the facial nerve were examined by using the technique of transganglionic tracing with horseradish peroxidase (HRP).

When each branch except the recurrent was treated with HRP, labeled fibers were observed in the facial sensory root and the descending trigeminal root (FIG. 2). The facial fibers of the mandibular barbel branch (B), maxillary barbel B, nasal barbel B and recurrent B, terminate, respectively, in the medial and intermediate 1, intermediate 2, lateral, and dorsal lobules. Those of mandibular B, maxillary B, and palatine B end in the medial, lateral, and ventral portions of the caudal FL. These results show that there is a clear topographical relation between the receptive field of the seven branches and their locus of representation in the facial lobe.

REFERENCES

1. SATO, M. 1937. On the barbels of a Japanese sea catfish, *Plotosus anguillaris* (Lacepede). Sci. Rep. Tohoku Imp. Univ. Ser. 4. **11:** 323-332.
2. FINGER, T. E. 1976. Gustatory pathways in the bullhead catfish 1. Connections of the anterior ganglion. J. Comp. Neurol. **165:** 513-526.

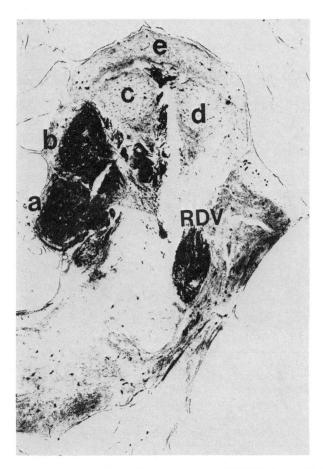

FIGURE 2. Distribution of labeled fibers following application of HRP to the mandibular barbel branch. Note the heavy labeling of the medial lobule (**a**), intermediate 1 lobule (**b**), and descending trigeminal root (**RDV**).

The Effects of Bilateral Sectioning of the Chorda Tympani and the Greater Superficial Petrosal Nerves and the Submaxillary and Sublingual Salivary Glands on the Daily Eating and Drinking Patterns in the Rat[a]

ROBIN F. KRIMM, MOHSSEN S. NEJAD, JAMES C. SMITH, AND LLOYD M. BEIDLER

Department of Psychology
The Florida State University
Tallahassee, Florida 32306

It was recently reported that rats, after bilateral sections of both the chorda tympani (CT) and the greater superficial petrosal nerves (GSP), showed a decrease in food intake and an alteration in the pattern of drinking and eating.[1] Bilateral sectioning of the CT alone resulted in a pattern of intake that was profound, but different from that observed with combined CT and GSP sections.

The CT nerve innervates the submaxillary and sublingual salivary glands (SMSL) as well as the anterior tongue. Stricker[2] has shown that sectioning of the CT alters eating behavior of deprived rats in short-term feeding tests in a manner similar to the altered behavior after removal of the SMSL glands. Therefore, it seems possible that the alterations in drinking and eating patterns observed by Krimm *et al.*[1] in animals with bilateral CT sections could be the result of partial desalivation. In order to better understand the effects of desalivation on the 23-hour daily drinking and eating behavior in nondeprived rats, comparisons were made between rats with the CT nerves bilaterally sectioned and rats that had the SMSL glands removed. The apparatus allowed for monitoring of drinking and eating by the rat during consecutive 30-second periods over a 23-hour day. Bilateral sectioning of CT resulted in changes in food, but not water, intake patterns. For example, the average time the rat's head was in the food jar increased even though there was no increase in food intake. Strip-chart records for an intact rat, a CT-sectioned rat, and a rat with the SMSL glands removed can be seen in FIGURE 1. We could not distinguish CT-sectioned rats from desalivated rats in terms of the water and food intake patterns. Furthermore, when CT-sectioned animals were subsequently desalivated, no additional changes in eating patterns occurred.

Ingestive patterns followed bilateral sections of only the GSP nerves were also measured resulting in significantly fewer water and food bouts and less water and food consumption. The pattern of licking of GSP-sectioned rats was more interrupted than that seen with CT-sectioned rats. They spent significantly less time out of each

[a]Supported in part by a grant from NIA (5R01AG04932).

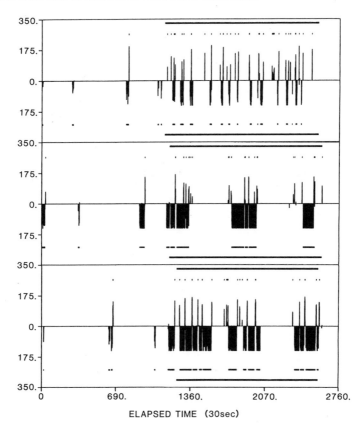

FIGURE 1. Twenty-three-hour records of drinking (top of each panel) and eating (bottom of each panel) are shown above for an intact rat (top panel), a CT-sectioned rat (middle panel), and a rat with the SMSL glands removed (bottom panel). The 23-hour day is divided into 2,760 thirty-second bins. The solid horizontal lines indicate periods when the room lights were off. The tic marks indicate drinking bouts.

TABLE 1. Mean Values for the Dependent Variables Indicated

Dependent Variables	Sham	CT	SMSL	GSP	CT + GSP
Water intake	37.12	29.24	37.12	9.90	31.63
Total licks	7162	6177	7746	2443	6992
No. of drinking bouts	23.0	22.0	22.8	11.6	37.5
Licks per bout	378.5	290.0	377.6	219.6	212.8
Interbout interval	65.2	59.3	55.5	141.8	31.6
Licks per 30-second bin	95.5	100.3	87.2	37.8	42.8
Food intake	23.9	23.4	28.2	5.9	7.18
Total time eating (min)	78.7	198.0	161.0	23.7	158.1
No. of eating bouts	14.8	11.8	13.2	8.43	24.2
Time per bout	4.7	17.6	12.3	2.79	9.08
Interbout interval	80.8	97.1	72.0	149.1	34.5
Eating efficiency	0.304	0.118	0.175	0.249	0.045
Time per 30-second bin	16.1	20.0	17.4	7.55	4.49

30-second period drinking and eating. When GSP sections and desalivations were performed in the same rat, an intake pattern emerged that was different from that previously seen with CT + GSP sections. A quantification of drinking and eating bouts for five of the surgically manipulated groups can be seen in TABLE 1.

REFERENCES

1. KRIMM, R. E., M. S. NEJAD, J. C. SMITH & L. M. BEIDLER. 1985. The effects of bilateral sectioning of the chorda tympani and the greater superficial petrosal nerves on feeding behavior and the sweet taste in rats. Chem. Senses **10:** 449.
2. STRICKER, E. M. 1970. Influence of saliva on feeding behavior in the rat. J. Comp. Physiol. **70:** 103-112.

Dynamic Conformational Changes of Receptor Domains in Gustatory Cell Membranes

KENZO KURIHARA, MAKOTO NAKAMURA,
AND YUSUKE KATOH

Faculty of Pharmaceutical Sciences
Hokkaido University
Sapporo 060, Japan

The effects of temperature and divalent cations on the gustatory nerve responses of the rat and the frog were examined, and the results obtained were interpreted in terms of dynamic conformational changes of the receptor domains.

TEMPERATURE DEPENDENCE OF THE TASTE NERVE RESPONSES

The responses to various stimuli showed maximal values around 30°C. The slopes of the curves for the response-temperature relationships below 30°C were increased in the following order: NaCl < KCl < HCl < quinine < sucrose = glycine.

The temperature dependence of the responses to various salts greatly varied with species of salts. The amiloride-insensitive component of the response to 10 mM NaCl had maximal value around 30°C and the amiloride-sensitive component had maximal value at 7°C. The responses to 100 mM NaCl and KCl were decreased above 30°C, but an response to 100 mM NH$_4$Cl was increased with an increase of temperature up to 43°C.

The concentration-response relationships were not greatly changed between 5 and 30°C except for the response to NaCl, suggesting that the equilibrium constants between stimuli and the receptor sites are little changed with temperature. In other words, the enthalpy changes in generation of the taste responses are not significant. The result that the temperature dependence of the responses greatly varies among different species of stimuli suggested that the temperature dependence does not come from the process of the transmitter release. Probably conformational changes of the receptor domains brought about by temperature changes lead to changes in the number of receptor sites available. In this connection, it is noted that the frog taste response to NaCl after removal of Ca^{2+} from the receptor membrane exhibited no temperature dependence between 2 and 20°C.

425

EFFECTS OF DIVALENT CATIONS ON THE FROG GUSTATORY RESPONSES TO VARIOUS STIMULI

The presence of divalent cations in the adapting solution for the frog tongue led to modification of the magnitude and the time course of the taste nerve responses. Adaptation of the tongue to solutions containing high concentrations of $CaCl_2$ (e.g. 7.5 mM) led to a great enhancement of the response to HCl. The adaptation also modified the time course of the response to quinine: The response to quinine after adaption to low-Ca solution had practically no tonic response, while adaptation of the tongue to high-Ca solution produced a large tonic response. In a previous paper,[1] we proposed that the initial process of chemoreception is expressed by the following equation: $S + A \rightleftarrows (SA)_{active} \rightleftarrows (SA)_{inactive}$. The presence of Ca^{2+} seems to prevent the conversion of $(SA)_{active}$ into $(SA)_{inactive}$.

The frog taste responses to various stimuli such as HCl, quinine, NaCl, and sucrose were greatly increased by adapting the tongue in $BaCl_2$ solution. The ability of the tongue to produce the enhanced responses to these stimuli held even in the absence of Ba^{2+} after the tongue was adapted to solutions containing high concentrations of Ba^{2+}, suggesting that the presence of Ba^{2+} in external medium is not always necessary.

The effects of Ca^{2+} and Ba^{2+} were not inhibited by Ca-channel blockers, suggesting that influx of the divalent cations into taste cells is not concerned with the modification of the responses. The divalent cations seem to expose the receptor domains or to change the efficacy of the stimulus binding into generation of the receptor potential.

REFERENCE

1. KAMO, N., T. KASHIWAGURA, K. KURIHARA & Y. KOBATAKE. 1980. J. Theor. Biol. 83: 111-130.

Purification and Chemical Structure of Taste Modifier

Ziziphin and Taste-Modifying Protein

YOSHIE KURIHARA

Department of Chemistry
Faculty of Education
Yokohama National University
Yokohama, Japan

KAZUYOSHI OOKUBO

Faculty of Agriculture
Tohoku University
Sendai, Japan

BRUCE P. HALPERN

Department of Psychology
Cornell University
Ithaca, New York

ZIZIPHIN

Leaves of *Ziziphus jujuba* contain ziziphin, which has anti-sweet activity. We have succeeded in purifying ziziphin and in determining its chemical structure.

The dried leaves of *Ziziphus jujuba* were extracted by organic solvents. The extracts were applied to a Sephadex LH-20 column and a fraction having anti-sweet activity was collected. This material (Zi-1) was subjected to high-pressure liquid chromatography (HPLC) using an RP-18 column. Among many peaks, only one peak had anti-sweet activity. To collect the active material more efficiently, Zi-1 was subjected to chromatography on a silica gel column, followed by an RP-2 column. The fraction having anti-sweet activity was subjected to HPLC using an RP-18 column. One peak with the anti-sweet activity was collected (Zi) and used for estimation of chemical structure.

FAB-MS of Zi as well as the above HPLC data showed that the Zi obtained was highly pure. The molecular weight of Zi was 980. Analysis of alditol acetate, which

427

was obtained by acid-catalyzed hydrolysis of Zi by GC, indicated that Zi contains two moles of L-rhamnose and one mole of L-arabinose. Methylation analysis showed that L-rhamnosyl-L-arabinose and L-rhamnose are linked to aglycone at two different positions. IR and ^{1}H NMR spectra of Zi indicated the presence of two acetyl groups. On acid hydrolysis, Zi gave ebelin lactone as a secondary aglycone. Actual aglycone of Zi was jujubogenin, which was converted to ebelin lactone.[1] On the basis of these data, the structure of Zi was established.

TASTE-MODIFYING PROTEIN

A number of methods on isolation of the active principle from miracle fruit have been reported. In previous studies the active protein, which could not be solubilized with water, was extracted with a carbonate buffer of pH 10.5. The extracted solution contained deep-colored materials that were difficult to eliminate. The extraction with alkaline solution also led to partial loss of the activity.

Therrasilp and Kurihara have found that the active protein can be extracted with a NaCl solution of neural pH or phosphate buffer of pH 8.5. The extracts were colorless and showed high activity. The extracts were subjected to $(NH_4)_2SO_4$ fractionation and the active fraction was applied to a CM-Sepharose column. Application of the active fraction to HPLC gave a single sharp peak. The molecular weight was about 25,000, which is smaller than that reported previously (40,000-48,000). The study on amino acid sequence of the active protein is now in progress.

REFERENCES

1. OKAMURA, N., T. NOHARA, A. YAGI & I NISHIOKA. 1981. Chem. Pharm. Bull. **29:** 676.
2. THERRASILP, S. & Y. KURIHARA. 1986. Proc. Jpn. Symp. Taste Smell **20:** 195.

Perceptual and Intraoral pH Measurements in Response to Oral Stimulation

D. B. KURTZ, J. C. WALKER, J. H. REYNOLDS,
AND D. L. ROBERTS

R. J. Reynolds Tobacco Company
Bowman Gray Technical Center
Winston-Salem, North Carolina 27102

S. L. YANKELL

University of Pennsylvania
School of Dental Medicine
Philadelphia, Pennsylvania 19104

Integrated Ionics, Inc.
Dayton, New Jersey 08810

Intraoral pH has been measured in people using sensors mounted on Hawley appliances.[1] In this study we have combined this measurement with the recording of perceptual ratings and the measurement of puffing behavior in order to study orosensory stimulation during smoking. In addition, the responses to simple aqueous tastant solutions, presented in a "sip and spit" fashion, were also recorded. Three electrodes, mounted on a Hawley appliance positioned on the maxilla, were used to measure changes in salivary pH in response to oral stimulation.

Control experiments demonstrated that the appliance itself did not alter perceptual ratings of either cigarettes or simple tastants[2] or puffing behavior on cigarettes. An example of the effect of a cigarette on oral pH is shown in FIGURE 1A. Each puff caused a rapid increase in pH which we attributed to the evaporative cooling of saliva by air brought into the mouth. The rapid increase in pH was followed by a slower partial return to the pH just before the puff. Subsequent puffs brought the salivary pH progressively closer to neutrality. The magnitude of the pH change was inversely related to starting oral pH and directly related to FTC "tar" delivery; that is, large pH changes toward neutrality were related to an acid starting condition in the mouth and puffs on a high "tar" cigarette. We hypothesize that the effects we observed were largely due to the increase in salivary flow as a result of the stimulation of chemoreceptors in the oral cavity.

Unlit or low "tar" cigarettes caused only small changes toward neutrality. FIGURE 1B shows the small shift in oral pH occurring during "puffs" on an unlit cigarette. Oral pH increased by approximately 0.5 unit when puffs were taken on an "ultra-low tar" cigarette; the magnitude of the change in pH was independent of the prestimulus pH.

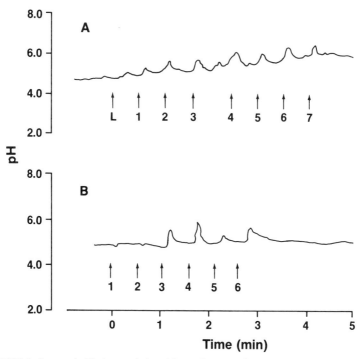

FIGURE 1. Intraoral pH changes induced by puffing on a lit cigarette (**A**) and puffing on an unlit cigarette (**B**).

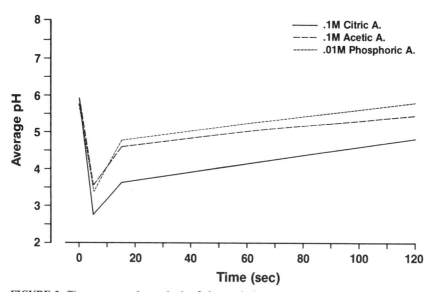

FIGURE 2. Time course and magnitude of changes in intraoral pH induced by several acids.

Solutions of sodium chloride or sucrose caused little change in oral pH. As shown in FIGURE 2, solutions of acids (phosphoric, acetic, and citric) caused large acidic excursions. These excursions peaked at approximately the pH of the acid solution by the end of the five-second sip. Recovery of oral pH to the pre-acid condition was relatively slow, being only 50% complete after expectoration and a one-minute wait. A distilled water rinse and an additional one-minute wait failed to return oral pH to the prestimulus condition for two of the three acids tested.

This method of monitoring salivary pH has the benefit that pH can be measured on a continuous basis. In addition, because of its inobtrusiveness, it serves as a valuable complement to psychophysical techniques.

REFERENCES

1. WALKER, J. C., D. B. KURTZ, D. J. JULIUS & S. L. YANKELL. 1986. Taste perception and the measurement of intra-oral pH. J. Dent. Res. (Abstr. # 945) **65:** 274.
2. YANKELL, S. L. & I. R. LAUKS. 1985. pH fluctuations in clinical plaque and saliva. J. Dent. Res. (Abstr. # 540) **64:** 235.

Levels of Immunoreactive β-Endorphin in Rostral and Caudal Sections of Olfactory Bulbs from Male Guinea Pigs Exposed to Odors of Conspecific Females[a]

JAY B. LABOV,[b,c] YAIR KATZ,[b] CHARLES J. WYSOCKI,[b] GARY K. BEAUCHAMP,[b] AND LINDA M. WYSOCKI [b]

[b]Monell Chemical Senses Center
Philadelphia, Pennsylvania 19104

[c]Department of Biology
Colby College
Waterville, Maine 04901

β-Endorphin (BE) exists in peripheral and central regions of both the main and accessory olfactory systems of rodents, but the relationship of this endogenous opioid to chemoreception is uncertain.[1] In rodents, stimulation of the vomeronasal organ (VNO) and accessory olfactory system by conspecific chemosignals appears to be innately reinforcing and may abet associative learning via the main olfactory system.[2,3] BE concentrations in other brain regions are altered during learning,[4] hence it is possible that changes in titers of BE in the accessory or main olfactory system may be associated with stimulation of the VNO. This study investigated whether BE titers would be altered in the rostral (main olfactory) or caudal (accessory olfactory) portions of the olfactory bulbs of male guinea pigs after brief exposures to female chemical cues of low volatility. When exposed to female chemical cues, male guinea pigs exhibit head bobbing, a stereotyped rostrocaudal movement of the head that may facilitate access of low-volatile molecules to the vomeronasal organ.[5] In this experiment, we recorded amounts of head bobbing and evaluated a possible correlation between this behavior and BE titers.

Subjects were permitted one-minute contacts with each of 10 glass plates that were smeared with either saline (controls) or secretions from the perineum of female conspecifics (experimentals). Subjects were observed during each trial for the number and duration of head-bobbing bouts and for the number of times and duration that the nose or mouth contacted the plate. Twelve minutes after stimulus presentations, subjects were killed with CO_2, their brains were rapidly removed, frozen in liquid

[a]Supported by National Science Foundation Grant BNS 83-16437 (CJW & YK), National Science Foundation Research Opportunity Award (JBL), National Institutes of Health Grant R01 NS 22623 (GKB), and a Sabbatical Leave Extension Grant from Colby College (JBL).

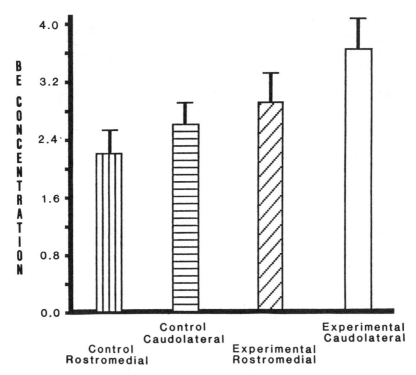

FIGURE 1. Concentrations (Mean ± SEM) of β-endorphin (pg/mg tissue) in sections of left olfactory bulbs from male guinea pigs exposed to saline or to perineal odors of conspecific females.

freon, and stored at $-80°C$. After warming to $-20°C$, the left olfactory bulbs were cut into rostromedial (main olfactory) and caudolateral (includes accessory olfactory) sections. Right olfactory bulbs and brains were returned to $-80°C$ for later assay. Immunoreactive BE in the rostral and caudal portions of both olfactory bulbs was measured using standard double-antibody RIA (Immuno Nuclear Corp., Stillwater, MN).

BE concentrations were significantly greater in left olfactory bulbs of experimentals than in controls. Caudolateral sections contained significantly higher titers of BE than did rostromedial sections (FIG. 1). A significant inverse correlation between BE concentration and tissue weight was observed for caudolateral sections from both controls and experimentals and from rostromedial sections from controls (TABLE 1A), suggesting that BE may be concentrated in localized portions of the bulbs.

Experimental males spent more time in contact with the plates and contacted the plates more often than controls, but mean time per contact with the plates did not differ statistically between the two treatments (TABLE 1B). Experimentals also exhibited significantly more head bobbing than controls (TABLE 1C). BE levels in caudolateral sections were significantly correlated with the mean duration of head-bobbing bouts for experimental, but not for control, subjects. There was no relationship between BE concentration and oronasal contact with the glass plates for either experimentals or controls. Replication using right olfactory bulbs confirmed significantly

TABLE 1A. Correlations between Titers of β-Endorphin (BE) in Rostral and Caudal Sections of *Left* Olfactory Bulbs Versus Tissue Weights and Behaviors

Section or Behavior	Controls		Experimentals	
	r	p	r	p
BE titers vs. tissue weight of rostral sections of bulb	−0.72	0.05	−0.15	N.S.
BE titers vs. tissue weight of caudal sections of bulb	−0.84	0.01	−0.79	0.01
BE titers vs. mean time per head-bobbing bout	0.48	N.S.	0.71	0.05
BE titers vs. mean time per contact with plate	−0.20	N.S.	0.34	N.S.

TABLE 1B. Total Time Spent in Contact with Plate, Number of Contacts with Plate, and Mean Time Per Contact with Plate for Control and Experimental Subjects

Behavior	Controls	Experimentals	p
Time in contact with plate (sec)	187.5 ± 49.2 (n = 10)	302.9 ± 45.1 (n = 10)	0.01
Number of contacts with plate	24.6 ± 3.6 (n = 8)[a]	31.9 ± 2.8 (n = 8)[a]	0.05
Mean time per contact with plate (sec)	7.8 ± 3.1 (n = 8)[a]	9.3 ± 2.0 (n = 8)[a]	N.S.

[a] Total number of head-bobbing bouts and number of contacts with the plate were not recorded for two control and two experimental subjects.

TABLE 1C. Total Time Spent Head Bobbing, Number of Head-Bobbing Bouts, and Mean Time per Head-Bobbing Bout for Control and Experimental Subjects

Behavior	Controls	Experimentals	p
Time spent head bobbing (sec)	13.5 ± 5.0 (n = 10)	125.8 ± 21.2 (n = 10)	0.01
Number of head-bobbing bouts	9.0 ± 3.4 (n = 8)[a]	24.1 ± 3.2 (n = 8)[a]	0.01
Mean time per head-bobbing bout	1.1 ± 0.3 (n = 8)[a]	4.7 ± 0.7 (n = 8)[a]	0.01

[a] Total number of head-bobbing bouts and number of contacts with the plate were not recorded for two control and two experimental subjects.

elevated BE titers in caudolateral sections, but a significant treatment effect was not observed. There were no significant correlations between BE concentrations in either rostromedial or caudolateral sections of right olfactory bulbs and mean duration of head bobbing.

Our results suggest that the accessory olfactory bulbs contain higher concentrations of immunoreactive BE than main olfactory bulbs. Data from the first experiment also suggest that BE titers in the accessory olfactory bulbs may increase in response to stimulation of the vomeronasal system; however, additional experimental manipulations and further replication are required.

REFERENCES

1. PERT, C. B. & M. HERKENHAM. 1981. From receptors to brain circuitry. *In* Biochemistry of Taste and Olfaction. R. H. Cagan & M. R. Kare, Eds.: 511-522. Academic Press. New York.
2. BEAUCHAMP, G. K., C. J. WYSOCKI & J. L. WELLINGTON. 1985. Extinction of response to urine odor as a consequence of vomeronasal organ removal in male guinea pigs. Behav. Neurosci. **99:** 950-955.
3. WYSOCKI, C. J., N. J. BEAN & G. K. BEAUCHAMP. 1986. The mammalian vomeronasal system: Its role in learning and social behaviors. *In* Chemical Signals in Vertebrates: Ecology, Evolution and Comparative Biology. D. Duvall, D. Müller-Schwarze & R. M. Silverstein, Eds. Vol. **4:** 471-485. Plenum Press. New York.
4. RILEY, A. L., D. A. ZELLNER & H. J. DUNCAN. 1980. The role of endorphins in animal learning and behavior. Neurosci. Biobehav. Rev. **4:** 69-76.
5. WYSOCKI, C. J., J. L. WELLINGTON & G. K. BEAUCHAMP. 1980. Access of urinary non-volatiles to the mammalian vomeronasal organ. Science **207:** 781-783.

Derivation of Power Law Exponents from Olfactory Thresholds for Pure Substances

PAUL LAFFORT AND FRANÇOIS PATTE

Laboratoire de Physiologie de la Chimioréception
C.N.R.S.
F-91190 Gif-sur-Yvette, France

EAG dose-response curves for 59 pure substances have been obtained in the honeybee by Patte *et al.*[1] These sigmoids in semilog coordinates fit the Hill model, which involves three parameters: the power law exponent (n), the maximal amplitude of response or plateau (Vm), and the concentration corresponding to the inflexion point (Cx). In addition, a fourth parameter (Co) can be easily derived from these three: the concentration corresponding to a response of 0.1 mV, which we consider analogous to an electrophysiological threshold. A significant positive correlation appears ($p < 0.02$) between log Co and n.[1] In other words, the following trend is observed: the lower the threshold, the smaller the power law exponent. This correlation approaches one (Spearman RHO = 0.87) if the threshold is expressed in terms of fraction of saturated vapor pressure instead of real concentration (FIG. 1).

From several sets of psychophysical data, a strong correlation between thresholds and power law exponents was also demonstrated by Laffort *et al.*[2] However, in this case the initial correlations are diminished when the concentrations are weighted by saturated vapor pressure and tend to be enhanced when they are weighted by a polar factor called "filter effect." The addition of physicochemical properties (polar and nonpolar) influences the initial correlations between thresholds and power law exponents inversely for these two sets of data: psychophysical and EAG on honeybee (FIG. 2). This could be due to the different means by which odorous molecules reach dendrites: aqueous mucus for vertebrates and pore tubules for insects.[3] In order to compare vertebrates and insects at the level of reception, an experiment is in progress, in cooperation with E. P. Köster and colleagues,[4] to obtain EOG dose-response curves in frog, for the same substances as those for which we recorded the EAG on honeybee.

REFERENCES

1. PATTE, F., M. ETCHETO & P. MARFAING. 1984. Etude comparative des réponses électroantennographiques de l'abeille à 59 substances. J. Physiol. Paris. **79:** 67 A.
2. LAFFORT, P., F. PATTE & M. ETCHETO. 1974. Olfactory coding on the basis of physicochemical properties. Ann. N.Y. Acad. Sci. **237:** 193-208.
3. KEIL, T. A. 1982. Contacts of pore tubules and sensory dendrites in antennal chemosensilla of a silkmoth: Demonstration of a possible pathway for olfactory molecules. Tissue Cell **14:** 451-462.

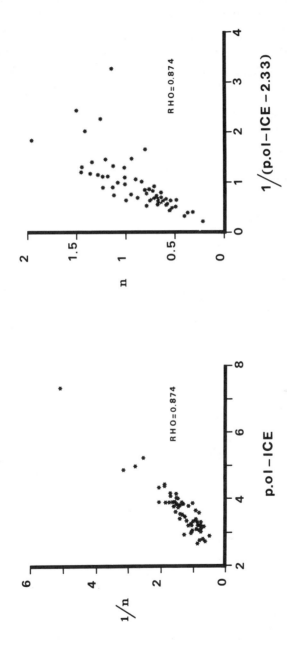

FIGURE 1. EAG on honeybee. Diagrams of power law exponents (*n* and 1/*n*) versus threshold values weighted by saturated vapor pressure (see definitions of *p.ol.* (olfactory power) and ICE (internal cohesive energy) in FIG. 2).

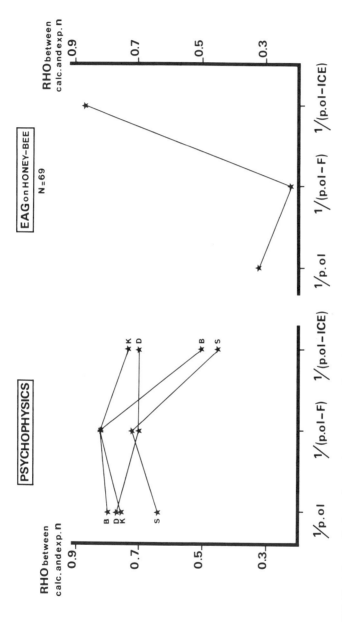

FIGURE 2. Tentative predictions of power law exponents n from thresholds for several sets of psychophysical data and for recent electroantennographic ones. B = Berglund et al.[5] ($n = 22$); K = Katz and Talbert[6] ($n = 34$); D = Dravnieks and Laffort[7] ($n = 31$); S = Standard[8] ($n = 82$). F = filter effect[9] ($0.6\epsilon + 0.9\pi$); ICE = saturated vapor pressure (nlmf); p.ol = olfactory power (nlmf threshold). nlmf means negative log of molar fraction; ϵ and π are from Patte et al.[9]

4. KÖSTER, E. P., H. J. KUIPER & W. VAN AS. 1986. (Psychological Laboratory, Utrecht University, Sorbonnelaan 16, 3584 CA Utrecht, the Netherlands.) Personal communication.
5. BERGLUND, B., U. BERGLUND & G. EKMAN. 1971. Individual psycho-physical functions for 28 odorants. Percep. Psychophys. **9:** 379-384.
6. KATZ, S. H. & E. J. TALBERT. 1930. Intensities of odors and irritating effects of warning agents for inflammable and poisonous gases. U.S. Dept. Comm. Bureau Mines. Technical paper. **480:** 1-37.
7. DRAVNIEKS, A. & P. LAFFORT. 1972. Physico-chemical basis of quantitative and qualitative odor discrimination in humans. *In* Olfaction and Taste IV. D. Schneider, Ed.: 142-148. Wissens-Verlag. MBH. Stuttgart, Germany.
8. PATTE, F., M. ETCHETO & P. LAFFORT. 1975. Selected and standardized values of suprathreshold odor intensities for 110 substances. Chem. Senses Flavor **1:** 283-305.
9. PATTE, F., M. ETCHETO & P. LAFFORT. 1982. Solubility factors for 240 solutes and 207 stationary phases in gas-liquid chromatography. Anal. Chem. **54:** 2239-2247.

Tactile and Taste Responses in the Superior Secondary Gustatory Nucleus of the Catfish

C. F. LAMB, IV AND J. CAPRIO

Department of Zoology and Physiology
Louisiana State University
Baton Rouge, Louisiana 70803-1725

The superior secondary gustatory nucleus (nGS) of teleosts receives ascending second-order neurons from the primary medullary taste centers, the facial (FL) and vagal (VL) lobes.[1] The nGS is located in the isthmic region of the hindbrain, ventral to the corpus cerebelli and lateral to the brachium conjunctivum (FIG. 1). Electrophysiological studies indicate that peripheral input into the FL[2] and VL[3] of the channel catfish, *Ictalurus punctatus,* is represented in a somatotopic and viscerotopic manner, respectively. FL and VL efferents remain segregated in the ascending secondary gustatory tract and terminate in separate regions of the nGS.[4] We studied the electrophysiological responses of nGS neurons to mechanical and chemical stimulation of their peripheral receptive fields (RFs) to determine if the neurons of the nGS are topographically arranged. Tactile stimulation was produced by stroking the surface of the fish with a fine-tipped glass rod, and chemical stimuli included filtered liver extract (10 g/1) and an amino acid mixture (L-ala, L-arg, L-pro, and betaine, each at 1.0 mM).

Phasic responses to mechanical stimulation of the oral and extraoral epithelium were found throughout the superior secondary gustatory nucleus of the channel catfish, but were more abundant in the rostral portion of the nucleus. Receptive fields of single units and receptive areas (RAs) of multiunit activity were greater than 100 mm^2, with many including the whole body surface and oropharyngeal cavity (FIG. 2). The large tactile RFs and RAs within the nGS did not indicate a topographical organization of nGS neurons.

Responses to chemical stimulation were encountered much less frequently than were tactile responses. Chemosensitive units were located in the central portion of the nGS, within the mechanosensitive areas. Chemical RFs included either the maxillary barbels or the oropharyngeal cavity. Four of the five chemosensitive units analyzed were bimodal, responding also to mechanical stimulation (FIG. 2). Tactile RFs of the bimodal units varied in size from the entire oropharyngeal cavity and mandibular barbels to the whole body surface. The chemoresponses of nGS neurons analyzed in this study were not topographically arranged. These results suggest that the neurons within the nGS do not display the precise topographical representation that is seen in the neurons of the facial and vagal lobes.

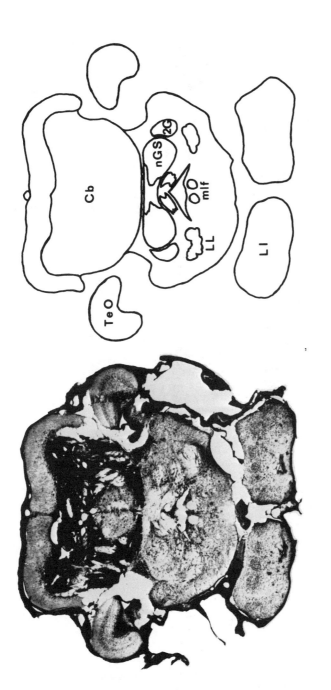

FIGURE 1. Transverse section through the catfish brain at the level of the rostral nGS. Photomicrograph of a Nissl-stained section (left) and an outline drawing of the same section (right) with the relevant areas indicated (Cb, cerebellum; 2G, ascending secondary gustatory tract; LI, inferior lobe; LL, lateral lemniscus; mlf, medial longitudinal fasciculus; nGS, superior secondary gustatory nucleus; TeO, optic tectum).

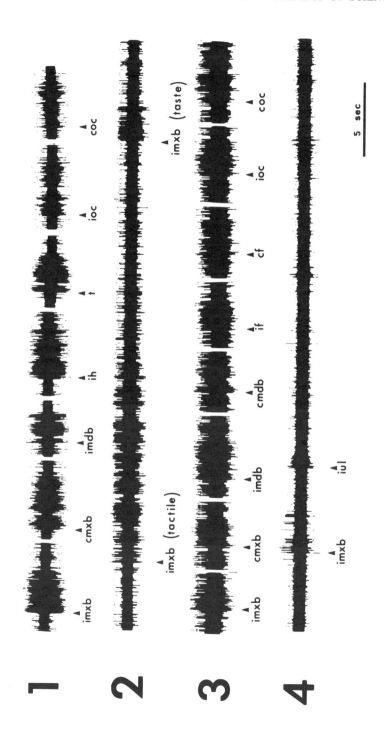

REFERENCES

1. HERRICK, C. J. 1905. Central gustatory paths in brains of bony fishes. J. Comp. Neurol. **15:** 375-456.
2. MARUI, T. & J. CAPRIO. 1982. Electrophysiological evidence for the topographical arrangement of taste and tactile neurons in the facial lobe of the channel catfish. Brain Res. **231:** 185-190.
3. KANWAL, J. & J. CAPRIO. 1984. Topographic arrangement and response properties of gustatory neurons in the vagal lobe of the catfish. Assoc. Chemorecept. Sci. Abstr. VI, #69.
4. FINGER, T. E. 1978. Gustatory pathways in the bullhead catfish. I. Connections of the anterior ganglion. J. Comp. Neurol. **180:** 691-706.

FIGURE 2. Electrophysiological records of mechanosensitive and chemosensitive responses within the nGS. Each trace (1-4) is recorded from a single location in a single fish. Traces 1 and 2 are from the central region of the nGS (750 μm lateral to the midline, 3.20 mm ventral to the surface of the cerebellum). Trace 1 shows multiunit responses to tactile stimulation of the whole body surface and oropharyngeal cavity. Trace 2 shows responses to tactile and taste (liver extract) stimulation of the ipsilateral maxillary barbel. Tactile stimulation of the rest of the body surface and the oropharyngeal cavity also produced responses from these units (not shown). Traces 3 and 4 are from the rostral nGS of different fish. Trace 3 shows responses in the rostromedial nGS (500 μm lateral to the midline, 2.80 mm ventral to the surface of the cerebellum) to mechanical stimulation of the whole body surface and the oropharyngeal cavity. Trace 4 shows limited responses of the rostrolateral nGS (1.00 mm lateral to the midline, 3.20 mm ventral to the surface of the cerebellum) to mechanical stimulation of the ipsilateral maxillary barbel and upper lip. (Abbreviations: cf, contralateral flank; cmdb, contralateral mandibular barbel; cmxb, contralateral maxillary barbel; coc, contralateral oropharyngeal cavity; if, ipsilateral flank; ih, ipsilateral head, imdb, ipsilateral mandibular barbel; imxb, ipsilateral maxillary barbel; ioc, ipsilateral oropharyngeal cavity; iul, ipsilateral upper lip; t, tail.)

Histogenesis of Pontine Taste Area Neurons in the Albino Rat[a]

PHILLIP S. LASITER [b] AND DAVID L. HILL [c]

Department of Psychology
University of Toledo
Toledo, Ohio 43606

Pontine taste area (PTA)[1,2] neurons respond to many, but not all, chemical stimuli in rats aged four to seven days. PTA neurons in rats older than 14 days respond to each of the four basic tastes. For instance, PTA neurons in rats aged four to seven days do not respond to 100 mM KCl and quinine hydrochloride (QHCl), whereas neurons in rats aged 14 days and older respond to 100 mM and 500 mM solutions of monochloride salts, to citric and hydrochloric acids, to sucrose and sodium saccharin, and to QHCl. Although PTA neurons respond to each of the four basic tastes by the second week of life, major developmental changes continue to occur with regard to response frequency: Response frequencies increase to all stimuli between 14 and 60 days of age. Thus, PTA neurons respond to a greater number of stimuli between the first and second week of life, and thereafter PTA neurons show greater response sensitivity.[3]

Quantitative Golgi studies were conducted to examine the neural correlates of developing taste responses in PTA neurons. To that end, a variant of the Golgi-Kopsch method[4] was used to impregnate neurons in rats aged 16-17 days (n = 26), 18-22 days (n = 29), 34-35 days (n = 23), and 72-131 days (n = 27). Planar morphometrics were subsequently performed on fully impregnated neurons confined to taste-responsive regions of the PTA in the developing rat.[5]

Neurons in the PTA of all age groups were classified as either multipolar or fusiform.[2] Multipolar neurons are ovoid or spherical in shape and possess three to five dendrites organized equidistantly around somata. Fusiform neurons are spindle-shaped and possess two to three large-diameter dendrites organized at major poles of somata (FIG. 1b). Between 40-48% of impregnated PTA neurons were multipolar, and the proportion of multipolar and fusiform neurons that were impregnated in each age group was statistically equivalent (p > 0.05; $\bar{\chi}$ = 45%). Mean somatic diameters of multipolar neurons ($\bar{\chi}$ = 13.39 μm; SEM = 0.31 μm) and fusiform neurons ($\bar{\chi}$ = 11.23 μm; SEM = 0.26 μm) also did not differ between age groups (p > 0.05).

FIGURE 1a shows concentric ring analyses[6] across age groups. Extensive radial growth of dendrites is observed between 16 and 35 days of age, yet no reliable difference in dendritic organization is observed between 35 days and 131 days. At 16-17 days dendrites extend maximally approximately 75 μm from somata; at 18-22 days 175 μm, at 34-35 days 225 μm, and at 72-131 days 250 μm. Thus, extensive radial dendritic growth occurs between 16 and 35 days.

[a] Supported by National Institutes of Health Grants NS11618, NS20538, and NS00964 and DeArce Memorial Endowment Grant 2131283675.

[b] Present address: Department of Psychology, Florida Atlantic University, Boca Raton, FL 33431.

[c] Present address: Department of Psychology, University of Virginia, Charlottesville, VA 22903.

FIGURE 1c shows concentric growth density between 16-35 days. FIGURE 1c depicts the mean intersection difference between 16-17 days and 18-22 days in addition to the mean intersection difference between 18-22 days and 34-35 days. Although radial dendritic growth is evident in concentric ring analyses for these age groups (see FIGURE 1a), a second and more subtle pattern of dendritic development occurs between 16 and 35 days. Concentric growth curves show that maximal growth density occurs

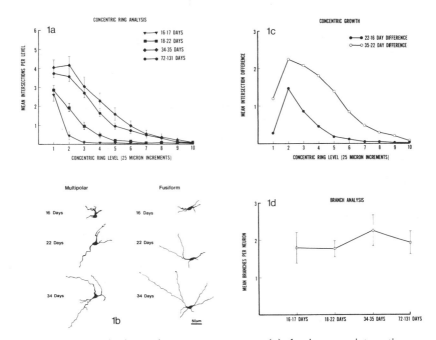

FIGURE 1. Concentric ring analyses across age groups (a) showing mean intersections per level and standard errors. Mean intersections per level differ reliably between rats aged 16-17 days, 18-22 days, and 34-35 days ($p < 0.05$; ANOVA). Mean intersections per level do not differ reliably between rats aged 34-35 days and 72-131 days ($p > 0.05$; ANOVA). Extensive radial growth of dendrites is evident from 16-35 days. Panel (b) shows typical multipolar and fusiform neurons observed in the PTA at 16 days, 22 days, and 34 days. Panel (c) shows concentric dendritic growth between 16 to 22 days and between 22 to 35 days. Concentric growth analyses show that throughout development PTA dendrites tend to elaborate at 50-75 μm surrounding somata. This effect is observed both for multipolar neurons and for fusiform neurons. Panel (d) shows branch analyses for developing PTA neurons. Means and standard errors are shown. Mean branches per neuron do not differ reliably between age groups ($p > 0.05$; ANOVA). Branch analyses indicate that dendritic elaboration within 50-75 μm of somata is not obviously related to the development of additional dendritic branches.

approximately 50-75 μm surrounding somata in rats aged 16-35 days. Dendritic growth in this region does not obviously reflect the proliferation of additional dendritic branches near cell bodies, because mean number of branches are statistically equivalent between age groups ($p > 0.05$) (see FIG. 1d). Rather, dendritic growth within 50-75 μm of somata appears to relate to an elaboration of short dendrites that are organized near somata on immature neurons (e.g., 16 days; see FIG. 1b).

These data show that the developmental frequency change in PTA neural responses are highly correlated with dendritic outgrowth and elaboration within the PTA. Such frequency changes may be due, in whole or in part, to the proliferation of additional functional synapses along developing dendrites.

REFERENCES

1. NORGREN, R. & C. M. LEONARD. 1973. J. Comp. Neurol. **150:** 217-238.
2. LASITER, P. S. & D. L. GLANZMAN. 1983. Brain Res. **253:** 299-304.
3. HILL, D. L. 1983. Soc. Neurosci. Abstr. **9:** 378.
4. BRAITENBERG, V., V. GUGLIELMOTTI & E. SADA. 1967. Stain Techn. **42:** 277-283.
5. HILL, D. L. 1987. J. Neurophysiol. **57:** 481-495.
6. SHOLL, D. A. 1956. *In* The Organization of the Cerebral Cortex. Methuen. London, England.

Olfaction Test in Laryngectomized Patients by the Artificial Tube Method

JEUNG GWEON LEE, MASARU OHYAMA, ETSURO
OBTA, KAZUYOSHI UENO, AND SHOKO KATAHIRA

Department of Otolaryngology
Faculty of Medicine
Kagoshima University
Kagoshima 890, Japan

After laryngectomy, olfactory acuity is disturbed with the abolished air current, but the power of olfaction does not change.[1] The authors noticed the occurrence of the nasal air current in the laryngectomized patients by connecting a tube between the tracheal stoma and the naris, so named this technique an artificial airway tube method. By this manner, the olfactory acuities of laryngectomized patients were evaluated by measuring the detection threshold for the smell of various odors such as β-phenyl ethyl alcohol, cyclotene and isovaleric acid. At the same time we also measured the latency and duration time for the sense of smell of thiamine propyl disulfide administered intravenously.

The results obtained indicate that the mean detection threshold in the laryngectomized patients is nearly equal to that obtained with normal persons (TABLE 1), and in the intravenous olfaction tests, inferior data were seen in the laryngectomized patients. We also observed that there are no significant differences in olfactory acuities between the groups, one with esophageal speech and the other without it.

Although repair of olfactory epithelium has been reported by many authors,[2,3] its change after laryngectomy has rarely been reported. In our experimental study using laryngectomized dogs, the morphological changes in olfactory epithelia were observed

TABLE 1. Perinasal Olfactory Test in Patients after Laryngectomy[a]

Odor Source	Our Study 60 cases (1975)	Our Study 15 cases (1986)	Normal, Takai 50 cases (1971)
β-Phenyl ethyl alcohol	3.8 ± 0.3	2.8 ± 0.4	3.8
Cyclotene	4.5 ± 0.2	3.9 ± 0.4	4.6
Isovaleric acid	5.0 ± 0.2	3.9 ± 0.4	5.2
Total	13.3 ± 0.6	10.6 ± 1.1	

[a] The unit is a number of a negative power of the diluted odor solution.

by light microscopy and scanning electron microscopy (FIG. 1). One week after laryngectomy, loss of cilia and swollen vesicles of olfactory cells were observed. Three months after the operation, although some variations in the degree of morphological alteration were found, the same patterns appeared as in the normal controls.

This study indicates that there are no visible morphological and functional changes in the olfactory epithelium in long-term laryngectomized subjects.

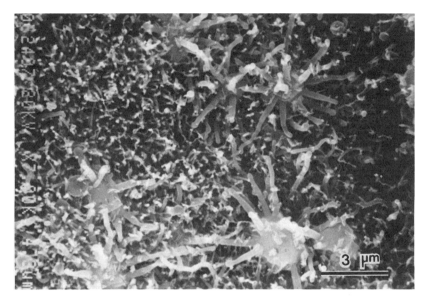

FIGURE 1. Olfactory epithelium of the dog three months after laryngectomy. Numerous olfactory cilia appear to be protruding from the irregularly shaped olfactory vesicle.

REFERENCES

1. RITTER, F. N. 1964. Fate of olfaction after laryngectomy. Arch. Otolaryngol. **79:** 169-171.
2. SCHULTZ, E. W. 1960. Repair of the olfactory mucosa. Am. J. Pathol. **37:** 1-19.
3. GRAZIADEI, G. A. M. & P. P. C. GRAZIADEI. 1979. Neurogenesis and neuronal regeneration in the olfactory system of mammals. II. Degeneration and reconstitution of the olfactory sensory neurons after axotomy. J. Neurocytol. **8:** 197-213.

Vomeronasal Chemoreception May Activate Reproduction in Reflex-Ovulating Prairie Voles[a]

JOHN J. LEPRI AND CHARLES J. WYSOCKI

Monell Chemical Senses Center
Philadelphia, Pennsylvania 19104

Unlike spontaneously ovulating Norway rats and house mice, the reproductive system of female prairie voles (*Microtus ochrgaster*) remains inactive until the female is exposed to a male. Growth of the ovaries and uterus can be stimulated by urinary chemosignals from males, but behavioral estrus and copulation-induced ovulation require physical contact with a male.[1] A single drop of male urine placed on the upper lip of a female prairie vole causes changes in the concentrations of norepinephrine and luteinizing hormone-releasing hormone in the caudal portion of the olfactory bulbs,[2] to which the sensory afferent neurons of the vomeronasal system project. We tested whether the surgical removal of the vomeronasal organ hinders male-induced activation of female reproduction.

Three types of adult female prairie voles were tested: vomeronasal organ surgically removed (VNX); surgical manipulation but vomeronasal organs left intact (SHAM); no surgical manipulation (NORMAL). Complete removal of the vomeronasal organs from VNX females was histologically verified by the lack of glomeruli in the accessory olfactory bulb. After recovery from surgery, the VNX females quickly found a piece of apple hidden in cage bedding, suggesting that the VNX procedure did not prohibit chemosensory foraging. Females from each of the three treatment groups were then either unexposed to males (zero hours), paired with stud males for 12 hours, or paired with stud males for 60 hours. Results demonstrate that uterine and ovarian weights were equally low in all females in the zero-hours group. Mean uterine weight of NORMAL and SHAM females approximately doubled after 12 hours and tripled after 60 hours of pairing with stud males. However, mean uterine weight of VNX females did not increase as much as that of NORMAL and SHAM females, even after 60 hours of pairing (FIG. 1). Furthermore, 8 of 9 NORMAL, and 10 of 13 SHAM females in the 60-hour group mated (sperm present in the vagina) whereas only 4 of 9 VNX females in the 60-hour group mated. Behavioral observations recorded during the first 10 minutes of interactions with stud males, including duration of naso-nasal and naso-anal sniffing, were similar for NORMAL, SHAM, and VNX females. The deficits in reproductive activation in the VNX females are a striking result of destroying a chemoreception system that has been traditionally labeled as "accessory." Indeed, for female voles, vomeronasal chemoreception of stimuli from males may facilitate activation of female reproduction.

[a]Supported by National Institutes of Health Grant 2T32NSO7176-07 (JJL) and National Science Foundation BNS 83-16437 (CJW).

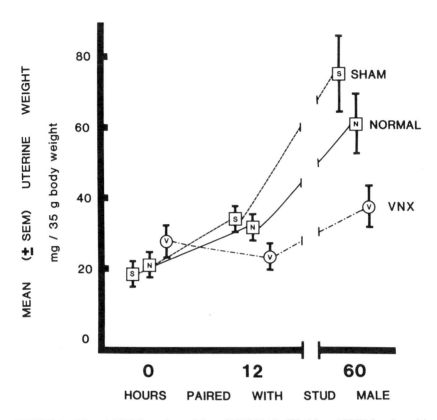

FIGURE 1. Mean (±SEM) uterine weights of NORMAL, SHAM, and VNX female prairie voles after zero, 12, or 60 hours of physical contact with stud males. Planned comparisons between surgical treatments in the 60-hour group resulted in statistically significant differences between SHAM and VNX females, and between NORMAL and VNX females, but not between the NORMAL and SHAM females.

REFERENCES

1. CARTER, C. S. & L. L. GETZ. 1985. Social and endocrine determinants of reproductive patterns in the prairie vole. *In* Comparative Neurobiology. R. Gilles & J. Balthazart, Eds.: 18-36. Springer-Verlag. New York.
2. DLUZEN, D. E., V. D. RAMIREZ, C. S. CARTER & L. L. Getz. 1981. Male vole urine changes luteinizing hormone-releasing hormone and norepinephrine in female olfactory bulbs. Science **212:** 573-575.

A Changing Density Technique to Measure Nasal Airflow Patterns [a]

DONALD A. LEOPOLD, DAVID E. HORNUNG,
ROBERT L. RICHARDSON, PAUL F. KENT,
MAXWELL M. MOZELL, AND
STEVEN L. YOUNGENTOB

*Departments of Otolaryngology, Physiology, and Radiology
SUNY Health Science Center at Syracuse
Syracuse, New York 13210*

To study the relationship between nasal airflow and olfactory function, a radiation detection technique was developed to quantify the airflow patterns through a uninasal model of the human air passageways. The theoretical basis of this technique is that for any point in the nasal model, the rate of displacement of one gas by another is proportional to the flow rate at that point. In these studies, the model was first filled with xenon gas, and then, as room air was drawn through the model, the xenon was displaced. To measure the changing xenon density, a collimated gamma ray source (200 millicuries of ^{125}I) was directed through the model toward a specially designed sodium iodide detector (FIG. 1). Because of the collimation, only a small cross section of the model (approx. 4 mm^2) was studied at any one time. This technique of measuring the changing xenon density takes advantage of the fact that the number of gamma rays received per second by the detector is dependent upon the density of gas inside the nasal model. Since xenon gas is much denser than air, the number of gamma rays received by the detector increased as the xenon gas was removed from the model. The detector output was fed into a multichannel scaler such that each succeeding channel contains the output for each succeeding 0.4 msec interval.

To analyze the data, the counts versus time least-squares regression equation was calculated. The slope of this equation is the rate of displacement of xenon by the air. As described above, the slope is, therefore, also directly proportional to the flow rate. To determine the reproducibility of the data, the slope of the counts versus time curve for each of four flow rates (2, 6, 15, 45 l/min) was determined for a single nasal position. The four randomly ordered flow rates defined a block of trials, and four such blocks were run. The two-way analysis of variance showed there was a significant effect of flow rate ($p < 0.01$), but not blocks ($p < 0.05$). Thus, it appears that this technique can give reproducible results.

In the second study three blocks at the above flow rates were run at each of two nasal positions. Position A was in the middle meatus near the center of the nasal fossa. Position B was just posterior to the nasal valve, midway between the nasal floor and dorsum. The results of this study (TABLE 1) show that at position A there was

[a] Supported by National Institutes of Health Grant #NS19658.

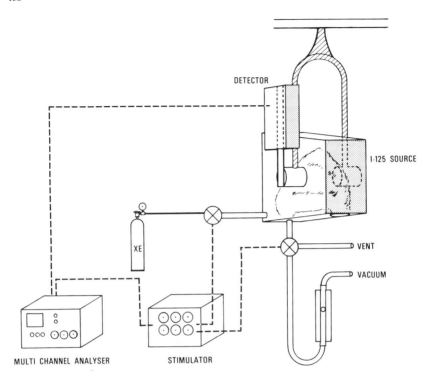

FIGURE 1. Experimental setup for the changing density airflow measuring technique. The source and detector, which are supported on a rigid yoke, can be accurately placed in any nasal position by using a micromanipulator.

an increase in the slope of the counts versus time curve as the airflow through the model was increased from 2 to 45 l/min. Likewise, at position B, there was an increase in slope as the velocity increased from 2 to 15 l/min; however, as the flow was increased beyond 15 l/min, the slope decreased. One explanation for this slope decrease may be that in this nasal region the flow characteristics changed from laminar to turbulent when the flow was increased from 15 to 45 l/min. That is, the turbulence would create a remixing of the dense and light gases, which would then produce the observed decrease in slope. This explanation is consistent with Masing's[1] observation

TABLE 1. Slopes of the Counts Versus Time Curve at Two Nasal Positions

Flow Rate in 1/min	Slope at Position A ± SE	Slope at Position B ± SE
2	4.16 ± 0.93	21.35 ± 8.51
6	9.42 ± 4.13	40.02 ± 5.02
15	17.90 ± 5.43	48.55 ± 10.13
45	44.01 ± 4.92	46.66 ± 7.92

that a turbulence is found just posterior to the nasal valve region (where position B is located). In contrast, position A is in a wider section of the nasal fossa where the flow is more likely to be laminar throughout the range of flows use in this study. Although these results should be considered very preliminary, this technique does seem to permit a noninvasive measurement of nasal airflow.

REFERENCE

1. MASING, H. 1967. Arch. Klin. Exp. Ohr. Nas. Kehkopf. **189:** 371-381.

Electrophysiological Properties of Immature and Mature Olfactory Receptor Cells[a]

M. S. LIDOW [b]

Department of Neurobiology and Physiology
Northwestern University
Evanston, Illinois 60201

R. C. GESTELAND, S. J. KLEENE, AND
M. T. SHIPLEY

Department of Anatomy and Cell Biology
University of Cincinnati Medical Center
Cincinnati, Ohio 45267

Because cells in the normal olfactory epithelium of adult animals are continuously turning over,[1,2] it contains not only mature receptor cells (which have established synapses within the olfactory bulb), but also a population of immature receptor cells (whose axons have not yet reached the olfactory bulb). We have conducted a comparative study of the electrophysiological properties of immature and mature receptor cells in order to understand their contributions to the electrophysiological properties of normal frog olfactory epithelium. We used a recently developed method of obtaining olfactory epithelia with developmentally synchronized populations of receptor cells.[3] To produce such epithelia, we ablated normal frog olfactory epithelia with $ZnSO_4$, allowed them to regenerate for 10 days (this is a minimal period required to obtain a population of receptor cells suitable for electrophysiological study), and then suppressed generation of new cells by continuous treatment with hydroxyurea. In one group of animals, hydroxyurea treatment lasted eight days and in the other group for 24 days. The oldest cells in these epithelia originated on the sixth day after ablation[3] and the youngest ones on the 10th day after ablation. Thus by the time of observation, the epithelia from the first group of animals contained receptor cells 8-12 days old, while the epithelia in the second group of animals contained receptor cells 24-28 days old. The anterograde WGA-HRP labeling of these epithelia indicated that axons of receptor cells 8-12 days old do not reach the olfactory bulb, while axons of receptor cells 24-28 days old do reach it.

[a]This work was supported by National Institutes of Health Grants NS18490, NS07223, and NS14663 and by National Science Foundation Grants BNS8117075, BNS8316827, and BNS8596011.

[b]Address for correspondence: Michael S. Lidow, Ph.D., Section of Neuroanatomy, Yale University School of Medicine, 333 Cedar St., New Haven, CT 06510.

No differences were found in the shape of electroolfactograms (EOGs) recorded from olfactory epithelia composed of immature or mature receptor cells. The sources of all EOG components are present in both types of epithelia; however, amplitudes of EOGs recorded from epithelia composed of mature receptor cells were generally higher than those from epithelia composed of immature cells. This indicates that mature cells are the main contributors to the EOG. The majority of immature olfactory receptor cells had low spontaneous activity or none at all. Mature olfactory receptor cells have a variety of frequencies of spontaneous activities ranging from less than one to more than 70 spikes/min. This study showed that, contrary to what had been observed in rat embryo,[4] olfactory receptor cells in regenerating epithelia of adult frogs do not go through a stage in which they respond to all odorants. If the response strength is not considered, immature olfactory receptor cells were no less odorant-selective than mature ones. Changes in the concentration of any tested stimulus did not produce dramatic changes in the response intensities of immature olfactory receptor cells. However, changes in the concentration of some stimuli did result in significant changes in response intensities of mature receptor cells.

REFERENCES

1. GRAZIADEI, P. P. C. & J. F. METCALF. 1971. Z. Zellforsch. **116:** 305-318.
2. MOULTON, D. G. 1974. Ann. N.Y. Acad. Sci. **237:** 52-61.
3. LIDOW, M. S., S. J. KLEENE & R. S. GESTELAND. 1987. Dev. Brain Res. **31:** 243-258.
4. GESTELAND, R. C., R. A. YANCEY & A. I. FARBMAN. 1982. Neuroscience **7:** 3127-3136.

Decreased Sensitivity to Bitter Solutions Following Chronic Opioid Receptor Blockade

WESLEY C. LYNCH,[a] CHARLES M. PADEN,[b] AND
SUSAN KRALL[b]

[a]Department of Psychology
[b]Department of Biology
Montana State University
Bozeman, Montana 59717

Numerous recent studies have implicated brain opioid systems in the regulation of feeding and taste-motivated behavior. In addition, we have shown that blockade of opioid receptors by naloxone inhibits intake of highly preferred solutions of saccharin[1] and that repeated daily naloxone in naive animals blocks normal preference acquisition.[2] Opioid receptors are found in high concentrations in numerous brain areas, particularly in sensory and limbic systems. Furthermore, chronic blockade of these receptors leads to receptor up-regulation in many of these areas (including taste and feeding areas) and to an increased sensitivity to morphine analgesia. Given the role of opioid systems in taste and feeding, the following study aimed to determine whether chronic naloxone might also modify taste preferences.

Ten adult male Sprague-Dawley albino rats were implanted with subcutaneous Alzet minipumps (Model 2002) containing either 150 mg/ml Naloxone-HCl ($n = 6$) or 0.9% NaCl ($n = 4$). An automated lick-monitoring system was used to measure consumption twice daily during two weeks of treatment and for one week following pump removal. At 9 A.M. daily (4 h after lights on) each animal was offered three calibrated bottles containing either sucrose (0.2 M and 0.7 M) or water. For this test animals were nondeprived. At 10 A.M. a six-hour period of water deprivation began and at 4 P.M. each animal was again offered three bottles, this time containing either quinine-HCl (0.00001 M or 0.00003 M) or water. Intake data were collected for the following six hours and these three bottles remained on the cages until 9 A.M. the following day. Additional tests for morphine analgesia were carried out during the week following pump removal.

In a previous experiment we have shown that chronic naloxone (NAL) treatment identical to that used in this study leads to the widespread up-regulation of brain opioid receptors that is both regionally and receptor subtype specific. In this study we found that this treatment also modifies intake of sucrose and quinine solutions. Contrary to expectations based on our previous work using repeated daily injections of NAL, we did not see an inhibition of sucrose intake during the period of exposure but instead intake tended to gradually increase over the two weeks of testing. This effect persisted after pump removal and was due both to a greater increase in drinking during the first few minutes of the session and to a delay in satiation in the naloxone-treated animals. Quinine drinking during the two weeks of drug exposure was not affected by naloxone. However, following pump removal, quinine intake increased in the naloxone-treated group for approximately four days and then returned to control levels. This effect on quinine intake was due almost entirely to an increase in consumption during the first 10 minutes of testing. Morphine analgesia, as reported previously by others, was also enhanced by NAL during the posttreatment period.

These results suggest that the qualitative aspects of gustatory events or the feedback from these events is influenced by chronic blockade of opioid receptors. This effect may be related to receptor up-regulation in specific brain areas, which has a similar time course. To our knowledge, this is the first report of a functional change in behavior other than sensitivity to pain that has been related to opioid receptor up-regulation. Future work will be needed to determine whether or not normal variations in the regulation of opioid receptors play a significant role in taste-motivated behavior.

REFERENCES

1. LYNCH, W. C. & L. LIBBY. 1983. Naloxone suppresses intake of highly preferred saccharin solutions in food deprived and sated rats. Life Sci. **33**(19): 1909-1914.
2. LYNCH, W. C. 1986. Opiate blockade inhibits saccharin intake and blocks normal preference acquisition. Pharmacol. Biochem. Behav. **24**(4): 833-836.
3. TEMPEL, A., E. L. GARDNER & R. S. ZUKIN. 1985. Neurochemical and functional correlates of naltrexone-induced opiate receptor up-regulation. J. Pharmacol. Exp. Ther. **232**(2): 439-444.

Gustatory Activity in the NTS of Chronic Decerebrate Rats

GREGORY P. MARK AND THOMAS R. SCOTT

*Department of Psychology and
Institute for Neuroscience
University of Delaware
Newark, Delaware 19716*

The hindbrain of the rat is involved in sophisticated ingestive functions, including those associated with taste acceptance or rejection, homeostasis, and conditioning. Some capacities, however, are disrupted by hindbrain isolation (decerebration). With so much data on taste-related behavior from the decerebrate rat now available, it becomes important to know how the taste system responds in this preparation.

We decerebrated rats at a supracollicular level and maintained them for 7-21 days. We then isolated the activity of 50 taste cells from the nucleus of the solitary tract (NTS). Taste stimuli were seven Na-Li salts (including four concentrations of NaCl), three sugars, HCl and citric acids, quinine, and two concentrations of Na saccharin. Responses were compared with corresponding data collected independently in intact rats by Chang.[1]

The characteristic broad sensitivity of taste cells was retained. Mean spontaneous activity was 6.5 ± 5.6 spikes/sec, 36% below the level in controls (insignificant). Evoked response rates were reduced by a corresponding amount (33%). But while reduction was a rather consistent phenomenon, there was a clear tendency for the cells of decerebrates to fall farther behind those of intact rats as higher discharge rates were demanded (FIG. 1, top).

To analyze intensity coding, we compared the concentration-response function to NaCl with that generated by Ganchrow and Erickson[2] (FIG. 1, bottom). At low concentrations (10 mM), both preparations gave similar responses. As intensity increased, however, activity from decerebrates fell progressively behind. Thus the greater the demands put on the taste cells of decerebrates, whether by the application of particularly effective taste qualities or by high concentrations, the more abnormal are their responses.

We employed a multidimensional scaling routine to examine relative similarities among stimulus profiles (FIG. 2). Within stimulus groups, coherence was maintained, but Na-Li salts moved away from the acids and quinine so that the space yielded three distinct groups: sugars, salts, and acid-quinine. The same three stimulus clusters have been identified by others in normal hamsters, however,[3] so we do not necessarily consider these results an anomaly of decerebration.

Taste responses in the NTS of decerebrate rats may be compared with those from intact animals as follows:

1. The characteristic broad range of responsiveness was retained.
2. Spontaneous activity was lower by 36%.

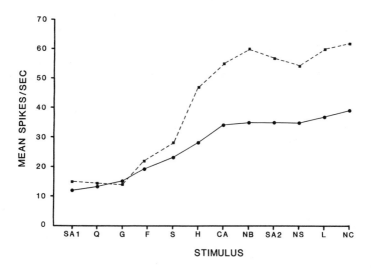

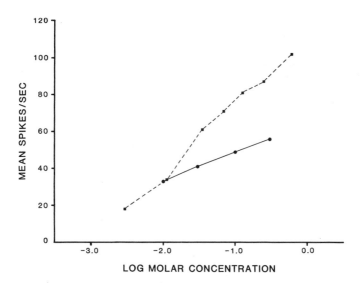

FIGURE 1. The relationship between evoked activity in chronic decerebrate rats versus intact controls across qualities and intensities. Top: Net (evoked minus spontaneous) spike rates across 50 NTS neurons in decerebrate (solid line) and intact rats. Data from intact rats are from Chang and Scott.[1] Abbreviations are: 0.0025 M Na saccharin (SA1), 0.005 M quinine (Q), 0.5 M glucose (G), 0.5 M fructose (F), 0.5 M sucrose (S), 0.01 M HCl (H), 0.01 M citric acid (CA), 0.1 M NaBr (NB), 0.25 M Na saccharin (SA2), 0.1 M Na$_2$SO$_4$ (NS), 0.1M LiCl (L), 0.1 M NaCl (NC). Bottom: Gross discharge rates evoked by various concentrations of NaCl in decerebrate (solid line) and intact rats. Data from intact rats are from Ganchrow and Erickson.[2]

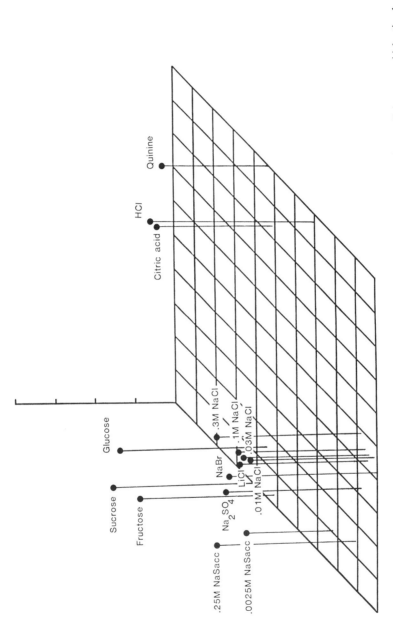

FIGURE 2. Multidimensional stimulus space representing relative similarities among taste qualities. The correlation coefficients on which stimulus positions are based are calculated from three seconds of net evoked activity from each of 50 NTS neurons.

3. Evoked activity was reduced in proportion to the spike rate demanded by virtue of the quality or the intensity of the stimulus.
4. Stimulus profiles maintained strong coherence within basic taste qualities, though there was some suggestion of stimulus realignment in decerebrates.

REFERENCES

1. CHANG, F-C. T. & T. R. SCOTT. 1984. Conditioned taste aversions modify neural responses in the rat nucleus tractus solitarius. J. Neurosci. **4:** 1850-1862.
2. GANCHROW, J. R. & R. P. ERICKSON. 1970. Neural correlates of gustatory intensity and quality. J. Neurophysiol. **33:** 768-783.
3. SMITH, D. V., R. L. VAN BUSKIRK, J. B. TRAVERS & S. L. BIEBER. 1983. Coding of taste stimuli by hamster brain stem neurons. J. Neurophysiol. **50:** 541-558.

Morphological and Behavioral Evidence for Chemoreception by Predaceous Stonefly Nymphs and Their Mayfly Prey

LEE ANNE MARTÍNEZ

Department of Entomology
Cornell University
Ithaca, New York 14853

To date, few studies have examined the use of chemosenses by stream-dwelling insects.[1] This research attempts to determine whether the immature stages of stoneflies (Plecoptera) and mayflies (Ephemeroptera) that inhabit fast-flowing "trout streams" are able to detect chemical cues. More specifically, the study focuses on the use of chemoreception in the interaction between these stonefly predators and their mayfly prey.

Two perlodid stoneflies from the Rocky Mountains of Colorado (*Megarcys signata* and *Kogotus modestus*) and five of their mayfly prey (*Ephemerella infrequens, Cinygmula* spp., *Epeorus* sp., *Baetis bicaudatus* and *B. tricaudatus*) were used in this study. Behavioral assays of two types were employed to test the responses of these insects to various chemical stimuli[2]: (1) Waterborne cues were assayed by placing the test organism in an in-stream observation box and giving it a choice between a channel of water flowing over a stimulus source and a channel of plain stream water. Its movement within the observation box was monitored for 10 minutes, and the total elapsed time spent in the stimulus channel was compared with the elapsed time spent in the same channel during a paired test in which no stimulus was presented. (2) Chemotactile cues were tested by placing organisms in an arena in which they must choose between filter paper treated with an ethanol extract of the stimulus source and filter paper treated with plain ethanol. For stoneflies, the movement of a single organism was monitored for 10 minutes, and the total elapsed time on treatment and control papers was compared to determine the stonefly's preference. For mayflies, six organisms were tested simultaneously by noting their location every 15 seconds for 10 minutes, and determining the cumulative number on the treatment versus the control paper over the observation period.

Morphological studies were also carried out in order to locate possible chemoreceptors.[3] Specimens were observed under scanning electron microscopy (SEM). Special attention was given to the cuticular fine structure of the nymphs' antennae and cerci.

In the waterborne cues assay, stoneflies were found to respond positively to the upstream presence of wounded prey (FIG. 1). This attraction appears to be limited to the immediate vicinity (< 10 cm) of the stimulus source. Such close-range response has also been shown for lobster, who live in a similarly turbulent environment.[4] The attraction also seems to be dependent on the wounding of prey, which probably increases the stimulus by releasing prey hemolymph into the water. In addition, these stoneflies are attracted to filter paper treated with an ethanol abstract of prey, while the smaller stonefly (*Kogotus*) avoids filter paper treated with an extract of stonefly competitor/potential predator (*Megarcys*) (FIG. 2). Like stoneflies, mayflies are attracted to an extract of their (algal) food source, yet they strongly avoid extracts of predatory stoneflies (FIG. 2). Both sets of observations indicate that chemical cues can be used to detect predators and/or prey.

Scanning electron microscopy (SEM) reveals that both stonefly and mayfly nymphs have a complex array of sensilla on their antennae, cerci, and other parts of their

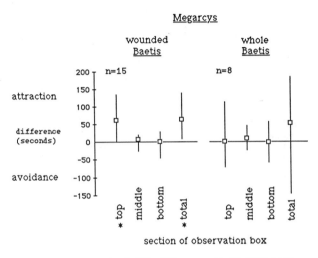

FIGURE 1. Representative data from waterborne cues assays show that the stonefly, *Megarcys*, spends significantly more time in the (top) section of the observation box immediately below the stimulus source, indicating that it is attracted to wounded prey (*Baetis*). No attraction is seen in the distal (bottom) section of the box, suggesting that the response is close-range (< 10 cm). In addition, no response is elicited unless prey are wounded. Similar results are seen in *Kogotus* trials ($n = 15$, $p = 0.015$ for the top section). Values represent mean differences (± 95% CIs) between the time spent in a section during the treatment versus the control (see text).

anatomy. As in a similar study on the stonefly *Paragnetina media*,[5] both *Megarcys* and *Kogotus* possess thick-walled sensilla basiconica, sensilla tricodea, and "sheaves" of coniform sensilla. The mayfly, *Epeorus*, has coeleconic pegs on the ventral side of its short antennae, while all mayflies (except *Ephemerella*) have various forms of long, flattened sensilla with constricted bases on their antennae and/or cerci. This flattening may be an adaptation to the fast-flowing water in which these insects live. In *Cinygmula*, pores are also apparent in the tips of its unusual arrowform flattened sensilla, suggesting that they play a role in chemoreception.[6] In addition, *Baetis* spp. have antennae that are virtually identical in cuticular fine structure to their cerci. The observation that *Baetis* nymphs often direct their cerci toward disturbances in the water suggests that the cerci may be used in detection. The sensillar complexity exhibited in the nymphs of both stoneflies (also see Kapoor[5]) and mayflies is lost in the nonfeeding terrestrial adults. While the gross morphology of sensilla cannot be used to determine function, the fact that the type and complexity of sensilla is lost in nonfeeding stages, and the similarity between some of these sensilla and known chemoreceptors of other insects, implicate these structures in the detection of chemical cues in the behavioral assays. Such results indicate that chemoreception may play an important role in prey detection and predator avoidance among stream insects.

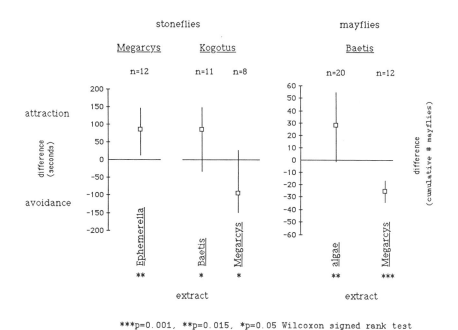

***p=0.001, **p=0.015, *p=0.05 Wilcoxon signed rank test

FIGURE 2. Data from the chemotactile assays show that both stoneflies prefer filter paper with an extract of prey (attraction). In contrast, *Kogotus* avoids an extract of its predator/competitor, *Megarcys.* The mayfly *Baetis* also shows an attraction towards extract of its food, while it avoids extract of its predator. Values represent mean differences (± 95% CIs) between time spent (stoneflies), or cumulative number of mayflies, on the treatment versus the control filter papers (see text).

REFERENCES

1. Martínez, L. A. 1987. Sensory cues used in the predator-prey interaction between stonefly nymphs and mayfly nymph prey. Ph.D. Dissertation. Cornell University.
2. Martínez, L. A. & B. L. Peckarsky. Behavioral responses of stonefly and mayfly nymphs to chemical stimuli: A case for chemoreception by stream-dwelling insects. Oecologia (Berlin), submitted.

3. MARTÍNEZ, L. A. A survey of the sensilla of a stream-dwelling stonefly nymph, *Kogotus modestus* (Plecoptera: Perlodidae), with comparisons to other perlodid nymphs. Can. J. Zool. Submitted.
4. ZIMMER-FAUST, R. K. & J. F. CASE. 1983. A proposed dual role of odor in foraging by the California spiny lobster, *Panulirus interruptus* (Randall). Biol. Bull. **167:** 341-353.
5. KAPOOR, N. N. 1985. External morphology and distribution of the antennal sensilla of the stonefly, *Paragnetina media* (Walker) (Plecoptera: Perlidae). Int. J. Insect Morphol. Embryol. **14(**5**):** 273-280.
6. SLIFER, E. H. 1960. A rapid and sensitive method for identifying permeable areas in the body wall of insects. Entomol. News **71:** 179-182.

Neural Connections of the Facial and Vagal Lobes in the Japanese Sea Catfish, *Plotosus anguillaris*

T. MARUI AND Y. KASAHARA

Department of Oral Physiology
Kagoshima University Dental School
Kagoshima 890, Japan

J. S. KANWAL AND J. CAPRIO

Department of Zoology and Physiology
College of Basic Science
Louisiana State University
Baton Rouge, Louisiana 70803

S. KIYOHARA

Biological Institute
College of Liberal Arts
Kagoshima University
Kagoshima 890, Japan

Gustation is the dominant sense for food searching behavior in Cyprinoid and Siluroid fishes. The gustatory sense in fish consists of two major dissociable components: the facial and vagal nerve systems.[1] The topology and taste-tactile responsiveness of neurons in the primary gustatory centers (facial and vagal lobes, FL and VL) have been studied with electrophysiological[2-4] and anatomical techniques.[5,6] However, the neural connectivities of these centers are relatively unknown, although cross-modality interactions may play an important role in recognizing and orienting toward biologically important objects. In this study, the Japanese sea catfish, which displays an extraordinary development of the gustatory nuclei, is used. To provide information for an accurate interpretation of continuous electrophysiological studies, neural tracing studies with horseradish peroxidase (HRP) were performed.

The facial lobe projects bilaterally to the posterior thalamic nucleus and the superior secondary gustatory nucleus. The FL has reciprocal connections with the nucleus lobobulbaris, the medial reticular formation of the rostral medulla, the funicular nuclei, and the descending trigeminal nucleus. Also, the FL receives inputs from the raphe nuclei, the pretectal region and perilemniscal neurons located adjacent to the ascending gustatory lemniscal tract. The vagal lobe projects bilaterally to the superior secondary gustatory nucleus, the lateral reticular formation, and ipsilaterally to the nucleus ambiguus. The VL has reciprocal connections with the ipsilateral lobobulbar nucleus, the medullary reticular formation, and the perilemniscal neurons as well. These anatomical findings with HRP tracing techniques are similar to the results for the bullhead catfish, *Ictalurus nebulosus*,[7] and may be critical for the interpretation of electrophysiological data obtained from the Japanese sea catfish.

REFERENCES

1. FINGER, T. E. & Y. MORITA. 1985. Two gustatory systems: Facial and vagal gustatory nuclei have different brainstem connections. Science **227:** 776-778.
2. MARUI, T. 1977. Taste responses in the facial lobe of the carp, *Cyprinus carpio* L. Brain Res. **130:** 287-298.
3. MARUI, T. & J. CAPRIO. 1982. Electrophysiological evidence for topographical arrangement of taste and tactile neurons in the facial lobe of the channel catfish. Brain Res. **231:** 185-190.
4. MARUI, T., J. CAPRIO, S. KIYOHARA & Y. KASAHARA. Topographical organization of taste and tactile neurons in the facial lobe of the Japanese sea catfish, *Plotosus anguillaris.* Brain Res. Submitted.
5. FINGER, T. E. 1976. Gustatory pathways in the bullhead catfish. I. Connections of the anterior ganglion. J. Comp. Neurol. **165:** 513-526.
6. MORITA, Y., T. MURAKAMI & H. ITOH. 1983. Cytoarchitecture and topograhic projections of the gustatory centers in the teleost, *Carassius carassius.* J. Comp. Neurol. **218:** 378-394.
7. MORITA, Y. & T. E. FINGER. 1985. Reflex connections of the facial and vagal gustatory systems in the brainstem of the bullhead catfish, *Ictalurus nebulosus.* J. Comp. Neurol. **231:** 547-558.

Covalent Modification of Schiff Base-Forming Proteins: *In Vitro* Evidence for Site Specificity and Behavioral Evidence for Production of Selective Hyposmia *in Vivo*[a]

J. RUSSELL MASON AND LARRY CLARK

Monell Chemical Senses Center and
Department of Biology
University of Pennsylvania
Philadelphia, Pennsylvania 19104

THOMAS HELLMAN MORTON[b]

Department of Chemistry
Harvard University
Cambridge, Massachusetts 02138

Chemical treatment of the olfactory epithelium can impair the sense of smell in two ways: production of a selective hyposmia or of general hyposmia. We have reported the former in tiger salamanders conditioned to respond to two odorants, cyclohexanone and dimethyldisulfide.[1-3] An example of the latter can be seen in the results of application of methylmercury hydroxide, CH_3HgOH (which rapidly attacks -SS- and -SH groups), directly to the receptor epithelium. As FIGURE 1A summarizes, responding to both odorants is impaired to an equal extent. By contrast, application of the sulfhydryl reagent iodoacetamide, ICH_2CONH_2 has, by itself, no effect on responding (FIG. 1B), nor does it enhance the effect of CH_3HgOH (FIG. 1C).

Our interest focuses on the blocker-fixer sequence, acetoacetic ester (CH_3COCH_2COOR) as blocker followed by sodium cyanoborohydride ($NaBH_3CN$) as fixer, which produces selective hyposmia to aldehydes and ketones, but does not affect responding to simple esters (e.g. ethyl butyrate), alcohols, or sulfides.[4,5] This two-step procedure is chemically specific for proteins with Schiff base-forming sites that bind electrically uncharged carbonyl compounds. *In vitro* experiments to probe this specificity were conducted with AAD (the bacterial enzyme acetoacetate decar-

[a] This work was supported by National Science Foundation Grant CHE 85-09557 and National Institutes of Health Grant NS 19424.

[b] Address for correspondence: T. H. Morton, Department of Chemistry, University of California, Riverside, CA 92521.

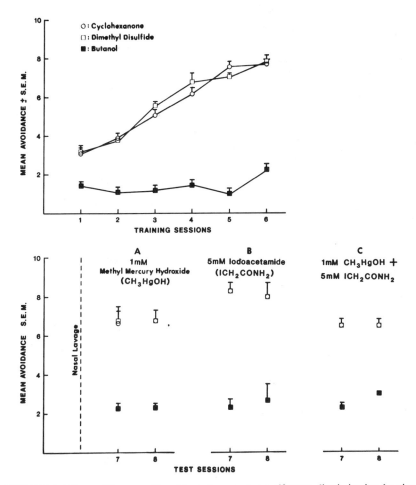

FIGURE 1. Effects of lavage with sulfhydryl reagents on olfactory discrimination by tiger salamanders. Top: acquisition of conditioned avoidance[1-3] to cyclohexanone and dimethyl sulfide (open symbols) with concurrent, unconditioned presentations of *n*-butanol (solid squares) ($n=11$). Bottom: mean performances for two postlavage test sessions for subgroups given bilateral lavage with **(A)** 100 μl methylmercury hydroxide ($n=3$); **(B)** 100 μl iodoacetamide ($n=3$); and **(C)** 100 μl methylmercury hydroxide followed by 100 μl iodoacetamide ($n=4$).

boxylase), a Schiff base-forming protein with a well-characterized active site. The following data provide evidence that a single active-site lysine undergoes covalent modification in this model system:

(1) Since all five hydrogens of CH_3COCH_2COOR exchange with solvent in the presence of AAD,[6] tritiated water can be used to incorporate radiolabel into the blocker and thence into the protein. A solution of 0.07 mM AAD and 3.4 mM ethyl 3-[^{14}C]acetoacetate (1 mCi/mmol) in tritiated water (0.3 Ci/ml) at pH 6 treated with 15 mM NaBH$_3$CN for one hour incorporated nondialyzable label into AAD in a ratio

of $\dfrac{\text{dpm }^{3}\text{H}}{\text{dpm }^{14}\text{C}} = 5$ (after correction for a control in which NaBH_3CN was omitted). In an experiment performed using *tert*-butyl acetoacetate, AAD incorporated comparable levels of ^{3}H: 0.056 μCi/mg protein from 3.2 mM $\text{CH}_3\text{COCH}_2\text{COOC(CH}_3)_3$ and 0.032 μCi/mg protein from 0.1 mM $\text{CH}_3\text{COCH}_2\text{COOC(CH}_3)_3$ [as opposed to 0.004 μCi/mg protein in a control experiment with 0.4 mM $\text{CH}_3\text{COCH}_2\text{COOC(CH}_3)_3$ in which NaBH_3CN was omitted].

(2) Since acetic anhydride specifically blocks the active-site lysine of AAD,[7] acetylation can be used to prevent subsequent covalent modification by the blocker-fixer sequence, as shown to the left in FIGURE 2. AAD with > 99% of its active sites acetylated shows a greatly reduced level of ^{3}H incorporation from 3.2 mM $\text{CH}_3\text{COCH}_2\text{COOC(CH}_3)_3$ plus NaBH_3CN in tritiated water, 0.014 μCi/mg protein. AAD that was acetylated with 1-[^{14}C]acetic anhydride (13.8 μCi/mmol) and subsequently treated with 3.5 mM ethyl acetoacetate in tritiated water incorporated a ^{3}H-to-^{14}C ratio of 13 when NaBH_3CN was omitted and a ^{3}H-to-^{14}C ratio of 14 under identical conditions when NaBH_3CN was included. We draw two conclusions: (1) Incorporation of radiolabel into AAD does not depend on the identity of alkyl group R in the acetoacetic ester; (2) the blocker-fixer sequence specifically modifies the active-site lysine, as depicted by the reaction sequence to the right in FIGURE 2.

Intranasal lavage with 0.5 mM ethyl acetoacetate followed by 50 mM NaBH_3CN produces a selective hyposmia not only in tiger salamanders, but in other species, as well. Results are tabulated below for cardiac conditioning of a starling *(Sturnus vulgaris)*. The bird was conditioned to respond to cyclohexanone and ethyl butyrate with an increase in heart-beat rate (with concurrent, unreinforced presentations of *n-*

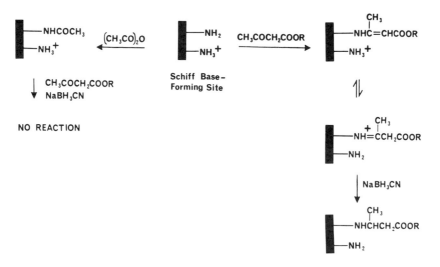

FIGURE 2. Site-selective covalent modification of a Schiff base-forming protein, as typified by the active site of acetoacetate decarboxylase (AAD), which contains two amine functions in close proximity. To the left, selective acetylation[7] blocks further modification by the blocker-fixer sequence of $\text{CH}_3\text{COCH}_2\text{COOR}$ followed by NaBH_3CN. To the right, a mechanism for selective covalent modification of the active site via a blocker-fixer sequence.

TABLE 1. Two-Step Lavage Abolishes Significant Responding to Cyclohexanone

Odorant	Prelavage		Postlavage	
	Prior	During	Prior	During
Cyclohexanone	7.2 Hz[a]	8.3 Hz[a]	6.8 Hz[b]	7.0 Hz[b]
Ethyl butyrate	6.9 Hz[a]	7.7 Hz[a]	6.9 Hz[a]	8.0 Hz[a]

[a] Meets significance criterion $p < 0.001$.
[b] Fails to meet significance criterion $p < 0.15$.

butanol, to which the bird did not respond). Average heartbeat rates for 15 seconds before and for 15 seconds during odorant presentation are given (means of 10 trials). Paired *t*-tests of postlavage performance show that two-step lavage abolishes significant responding to cyclohexanone (TABLE 1).

Starlings recover intact olfactory function within one to two hours after lavage, much more quickly than do tiger salamanders (which take several days). This result is consistent with replacement of modified receptors via new protein synthesis, which, we have argued,[4,5] is required for recovery from the agonistic effect of irreversible covalent modification.

REFERENCES

1. MASON, J. R. & T. H. MORTON. 1982. Physiol. Behav. **29:** 709-714.
2. MASON, J. R. & T. H. MORTON. 1984. Tetrahedron **40:** 483-492.
3. MASON, J. R., L. CLARK & T. H. MORTON. 1984. Science **226:** 1092-1094.
4. MASON, J. R., F.-C. LEONG, K. W. PLAXCO & T. H. MORTON. 1985. J. Am. Chem. Soc. **107:** 6075-6084.
5. MASON, J. R., K. K. JOHRI & T. H. MORTON. 1987. J. Chem. Ecol. **13:** 1-18.
6. HAMMONS, G., F. H. WESTHEIMER, K. NAKAOKA & R. KLUGER. 1975. J. Am. Chem. Soc. **97:** 1568-1572, 4152.
7. O'LEARY, M. & F. H. WESTHEIMER. 1968. Biochemistry **7:** 913-917.

Skin Lipids of Garter Snakes Serve as Semiochemicals

ROBERT T. MASON,[a] JOHN W. CHINN,[b] AND
DAVID CREWS [a]

[a]Institute of Reproductive Biology
Department of Zoology
[b]Department of Chemistry
University of Texas at Austin
Austin, Texas 78712

Chemical communication in the Reptilia has not been widely studied despite the fact that many species of lizards and snakes rely heavily on chemical cues obtained by tongue-flicking. As early as the 1930s, G. K. Noble[1] hypothesized that integumental chemical cues play a powerful role in garter snake reproductive behavior.

The red-sided garter snake, *Thamnophis sirtalis parietalis*, is the most northerly living reptile in North America and perhaps the world. In the spring, garter snakes leave underground hibernacula where they have hibernated the previous winter. In Manitoba, Canada where our studies are conducted, these animals hibernate together in large groups of up to several thousand animals.[2] Upon emergence, male garter snakes congregate around the entrances to the dens waiting for females that emerge singly or in small groups. When a male encounters a newly emerged female he will investigate her, tongue-flicking repeatedly. All newly emerged females are sexually attractive and elicit courtship behavior from males. Soon, groups of males begin to court females when the males detect a pheromone on the female's dorsal surface. Courtship ensues as the males press their chins onto the female's back. "Chin-rubbing" is accompanied by rapid tongue-flicking by the male. This tongue-flicking behavior serves to deliver chemical cues from the female's dorsal surface to the male's vomeronasal organ.[3]

We have been investigating the chemical communication used by garter snakes in conjunction with their reproductive behavior. Previous work[4] has demonstrated that chemical cues removed from the female's integument with hexane elicit courtship behavior from male garter snakes. These results suggest that nonpolar lipids are serving as semiochemicals in these snakes.

Our studies to date have focused on isolating and identifying the sex-recognition pheromone of the red-sided garter snake. We have used the techniques of thin-layer chromatography (TLC) and gas chromatography/mass spectrometry (GC/MS). During the breeding season, hexane washes of the skin of males and females show different components on charred TLC plates. These results suggest a sex difference in the skin lipids of males and females. Analysis of the same samples using GC/MS confirmed this hypothesis. FIGURE 1 shows a clear sex difference in the GC trace of males as compared to females. Mass spectral data identified primarily long-chain fatty acids in the male washes. The female washes were comprised mostly of small hydrocarbons as well as cholesterol and two other steroids similar in structure to androstanediones. Both males and females possess cholesterol in their skin lipids.

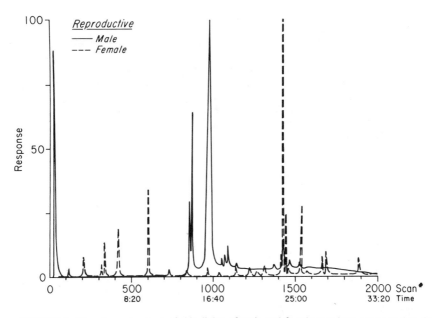

FIGURE 1. Gas chromatograms of skin lipids of male and female red-sided garter snakes, *T. s. parietalis* acquired during the breeding season (retention times in min).

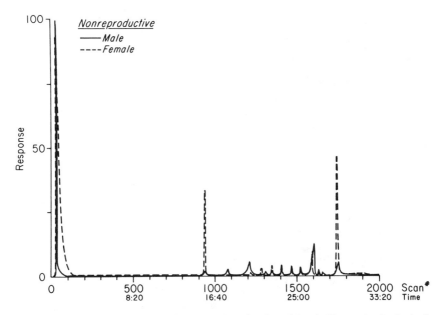

FIGURE 2. Gas chromatograms of the skin lipids of male and female *T. s. parietalis* obtained during the nonbreeding season (retention times in min).

During the nonbreeding season the GC traces of males and females are quantitatively as well as qualitatively different from the washes acquired during the breeding season (FIG. 2). The fatty acid peaks are absent from the male trace as are the two steroid peaks from the female trace. Cholesterol is still found in both the male and female washes. These results lend support to the theory that cholesterol is playing a role in the regulation of cutaneous water loss in reptiles. Indeed, Schell and Weldon[5] found cholesterol in the skin lipids of all 14 species of snakes they examined.

In summary, there is a distinct sex difference in the skin lipids of garter snakes. In addition, there is a pronounced seasonal difference. This seasonal difference in garter snake skin lipids may be responsible for the extinction of male courtship after the brief three- to four-week breeding season.

REFERENCES

1. NOBLE, G. K. 1937. The sense organs involved in the courtship of Storeria, Thamnophis, and other snakes. Bull. Am. Mus. Nat. Hist. 73: 673-725.
2. GREGORY, P. T. 1974. Patterns of spring emergence of the red-sided garter snake, Thamnophis sirtalis parietalis in the Interlake region of Manitoba. Can. J. Zool. 52: 1063-1069.
3. KUBIE, J. L. & M. HALPERN. 1978. Garter snake trailing behavior: Effects of varying prey-extract concentration and mode of prey-extract presentation. Comp. Phys. Psych. 92: 362-373.
4. MASON, R. T. & D. CREWS. 1985. Female mimicry in garter snakes. Nature 316: 59-60.
5. SCHELL, F. M. & P. J. WELDON. 1985. [^{13}C]NMR analysis of snake skin lipids. Agric. Biol. Chem. 49: 3597-3600.

Changes in Excitable Properties of Olfactory Receptor Neurons Associated with Nerve Regeneration[a]

LEONA M. MASUKAWA, BRITTA HEDLUND, AND
GORDON M. SHEPHERD

Section of Neuroanatomy
Yale University School of Medicine
New Haven, Connecticut 06510

Analysis of the physiological properties of olfactory receptor neurons following nerve transection began with the pioneering studies of Simmons and Getchell[1] using extracellular recording techniques. We have carried out intracellular voltage recordings from cells in *in vitro* epithelial preparations of the normal tiger salamander (land phase), *Ambystoma tigrinum,* and at two and four weeks post-nerve transection (for methods, see Masukawa *et al.*[2]) in order to examine the action potential generating properties of olfactory neurons under these conditions. We summarize the results of these studies, and some insights they have provided into the membrane mechanisms related to the generation of new neurons.

Several specific changes have been observed in the excitable properties of olfactory receptor neurons in the two- and four-week period after olfactory nerve transection. The threshold for generation of the initial impulse response to injected current is significantly higher compared to normal receptor neurons (see TABLE 1 and FIG. 1B). There is a decreased frequency sensitivity to injected current (see TABLE 1, four weeks). Small spikelike potentials occur during the response to current injection at two weeks post-transection (FIG. 1Ae). Normally only one action potential is generated even with high current intensities at this time period. Following the action potential, a greatly reduced after-hyperpolarization is present (compare FIG. 1A, a, b, and c).

Most neurons with transected axons die within the first week,[3,4] so it may be concluded that the changes at two weeks occur in newly generated neurons. We postulate that the decrease in excitability, as seen in the several types of properties summarized above, may reflect changes in the expression of K channels. A possible change involving K channels is the lack of transient A current channels at two weeks post-transection. In support of this possibility is the observation that the removal of this current by exposure to 4-AP in dissociated mature olfactory cells leads to the generation of single action-potential responses to current steps.[1] Subsequent changes at four weeks post-transection may be related to the sequence of outgrowth of the axon, contact by the axonal growth cone with the olfactory bulb, and establishment there of synaptic contacts with bulbar neuronal dendrites in the olfactory glomeruli. Maturation of low-threshold, slowly adapting impulse responsiveness must therefore

[a]This work was supported by Research Grants NS-07609 and NS-10174 (to G.M.S.) from the National Institute for Neurological and Communicative Disorders and Stroke, and by a James Hudson Brown-Alexander B. Coxe Fellowship (to B.H.).

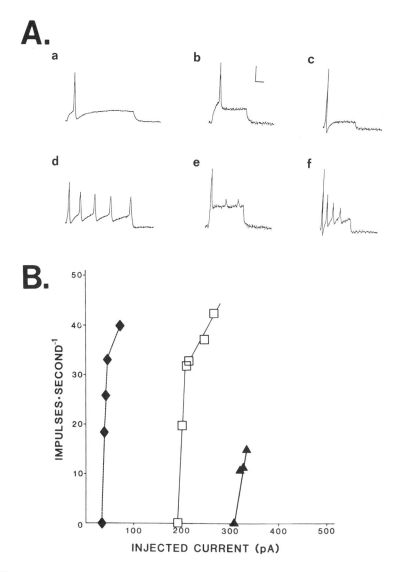

FIGURE 1. (A) Impulse responses of receptor cells to current injection in control epithelium (a and d), two weeks (b and e) and four weeks (c and f) post-nerve transection. Intracellular responses of receptor cells to depolarizing current injections of two different intensities; top traces are to the lower intensity. Resting membrane potentials were at least −45 mV, and the input resistances were at least 160 Mohm. (B) Firing frequency of the second action potential (inverse of the time interval between the first and second action potential in the train) plotted against amount of injected current. Different symbols refer to three cells for which data are presented in A (control (◆); two weeks (▲), and four weeks (□). The threshold currents for eliciting the first action potential in each cell are indicated along the abscissa. The frequency response is representative of the mean response under each condition. Calibrations are 40 mV and 20 msec (b, c, e, and f) and 40 mV and 10 msec (a and d).

involve both anterograde signals from the receptive dendritic sites as well as retrograde signals from the axonal synaptic sites, in order to control, in a coordinated manner, the expression and/or insertion of channel proteins in their appropriate numbers and sites.

TABLE 1. Summary of Properties of Olfactory Receptor Neurons in Control Epithelium, Two and Four Weeks after Olfactory Nerve Transection

	Resting Membrane Potential (mV)	Cell Input Resistance (Mohm)	Threshold for First Spike (pA)	Current for Second Spike (pA)	Rate of Increase in Firing Freq. (imp. sec^{-1}pA^{-1})
Control	-57 ± 21	259 ± 159	74 ± 46	87 ± 56	1.34 ± 0.80 ($n = 9$)
Two weeks	-48 ± 4	329 ± 96	176 ± 41	251 ± 63	
Four weeks	-39 ± 7	258 ± 145	191 ± 71	254 ± 80	0.49 ± -0.28 ($n = 6$)

REFERENCES

1. MASUKAWA, L. M., B. HEDLUND & G. M. SHEPHERD. 1985. Changes in the electrical properties of olfactory epithelial cells in the tiger salamander. J. Neurosci. **5:** 136-141.
2. GRAZIADEI, P. P. C. & G. A. MONTI GRAZIADEI. 1978. Continuous nerve cell renewal in the olfactory system. *In* Handbook of Sensory Physiology, Vol. IX. Development of Sensory Systems. M. Jacobson, Ed.: 55-83. Springer Verlag. Berlin, Heidelberg, New York.
3. SIMMONS, P. A. & T. V. GETCHELL. 1981. Physiological activity of newly differentiated olfactory receptor neurons correlated with morphological recovery from olfactory nerve section in the salamander. J. Neurophysiol. **45:** 529-549.
4. FIRESTEIN, S. Personal communication.

Blocking Learned Food Aversions in Cancer Patients Receiving Chemotherapy[a]

RICHARD D. MATTES

Monell Chemical Senses Center
Philadelphia, Pennsylvania 19104

CATHY ARNOLD AND MARCIA BORAAS

Fox Chase Cancer Center
Fox Chase, Pennsylvania 19111

Aversions that form toward foods or beverages following a pairing of their ingestion with a period of malaise are termed learned food aversions. Nutritious foods are often targeted,[1] and this phenomenon has been implicated in the anorexia of cancer.[2] Such aversions may also adversely affect the quality of patients' lives since problematic foods may be highly preferred, regularly consumed items. The purpose of this project was to evaluate whether chemotherapy patients surreptitiously exposed to a nutritionally inconsequential "scapegoat" food (fruit-flavored beverage) in the one-hour period before their first course of treatment would direct any newly formed aversion towards the scapegoat, and thereby spare previously acceptable, nutrient-dense items from aversions.

Patients were 25-76 years of age with histologically confirmed cancers primarily of the breast or lung. All were tested before receiving their first course of chemotherapy and none received radiotherapy during their time of participation. Forty-five patients were randomly assigned to receive scapegoat exposure while 29 patients served as controls. The scapegoat was either grape or lemon-lime flavored beverage (presented in a counter-balanced order) served as a 150 ml sample that subjects evaluated on selected sensory properties to surreptitiously provide exposure. An aversion to the scapegoat was defined as present when, in a two-choice condition, one-third or less of total beverage ingestion was derived from the scapegoat beverage. Aversions to foods in the patient's diet were monitored by open-ended questionnaires and nine-point food action rating scale responses to the foods ingested in the 48-hour period surrounding the first treatment session. These assessments are conducted one, two, four, and six months following the initiation of treatment.

Following exposure to the scapegoat beverage, 40% (18/45) of patients reported an aversion to other foods during the follow-up period. The incidence in unexposed patients was 52% (15/29). However, in the subset of patients actually forming an aversion to the scapegoat, the incidence of new aversions was reduced to approximately

[a]Supported by Grant #RO1-CA37298 from the National Cancer Institute.

478

10% (2/18). The efficacy of the blocking approach appeared comparable across cancer types. The amount of scapegoat beverage ingested was significantly related to its efficacy in limiting aversions to other foods. ($p < 0.004$).

These findings suggest that exposure to a nutritionally inconsequential food just before a course of chemotherapy blocks the formation of aversions to other nutrient-dense foods in the diet and may therefore be a valuable therapeutic aid in the management of cancer patients receiving chemotherapy.

REFERENCES

1. MIDKIFF, E. E. & I. L. BERNSTEIN. 1985. Targets of learned food aversions in humans. Physiol. Behav. **34:** 839-841.
2. BERNSTEIN, I. L. 1985. Learned food aversions in the progression of cancer and its treatment. Ann. N.Y. Acad. Sci. **443:** 365-380.

Association between Anosmia and Anorexia in Cats

KIMBERLY MAY

Department of Physiology and Pharmacology
Auburn University
Auburn, Alabama 36849

Anorexia is a common disorder in the feline species. Anosmia, the absence of the sense of smell, has been suggested by current literature to specifically contribute to the condition of anorexia. Therefore, an attempt was made to associate the anosmia with a decrease in food intake.

In this study the olfactory system was treated to induce anosmia. Treatment groups included the following: (1) control (no treatment); (2) 5% zinc sulfate ($ZnSO_4$) deposited intranasally; (3) sham zinc sulfate (substitution of sterile physiological saline for $ZnSO_4$); (4) olfactory bulb transection; (5) sham surgery (surgical exposure of olfactory bulbs but with neither transection nor ablation); and (6) olfactory bulb ablation.

Each animal's food intake was monitored daily pre- and post-treatment for at least 27 days. Analysis of these data revealed significant differences in groups 4 (transection) and 6 (ablation), as reported in TABLE 1. In addition, all animals, with the exception of cats R-16 and R-24, were tested seven days posttreatment for anosmia; the latter members were tested 29 days posttreatment. All the cats in groups 2, 4, and 6, with

TABLE 1. The Pre- and Posttreatment Daily Average Food Intake Data for Each Group

Treatment Group	Pre/Post	Average Food Intake (g)	Average Days Observed
(1) Control	Pre	76	35
	Post	70	29
(2) $ZnSO_4$	Pre	74	35
	Post	78	28
(3) Sham $ZnSO_4$	Pre	72	38
	Post	76	28
(4) Transection	Pre	78	39
	Post	89[a]	30
(5) Sham surgery	Pre	93	39
	Post	90	28
(6) Ablation	Pre	88	42
	Post	57[a]	34

[a] Indicates a significant difference ($p < 0.01$ for group 4 and $p < 0.0001$ for group 6) between pre- and posttreatment daily food intake.

480

TABLE 2. Presence of Degeneration in the Olfactory System of Each Cat in Their Treatment Groups

Treatment Groups	Lab ID#	Area Observed for Degeneration		
		Inner	Middle	Outer
(1) Control	C-5	−	−	−
	C-10	−	−	−
	C-19	−	−	−
	R-1	−	−	−
	R-12	−	−	−
(2) ZnSO₄	C-4	+	+	−
	C-9	−	−	−
	C-12	−	−	−
	C-16	−	−	−
	R-5	−	−	−
	R-11	+ +	+ +	−
(3) Sham ZnSO₄	C-8	−	−	−
	C-13	−	−	−
	C-18	−	−	−
	R-4	−	−	−
	R-9	−	−	−
(4) Transection	C-2	+ + +	+ + +	−
	C-7	+ + +	+ + +	−
	C-20	+ + +	+ + +	−
	R-3	+ + +	+ + +	−
	R-7	+ + +	+ + +	−
	R-16	+ + +	+ + +	−
(5) Sham surgery	C-1	−	−	−
	C-6	−	−	−
	C-17	−	−	−
	R-2	+	+	+
	R-10	−	−	−
		Olfactory Tracts	Rostral Commissure	Cerebral Cortex
(6) Ablation	R-17	+ + +	+ + +	+ +
	R-20	+ + +	+ + +	+ +
	R-21	+ + +	+ + +	+ +
	R-22	+ + +	+ + +	+ +
	R-23	+ + +	+ + +	+ +
	R-24	+ + +	+ + +	+ +

the exception of cat R-16, were judged anosmic. After termination, the olfactory bulb, tracts, and basal frontal cerebral cortex was extracted and prepared by the Nauta-Gygax technique[2] for examination of olfactory nerve degeneration. In analyzing the slides, three regions of the olfactory bulb were observed: (1) the "inner" or the periventricular layer, (2) the "middle" or the internal plexiform and mitral cell layer, and (3) the "outer" or the external plexiform and glomerular layer. Degeneration was subjectively ranked according to its magnitude: + indicated slight degeneration, + + indicated moderate degeneration, and + + + indicated massive degeneration (TABLE 2).

In the animals in group 6 (ablation), which were anosmic due to absence of an olfactory bulb, a correlation was determined to be significant with a decrease in food intake. In group 4 (transection) the animals exhibited an increase in food intake even though they appeared anosmic, and histological preparations showed massive degeneration (though not complete) of olfactory input to the central nervous system. The last olfactory treatment group 2 ($ZnSO_4$) had no change in food consumption. Hence, the original hypothesis that elimination of olfactory function produces a decrease in food intake was confirmed only in the ablation-treated group. However, even though behavioral evidence would suggest anosmia in the other two groups, partial loss, as demonstrated by histological preparations, resulted in no change ($ZnSO_4$ group) or increased food intake (transected group).

REFERENCES

1. MORRIS, M. L., S. M. TEETER & G. G. DOERING. 1984. Anorexia; A Commentary on Nutritional Management of Small Animals. Mark Morris Associates. Topeka, KS.
2. NAUTA, W. J. H. & P. A. GYGAX. 1954. Silver impregnation of degenerating axons in the central nervous system: A modified technique. Stain Tech. **29**(2): 91-93.

A Linkage between Coding of Quantity and Quality of Pheromone Gland Components by Receptor Cells of *Trichoplusia ni*

M. S. MAYER AND R. W. MANKIN

*United States Department of Agriculture, ARS
Insect Attractants, Behavior, and Basic Biology Research Laboratory
Gainesville, Florida 32604*

The responses elicited in two specialized pheromone receptor cells of *T. ni* by six pheromone gland components link the coding of pheromone quality in the central nervous system (CNS) inextricably with the coding of pheromone quantity.

In a study of quality and quantity coding of *T. ni* sex pheromone components, we have used a combination of various methods, including gas-liquid chromotography, electroantennogram, and radiolabeling, to measure the emission rates of the pheromone gland components from glass and rubber septum dispensers. A newly developed stimulus delivery system controlled the stimulus concentration and duration. Neuronal responses were recorded from a tungsten electrode inserted at the base of a sensillum that contained the receptor neurons. In *T. ni* there are two morphologically distinct sensilla, HS and LS, that contain two or more neurons, designated (a) and (b), whose responses can be distinguished by spike amplitude.[1,2]

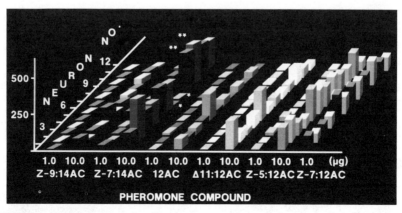

FIGURE 1. HS(a) impulses during three-second stimulus interval. *1.0 µg of new sample; **0.1 µg of new sample.

We found that the HS(a) neuron responded most sensitively to Z-7:12AC, with action potentials elicited at concentrations lower than 1×10^{-11} μmole/cm^3 (FIG. 1). The next most stimulatory gland component for this neuron was Z-7:14AC. Dodecan-1-ol acetate failed to elicit reliable responses until the concentration exceeded about 6×10^{-5} μmole/cm^{-3}. Another neuron in this sensillum, HS(b), responded only to Z-7:12OH at or above such concentrations. A neuron in the other sensillum, LS(b), was most sensitive to Z-7:14AC, with a midrange response at a concentration of 1×10^{-8} μmole/cm^3 (FIG. 2). The LS(a) neuron was not stimulated by any of the pheromone components.

The linkage between quantity to quality coding is evident from the responses where the stimulating doses were increased just one magnitude. Both of the neurons that responded selectively at low doses lost selectivity at the high dose; both the HS(a) and LS(b) neurons responded to all pheromone gland components except Z-7:12OH at stimulus concentrations above about 1×10^{-6} μmole/cm^{-3}. Because HS(a) and LS(b) neurons responded to six of the seven pheromone gland components at these elevated concentrations, it is not clear how the individual components are discriminated by the CNS.

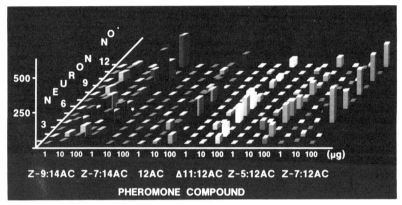

FIGURE 2. LS(b) impulses during three-second stimulus interval. * indicates new sample 0.1 μg dose.

It is evident from FIGURES 1 and 2 that the selectivity of the response to different pheromone components decreases as the stimulus intensity increases. We conclude that neither of these neurons in *T. ni* can be defined as a specialist cell in the sense defined for the bombykol receptor neuron in *Bombyx mori* because of the linkage between quantity and quality. Whether this linkage is unique or general remains to be determined.

REFERENCES

1. O'CONNELL, R. J., A. J. GRANT, M. S. MAYER & R. W. MANKIN. 1983. Morphological correlates of differences in pheromone sensitivity in insect sensilla. Science **220:** 1408-1410.
2. MAYER, M. S. & R. W. MANKIN. 1985. Neurobiology of pheromone perception. *In* Comprehensive Insect Physiology, Biochemistry and Pharmacology. G. A. Kerkut & L. I. Gilbert, Eds. Vol. **9:** 95-144. Pergamon Press. London, England.

Locations of Taste-Evoked Activity Are Not Coincident with Chorda Tympani Afferent Endings in the Nucleus of the Solitary Tract in the Hamster[a]

MARTHA McPHEETERS, THOMAS P. HETTINGER,
MARK C. WHITEHEAD, AND MARION E. FRANK

Department of BioStructure and Function
University of Connecticut Health Center
Farmington, Connecticut 06032

The medullary solitary nucleus is the first site of neural processing in the mammalian gustatory system. Neurophysiological experiments[1,2] in the hamster discovered responses to gustatory stimulation of the anterior tongue in the lateral part of the rostral pole of the nucleus of the solitary tract (NTS). Neuroanatomical observations and tracing experiments[3] divided the nucleus into cytoarchitectonic subdivisions and located the terminals of afferent fibers within various subdivisions of the NTS. Afferent endings of the chorda tympani are predominately in the rostral third of the central subdivision (FIG. 1). We attempted to correlate neurophysiological findings with these neuroanatomical results in the same experiment in the hamster brainstem.

Platinum-iridium microelectrodes (0.5-2.0 megohms) comparable to the more common tungsten electrodes were used to record multiunit activity from the hamster NTS and surrounding areas while the anterior tongue was stimulated with a search solution (0.03 M NaCl + 0.1 M sucrose + 0.1 M KCl). Locations responding to the search solution were sequentially stimulated with the individual search components, 0.003 M HCl, 0.001 M quinine hydrochloride, hot water (45°C) and cold water (0°C). Responses were recorded on cassette tape and played back through a window discriminator into a computer that produced histograms of frequency (spikes per second) versus time. At the close of a recording session, small electrolytic lesions were made above and below the NTS at known locations. Later these histologically verified lesions were used to locate the electrode tracks and the locations of each taste response.

Response profiles showed a great deal of variation from location to location and many histologically identified sites of gustatory response were outside of the NTS. At locations within the NTS, the search solution tended to be the most effective stimulus and outside of the NTS either hot water, cold water, or HCl tended to give the maximal response, although all locations responded to the search solution.

The sites of taste-elicited activity are found around the solitary tract and/or the

[a]Supported by National Institutes of Health Grant NS 16993.

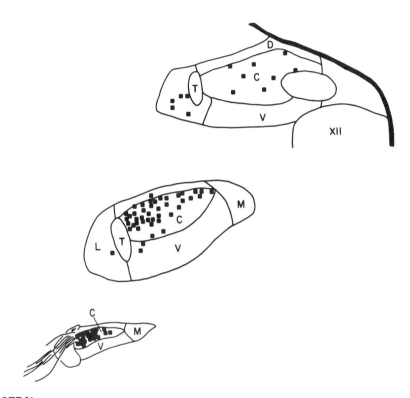

ROSTRAL

FIGURE 1. Schematic diagram of the rostral half of the hamster NTS showing afferent endings from the chorda tympani (■ , solid squares) and cytoarchitectonic subdivisions. Adapted from Figure 2 in Whitehead and Frank.[3] C, central; D, dorsal; L, lateral; M, medial; T, solitary tract; V, ventral; XII, hypoglossal nucleus.

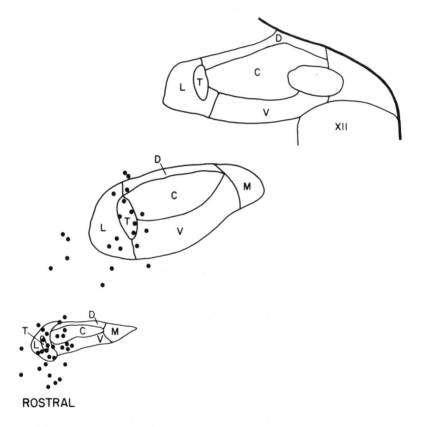

ROSTRAL

FIGURE 2. Schematic diagram of the rostral half of the NTS showing locations of taste-elicited activity (●, solid circles). Abbreviations as in FIGURE 1.

lateral subdivision of the NTS as shown in FIGURE 2. This pattern of locations is a marked contrast to the pattern of afferent endings shown in FIGURE 1. This curious finding, which replicates that of previous workers, raises questions about where neural processing of first-order gustatory information occurs and alternatively whether the electrodes may be recording responses from afferent fibers as they exit the tract rather than second-order cells. This experiment provides evidence that the locations of easily recorded responses to taste stimulation of the anterior tongue are not coincident with the locations of chorda tympani afferent endings on brain stem neurons.

REFERENCES

1. PFAFFMANN, C., R. P. ERICKSON, G. P. FROMMER & B. P. HALPERN. 1961. Gustatory discharges in the rat medulla and thalamus. In Sensory Communication. W. A. Rosenblith, Ed.: 455-473. Wiley. New York.
2. TRAVERS, J. B. & D. V. SMITH. 1979. Gustatory sensitivities in neurons of the hamster nucleus tractus solitarius. Sens. Proc. 3: 1-26.
3. WHITEHEAD, M. C. & M. E. FRANK. 1983. Anatomy of the gustatory system in the hamster: Central projections of the chorda tympani and the lingual nerve. J. Comp. Neurol. 220: 378-395.

Relationships between and among Selected Measures of Sweet Taste Preference and Dietary Habits[a]

D. J. MELA AND R. D. MATTES

Monell Chemical Senses Center
Philadelphia, Pennsylvania 19104

Previous studies from this laboratory[1,2] and others have failed to elucidate any clear relationships between individual measures of gustatory function (threshold sensitivity, suprathreshold intensity ratings, preferences) and aspects of dietary intake. However, the strongest associations were noted with measures of taste preference, and this prompted the hypothesis that a profile developed from an array of taste preference procedures, using different response scales, test media, and contexts, might facilitate the detection of diet-taste relationships. This report focuses on selected measures of sweet taste and intake.

Preference measures include (1) questionnaires assessing liking and frequency of consumption of selected sweet and nonsweet foods, (2) "adjustment tasks" wherein subjects modified beverage samples to their optimal preferred sweetness level, and (3) hedonic ratings of oatmeal and coffee samples prepared with a range of sweetener levels. These samples were assessed under several conditions: expectorated and swallowed, rated on a visual analogue and a verbal nine-point category scale, and viewed in the context of either a prepared "food" or an isolated laboratory "sample." Dietary records were collected throughout the week of sensory testing and, in addition to food descriptions and portion estimates, subjects recorded the perceived predominant taste of each food consumed.[1] Subjects were twenty-five nonsmoking males aged 17 to 34.

Variations in the conditions for evaluation of the oatmeal and coffee samples had little effect on the preferred sweetener levels, although over a one-week period responses following expectoration of samples were found to generate more repeatable results than those obtained after the samples were swallowed. The preferred sweetener level identified for prepared coffee under all conditions was significantly correlated with the preferred sweetener level as produced by the subjects in the adjustment task ($r = 0.55$ to 0.73, all $p < 0.01$).

Correlations between single taste measures and intake variables demonstrated no consistent relationships except a positive correlation between preferred oatmeal sweetener level and percent of calories from foods identified as sweet ($p < 0.01$). The various preference measures were loaded into a discriminant analysis program to establish whether they could be used to classify subjects into upper and lower tertiles

[a] This work has been published in full as: MATTES, R. D. & D. J. MELA. 1986. Relationships between and among selected measures of sweet-taste preference and dietary habit. Chem. Senses **11:** 523-539.

for selected intake measures. Initial analyses identified no particular advantage for any one of the several methodological approaches to testing, but from these initial analyses, "best predictor" discriminant functions were developed. As shown in TABLE 1, preference measures yielded significant discriminant functions for percent of calories from carbohydrates and from sweet foods, and the percent of food items identified as sweet.

TABLE 1. Sweet Taste Preference Profiles

Intake Measure	Canonical Correlation	Percent of Subjects Correctly Classified	p
% Carbohydrate calories	0.82	100	0.006
% Calories from sweet foods	0.80	94	0.0007
% Sweet food items	0.81	100	0.0004

These results reported in TABLE 1 demonstrate that sweet taste preference profiles may be predictive of dietary habits related to sweet food and carbohydrate intake. Whether such relationships exist for other tastes and intake measures will be the focus of future studies.

REFERENCES

1. MATTES, R. D. 1985. Gustation as a determinant of ingestion: Methodological issues. Am. J. Clin. Nutr. **41:** 672-683.
2. MATTES, R. D. 1985. Gustation and nutrition. Chem. Senses **10:** 456-457.

A Freeze-Fracture Study on the Prenatal Development of Ciliated Surfaces in Rat Olfactory Epithelia[a]

BERT PH. M. MENCO

Department of Neurobiology and Physiology
Northwestern University
Evanston, Illinois 60201

The purpose of this study was to investigate membrane transformations accompanying outgrowth of receptor cell cilia and supporting cell microvilli in the olfactory epithelium during normal development.[1,2] As yet there has been only one report on this subject.[3] Fixed, cryoprotected samples of rat embryos of intrauterine days E14, E16, E18, E19 (E1 = day that the dams were sperm-positive), and of adults were fractured in a Cressington Freeze-Fracture Apparatus at −170°C and at a vacuum of at least

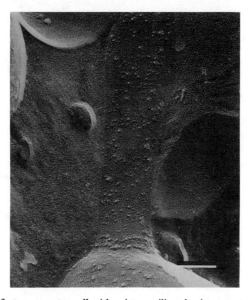

FIGURE 1. E16 olfactory receptor cell with primary cilium having two necklace strands.

[a]Supported by National Institutes of Health Grant #NS 21555.

10^{-6}mbar. Platinum-carbon rotary replication (about 600 rpm) was carried out at 20° angles, and the replicas were reinforced with carbon evaporated from above (see Menco[4] for technical details and nomenclature). Endings with single cilia or none had lower densities of intramembranous particles than endings with more than one cilium, but in both instances particle densities increased with development. Increases in endings with one cilium or none ranged from 220 particles/μm^2 at E14 to 900 particles/μm^2 at E19 for the P-faces. In endings with more than one cilium, values ranged from 550 (E16) to 1,000-1,200 particles/μm^2 (from E18 on). E-face particle densities were around 200-300 particles/μm^2, except for E14, in which the densities were about 100 particles/μm^2. Necklaces of primary cilia of receptor and supporting

FIGURE 2. E16 supporting cell apex with rod-shaped particles and primary cilium with three necklace strands. Magnification bars in both figures indicate 0.1 μm.

cells of main and vomeronasal olfactory organs had about three strands (FIGS. 1, 2). Necklaces of secondary olfactory cilia increased from four (E16) to five-six (E18, E19) and seven (adult). Two types of supporting cells were distinguished, those without (from E14 on) and with dimer or rod-shaped particles (from E16 on; FIG. 2) in their apical membrane structures. Those with rod-shaped particles had higher densities than those without. Highest densities were encountered in adults. In conclusion, various membrane transformations accompany outgrowth of cilia and microvilli during embryonic development of the olfactory epithelium, predicting that the cells undergo changes in their biochemical and physiological properties during development.[5]

REFERENCES

1. MENCO, B. PH. M. & A. I. FARBMAN. 1985. Genesis of cilia and microvilli of rat nasal epithelia during pre-natal development. I. Olfactory epithelium, qualitative studies. J. Cell Sci. **78:** 283-310.

2. MENCO, B. PH. M. & A. I. FARBMAN. 1985. Genesis of cilia and microvilli of rat nasal epithelia during pre-natal development. II. Olfactory epithelium, a morphometric analysis. J. Cell Sci. **78:** 311-336.
3. KERJASCHKI, D. 1977. Some freeze-etching data on the olfactory epithelium. *In* Olfaction and Taste. J. Le Magnen & P. MacLeod, Eds. Vol. **6:** 75-85. Information Retrieval Ltd. London.
4. MENCO, B. PH. M. 1986. A survey of ultra-rapid cryo-fixation methods with particular emphasis on application to freeze-fracturing, freeze-etching, and freeze-substitution. J. Electron Microsc. Techn. **4:** 177-260.
5. GESTELAND, R. C., R. A. YANCEY & A. I. FARBMAN. 1982. Development of olfactory receptor neuron selectivity in the rat fetus. Neuroscience **7:** 3127-3136.

Mate Recognition by an Antarctic Isopod Crustacean[a]

WILLIAM C. MICHEL [b]

Department of Biology
University of California, Los Angeles
Los Angeles, California 90024

Precopulatory mate guarding is common among pericarid crustaceans (reviewed by Ridley[1]). Male amphipods identify receptive females from olfactory[2,3] or contact chemosensory information (see Hartnoll and Smith[5] for other references). Vandel[6] reports that male *Asellus aquaticus* isopods identify females by contact, but the chemosensory basis for mate recognition by isopods has not been thoroughly examined.

The Antarctic isopod *Serolis polita* is known to maintain precopulatory pairs for extended periods (Luxmoore[7]), but the sensory basis for mate selection was not examined. An olfactory and contact chemosensory basis for male recognition of receptive females is described here. The reproductive and foraging behavior of this species will be considered in detail elsewhere.

MATERIALS AND METHODS

Serolis polita were collected by scuba divers with hand nets between April and November 1985 (austral winter) near Palmer Station, Anvsers Island, Antarctica (64°46′S., 64°03′W.). Nongravid females, collected from precopulatory pairs (hence, considered receptive), were segregated and used in subsequent experiments. Further details of experiments can be found in figure captions.

RESULTS

When searching, male *Serolis polita* sweep their antennae forward and after flagellar contact, lunge onto the dorsal surface to evaluate the prospective mate. Significantly more males displayed antennal alert, postural shifts, and locomotion to FNG aquaria

[a] Supported by National Science Foundation Grant #DPP85-40817 and DPP83-02852 to W. M. Hamner.

[b] Current address: Dept. of Zoology and Physiology, Louisiana State University, Baton Rouge, LA 70803.

water than to SW controls at a 10^{-1} dilution (FIG. 1). Neither the 10^{-3} dilution of FNG nor either dilution of FG or M aquaria waters were significantly more effective than SW.

During the first hour of choice experiments, total contact/inspections and subsequent precopulatory pairings were not FNG biased (FIG. 2A). After 24 hours significantly more FNG were selected for precopula than either FG or immature males (MI). When choice animals did not include a FNG, the incidence of male contact/inspections and precopula pairings during the first hour did not change and inappropriate mate selections after 24 hours were similar (FIG. 2B & 2C).

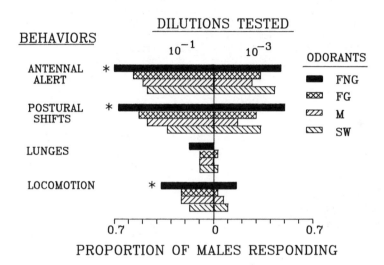

FIGURE 1. A summary of the *Serolis polita* males displaying behaviors after exposure to 10 ml (from 500 ml total) of 5-micron filtered aquaria water that held either 50 mature males (M), 50 gravid females (FG), or 60 non-gravid females (FNG; 10 extra to correct for total weight) or seawater (SW; as a control) for the previous 24 hours. The odorants were presented in a random, blind fashion. Each male (30 tested) was observed for three minutes before and after stimulus introduction and tested only once. An asterisk indicates a significant difference in the proportion responding to an odorant versus SW control (Chi-square test; $p < 0.05$).

DISCUSSION

FNG, FG, and M aquaria water all evoked more search behavior than SW (FIG. 1) but the waterborne factor released by FNG is not essential for initiation of mate search, since male search behavior is unchanged when FNG are absent (FIG. 2). Antennal contact is not sufficient to confirm the female is receptive. Only after the male contacted the dorsal surface did rejection of prospective mates occur. Apparently, the plumose setae on the second periopod must contact the female to confirm her reproductive status (Michel and Stretch, in preparation). These results suggest that both olfactory and contact chemosensory mechanisms are involved in mate recognition.

Olfaction may serve to aggregate *Serolis polita* before heterosexual pairings, however, the ultimate recognition of a receptive female requires contact.

SUMMARY

The proportion of male *Serolis polita* (Pfeffer; Crustacea, Isopoda, Flabellifera) initiating search behavior upon exposure to aquaria water that previously held non-gravid females (FNG) > gravid females (FG) > males (M) > seawater (SW). Only the FNG aquaria water evokes significantly more search behavior than SW (FIG. 1). Provided with choice animals, males preferentially select the FNG but contact and inspect FNG, FG, and M equally (FIG. 2). The absence of a FNG as a choice did not alter male search.

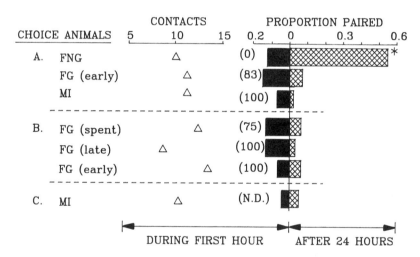

FIGURE 2. Contact/inspections and precopula pairs observed during the first hour and pre-copula pairs observed after 24 hours. The proportions are based on 40 trials conducted during experiment A and 30 trials each during experiments B and C. The numbers in paratheses to the right of the pairing bars indicate the percentage of males changing precopula choices between the one-hour observation and the final check at 24 hours. The early, late, and spent in paratheses next to the FG refers to the approximate brood stage. Immature males (MI) were used to eliminate the potential confusion of a competitor male. N.D. = not determined; all other abbreviations as in FIGURE 1.

REFERENCES

1. RIDLEY, M. 1983. The explanation of organic diversity: The comparative method and adaptations for mating. Clarendon Press. Oxford, England. 272 pp.
2. BOROWSKY, B. 1984. Effects of receptive females' secretions on some male reproductive behaviors in the amphipod crustacean *Microdeutopus gryllotalpa.* Mar. Biol. **84**: 183-187.

3. BOROWSKY, B. 1985. Responses of the amphipod crustacean *Gammarus palustris* to water-borne secretions of conspecifics and congenerics. J. Chem. Ecol. **11:** 1545-1552.
4. DAHL, E., H. EMANUELSSON & C. VON MECKLENBURG. 1970. Pheromone reception in the males of the amphipod *Gammarus duebeni,* Lilljeborg. Oikos **31:** 42-47.
5. HARTNOLL, R. G. & S. M. SMITH. 1980. An experimental study of sex discrimination and pair formation in *Gammarus duebenii* (Amphipoda). Crustaceana **38:** 253-264.
6. VANDEL, A. 1926. La reconnaissance asexuelle chez les asellus. Bull. Soc. Zool. Fr. **51:** 163-172.
7. LUXMOORE, R. A. 1982. The reproductive biology of some serolid isopods from the Antarctic. Polar Biol. **1:** 3-11.

Generalization of Conditioned Taste Aversion to Sodium Chloride in Fischer-344 and Wistar Rats

ELEANOR E. MIDKIFF[a] AND ILENE L. BERNSTEIN

Department of Psychology
University of Washington
Seattle, Washington 98195

The voluntary ingestion of NaCl solutions by sodium-replete rats has been accepted as a generic behavior of that species since Richter[1] reported it. However, we recently reported[2] that Fischer-344 rats, unlike any strain previously described, failed to prefer NaCl over water at any concentration, and indeed, avoided NaCl solutions that appeared highly palatable to other strains. Two possible explanations for this are proposed: F-344 rats may be insensitive to NaCl at low concentrations while experiencing an abnormally exaggerated growth of sensation as concentration increases ("recruitment"); or NaCl may taste qualitatively different to F-344 rats than it does to other strains. These studies investigated whether F-344 rats are able to form aversions to dilute NaCl solutions, and whether NaCl aversions generalize to taste solutions that are qualitatively different.[3]

METHODS

Experiment 1: Inbred F-344 and outbred Wistar rats ($n = 8$ per group) were injected with 0.15 M lithium chloride (LiCl; 1% BW) after drinking 0.033 M NaCl. Control subjects were saline-injected. After two recovery days, water-deprived animals drank 0.033 M NaCl for 30 minutes, and intakes were measured.

Experiment 2: Animals from Experiment 1 were reconditioned by pairing consumption of 0.1 M NaCl with another LiCl treatment (1.2% BW). Animals were tested for generalization of the aversion to quinine sulfate and citric acid. They were then tested for aversions to NaCl to be certain that the NaCl aversion had not extinguished.

Experiment 3: New groups of animals were conditioned to 0.1 M NaCl and tested for generalization of aversions to other monochloride salts, that is, KCl and NH$_4$Cl. They were then tested with 0.1 M NaCl.

[a] Address for correspondence: Eleanor E. Midkiff, Department of Psychology, Pacific University, 2043 College Way, Forest Grove, OR 97116.

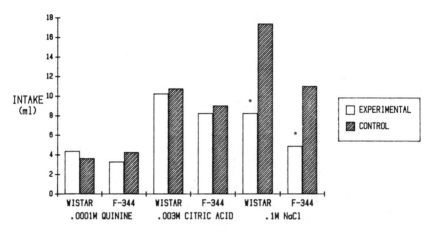

FIGURE 1. Suppression of intake of quinine sulfate, citric acid, and 0.1 M NaCl by Fischer-344 and Wistar rats with conditioned aversions to 0.1 M NaCl. Asterisks (*) indicate experimental animals were significantly different from controls of the same strain ($p < 0.05$).

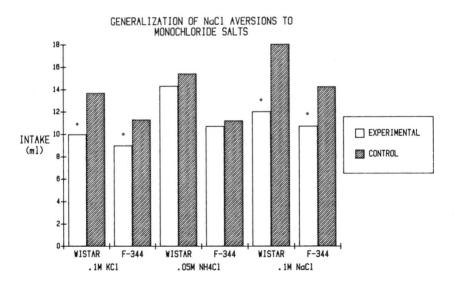

FIGURE 2. Suppression of intake of potassium chloride, ammonium chloride, and 0.1 M NaCl by Fischer-344 and Wistar rats with conditioned aversions to 0.1 M NaCl. Asterisks (*) indicate experimental animals were significantly different from controls of the same strain ($p < 0.05$).

RESULTS

Experiment 1: Wistar and F-344 rats treated with LiCl showed significantly lower intake of 0.033 M NaCl than controls of the same strain. Percent suppression was approximately 50% for both strains.

Experiment 2: Neither F-344 nor Wistar rats showed a generalization of CTAs to $QuSO_4$ or citric acid (FIG. 1). Significant aversions to NaCl were evident that did not differ between strains.

Experiment 3: There were again no significant strain differences in generalization of NaCl aversions to monochloride salts (FIG. 2). Neither strain showed a suppression of intake of NH_4Cl, and both strains showed some suppression of KCl.

DISCUSSION

In these studies, F-344 rats (which fail to prefer NaCl to water at any concentration) and outbred Wistar rats (which show typical NaCl preference) were *not* different in their patterns of development and generalization of NaCl CTAs. Thus, F-344 rats are able to detect NaCl at concentrations as dilute as 0.033 M. Furthermore, F-344 rats, like Wistars, did not generalize NaCl aversions to $QuSO_4$, citric acid, or NH_4Cl. Both Wistar and F-344 rats generalized NaCl aversions to KCl. These results fail to provide evidence that NaCl tastes qualitatively different to F-344 rats than to Wistars. Since studies of sodium metabolism of F-344 rats do not reveal differences between them and Wistars,[1] physiological effects of NaCl ingestion do not appear to be responsible for their behavior.

REFERENCES

1. RICHTER, C. P. 1959. Salt appetite in mammals: Its dependence on instinct and metabolism. *In* L'Instinct dans le Comportement des Animaux et de l'Homme. M. Autori, Ed.: 577-629. Masson. Paris, France.
2. MIDKIFF, E. E., D. A. FITTS, J. B. SIMPSON & I. L. BERNSTEIN. 1985. Absence of sodium chloride preference in Fischer-344 rats. Am. J. Physiol. **249** (Regulatory Integrative Comp. Physiol. 18): R438-R442.
3. NOWLIS, G. H., M. E. FRANK & C. PFAFFMANN. 1980. Specificity of acquired aversions to taste qualities in hamsters and rats. J. Comp. Physiol. Psychol. **94**: 932-942.

Human Fungiform Taste Bud Density and Distribution[a]

INGLIS J. MILLER, JR.

Department of Anatomy
Bowman Gray School of Medicine
Wake Forest University
Winston-Salem, North Carolina 27103

Human taste sensitivity varies among individuals, and there is a relationship between taste sensitivity and the number of stimulated receptors. Smith[1] showed that taste intensity in normal subjects is proportional to the number of fungiform papillae that are stimulated. Arvidson and Freiberg[2] stimulated individual papillae with chemicals to elicit sensations, excised each papilla and examined them microscopically to count taste buds. They found that more taste qualities could be elicited from papillae with greater numbers of taste buds. Because of sampling problems, it is difficult to assess potential differences in receptor density among subjects by excising a few papillae. We reported recently[3] that tongues from male cadavers averaged 116.3 taste buds/ cm^2 on the tip and 25.2 taste buds/cm^2 on the midregion. We now have observations from eight female cadaver tongues that permit a comparison of taste bud density between the two sexes and provide a sample of 18 subjects to see if human taste bud density varies with age.

The specimens came from 57 human cadavers ranging in age from premature infants to adults 95 years old. Photography was used to document morphological features of the tongue surface. Areas of the tongue measuring 1 cm^2 were sectioned serially with 20-μm frozen sections, mounted on slides, stained with H & E, and examined by light microscopy. Taste buds were found on classical fungiform papillae as well as on short, atypical papillae, especially on the lateral margin of the midregion. Taste bud density (per cm^2 of surface), gustatory papilla density and the number of taste buds per papilla were counted on two regions of the tongue, tip and midregion. Data are reported for eight females and ten males aged 22-90 years in TABLE 1.

Mean taste bud density on the tongue tip was 115.1 TB/cm^2 with no significant differences between females (113.6) and males (116.3). For both males and females, the range of taste bud density among subjects was over 2 log units. On the tip, taste bud densities divided the subjects into three groups: 3-75 TB/cm^2 ($n = 9$), 113-170 TB/cm^2 ($n = 7$) and 322-514 TB/cm^2 ($n = 2$). The mean density of taste buds on the midregion was 23.0 TB/cm^2 (range 0-85.9) for all subjects without significant differences between sexes. For both regions, taste bud density was proportional to the density of gustatory papillae (those bearing taste buds). The mean numbers of taste buds per papilla were 3.37 on the tip and 2.57 on the midregion, which were statistically different ($p < 0.05$, paired t-test). The range of taste buds per gustatory papilla was 1-18 on the tip and 1-9 on the midregion.

[a] This work was supported by National Institutes of Health Grant NS 20101 from the National Institute of Neurological and Communicative Diseases and Stroke.

TABLE 1. Taste Bud Distribution on the Tongue Tip and Midregion

Subject			Tip Region			Midregion		
Tip Rank	Age	Sex	TB[a]/cm²	Pap[b]/cm²	TB/Pap	TB/cm²	Pap/cm²	TB/Pap
1	62	F	3.0	2.0	1.5	1.8	1.0	2.0
2	22	M	3.6	2.4	1.5	22.9	9.3	2.5
3	53	M	4.9	3.3	1.5	3.0	1.5	2.0
4	50	F	10.6	5.6	1.9	0	0	0
5	22	M	18.7	3.7	5.0	0	0	0
6	75	M	23.2	9.5	2.4	2.0	2.0	1.0
7	76	F	25.9	7.6	3.4	12.0	5.3	2.2
8	80	M	44.0	26.8	1.6	17.7	11.3	1.6
9	56	M	74.5	27.3	2.7	8.4	3.2	2.7
10	90	F	113.6	30.9	3.7	32.3	11.4	2.8
11	26	F	129.5	55.7	2.3	45.4	16.7	2.7
12	63	F	140.0	58.3	2.4	11.4	6.3	1.8
13	79	M	148.5	32.0	4.6	68.7	17.3	4.1
14	53	M	160.7	38.6	4.0	19.0	4.8	4.0
15	36	F	163.6	44.5	3.7	21.4	7.1	3.0
16	29	M	170.8	21.4	8.0	24.5	4.4	5.6
17	58	F	322.8	80.7	4.0	37.1	6.4	5.8
18	70	M	514.0	80.0	6.4	85.9	28.2	3.0
Mean	55.6	n = 18	115.1	29.4	3.37	23.0	7.6	2.57
SD	21		131	26	1.8	24	7.3	1.6
Females	57.6	n = 8	113.6	35.7	2.86	20.2	6.8	2.48
SD	20.6		105	29	0.9	17	5.4	1.6
Males	51.5	n = 10	116.3	24.5	3.8	25.2	8.2	2.6
SD	22.8		154.4	23.5	2.2	29.1	8.8	1.5

[a] TB = taste buds.
[b] Pap = gustatory papillae.

In studies where single papillae have been excised, it has been easy to determine the proportion of papillae that contained taste buds. In blocks of tissue, determining the proportion of fungiform papillae that contain taste buds is complicated. We found a wide variation in the structure of gustatory papillae from the tall classical variety to short structures more like the fungiform papillae in rodents. It is difficult to determine which papillae without taste buds to include in the calculation.

There is no significant difference in the ages of the subjects with the highest and lowest taste bud densities. By dividing the subject population in halves ($n = 9$) according to rank of taste bud densities on the tip (column 1 of the table), the "low density half" (3-75 TB/cm^2) has a mean age of 55.1 $\pm$ 22 yr, while the "high density half" (114-514 TB/cm^2) has an equivalent age distribution (mean = 56 $\pm$ 22 yr). While we cannot determine the effect of aging on taste bud densities, we found a similar distribution of high-density and low-density subjects at both ends of the age range (22-90 yr).

In conclusion, we found a 2 log unit difference in fungiform taste bud densities among human cadaver tongues and a fivefold difference in taste bud densities between the tongue tip and midregion. These differences may account for some of the variation in taste sensitivity between subjects and regions of the tongue.

ACKNOWLEDGMENTS

We acknowledge the technical efforts of Elise Williams, Shiva Jarrahi, and Derrick Brown. This work would not have been possible without anatomical gifts from posthumous donors.

REFERENCES

1. SMITH, D. V. 1971. Taste intensity as a function of area and concentration: Differentiation between compounds. J. Exp. Psychol. **87:** 163-171.
2. ARVIDSON, K. & U. FREIBERG. 1980. Human taste response and taste bud number in fungiform papillae. Science **209:** 807-808.
3. MILLER, JR., I. J. 1986. Variation in human fungiform taste bud densities among regions and subjects. Anat. Rec. **216:** 474-482.

Relation of Receptive Field Size and Salt Taste Responses in Chorda Tympani Fibers during Development[a]

CHARLOTTE M. MISTRETTA,[b] TAKATOSHI NAGAI,
AND ROBERT M. BRADLEY

Department of Oral Biology
School of Dentistry and
Center for Nursing Research
University of Michigan
Ann Arbor, Michigan 48109

During development nerve fibers must establish precise connections with sensory organs. For example, in the adult mammal a single chorda tympani nerve fiber innervates receptive fields on the tongue that contain from one to several fungiform papillae and associated taste buds. However, there has been no developmental study to determine how taste receptive fields are established. We hypothesized that previously observed differences in development of taste responses to various salts[1] might relate to development of receptive fields. Therefore, the size and chemical response properties of receptive fields for single chorda tympani nerve fibers in sheep fetuses, perinatal animals, and lambs were determined.

Seventy-four fibers were studied in 44 animals from three age groups: fetuses about 130 days of gestation (term = 147 days); perinatal animals (about one week before or after birth); lambs (about one to three months postnatal). A single chorda tympani fiber was dissected and the tongue was stimulated with 0.5 M NH_4Cl, $NaCl$, and KCl to record salt taste responses. Then individual fungiform papillae were stimulated electrically with 5 microamps anodal current from a fine platinum probe to determine number of papillae innervated by the fiber.

There was a broad range of receptive field sizes across age groups: for fetuses the range was 3 to 40 papillae per field, mean (SD) = 13 (10); for perinatal animals the range was 1 to 33 papillae, mean (SD) = 15 (9); for lambs the range was 2 to 29 papillae, mean (SD) = 11 (7). There was a significant difference in receptive field size across age groups with a median test ($p < 0.10$), and posttests indicated that lamb fields were smaller than perinatal fields. Not only were small fields observed more frequently in fibers in older animals, but also there was an increase in the proportion of fibers that responded with highest frequency to $NaCl$ compared to NH_4Cl ($p = 0.06$). Furthermore, receptive field size correlated *negatively* with the $NaCl : NH_4Cl$ response ratio. That is, smaller receptive fields had higher frequency

[a] Supported by National Science Foundation Grant BNS 83-11497 to CMM and RMB.

[b] Address for correspondence: Dr. Charlotte Mistretta, Department of Oral Biology, School of Dentistry, University of Michigan, Ann Arbor, MI 48109.

responses to NaCl than to NH_4Cl ($p = 0.05$). Receptive field size correlated positively with NH_4Cl and KCl response frequencies ($p = 0.01$).

These results demonstrate that there is already extensive branching of peripheral taste nerve fibers in fetuses providing innervation for large and small receptive fields. However, fibers with small receptive fields are encountered more frequently in older animals, and these tend to be more responsive to NaCl than other salts. This suggests a developmental reorganization of peripheral innervation with possible reduction of fiber branching in some large receptive fields, or an addition of fibers that innervate small, Na-responsive receptive fields.

REFERENCE

1. MISTRETTA, C. M. & R. M. BRADLEY. 1983. Neural basis of developing salt taste sensation: Response changes in fetal, postnatal, and adult sheep. J. Comp. Neurol. **215:** 199-210.

Contribution of Cations to Generation of Salt-Induced Receptor Potential in Frog Taste Cell[a]

TAKENORI MIYAMOTO,[b] YUKIO OKADA, AND
TOSHIHIDE SATO

Department of Physiology
Nagasaki University School of Dentistry
Nagasaki 852, Japan

We have reported that the permeability change of the basolateral membrane of a frog taste cell plays an important role in generation of salt-induced receptor potential.[1] We reexamined the effects of various modified salines substituted for either superficial fluid (SF) or interstitial fluid (ISF) on the receptor potential of a taste cell in response to salty taste stimuli.

Adult bullfrogs anesthetized with urethane were used. The tongue surface was usually adapted to normal saline (115 mM NaCl, 2.5 mM KCl, 1.8 mM CaCl$_2$, 5 mM HEPES, pH 7.2). Arterial perfusion was performed to modify the ionic composition of ISF. The taste stimuli used were 0.5 M NaCl and 0.5 M KCl.

The amplitude of receptor potential evoked by 0.5 M KCl (V_K) was usually larger than that evoked by 0.5 M NaCl (V_{Na}) (FIG. 1). Reversal potential of V_{Na} was obviously different from that of V_K in a single taste cell (FIG. 2).

After only Na$^+$ or both Na$^+$ and Ca^{2+} in ISF were totally replaced with any of K$^+$, Li$^+$, choline$^+$, tetramethylammonium$^+$, and tetraethylammonium (TEA)$^+$, V_{Na} and V_K reduced to 30-70% of those controls. Replacement of CaCl$_2$ in ISF with MgCl$_2$ significantly reduced only V_K, whereas total replacement of ISF with isotonic CaCl$_2$ or MgCl$_2$ markedly potentiated V_{Na}. Addition of 3 μM tetrodotoxin or 5 mM Co^{2+} to ISF did not affect either receptor potential significantly.

After adapting the tongue surface to Ca-free saline, both V_{Na} and V_K reduced significantly with the depolarization of resting membrane potential.

These results suggest that the generation of salt-induced receptor potentials in the frog taste cell depends on the presence of Na$^+$ and Ca^{2+} in ISF but only Ca^{2+} in SF. V_{Na} and V_K are evoked by a somewhat different mechanism. Voltage-dependent Na or Ca channels[2] may not contribute to them intensively.

[a]This work was supported by a Grant-in-Aid for Scientific Research (No. 60771522) from the Ministry of Education, Science and Culture of Japan.

[b]Address for correspondence: Dr. Takenori Miyamoto, Department of Physiology, Nagasaki University School of Dentistry, 7-1 Sakamoto-machi, Nagasaki 852, Japan.

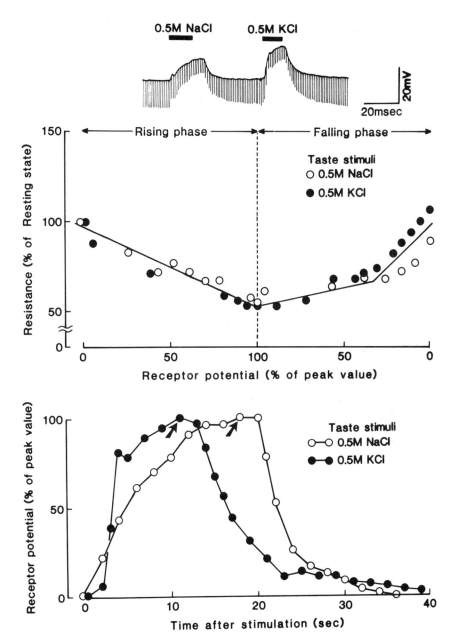

FIGURE 1. Time courses of membrane resistance and receptor potential to taste stimuli during adaptation to normal saline.

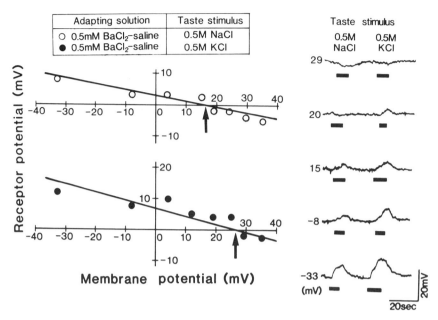

FIGURE 2. Reversal potentials of V_{Na} and V_K in a single taste cell during adaptation to saline containing 0.5 mM BaCl$_2$.

REFERENCES

1. SATO, T., K. SUGIMOTO & Y. OKADA. 1982. Ionic basis of receptor potential in frog taste cell in response to salt stimuli. Jpn. J. Physiol. **32:** 459–462.
2. KASHIWAYANAGI, M., M. MIYAKE & K. KURIHARA. 1983. Voltage-dependent Ca^{2+} channel and Na$^+$ channel in frog taste cells. Am. J. Physiol. **244:** C82–C88.

Trout Olfactory Receptors Degenerate in Response to Water-Borne Ions

A Potential Bioassay for Environmental Neurotoxicology[a]

DAVID T. MORAN AND J. CARTER ROWLEY

Rocky Mountain Taste & Smell Center
Department of Cellular and Structural Biology
University of Colorado School of Medicine
Denver, Colorado 80262

GEORGE AIKEN

United States Geological Survey
Arvada, Colorado 80002

During the course of our research on the ultrastructure of the trout olfactory system, we observed that wild brown trout (*Salmo trutta*) experienced complete loss of their olfactory receptors after spending two days in a large, 250-gallon aquarium in our aquatic facility. When these same fish were returned to the North Fork of the South Platte River—their home stream—their olfactory receptors were found to have *regenerated* within a period of eight days. When these same fish were once again reintroduced into our laboratory aquarium, their receptors degenerated, once again, within two days. Comparative chemical analysis of the water from the South Platte River and the laboratory aquarium revealed some striking differences. In the aquarium water, the levels of four ions, cadmium (Cd), cobalt (Co), copper (Cu), and zinc (Zn), were present at significantly higher levels than they were in stream water. Suspecting one or more of these ions might be associated with the loss of trout olfactory receptors, a pilot study was done in which trout were placed in separate glass containers. Each container was filled with stream water plus one of the ions named above set at the concentration levels found in the water from the 100-gallon laboratory aquarium. After trout had been in the "spiked" containers for three days, biopsies of their olfactory rosettes were taken and investigated by transmission electron mi-

[a] Supported by National Institutes of Health Program Project Grant NS20486 and National Science Foundation Research Grant BNS-821037.

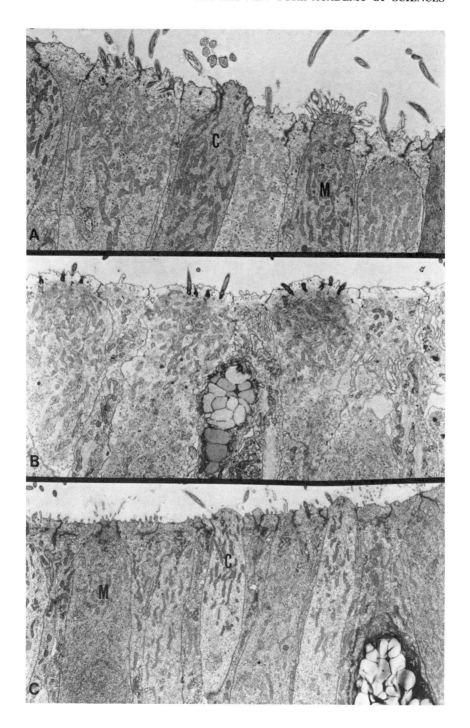

croscopy. All fish noses appeared normal with the exception of the experimental animal that had lived in the stream water "spiked" with copper ion: its nose had lost all of its olfactory receptors (FIG. 1). These results not only suggest that judicious regulation of ion content may provide a technique for noninvasive chemical olfactotomy, but also suggest that the trout nose may serve as a sensitive bioassay for environmental toxicology—especially in waters where the biological impact of introduced metal contaminants is in question.

◄ **FIGURE 1.** (A) Transmission electron micrograph (TEM) of normal trout olfactory epithelium showing ciliated (C) and microvillar (M) olfactory neurons. Magnification: 6,200×.

(B) TEM of olfactory epithelium of trout taken from ion-contaminated aquarium. Note absence of olfactory receptors. Magnification: 4,100×.

(C) TEM of olfactory epithelium of the same trout shown in (B) after the trout had been returned to, and spent eight days in, its native stream. Note the ciliated (C) and microvillar (M) receptors have regenerated. Magnification: 5,100×.

Degeneration-Regeneration of the Olfactory Neuroepithelium Following Bulbectomy

An SEM Study[a]

EDWARD E. MORRISON,[b] PASQUALE P. C. GRAZIADEI,[c] AND RICHARD M. COSTANZO [b]

[b]*Department of Physiology*
Medical College of Virginia
Richmond, Virginia 23298

[c]*Department of Biological Sciences*
Florida State University
Tallahassee, Florida 32306

The olfactory neuron is unique in the vertebrate nervous system in its ability to replace itself normally and when injured. The replacement process is assured by the presence of a persistent stem cell population (basal cells) located within the olfactory neuroepithelium.[1] Removal of the olfactory bulb (bulbectomy) results in a rapid retrograde degeneration of mature olfactory neurons, followed by an increase in basal cell mitotic activity.[2] Newly differentiated cells mature into olfactory neurons by developing an apical dendrite and an axonal process that grows centrally to the brain. In this study we have examined the olfactory neuroepithelium using the scanning electron microscope (SEM), following complete unilateral bulbectomy. The SEM allowed us for the first time to study detailed cellular surface morphology of the degeneration-regeneration process of this sensory neuron.

Adult hamsters ($n = 40$) were unilaterally bulbectomized and nasal turbinates and strips of septal mucosa were processed for SEM examination. The olfactory epithelium was then examined at recovery times of 0-120 days. During the degeneration period (0-4 days), the neuroepithelium consisted of supporting cells, basal cells, degenerating olfactory neurons and a few maturing olfactory neurons, whose axons were spared at the time of bulbectomy. The absence of mature olfactory neurons during the degeneration period allowed for the detailed examination of the supporting cells (FIG. 1). The columnar supporting cells extend to the basal lamina where they terminate in footlike processes. We observed adjacent supporting cells to be attached throughout their length by thin intercellular bridges. The epithelial surface lacked the characteristic dense sensory cilia blanket and consisted of supporting cell microvilli.

[a]Supported by VCU Grants in Aid (EEM) and National Institutes of Health, Grant NS 16741 (RMC).

512

The early recovery period (5-25 days) was marked by an increase in the number of developing olfactory neurons located primarily within the lower region of the epithelium. Occasionally, isolated regions of increased basal cell activity (active zones) were observed. Dendrites of maturing olfactory neurons grew toward the apical surface between and along the edges of supporting cells that appeared to surround and envelop the developing process. Olfactory axons arising from the basal region of olfactory neurons grew along and through the basal lamina. At later stages of recovery (35-120 days) the epithelial surface was covered with a dense blanket of olfactory cilia and

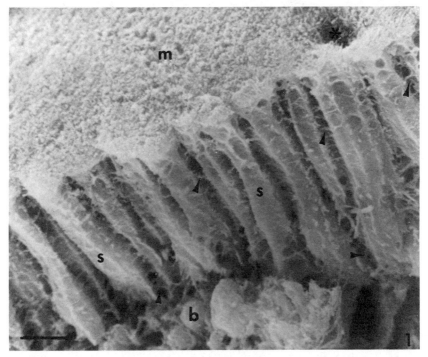

FIGURE 1. SEM of the olfactory neuroepithelium following degeneration of receptor neurons (day 3). The epithelium consists of supporting cells (s) covered with microvilli (m), interconnected by cellular bridges (arrows), and basal cells (b). Star (*) indicates duct opening of Bowman gland. Bar = 5μ.

the neuroepithelium contained numerous mature and developing olfactory neurons (FIG. 2). Sensory axons fasciculated into large bundles within the lamina propria and grew centrally.

The results of this study provide additional information on the process of degeneration-regeneration of sensory neurons. The SEM revealed, for the first time, the detailed cellular surface morphology of the olfactory neuroepithelium. We observed the development of the dendritic and axonal processes of maturing olfactory neurons. In addition, we examined the footlike process and intercellular bridges among the supporting cells, as well as the intricate structural relationship between supporting cells and olfactory neurons.

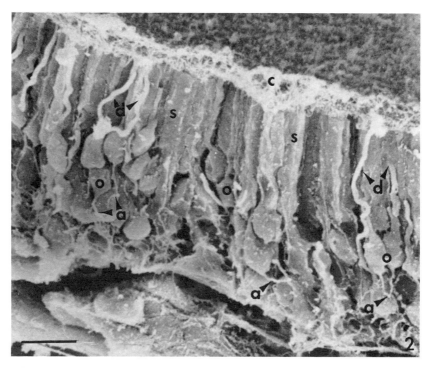

FIGURE 2. SEM of olfactory neuroepithelium following recovery (day 120). The neuroepithelium consists of supporting cells (s) and fully developed olfactory neurons (o) with dendritic (d) and axonal (a) processes. Olfactory cilia (c) are observed covering the epithelial surface. Bar = 10μ.

REFERENCES

1. MONTI-GRAZIADEI, G. A. & P. P. C. GRAZIADEI. 1979. Neurogenesis and neuron regeneration in the olfactory system of mammals. II Degeneration and reconstitution of the olfactory sensory neurons after axotomy. J. Neurocytol. **8:** 197-213.
2. COSTANZO, R. M. & P. P. C. GRAZIADEI. 1983. A quantitative analysis of changes in olfactory epithelium following bulbectory in hamsters. J. Comp. Neurol. **215:** 370-381.

Effects of Age and Biochemical Status on Preference for Amino Acids[a,b]

CLAIRE MURPHY [c]

Department of Psychology
San Diego State University
San Diego, California 92182

A primary reason for interest in age-associated chemosensory dysfunction has been the assumption that such changes are linked to changes in nutritional status in the elderly. Certainly, there are age-associated chemosensory losses (See Murphy[1] and Schiffman,[2] for reviews), and there is an age-associated decline in nutritional status.[3,4] However, empirical evidence linking chemosensory variables with lower nutritional status in the elderly has been lacking. The following studies provide evidence that demonstrates that both age and nutritional status affect the preference for a nutritionally significant chemosensory stimulus.

EXPERIMENT 1

Experiment 1 investigated the effects of aging and biochemical status on preference for casein hydrolysate, a mixture of the essential amino acids. The hypotheses were that (1) older participants would rate high concentrations of the amino acid mixture as more pleasant than young participants would, and (2) that participants with lower biochemical status would prefer higher concentrations of the amino acid mixture than would those with better biochemical status.

Ten young persons (M = 22.8 yr.) and 16 elderly persons (M = 84.0 yr.) had blood drawn to permit assay of total protein, albumin, and BUN. All rated the pleasantness of six concentrations (0, 1, 2, 3, 4, and 5% wt/vol) of the amino acid mixture in an amino acid deficient soup base. They marked a bipolar line scale to indicate degree of pleasantness or unpleasantness.[5]

Analysis of variance on the peak preferred concentration (PPC) indicated that older participants preferred higher concentrations of the amino acid mixture (M = 3.0%) than did younger participants (M = 0.5%). Similar analyses showed that subjects with lower biochemical status (above the median for BUN and below the

[a] This research was supported by Grant AG04085 from the National Institute on Aging.

[b] Complete details of this study will be found in J. Gerontol. 1987. **42**: 73-77.

[c] Address for correspondence: Dr. Claire Murphy, Department of Psychology, San Diego State University, San Diego, CA 92182-0350.

median for serum albumin) also preferred higher concentrations of the amino acid mixture than did subjects with higher biochemical status. There was no statistically significant difference in PPC associated with serum protein level. TABLE 1 shows ANOVA results.

EXPERIMENT 2

In a second study, we attempted to replicate the effects of age and biochemical status on preference for amino acids. To test the possibility that the differences in preference could simply be due to reduced input to the olfactory and taste systems of the elderly, subjects also rated the stimuli for intensity using the method of magnitude matching. A composite index of the biochemical indices of total protein, albumin, and blood urea nitrogen (BUN) was used to objectively define nutritional status.

TABLE 1. Significant Effects of Analyses of Variance[a]

Independent Variable	Dependent Variable	Mean (SD)		df	F
Experiment 1					
		Young	Elderly		
Age	Protein	6.95 (0.52)	6.56 (0.44)	1,24	4.28[b]
Age	Albumin	4.50 (0.23)	3.99 (0.30)	1,24	20.92[c]
Age	BUN	11.60 (3.60)	23.94 (5.25)	1,24	42.47[c]
Age	PPC	0.50 (0.85)	3.00 (1.67)	1,24	19.03[d]
		Low	High		
Protein	PPC	2.15 (1.82)	1.70 (2.00)	1,21	.32
Albumin	PPC	2.93 (1.71)	.44 (0.88)	1,22	16.26[d]
BUN	PPC	0.56 (0.88)	2.82 (1.78)	1,24	12.81[d]
Experiment 2					
		Young	Elderly		
Age	Protein	7.20 (0.41)	6.72 (0.33)	1,38	16.48[c]
Age	Albumin	4.56 (0.31)	4.21 (0.22)	1,38	16.48[c]
Age	BUN	12.85 (4.21)	15.90 (3.93)	1,38	5.62[b]
Age	PPC	0.45 (0.69)	1.35 (1.27)	1,32	48.43[c]
		Low	High		
Blood index	PPC	1.50 (1.11)	0.41 (0.49)	1,32	13.50[d]
Intensity	PPC	1.16 (0.76)	0.79 (0.82)	1,32	1.76

[a] "PPC" is the peak preferred concentration of casein hydrolysate (0, 1, 2, 3, 4, or 5%). "Intensity" is the geometric mean of intensity estimates at the six concentrations of casein hydrolysate. Higher values of protein and albumin and lower values of BUN are associated with better nutritional status.
[b] $p < 0.05$.
[c] $p < 0.0001$.
[d] $p < 0.001$.

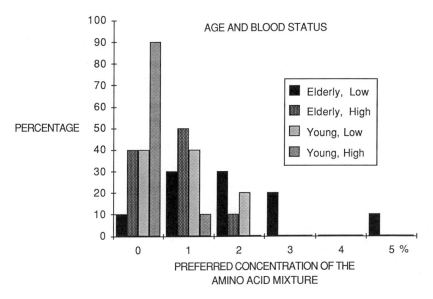

FIGURE 1. The percentage of subjects within each indicated group who chose a given concentration as most preferred.

The procedure was as in Experiment 1 except that the 20 young people ($M = 20.15$) and the 20 elderly people ($M = 70.75$) also rated the intensity of the six amino acid stimuli and of a series of six auditory stimuli, using the method of magnitude matching.[6]

While older participants made lower intensity estimates for amino acids (geometric mean $= 3.3$) than did young participants (geometric mean $= 5.4$), all participants successfully tracked increases in concentration. The psychophysical intensity function for the elderly was displaced downward, but was parallel to the function for the young subjects. In separate analyses, biochemical status as measured by the previously described composite biochemical index (grouped above and below a participant's age group median) showed no significant relationship to perceived intensity.

In order to test in an ANOVA the effects of age, biochemical status, and perceived intensity on preferred concentration (PPC), a geometric mean intensity rating was computed for each participant across the six concentrations. Both the geometric mean intensity rating and the previously described composite biochemical index were grouped above and below appropriate age group medians.

The three-way ANOVA showed that age and composite blood status were significantly related to PPC, but perceived intensity was not (see TABLE 1). Elderly participants preferred higher concentrations of casein hydrolysate ($M = 1.4\%$) than did young participants ($M = 0.4\%$). Across age, participants with lower composite biochemical indices also preferred higher concentrations of amino acids ($M = 1.5\%$) than participants with higher biochemical status ($M = 0.4\%$). FIGURE 1 shows PPC as a function of both age and composite biochemical status.

Two of the initial hypotheses in Experiment 2 were confirmed: (1) older participants preferred higher concentrations of the amino acid mixture, and (2) participants with lower values for the combined biochemical index preferred higher concentrations of the amino acid mixture. Finally, this study suggests that the elderly participants' higher preferred concentration of casein hydrolysate is not simply due to generally lower perceived intensity, since a participant's overall mean perceived intensity did not predict his preference. Thus, these studies revealed individual differences in preference for a nutritionally significant chemosensory stimulus, which varied independently with both biochemical status and age.

ACKNOWLEDGMENTS

I am grateful to Kelly Dozois, Elizabeth Konowal, Michele J. Reed, Jeanne Withee, and Ann Woltjen for excellent technical assistance and the San Diego State University Student Health Services Center for phlebotomy services. I am grateful to Gary K. Beauchamp for supplying the soup base and to Karen Knauff of Johnson and Johnson for supplying the casein hydrolysate.

REFERENCES

1. MURPHY, C. 1986. Taste and smell in the elderly. *In* Clinical Measurement of Taste and Smell. H. L. Meiselman & R. S. Rivlin, Eds.: 343-371. Macmillan. New York.
2. SCHIFFMAN, S. S. 1986. Age-related changes in taste and smell and their possible causes. *In* Clinical Measurement of Taste and Smell. H. L. Meiselman & R. S. Rivlin, Eds.: 326-342. Macmillan. New York.
3. YEARICK, E. S., M. L. WANG & S. J. PISIAS. 1980. Nutritional status of the elderly: Dietary and biochemical findings. J. Gerontol. **35:** 663-671.
4. JANSEN, C. & I. HARRILL. 1977. Intakes and serum levels of protein and iron for 70 elderly women. Am. J. Clin. Nutr. **30:** 1414-1422.

Electrophysiological and Innate Behavioral Responses of the Dog to Intravenous Application of Sweet Compounds

LAWRENCE J. MYERS, RANDY BODDIE, AND
KIMBERLY MAY

Department of Physiology and Pharmacology
Auburn University
Auburn, Alabama 36849

The ability to taste intravenously administered compounds has been used clinically in humans, as the intravenous saccharin circulation test and the phenomenon of intravenous taste has been studied both electrophysiologically and behaviorally.[1-3] Thirteen mixed-breed, mesatocephalic dogs were selected for examination of electro-physiological and innate behavioral responses to intravenously applied, ascending log dilution series of sucrose, fructose, glucose, and saccharin in Ringer's solution. Animals were lightly restrained in lateral recumbency, blindfolded, and a catheter was placed in the left cephalic vein. Electrodes were placed subcutaneously in the standard veterinary electroencephalographic (EEG) montage, and in the splenius muscles.[4]

In the first series of experiments, serial dilutions of each of the selected compounds were infused via the cephalic catheter at a constant rate of 5 ml/min for two minutes per dilution. Concentration of solutions ranged from 10^{-15} to $10^{-6} M$. In the second series the serial dilutions were infused at the rate of 5 ml/sec for two seconds at two-minute intervals. Two sets of solutions were used at this latter rate. The first ranged from 10^{-15} to $10^{-6} M$, and the second ranged from 10^{-6} to $10^{-1} M$.

Behavior and electroencephalographic-electromyographic (EMG) activity in response to stimulus application were recorded. The behavioral response used as evidence of perception of taste stimulus by the dogs was an alteration of orofacial expression principally evidenced by licking or swallowing. The criteria for EEG-EMG response were visually detectable alterations of frequency and amplitude during stimulus presentation. That the responses were mediated by taste was demonstrated by the abolition of response to sucrose following topical application of a tea derived from *Gymnema sylvestre* leaves. Behavior was found to be more reliable than visual analysis of EEG-EMG in establishing response to intravenous taste stimuli in this study. Reliability of EEG-EMG response in a single-blind evaluation of 20 records yielded only an 85% agreement of results.

Relatively stable thresholds were obtained for each compound (TABLE 1). Rapid injection of stimuli resulted in threshold values markedly higher than obtained with 5 ml/min stimulus rates. These thresholds ranged from 10^{-1} to $10^{-5} M$ for saccharin with a mean threshold value for EEG-EMG response of $10^{-3.38}$, and for a behavioral response of $10^{-3.75}$. These figures reflect positive response only, excluding one trial in which no EEG-EMG response was observed and ten trials in which no behavioral response was observed. Sucrose was ineffective in evoking the EEG-EMG or behavioral response in all 15 attempts.

The data obtained from this preliminary investigation suggest that the effects of some intravenously applied taste stimuli are cumulative over time or dependent on rapid dynamics of exposure of taste receptors to flavor compounds, and that the difference between thresholds at the two rates of application does not seem to be determined entirely by the total amount of flavor compound delivered to the animal.

TABLE 1. Comparisons of Thresholds between Compounds

Compound	Behavioral Threshold (Negative log) 5 ml/min $n = 24$	10 ml/sec	EEG Threshold (Negative log) 5 ml/min $n = 24$	10 ml/sec $n = 15$
Saccharin	11.20^a	3.75^b	11.59	3.38^c
(SEM)	(0.52)		(0.50)	
Glucose	11.19	—	11.38	—
(SEM)	(0.44)		(0.37)	
Sucrose	10.95	no resp.	11.38	no resp.
(SEM)	(0.46)		(0.41)	
Fructose	9.93^a	—	10.59	—
(SEM)	(0.70)		(0.54)	

^{a} $p < 0.05$ for differences.
^{b} 5 responses in 15 trials.
^{c} 14 responses in 15 trials.

REFERENCES

1. FISHEBERG, A. M., W. M. HITZIG & F. H. KING. 1933. Measurement of the circulation time with saccharin. Proc. Soc. Exp. Biol. Med. **30:** 651-652.
2. BRADLEY, R. M. 1973. Electrophysiological investigations of intravascular taste using perfused rat tongue. Am. J. Physiol. **224:** 300-304.
3. BRADLEY, R. M. & C. M. MISTRETTA. 1971. Intravascular taste in rats as demonstrated by conditioned aversion to sodium saccharin. J. Comp. Physiol. Psychol. **75:** 186-189.
4. REDDING, R. W. & C. D. KNECHT. 1984. Atlas of Encephalography in the Dog and Cat. Praeger Publishers. New York. 387 pp.

Sweet Intensity, Persistence, and Quality of Mixtures Containing Neohesperidin Dihydrochalcone and Bitter Stimuli[a]

MICHAEL NAIM,[b] EMMANUEL DUKAN, LYAT
YARON, AND URI ZEHAVI

Department of Biochemistry and Human Nutrition
The Hebrew University of Jerusalem
Rehovot 76100, Israel

The sweetness of some intensive sweeteners, such as neohesperidin dihydrochalcone (NHD) is characterized by undesirable sensory properties, such as slow taste onset and lingering aftertaste. NHD has been chemically modified in an attempt to improve the sweet intensity-time relationship.[1] This relationship follows a negative exponential function: $Ip = Ip_{max} \times e^{-t/T}$ where Ip is the perceived sweet intensity, Ip_{max} is a maximal intensity at an initial time (t) and T is the persistence time constant, that is, the length of time it takes for Ip to reach $1/e$ of its Ip_{max}.[2] The high sweet intensity of NHD is, perhaps, due to an interaction between two glucophores in the NHD molecule[1] and two sweet receptors. Previous data suggested that the sensation of sweet and bitter tastes are functionally related and that there is proximity between sweet

TABLE 1. Maximal Perceived Sweet Intensity (Ip_{max}) and Persistence Constant (T) of Mixtures Containing Neohesperidin Dihydrochalcone (NHD) with Either Naringin (NAR) or Sucrose Octaacetate (SOA)[a]

Sweet Source	Ip_{max}	T (sec)
0.03% NHD	27	82
0.03% NHD + 0.03% NAR	15[b]	39[b]
0.03% NHD + 0.005% SOA	15[b]	68

[a] Percentage concentrations are expressed as weight per volume. Ip_{max} units are related to an intensity level of 30 given to a 10% sucrose solution.

[b] Indicates significantly ($p < 0.05$) lower value than that for 0.03% NHD.

[a] This study was supported by a grant from the Jaf-Ora Company, Rehovot, Israel.

[b] Address for correspondence: Dr. Michael Naim, Department of Biochemistry and Human Nutrition, Faculty of Agriculture, The Hebrew University of Jerusalem, P.O. Box 12, Rehovot 76100, Israel.

and bitter receptors in the membranes of taste cells.[3] This research deals with the effect of bitter compounds on NHD sweet intensity and persistence and investigates whether the persistence of NHD is related to its sweet intensity.

Psychophysical functions for sweet intensity of mixtures have been previously suggested;[4,5] here we describe the Ip_{max} of mixtures determined for various concentrations while maintaining a constant ratio between the two solutes. The results indicated that both Ip_{max} and T of NHD were reduced in a mixture containing naringin (NAR), a bitter flavone analogue of NHD (TABLE 1). When equal bitter doses of sucrose octaacetate (SOA) were used in a mixture with NHD, Ip_{max} was reduced with no significant reduction in T value. Linear regression analyses performed on the Ip_{max} data of NHD+NAR (log Ip_{max} vs. log concentration) or on those of NHD+SOA produced lower slope values than that observed for NHD alone, suggesting that the sweet quality of the obtained mixtures is poor. Moreover, both three-dimensional scaling (MDS) and hierarchical tree structure (HTS) expressions (FIG. 1) of taste similarity indicated that the mixture of NHD+NAR was apart from all other taste stimuli that become condensed in the space diagram, thus suggesting that such a mixture produces a sweet quality inferior to that of NHD. It is concluded that the reduction of T value of NHD by NAR was related to the reduced Ip_{max} level. Sweet intensity and persistence of NHD are apparently related phenomena.

FIGURE 1. Hierarchical tree structure (HTS) of taste similarity data. Sucrose (Suc, 8.9%); fructose (Fru, 7.1%); glucose (Glc, 13.0%); aspartame (ASP, 0.08%); NHD, 0.012%; sodium saccharin (Sacc, 0.033%); and a mixture of neohesperidin dihydrochalcone and naringin (NHD + NAR, 0.031% + 0.031%).

REFERENCES

1. DUBOIS, G. E., G. A. CROSBY & R. A. STEPHENSON. 1981. Dihydrochalcone sweeteners. J. Med. Chem. **24:** 408-428.
2. NAIM, M., E. DUKAN, U. ZEHAVI & L. YARON. 1986. The water-sweet aftertaste of neohesperidin dihydrochalcone and thaumatin as a method for determining their sweet persistence. Chem. Senses **11:** 361-370.
3. BIRCH, G. G. & A. R. MYLVAGANAM. 1976. Evidence for the proximity of sweet and bitter receptor sites. Nature **260:** 632-634.
4. MOSKOWITZ, H. R. & C. DUBOISE. 1977. Taste intensity, pleasantness, and quality of aspartame, sugars, and their mixtures. J. Inst. Can. Sci. Technol. Aliment. **10:** 126-131.
5. CURTIS, D. W., D. A. STEVENS & H. T. LAWLESS. 1984. Perceived intensity of the taste of sugar mixtures and acid mixtures. Chem. Senses **9:** 107-120.

Taste Responses of the Cross-Regenerated Greater Superficial Petrosal and Chorda Tympani Nerves of the Rat

MOHSSEN S. NEJAD AND LLOYD M. BEIDLER

Department of Biological Science
The Florida State University
Tallahassee, Florida 32306

The rat greater superficial petrosal (GSP) nerve, contrary to the chorda tympani (CT), is highly responsive to sucrose and reverse is true for NaCl.[1] We wanted to know how cross regeneration would affect the gustatory afferent neural responses of the GSP and CT nerves in the rat.

In two groups of male Sprague-Dawley rats, cross-union anastomoses between the GSP and CT nerves in the middle ear were unilaterally made in such a manner that: in group I ($n = 6$), GSP nerve grew into the front of the tongue ($GSP_c = CT_p$) and in group II ($n = 5$), CT nerve grew into the palatal regions of the oral cavity ($CT_c = GSP_p$). In addition in a third group of animals ($n = 4$), the chorda tympani nerve was transected, then its central and peripheral ends were anastomosed for regeneration ($CT_c = CT_p$). After 16–24 weeks, integrated neural responses were recorded from the regenerated nerves in responses to several tastants (0.1 M chloride salts of Na$^+$, Li$^+$, K$^+$, NH$_4^+$, Ca^{2+}, 0.5 M sucrose, 0.01 M quinine-HC1, 0.05 M citric acid, and 0.02 M Na-saccharin). In addition, response-concentration functions for NaCl and sucrose were also measured. The relative integrated responses from the experimental nerves, control nerves (contralateral), and nerves of normal animals were compared (FIGS. 1, 2).

Electrophysiological and histological studies suggested that after cross-union regeneration of the GSP and CT nerves, the palatal and anterior tongue taste buds (ipsilateral) were reformed and were functional. The integrated neural response profile of the regenerated GSP into the anterior part of the tongue resembled that of the normal CT nerve. The integrated neural response profile of the regenerated CT into the palatal regions resembled that of the normal GSP nerve. These findings are in concordance with the CT and IX nerve cross-regeneration studies by B. Oakley.[2] Afferent neural activities of the cross-regenerated GSP and CT appeared to be influenced by their new peripheral gustatory receptor populations. Apparently, central neural influence did not maintain the original afferent neural profiles of the two contrasting GSP and CT nerves in response to the tested chemicals (assuming no neural rearrangement in the geniculate ganglion). Quantitative analysis of the response-concentration functions[3,4] of sucrose and NaCl showed that in both cross-regenerated GSP and CT nerves, the binding strengths (K) remained the same as control nerves (sodium chloride K values ~10 and sucrose K values ~7), whereas

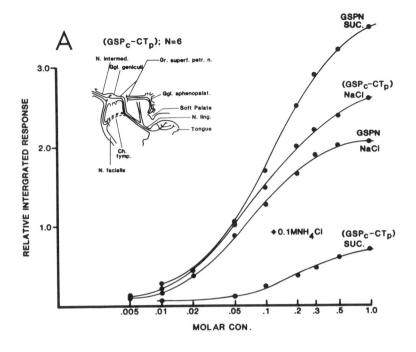

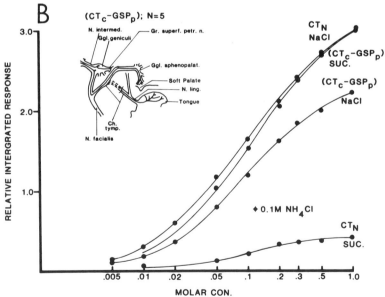

FIGURE 1. Response-concentration functions of sucrose and sodium chloride in the normal GSP and (GSPc-CTp) nerves (**A**) and the normal CT and (CTc-GSPp) nerves (**B**). All responses were normalized to responses to 0.1 molar ammonium chloride = 1.0.

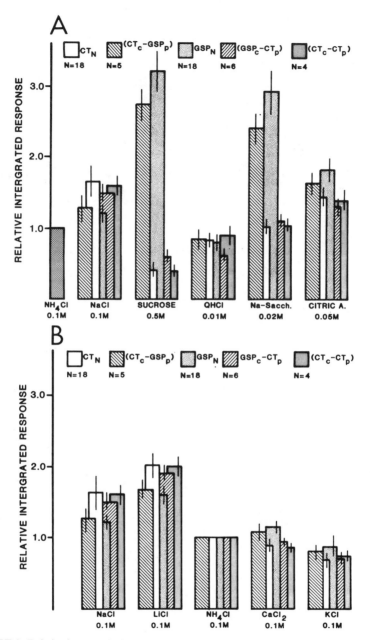

FIGURE 2. Relative integrated response to 0.1 molar sodium chloride, 0.5 molar sucrose, 0.01 molar quinine-HCl, 0.02 molar Na-saccharin and 0.05 molar citric acid (**A**) and to chloride salts of sodium, lithium, ammonium, calcium, and potassium (**B**). All responses were normalized to responses to 0.1 molar ammonium chloride = 1.0.

the maximum response (R_s) was changed (sodium chloride R_s values 2.7-3.2 and sucrose R_s values 0.46-3.9).

REFERENCES

1. NEJAD, M. S. 1986. The neural activities of the greater superficial petrosal nerve of the rat in response to chemical stimulation of the palate. Chem. Senses **11.**
2. OAKLEY, B. 1967. Altered temperature and taste responses from cross-regenerated sensory nerves in the rat's tongue. J. Physiol. **188:** 353-371.
3. BEIDLER, L. M. 1954. A theory of taste stimulation. J. Gen. Physiol. **38:** 133-139.
4. BEIDLER, L. M. 1958. Biophysics and chemistry of taste. *In* Handbook of Perception. E. Carterette & M. Friedman, Eds. Vol. **VIA:** 21-49. Academic Press. New York.

Genetics of the Ability to Perceive Sweetness of D-Phenylalanine in Mice

YUZO NINOMIYA,[a] TETSUICHIRO HIGASHI,
TSUNEYOSHI MIZUKOSHI, AND MASAYA
FUNAKOSHI

Department of Oral Physiology
School of Dentistry
Asahi University
Hozumi, Motosu
Gifu 501-02, Japan

D-Phenylalanine is an amino acid that tastes sweet to man and is preferred to water by some mammalian species. In mice, there are prominent strain differences in behavioral responses to this amino acid. In the C57BL/6CrSlc strain, a taste aversion conditioned to D-phenylalanine generalizes to sugars and sacharin Na but not L-phenylalanine, whereas the opposite is true for BALB/cCrSlc and C3H/HeSlc strains. This suggests that D-phenylalanine presumably tastes sweet to C57BL mice but not to BALB and C3H mice.[1] These strain differences correspond quite well with those observed for the responses of sucrose-sensitive chorda tympani fibers to D-phenylalanine among the three strains of mice.[2]

This study provides evidence for single-locus control of the difference in the ability to taste D-phenylalanine between mouse strains C57BL/6CrSlc and BALB/cCrSlc. Generalization patterns of the conditioned aversion from D-phenylalanine to the other compounds among C57BL, BALB, and their F_1 and F_2 hybrid mice were examined. Then, mice were classified into S- and non-S-tasters on the basis of whether D-phenylalanine generalized to sucrose (S-taster) or not (non-S-taster). This criterion produced no overlap between the two inbred strains, C57BL (S-taster) and BALB (non-S-taster). All 42 F_1 mice responded like C57BL mice, indicating that they were S-tasters. As is shown in TABLE 1, 93 F_2 mice were classified into 71 S-tasters (76.34%) and 22 non-S-tasters (23.66%). The proportion of 71 to 22 was statistically compatible with the expected three to one simple Mendelian ratio for the single-locus model. We tentatively designated the possible gene as *dpa* that would have a major effect on the ability to taste D-phenylalanine in mice. Linkage tests, performed between the gene, *dpa,* and the three fur-color genes, nonagouti (*a,* chromosome 2), brown (*b,* chr. 4), and albino (*c,* chr. 7; see TABLE 1), showed that there is a genetic linkage between the genes, *dpa* and *b* with a recombination value of about 11%, suggesting the possibility that the gene, *dpa,* is located at about 11% recombination distance from the fur-color gene, *b* (brown), on chromosome 4.

[a] Address for correspondence: Yuzo Ninomiya, Department of Oral Physiology, School of Dentistry, Asahi University, Hozumi, Motosu, Gifu 501-02, Japan.

TABLE 1. Numbers of S- and Non-S-tasters of F_2 Segregants from a Cross between C57BL/6CrSlc(S-taster) and BALB/cCrSlc (non-S-taster) Mice[a]

	Agouti (ABC)	Cinnamon (AbC)	Black (aBC)	Brown (abC)	Albino (—c)	Total
S-taster	36	4	15	1	15	71
Non-S-taster	2	11	1	3	5	22
Total	38	15	16	4	20	93

[a] F_2 mice were divided into five groups according to their fur colors, agouti, cinnamon, black, brown, and albino. Fur color genes of which each of five groups possesses are presented in parentheses (a: nonagouti; b: brown; c: albino).

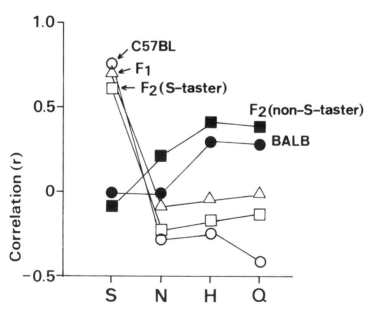

FIGURE 1. Correlation taste profiles of D-phenylalanine across the four basic stimuli (S: sucrose; N: NaCl; H: HCl; Q: quinine-HCl) in C57BL, BALB, F_1, and S-, and non-S-tasters of F_2 mice. The r's plotted are across-neuron correlation[3] coefficients between responses to D-phenylalanine and to one of the four basic stimuli obtained from 30, 30, 32, 30, and 32 chorda tympani fibers of C57BL, BALB, F_1, and S- and non-S-tasters of F_2 mice, respectively.

Responses of single chorda tympani fibers to D-phenylalanine and the four basic taste stimuli (sucrose, NaCl, HCl, and quinine-HCl) were compared among C57BL, F_1, and F_2 S-tasters and BALB and F_2 non-S-tasters. In general, chorda tympani fibers of S-tasters showing high sensitivities to sucrose responded to D-phenylalanine as well, but those of non-S-tasters did not. As appears in FIGURE 1, high correlations (across-neuron correlation[3]) between responses to D-phenylalanine and sucrose are obtained commonly in C57BL, F_1, and F_2 S-tasters, whereas in BALB and F_2 non-S-tasters responses to D-phenylalanine are positively correlated with those to HCl or quinine-HCl but not with responses to sucrose. These results suggest that differences observed in behavioral responses to D-phenylalanine between S- and non-S-tasters are based on their differential peripheral neural responsiveness to the stimulus. Therefore, it is propable that the site of action of the possible gene, *dpa,* is in taste cell membrane.

REFERENCES

1. NINOMIYA, Y., T. HIGASHI, H. KATSUKAWA, T. MIZUKOSHI & M. FUNAKOSHI. 1984. Qualitative discrimination of gustatory stimuli in three different strains of mice. Brain Res. **322:** 83-92.
2. NINOMIYA, Y., T. MIZUKOSHI, T. HIGASHI, H. KATSUKAWA & M. FUNAKOSHI. 1984. Gustatory neural responses in three different strains of mice. Brain Res. **302:** 305-314.
3. ERICKSON, R. P., G. S. DOETSCH & D. A. MARSHALL. 1965. The gustatory neural response function. J. Gen. Physiol. **49:** 247-263.

Axons Do Not Necessarily Compete for Targets[a]

BRUCE OAKLEY

Department of Biology
Neuroscience Laboratory Building
University of Michigan
Ann Arbor, Michigan 48109

Competition exists if A receives less because of the presence of B. Synergism occurs when the effects of A and B acting together exceed the sum of their separate effects. It is in vogue to assume that interactions between axons are competitive; however, axons might be expected to interact synergistically in instances where they exert trophic effects upon their targets. Possibilities include trophic effects upon glial cells, promotion of the outgrowth of dendrites or dendritic spines, or prevention of cell death in neurons receiving axonal projections. In some of these circumstances, innervation of the target cell by axon A might increase, rather than decrease, the probability of synapse formation by axon B.

Our recent observations indicate that taste axons in the rat vallate papilla normally interact in a noncompetitive, synergistic manner to augment the number of taste buds induced during development by more than 150. The vallate papilla of the rat is normally innervated by the right and left IXth nerves. One IXth nerve alone forms a mean of 228 vallate taste buds.[1] Two IXth nerves form, not twice 228, but 610 taste buds.[2] Thus, more than 150 additional taste buds are added because of synergistic interactions between the right and left IXth nerves.

Taste buds in the vallate papilla are neurally induced during a sensitive period that is maximal from 0 to 10 days postpartum.[3] After one IXth nerve is removed and the other crushed at three days of age, only 30 of the 610 vallate taste buds eventually develop. However, after both IXth nerves are crushed at three days, 144 vallate taste buds develop. Thus, in these experiments with crushed IXth nerves, the synergism among axons is even greater, as doubling the nerve supply (two crushed nerves) nearly quintuples the number of taste buds. The surviving axons of one crushed nerve (mean = 835 myelinated axons vs. 1897 in a normal nerve[1]) may be so dispersed in the papilla that the innervation density is insufficient to form more than a few (30) taste buds. The addition of a second IXth nerve may increase the innervation density above the threshold level necessary to form taste buds at many sites in the gustatory epithelium. This implies that the number of bilaterally innervated taste buds should increase according to the number of axons in the right and left IXth nerves. This inference is supported by a stochastic analysis that predicted to a first approximation the number of bilaterally innervated taste buds, given the number of axons in the right and left IXth nerves after nerve crush.[1]

[a] This research was supported in part by National Institutes of Health Grant NS-07072.

During development the IXth nerve forms 75% and the chorda tympani about 30% of the 121 foliate taste buds. Since the total is close to 100%, there is no evidence that some taste buds require dual innervation by both the IXth and chorda tympani nerves for development. However, in adults the IXth nerve will maintain all 121 foliate taste buds, including all of those formed by the chorda tympani. Hence, sometime before 90 days postpartum, the IXth nerve axons must provide enough innervation to maintain the chorda tympani pool of taste buds.

Why does synergism occur in the vallate papilla between the right and left IXth nerve but is not evident in the foliate papilla between the chorda tympani and IXth nerves? The chorda tympani seems to form its own pool of taste buds. Indeed, the chorda tympani induces about 35 taste buds whether the IXth nerve is present or absent during development.

Underlying the interactions between two nerves during taste bud induction are three possible models of dual innervation in which the innervation density might be: (a) so high that either nerve could have formed the given taste bud (logical "or"), (b) so low that both nerves are required (logical "and"), or (c) great enough for one nerve to form the taste bud, but not the other nerve. Model b is present in the vallate papilla and model c may be present in the foliate papilla.

The existence of innervation of taste papillae by two nerves has provided a convenient way to examine interactions among taste axons. Examples of synergistic, noncompetitive interactions among axons have been established for bilaterally induced taste buds in the vallate papilla of the rat. Rather than assuming that competition is *the* rule of interaction among axons, one should consider the possibility of synergism in those situations where axons have neurotrophic effects upon neuronal or nonneuronal targets.

REFERENCES

1. HOSLEY, M. A., S. E. HUGHES & B. OAKLEY. 1987. Neural induction of taste buds. J. Comp. Neurol. **260:** 224-232.
2. HOSLEY, M. A. & B. OAKLEY. 1987. Postnatal development of the vallate papilla and taste buds in rats. Anat. Rec. **218:** 216-222.
3. HOSLEY, M. A., S. E. HUGHES, L. L. MORTON & B. OAKLEY. 1987. A sensitive period for the neural induction of taste buds. J. Neurosci. **7:** 2075-2080.

Response Properties of Thalamocortical Relay Neurons Responsive to Natural Stimulation of the Oral Cavity in Rats

HISASHI OGAWA[a] AND TOMOKIYO NOMURA

Department of Physiology
Kumamoto University Medical School
Kumamoto 860, Japan

The parabrachial nucleus, the second taste relay nucleus, sends afferents to a parvicellular part of the posteromedial ventral nucleus (VPMpc) and center median-parafascicular nucleus complex (CM-PF) in the thalamus,[1] which in turn send efferents to the cortex, particularly the VPMpc sends efferents to the cortical taste area (CTA).[2,3] Previously, we recorded taste neurons in the two nuclei.[4] In this study, we aimed to examine the response properties of the thalamocortical relay (TC) and non-TC neurons in these two nuclei in rats.

Recordings were made from single thalamic neurons in amobarbital-anesthetized Sprague-Dawley rats, while taste and mechanical stimulation was being continuously applied to the whole oral cavity. Taste stimuli used were 0.1 M NaCl, 0.5 M sucrose,

TABLE 1. Thalamocortical Relay and Nonthalamocortical Neurons Responsive to Taste and Mechanical Stimulations in the VPMpc and CM-PF[a]

Neuron Type	VPMpc	CM-PF
Total neurons	61	29
Taste neurons	25	7
TC relay	15 (60%)	0
non-TC		
O type	5 (20%)	5 (71.4%)
N type	5 (20%)	2 (28.5%)
Mechanoreceptive neurons	36	22
TC relay	7 (19.4%)	2 (9.1%)
non-TC		
O type	14 (38.9%)	11 (50%)
N type	15 (41.7%)	9 (40.9%)

[a] Results expressed in number of neurons.

[a] Address for correspondence: Dr. H. Ogawa, Department of Physiology, Kumamoto University Medical School, Honjo 2-2-1, Kumamoto 860, Japan.

532

0.01 N HC1, and 0.02 M quinine-HC1. Mechanical stimulations were applied by stroking or pressing the tissue with a glass rod or by pinching them with a pair of nonserrated forceps. To antidromically activate neurons, 3-10 monopolar metal electrodes were placed ipsilaterally at the CTA and cathodal current pulses (maximum 400 μA 0.02 msec) were applied to them. When the neurons under study produced

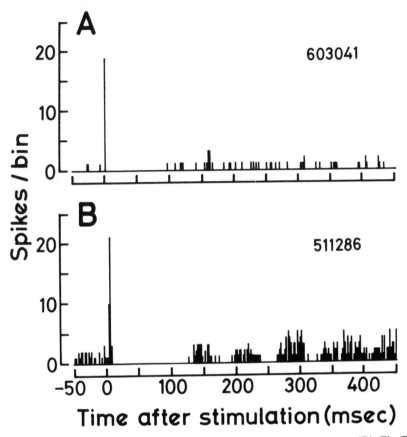

FIGURE 1. PST histograms of TC taste neuron (**A**) and non-TC taste neuron (**B**). The TC neuron in A was identified by a collision test. Two times the threshold current (150 μA) was used to stimulate the CTA, but the neuron produced spikes at ca. 40% of stimulation rate. The non-TC neuron in B was orthodromically excited from the CTA. Bin width: 2 msec, 50 trials.

spikes at an invariant latency, followed by double shock with an interval of 5 msec or less, and passed the collision test, they were assumed to be TC neurons.

A total of 91 neurons were recorded; 61 of these were obtained from the VPMpc and 30 from the CM-PF. A majority of the gustatory neurons (G neurons) and a small ration of mechanoreceptive (M) neurons in the VPMpc were of the TC type (TABLE 1), with an antidromic latency of 1-4 msec. In the CM-PF, however, only two M neurons were of the TC type, having an antidromic latency of 2.4 and 16 msec, respectively. Many M neurons in the VPMpc and a majority of the G and M neurons in the CM-PF were non-TC neurons, either orthodromically activated in a latency of 2.2-16 msec (O-type non-TC) or not activated in a latency of 20 msec or less (N-type non-TC). Both the TC neurons and O- and N-type non-TC neurons showed a period of suppressing spontaneous discharges following the anti- or orthodromic excitation or the onset of stimulation (FIG. 1). The duration of the suppression period was much shorter in the TC neurons (mean = 93.0 msec) than in the non-TC neurons (140.8 msec). Such a long-duration suppression period was not previously observed in the VPMpc.[3] No neuron was found to discharge repetitively during the suppressive period in response to CTA stimulation. Both types of neurons in the two nuclei showed rebound responses after the suppression period. No significant difference was found in the receptive properties between the TC and non-TC G neurons nor between the TC and non-TC M neurons in each of the two nuclei. But some differences were noticed in properties of the G and M neurons between the two nuclei. G neurons in the CM-PF produced smaller response magnitudes to NaCl and quinine on the average than those in the VPMpc, and the former tended to produce phasic responses in comparison with the latter. A high ratio of M neurons in the CM-PF tended to respond to heavy mechanical stimulation but neurons responding to light and heavy stimulations were equally observed in the VPMpc.

These findings showed that the VPMpc sends tonic information of taste to the CTA as the taste relay nucleus, but that some neurons in the CM-PF respond phasically to taste or heavy mechanical stimulation of the oral cavity and projects mainly to the cerebral cortex other than the CTA.

REFERENCES

1. NORGREN, R. & C. LEONARD. 1973. Ascending central gustatory pathways. J. Comp. Neurol. **150:** 217-237.
2. GANCHROW, D. & R. P. ERICKSON. 1972. Thalamocortical relations in gustation. Brain Res. **36:** 289-305.
3. YAMAMOTO, T., R. MATSUO & Y. KAWAMURA. 1980. Corticofugal effects on the activity of thalamic taste cells. Brain Res. **193:** 2285-292.
4. NOMURA, T. & H. OGAWA. 1985. Responses of neurons in the thalamic parafascicular nucleus and posterior nuclear complex to natural stimulation of the oral cavity in rats (in Japanese). J. Physiol. Soc. Jpn. **47:** 154.

Scanning Electron Microscopic and Histochemical Studies of Taste Buds

MASARU OHYAMA, KEI OGAWA, KOZO FUKAMI,
ERIKO TABUCHI, KAZUYOSHI UENO, AND
KAZUYO TANAKA

Department of Otolaryngology
Faculty of Medicine
Kagoshima University
Kagoshima, Japan

The purpose of this study was to determine if the use of both the secondary electron (SE) and the backscattered electron (BSE) images with zinc-iodine-osmium (ZIO) and ruthenium red (Rr) staining would show histochemical similarities among the chemical fine structures of rabbit taste buds. Additionally, the distribution of binding sites of HRP-labeled lectins, such as PNA, WGA, TCA, and UEA-1, were also studied in monkey taste buds using these staining methods.

RESULTS AND DISCUSSION

In SE and BSE images, the dense substance in the taste pore and the dark granules of the type I taste cell could be seen in the SE image. However, it was found that the BSE image produced by the ZIO method provided a contrast for both dense substances and dark granules superior to that of the conventional technique. Also a distinct BSE image could be obtained from the Golgi vesicles and DG near the Golgi apparatus in only the type I cell. Rr strongly stained the substance present along the membrane of the microvilli and some of the pore vesicles.

The characteristic distribution of binding sites of lectins with different specificities for glycoconjugates was found. Especially, binding sites for PNA were distributed more widely in not only the cell membranes of the intermediate layer in squamous epithelium but in those of glandular ducts and in the cytoplasm of nerve cells and fibers. An interesting result is that dense substances and dark granules exhibit the same strong BSE image; it strongly suggests that the dark granules originated from the Golgi apparatus in the perinuclear areas. It has been reported that ZIO reacts with intravesicular proteins containing SH-groups and that highly ZIO-positive vesicles represent mature transmitter vesicles available for release at the neuromuscular junction. This suggests that dense substances, which probably consist of proteins containing SH groups, may play an important role in facilitating and modifying the response of

535

taste cells to taste stimuli. Therefore, the primary role of the type I cell may be in the synthesis and secretion of a pore-modifying substance rather than in neurotransduction. It appears that utilizing the BSE image in conjunction with the SE image is very useful in determining the three-dimensional localization of cytochemical components of taste buds, in addition to showing a characteristic appearance of binding sites for different kinds of lectins.

Taste Concepts and Quadranomial Taste Description

M. O'MAHONY, R. ISHII, AND D. SHAW

Department of Food Science and Technology
University of California
Davis, California 95616

Taste description is a widely used psychophysical tool. For accurate communication it is important to arrange that subjects and experimenters use concepts held in common, with agreed labels. It thus becomes important to study the mechanisms of concept formation. Sensory concept formation is easily understood in reference to color. A concept like "redness" is formed, according to current theory,[1] by a two-part process: abstraction and generalization. For the first part, the concept of redness is abstracted from red and nonred stimuli. For the second, this concept is generalized or broadened beyond those sensations used in the abstraction process. Then, stimuli colored shades of red that have not been seen before can be categorized as falling within the concept; they have "redness." The concept is given the label "red" for purposes of communication between those who share the concept. The same reasoning can be applied to the formation of taste concepts with labels like "sweet," "salty," or "umami."

Because taste description is the commonly used psychophysical tool for studying taste quality, it is worth considering some basic requirements for such a tool. In terms of concept formation, taste description is essentially the use of labels to communicate taste concepts, presumably held in common. A necessary, but insufficient, requirement of any such descriptive method is that each concept should have a separate label. If they did not, separate concepts given the same label would be confused. Another way of stating this is that the number of labels should equal the number of concepts. If this were so and if tastes were sorted into their conceptual groupings, the number of labels would equal the number of groupings.

This requirement was tested for a method of taste description that incorporated two assumptions common in current taste-descriptive procedures. The first was that the labels for the taste concepts need be limited only to the 31 possible combinations of the words "sweet," "sour," "salty," "bitter," and "other." The second assumption was that everyday language concepts were sufficient for taste description, no preexperimental procedure being necessary to define the concepts with standard stimuli.

One hundred seventy American subjects were required first to taste and sort filter paper stimuli, which had been previously soaked in solutions of various concentrations of NaCl, LiCl, MSG, IMP, GMP, citric acid, caffeine, sucrose, aspartame, sodium benzoate, and KCl, into groupings according to their conceptual categories. After sorting, they named each separate taste grouping, using the descriptive method under investigation. The limitation in the number of allowable descriptive terms available resulted in there being fewer separate labels than categories for some subjects. This occurred for approximately a third of the subjects (56/170), with a mean shortfall of 1.7 labels less than categories. The experiment was repeated with 209 Japanese subjects, because prior research had indicated that Japanese exhibited differences from

Americans in naming strategy.[2] Here, the shortfall occurred for approximately half of the subjects (97/209) with a mean shortfall of 1.6.

For a large proportion of subjects, the basic requirement of a descriptive method, that concepts should be given separate labels, was not fulfilled by the descriptive system. Furthermore, variations in the sorting pattern indicated that everyday language taste concepts vary between subjects; for scientific measurement it is necessary to align their concepts, presumably by training with standards. These findings have important implications for current psychophysical measures of taste quality.

REFERENCES

1. MILLER, G. A. & P. N. JOHNSON-LAIRD. 1976. Language and Perception. Cambridge University Press. London, England.
2. O'MAHONY, M. & R. ISHII. 1986. A comparison of English and Japanese taste languages: Taste descriptive methodology, codability and the umami taste. Br. J. Psychol. 77: 161-174.

The Size of Mitral Cells Depends on the Age at Which Continuous Odor Exposure Commences

H. PANHUBER AND D. G. LAING

CSIRO Division of Food Research
North Ryde, N.S.W.
Australia 2113

A. MACKAY-SIM

Department of Physiology
University of Adelaide
Adelaide, S.A., Australia 5000

Exposing rats to one predominant odor for two months leads to odor-specific changes in the size of mitral cells, the relay cells of the olfactory bulb. The distribution and extent of these changes depends on the odor used and the age of the rats when exposure is commenced. Rats were continuously exposed to three different olfactory environments commencing at three ages: 1 day old, 14 days old, and 13 weeks old (adult). The three olfactory environments were: deodorized air containing a moderate concentration of cyclohexanone (an almond-mint-like odor), deodorized air, and the normal rat colony odors (controls). After two months' exposure, rats were perfused with a 4% formaldehyde, 1% glutaraldehyde solution; the olfactory bulbs were removed, embedded in Epon and sectioned at 5 μm. The cell areas of 3% of all mitral cells were measured using a computer-assisted technique.[1] The mitral cell layer was then divided into 40 equally spaced regions and the average cell area for each region was calculated. As the distribution of cell area in the coronal plane was consistent along the rostro-caudal axis of the bulb, the 40 regional means were averaged over all the coronal sections.[1]

The results are summarized in FIGURE 1 by schematic representations of coronal sections of the olfactory bulb. Regions of the mitral cell layer with a mean cell area significantly larger than that from the equivalent regions of control rats are shown as hatched (least significant difference calculated from the ANOVA, $p < 0.05$), and the regions significantly smaller than the controls are shown in black. The number of regions with cells smaller than controls increased with the age at which exposure was commenced; Day 1 had 9 regions, day 14 had 22, and adults had 34 regions smaller than controls.

The results show that adult rats are particularly vulnerable to prolonged odor exposure and show more marked changes than neonates. The severe cell shrinkage observed in adult rats exposed to an odor could be due to the combined effects of inhibition and lack of stimulation. With rats exposed from one day old, the inhibitory

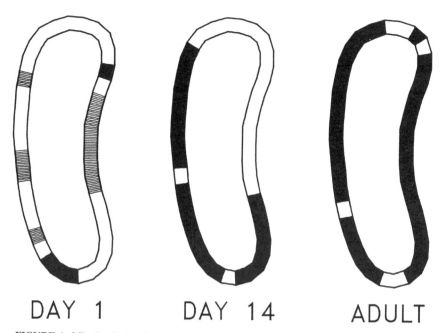

FIGURE 1. Mitral cell size changes for rats exposed to cyclohexanone at three different ages.

and centrifugal circuitry is undeveloped, and the changes in cell size induced by odor exposure are probably due to the differences in the level of stimulation received by different mitral cells. The development of the inhibitory and centrifugal circuitry during the initial three weeks of life may also be altered by the exposure environment when rats are exposed from day one and to a lesser extent when exposed from day 14.

REFERENCE

1. PANHUBER, H., D. G. LAING, M. E. WILLCOX, G. K. EAGLESON & E. A. PITTMAN. 1985. The distribution of the size and number of mitral cells in the olfactory bulb of the rat. J. Anat. **140:** 297-308.

Brief Taste Stimuli

Cued and Uncued Magnitude Estimates and Their Reaction Times[a]

TERESA PANTZER, STEVEN T. KELLING, AND
BRUCE P. HALPERN[b]

Department of Psychology, Field of Physiology, and
Section of Neurobiology and Behavior
Cornell University
Ithaca, New York 14853-7601

Taste intensity changed over time for stimulus durations close to or greater than one second,[1] and for 50-msec[2,3] to 100-msec durations.[4] Time-intensity experiments (see Halpern[5] for references) found rapid taste intensity increases to a flat maximum, then a decline. However, the sustained gustatory stimulation of the latter experiments prevented distinctions between central nervous system processing versus continuous peripheral input as the dominant factor in the reported time-intensity profile.

We used a closed-flow taste stimulus apparatus,[6] and measured vocal taste magnitude estimation reaction times,[3,5] using an electret throat microphone, or taste intensity tracking, using a single-axis joystick. For cued judgments, subjects judged and stated the taste intensity as soon as they heard a judgment signal; for uncued, as soon as they thought it appropriate (TABLE 1).

The uncued experiments confirmed increased taste intensity as stimulus duration increases (FIG. 1). However, in the cued magnitude estimation investigations, there were no significant changes in the intensity ratings across the five judgment cue times (TABLE 1). The latter was unexpected. The problem might be the cued vocal magnitude estimation method, or the absence of an accessible cumulation process at brief taste stimulus durations. To test this, time-intensity tracking was done with 50-msec, 100-msec, 300-msec, 1,000-msec, and 2,000-msec durations of 2 mM Na-saccharin, randomized with distilled water. Preliminary single-axis joystick results indicate that brief taste stimulus durations can be tracked, that resulting indications of intensity increase in magnitude and temporal extent as stimulus duration increases, and that, for durations of 300 msec or less, taste stimulus presentation and removal precedes the start of the joystick-produced intensity indication.

[a] Supported by National Science Foundation Grants BNS 8213476 and 8518865.

[b] Address for correspondence: Dr. Bruce P. Halpern, Department of Psychology and Section of Neurobiology and Behavior, Uris Hall, Cornell University, Ithaca, NY 14853-7601.

TABLE 1. Judgments of 300-msec Duration Stimuli[a]

Response Measure	Time of Judgment Signal Onset (msec) Cued						Uncued
	730	960	1240	1440	1640		
2 mM NaSac[b]							
Magnitude estimate	7 (.6)	8 (.7)	7.5 (.6)	8 (.7)	7.5 (.7)	ns[c]	10[d] (1)
Reaction time	1138 (15)	1283 (15)	1550 (17)	1709 (16)	1933 (11)	s[e]	1638 (83)
214 mM MSG[f]							
Magnitude estimate	6 (.6)	6 (.6)	6 (.7)	8 (1.2)	6.5 (.7)	ns	6[g] (.7)
Reaction time	1086 (34)	1349 (24)	1557 (25)	1723 (15)	1905 (23)	s	1217 (83)

[a] Cued (300-msec duration, 1200 Hz, 74 db tone) and uncued vocal (5) median (standard error) magnitude estimates (standard = 20 for 2,000-msec stimulus solution duration), and their vocal reaction times (in msec), for 300-msec duration (6) pulses of 2 mM sodium saccharin (NaSac) or 214 mM monosodium glutamate (MSG), at 10 ml/sec flow rate, randomized with distilled water control stimuli.

[b] Five subjects.

[c] ns: No significant difference in cued magnitude estimates across the five judgment signal onset times, Friedman nonparametric analysis of variance (NONOVA), $p > 0.05$.

[d] Significantly different from the cued magnitude estimates for 2 mM NaSac at the 730-msec and 1240-msec judgment signal onset times, Mann-Whitney U, two-tailed, $p < 0.05$.

[e] s: Significant difference in cued reaction time across the five judgment signal onset times, NONOVA, $p \leq 0.05$.

[f] Four subjects.

[g] No significant differences between uncued and cued magnitude estimates at any judgment signal onset time, U, $p > 0.1$.

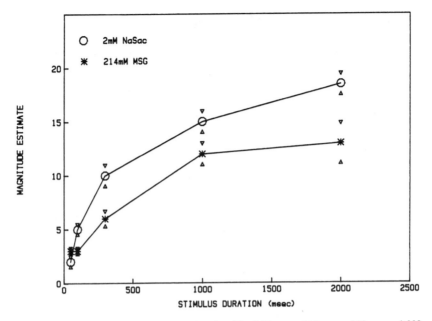

FIGURE 1. Uncued vocal magnitude estimations[3,5] of 50-msec, 100-msec, 300-msec, 1,000-msec, and 2,000-msec duration stimulus pulses[6] of 2 mM sodium saccharin (NaSac) or 214 mM monosodium glutamate (MSG), detected with a throat microphone.[5] Medians, plus and minus the standard error of the median, are shown. Subjects (NaSac = five; MSG = four) judged in proportion to a 2,000-msec duration identified standard stimulus of NaSac or MSG that was assigned an intensity of 20. During each of five data-collecting sessions, the control stimulus (distilled water) was presented twice at each duration, in random order, randomized with the stimulus solution presentations.

REFERENCES

1. BUJAS, Z. & A. OSTOJCIC. 1939. L'evolution de la sensation gustative en fonction du temps d'excitation. Acta Inst. Psychol. Zagreb **3:** 1-24.
2. KELLING, S. T. & B. P. HALPERN. 1983. Effects of stimulus duration and concentration on taste perception of MgSO₄ and HCl. Presented at the Fifth Annual Meeting of the Association for Chemoreception Sciences. Sarasota, FL.
3. KELLING, S. T. & B. P. HALPERN. Taste judgments and stimulus duration: Taste quality, taste intensity, and reaction times. In preparation.
4. KELLING, S. T. & B. P. HALPERN. 1983. Taste flashes: Reaction times, intensity, and quality. Science **219:** 412-414.
5. HALPERN B. P. 1986. Constraints imposed on taste physiology by human taste reaction time data. Neurosci. Biobehav. Rev. **10:** 135-151.
6. KELLING, S. T. & B. P. HALPERN. 1986. The physical characteristics of open flow and closed flow taste delivery apparatus. Chem. Senses **11:** 89-104.

Cytochrome Oxidase Staining in the Olfactory Epithelium and Bulb of Normal and Odor-Deprived Neonatal Rats[a]

P. E. PEDERSEN, G. M. SHEPHERD, AND
C. A. GREER

Sections of Neuroanatomy & Neurosurgery
Yale University School of Medicine
New Haven, Connecticut 06510

Cytochrome oxidase (CO), recently introduced as a marker of central nervous system (CNS) activity,[1] stains heterogeneously and laminae dependent in the olfactory bulb (OB).[2] CO decreases in intensity following transection of the olfactory nerve (ON) in the adult.[3] As part of continuous studies on the neonatal olfactory system, we assessed CO staining in the olfactory epithelium (OE) and OB following neonatal unilateral naris closure. We wished to: (1) establish the suitability of the CO procedure for simultaneously indexing activity within the OE and OB; (2) determine the effect of odor deprivation on CO staining in the developing OE and OB; and (3) compare the results obtained with the 2DG and CO procedures.

Sprague-Dawley rats, two to eight days postnatal, had one naris occluded by electrocautery and were sacrificed 2-40 days later. The skull was decalcified and the OE and OBs serially sectioned and processed for CO. Littermates were injected with 2DG, exposed to amyl acetate and processed with autoradiographic procedures. Planimetry determined the respective areas of each OB.

The pattern of CO staining in the nondeprived OB was consistent with prior reports (FIG. 1A, open arrow).[2] In general, CO staining was heaviest in areas of neuropil with little staining of cell bodies. Low levels of stain were present in the ON layer whereas among glomeruli the density of stain ranged from low to high. The external plexiform layer had a trilaminar pattern of staining; the granule cell layer ranged from high to low levels of staining from the superficial to deep granule cell layer, respectively. Unexpectedly, the pattern and intensity of staining in the deprived OB (FIG. 1A, closed arrow) was indistinguishable from the control OB even though the average area of the deprived OB was 25% smaller. This was in contrast to the 2DG studies in which a marked reduction of uptake was apparent in the deprived OB (FIG. 2).

The CO histochemical procedure permitted visualization and localization in the OE. The nondeprived OE exhibited heterogeneous CO staining. A bilaminar pattern

[a]Supported in part by NICHHD HD20994, NINCDS NS19430, and NINCDS NS07609 of the National Institutes of Health and the Fragrance Research Fund and March of Dimes 5-420.

was apparent with the heaviest staining in the superficial dendritic zone and less intense in the somata of the receptor cells (FIG. 1B). In sharp contrast to the results in the OBs, the deprived OE (FIG. 1C) exhibited marked reductions in the intensity of CO staining. Although a bilaminar pattern could be discerned, staining was more generally homogeneous across the deprived OE.

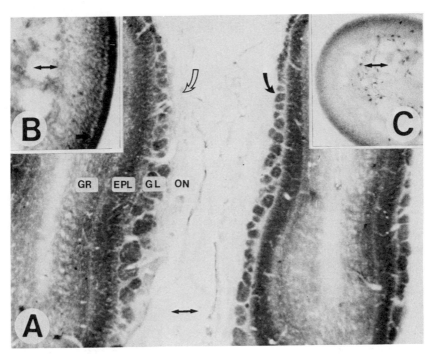

FIGURE 1. Cytochrome oxidase staining three weeks post-unilateral naris closure. In (**A**), the nondeprived (open arrow) and contralateral deprived (closed arrow) olfactory bulbs are shown. The darkened areas correspond to variations in the intensity of CO staining. Note that there are no apparent differences between the two bulbs in the intensity of the CO stain. However, a decrease in the size of the olfactory bulb laminae is apparent on the deprived side. Optical density measurements confirmed the equivalence of CO staining in the olfactory bulbs. In (**B**) and (**C**) examples of olfactory epithelium for nondeprived and deprived sides, respectively, are shown. Note in (**B**) the bilaminar pattern of staining. The arrow indicates the dendritic zone containing the heaviest staining. Immediately deep to the dendritic zone, the unstained nuclei of supporting cells are visible. In contrast the staining of deprived epithelim (C) was far less intense. Nevertheless, staining in the dendritic zone appeared to exceed background levels. Abbreviations: ON, olfactory nerve; GL, glomerular layer; EPL, external plexiform layer; Gr, granule cell layer. Calibration bars: A = 300 μm; B = 25 μm; C = 50 μm.

The effects of naris closure, as measured by planimetry and 2DG, are consistent with odor deprivation. Unlike other sensory systems, however, we did not observe any alterations in CO staining in the OB following chronic odor deprivation.[1,4] CO staining in the OB appears more susceptible to changes induced by loss of ON fibers than those changes accompanying odor deprivation. The differential effect of odor

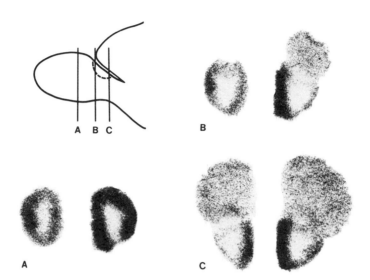

FIGURE 2. Autoradiographs of the olfactory bulbs three weeks after electrocautery occulsion of the left naris. In the schematic the levels of each section (**A-C**) are indicated. The darkened areas correspond to high levels of 2-deoxyglucose uptake. Note the absence of high levels of uptake in the deprived olfactory bulb relative to the contralateral control. However, frontal cortex, as shown in (C) exhibits comparable 2-deoxyglucose uptake on both sides suggesting that the differences observed between the olfactory bulbs are functionally related to the presence or absence of odor input.

deprivation on the OE suggests that CO may be useful for studying the functional organization at the receptor level.

REFERENCES

1. WONG-RILEY, M. 1979. Changes in the visual system of monocularly sutured or enucleated cats demonstrable with cytochrome oxidase histochemistry. Brain Res. **171:** 11-28.
2. SHIPLEY, M. & S. VAN OOTEGHEM. 1984. Cytochrome oxidase activity in the olfactory system: Normal distribution. Am. Chem. Soc. Abstr. **6:** 116.
3. COSTANZO, R., M. SHIPLEY & S. VAN OOTEGHEM. 1984. Cytochrome oxidase activity in the olfactory system: Ontogeny and response to peripheral deafferentation. Am. Chem. Soc. Abstr. **6:** 35.
4. WONG-RILEY, M. & C. WELT. 1980. Histochemical changes in cytochrome oxidase of cortical barrels following vibrissal removal in neonatal and adult mice. Proc. Natl. Acad. Sci. USA **77:** 2333-2337.

Isolation and Characterization of an Odorant-Binding Protein[a]

JONATHAN PEVSNER, PAMELA B. SKLAR, AND
SOLOMON H. SNYDER

*Departments of Neuroscience, Pharmacology and Experimental
Therapeutics, Psychiatry, and Behavioral Sciences
The Johns Hopkins University School of Medicine
Baltimore, Maryland 21205*

We have examined the binding of tritiated odorants to homogenates of bovine and rat nasal mucosa. Using the potent bell-pepper odorant [³H]2-isobutyl-3-methoxypyrazine ([³H]IBMP) we[1] and others[2] measured specific and saturable binding to homogenates from bovine and rat nasal epithelia. In each species, specific binding is detected in soluble extracts from nasal epithelia but not in membrane-bound fractions nor in a variety of other tissues including olfactory bulbs, brain, and liver.

An odorant-binding protein (OBP) has been purified from bovine and rat nasal epithelium.[1,3] The epithelia are homogenized, centrifuged (30,000 × g, 20 min) and the supernatant is applied to a DEAE-cellulose ion-exchange column eluted with a linear gradient of NaCl. Fractions assayed for [³H]IBMP binding activity reveal a single peak. Protein from this peak is applied to a reverse-phase HPLC column (FIG. 1) eluted with a linear gradient of 0-100% acetonitrile : n-propanol (2 : 1). The resulting peak contains odorant binding activity and is pure OBP. Twelve percent

TABLE 1. Binding Constants for Odorants to Purified Bovine OBP and Rat Mucus and Tears[a]

Odorant	Bovine OBP K_D (μM)	B_{max}	Rat Mucus K_D (μM)	B_{max}	Rat Tears K_D (μM)	B_{max}
[³H]amyl acetate	68	22	> 1,000	n.d.	> 1,000	n.d.
[³H]DMO	0.3	30	44	1.6	100	15.7
[³H]IBMP	3	27	20	2.3	11	1.7
[³H]MDHJ	8	18	35	0.3	25	0.3
[³H]methoxy-pyrazine	> 1,000	n.d.	> 1,000	n.d.	> 1,000	n.d.

[a] Binding assays were performed in triplicate using bovine OBP, rat mucus, or rat tears and 5–15 n*M* of each tritiated odorant. B_{max} values are nanomole bound per milligram protein. No specific binding was detected for the binding of 10 n*M* [³H]isovaleric acid to bovine OBP or rat mucus. Less than 5 fmol/mg protein of [³H]IBMP or [³H]DMO binding to rat saliva was detected. n.d., not determined. From Pevsner *et al.*[4]

[a] This work was supported by International Flavors and Fragrances, Union Beach, NJ.

polyacrylamide gels electrophoresed in the presence of SDS reveal a single band of 19,000 daltons (bovine OBP) or 20,000 daltons (rat OBP; FIGURE 1, inset). Based on data from gel filtration and sucrose gradients, the molecular size of bovine OBP is 37,000 daltons, suggesting the protein is a homodimer. In addition to IBMP, OBP binds the odorants [³H]methoxypyrazine, [³H]3,7-dimethyloctan-1-ol (DMO), [³H]methyldihydrojasmonate, and [³H]amyl acetate with micromolar to millimolar affinities (TABLE 1).[4]

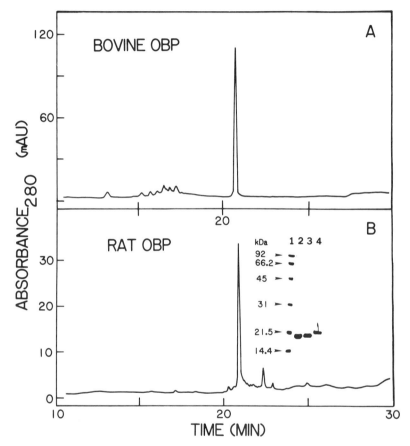

FIGURE 1. Reverse-phase HPLC analysis of bovine OBP and rat mucus, following fractionation on a DEAE-cellulose column. (**A**) Bovine OBP (50 μg; retention time, 20.58 min); (**B**) Rat OBP (40 μg; retention time, 20.88 min). (Inset) SDS/PAGE analysis of OBP. Lanes: 1, molecular size markers; 2, bovine OBP purified as described;[1] 3, the major peak from A; 4, the major peak from B.

A polyclonal antiserum to bovine OBP has been produced in rabbits.[4] The specificity of the antiserum for OBP was confirmed by immunoblotting, and its titer, determined by radioimmunoassay, was 1 : 8100. Eight-micrometer frozen sections of bovine olfactory and respiratory epithelium were immunohistochemically stained with a 1 : 20,000 dilution of antiserum. OBP-like immunoreactivity is detected in mucus-secreting glands of the lamina propria in both olfactory and respiratory epithelium. Preadsorption of the antiserum with pure OBP eliminates staining of the glands.

Given the localization of OBP to secretory glands of the nasal epithelium, we examined secretions of the rat for OBP. Rats were anesthetized with sodium pentobarbital and then injected with isoproterenol intraperitoneally to induce secretions. Tears, nasal mucus, and saliva were collected and assayed for the binding of tritiated odorants. [^{3}H]IBMP, [^{3}H]DMO, and [^{3}H]methyldihydrojasmonate bind to mucus and tears (TABLE 1), and OBP has been purified to homogeneity from those sources. No specific odorant binding is detected in saliva.

Other experiments suggest that odorants inhaled by rats *in vivo* can bind to OBP. The function of OBP may be to transport odorants within the nasal epithelium, either delivering them to receptors on cilia of olfactory neurons or eliminating them from the mucosa after they have been smelled.

REFERENCES

1. PEVSNER, J., R. R. TRIFILETTI, S. M. STRITTMATTER & S. H. SNYDER. 1985. Isolation and characterization of an olfactory receptor protein for odorant pyrazines. Proc. Natl. Acad. Sci. USA **82:** 3050-3054.
2. PELOSI, P., N. E. BALDACCINI & A. M. PISANELLI. 1982. Identification of a specific olfactory receptor for 2-isobutyl-3-methoxypyrazine. Biochem. J. **201:** 245-248.
3. BIGNETTI, E., A. CAVAGGIONI, P. PELOSI, K. C. PERSAUD, R. T. SORBI & R. TIRINDELLI. 1985. Purification and characterization of an odorant-binding protein from cow nasal tissue. Eur. J. Biochem. **149:** 227-231.
4. PEVSNER, J., P. B. SKLAR & S. H. SNYDER. 1986. Odorant-binding protein: Localization to nasal glands and secretions. Proc. Natl. Acad. Sci. USA **83:** 4942-4946.

Multiple Bitter Receptor Sites in Hamsters[a]

CARL PFAFFMANN AND M. SCOTT HERNESS

Laboratory of Neurobiology and Behavior
The Rockefeller University
New York, New York 10021

We have been studying the hamster's conditioned taste aversion (CTA) to sodium picrate, sodium cholate, and sodium nitrobenzoate, which taste bitter to humans and which hamsters avoid in two-bottle preference tests.[1] The degree of generalization, if any, to the basic four taste stimuli, 0.1 M sucrose, 0.1 M NaCl, 0.01 M HCl, and 0.001 M quinine hydrochloride, was determined. The CTA was established by the i.p. injection of apomorphine hydrochloride (Lilly, 30 mg/kg dissolved in H_2O) after the first exposure to the experimental solution. A mean suppression score was calculated from the amount consumed by an experimental group ($n = 5$) divided by the amount consumed by hamsters whose CS was water. This ratio subtracted from unity was multiplied by 100. Complete generalization is indicated by a score of 100, no generalization by a score of zero. A "suppression" is a score that differs from zero by more than two SE, a weak suppression differed from zero by 1.5 to 2.0 SE following the Nowlis *et al.*[2] procedure.

Groups of hamsters conditioned against the four basics and one of the three putative bitters showed good generalization of each of the four basics to itself. To the cholate, picrate, and nitrobenzoate sodium salts, two groups show generalization to NaCl, one no generalization; however, none suppressed to quinine.

We reconditioned this group of hamsters against their previous CSs and tested for generalization to additional stimuli, 0.15 M L-phenylalanine, 0.001 M strychnine sulfate, 0.5 urea, and 0.1 $MgCl_2$. Sucrose, NaCl, HCl, and quinine CS groups all produced strong suppression against themselves (FIG. 1), but now quinine gave strong suppressions (35% and 40%) to phenylalanine (Pa) and to strychnine (S9) and weak suppression (12%) to magnesium (Mg). Na cholate suppressed to strychnine plus Na, Na picrate only Na, and Na benzoate (NaNBA) was without effect.

Four groups of these same animals were reconditioned on their same stimuli, the quinine, Na picrate, Na cholate, and Na nitrobenzoate animals. But now all of these conditioning solutions were used as test stimuli as well as training stimuli. Previously only quinine had been tested against itself. FIGURE 2 shows the results.

The four tested CS groups all showed avoidance of the CS when presented as a test stimulus; thus, quinine as CS elicited a 62% suppression of quinine, Na cholate, a 69% suppression to itself as a test stimulus, Na picrate showed 49% suppression, and NaNBA, a 79% suppression to itself. Thus, a failure of the three putative bitter

[a] This study was supported in part by a grant from the National Science Foundation, BNS 8111816 and in part by BRSG S07 RR07065 awarded by the Biomedical Grant Program, Division of Research Resources, National Institutes of Health.

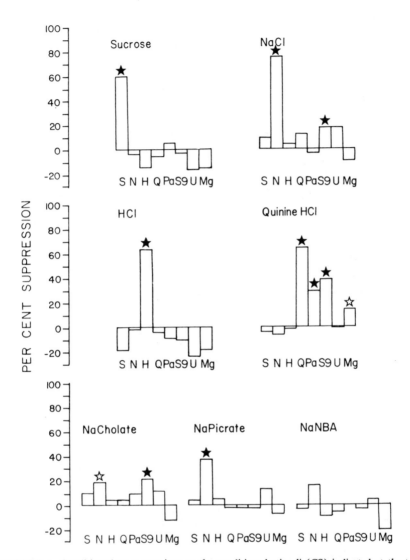

FIGURE 1. Conditioned taste aversions to the conditioned stimuli (CS) indicated at the top of each histogram, 0.1 *M* sucrose (S), 0.1 *M* NaCl (N), 0.01 *M* HCl (H), 0.001 *M* quinine HCl (Q), 0.15 *M* L-phenylalanine (Pa), 0.0001 *M* strychnine sulfate (S9), 0.5 *M* urea (U), 0.1 *M* MgSO$_4$ (Mg), 0.01 *M* Na cholate, 0.01 *M* Na picrate, 0.1 *M* Na NBA (Na benzoate). Percent suppression is the mean suppression score calculated by comparing drinking in milliliters of test animals with control animals for which the CS is distilled water. A solid star over a bar indicates a strong suppression, a mean suppression score greater than 2.0 standard error (SE). An open star indicates weak suppression, an SE from 1.5 to 2.0.

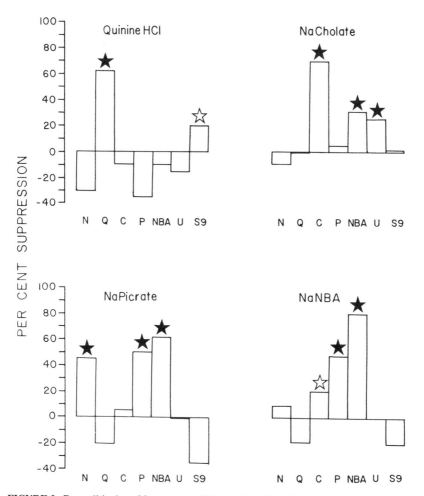

FIGURE 2. Reconditioning of four groups of FIGURE 1 each on their same CS, 0.001 M quinine HCl (Q), 0.01 M Na cholate (C), 0.01 M Na picrate (P), or 0.1 M NaNBA (NBA), and tested on these same stimuli plus 0.1 M NaCl (N), 0.5 M urea (U), and 0.0001 M strychnine sulfate (S9). All CS groups generalized to themselves (i.e., cholate as CS to cholate as test stimulus, TS). In addition NBA and picrate showed symmetrical generalizations to each other. A weaker NBA-cholate generalization is evident. Na picrate again demonstrates a generalization to NaCl.

compounds to generalize to certain other bitter compounds cannot be attributed to ineffective taste aversion conditioning or inadequate testing procedure.

Additionally, 0.01 M Na picrate and 0.1 M NaNBA generalized to one another. The Na picrate group suppressed drinking of NaNBA by 63% and, conversely, the NaNBA group suppressed drinking of Na picrate by 47%. Na cholate (0.01 M) and NaNBA (0.1 M) also produced symmetrical generalizations to one another. (The 0.01 M Na cholate suppressed drinking of 0.1 M NaNBA by 30%, a strong suppression, and 0.1 M NaNBA suppressed drinking of 0.01 M Na cholate by 20%, a weak suppression.) While Na picrate and Na cholate both produced symmetrical generalizations with NaNBA, there did not appear to be any significant generalizations between Na picrate and Na cholate.

Thus there appear to be at least two domains of bitter sensitivity sampled by our stimuli, one clustered around quinine, phenylalanine, and strychnine, the second grouping consisted of cholate and benzoate with which picrate and urea partially overlap.

Prior data from humans, rats, hamsters, mice, and frogs support the idea of separate or multiple sites for bitter stimuli depending on the solutions used, species or even strains being studied. Our results imply behaviorally distinct bitter sensitivities in hamsters, and thus more than one receptor site for bitters.

REFERENCES

1. HERNESS, M. S. & C. PFAFFMANN. 1986. Generalization of conditioned taste aversions in hamsters: Evidence for multiple bitter receptor sites. Chem. Senses 11(3).
2. NOWLIS, G. H., M. E. FRANK & C. PFAFFMANN. 1980. Specificity of acquired aversions to taste qualities in hamsters and rats. J. Comp. Physiol. Psychol. 94(5): 932-942.

Zigzag Flight as a Consequence of Anemotactical Imprecision

R. PREISS AND E. KRAMER

Max-Planck-Institut für Verhaltensphysiologie
D-8131 Seewiesen
Federal Republic of Germany

Moths orienting in a pheromone plume use visual information from ground pattern movement to proceed upwind along a zigzagging flight track. Groundspeed and track angles left and right of the wind direction appear to be kept constant if the wind varies. In the concept of "reversing anemomenotaxis,"[1,2] it was supposed that moths compensate for the differences in winddrift by adjusted alterations of their course, and the zigzag structure of this menotaxis was claimed to be caused by an internal—"self-steered"—program of "counterturning."

After having analyzed the orientation behavior of male gypsy moth using a "flight simulator"[3] we found an alternative explanation for the above peculiarities.[4] In this simulator the forces exerted by the moth during tethered flight are sensed and used to control speed, direction of motion and size of a projected ground pattern. The visual situation of the moth then corresponds to free flight with respect to all three axes of translation and their yaw-axis of rotation. Wind can be simulated by superimposing a translatory component to the moving pattern.

In our experiments, the course angle at all wind speeds was scattered unimodally around due upwind, which, in fact, is incompatible with menotaxis. Nevertheless, the distributions of the corresponding track angles were always clearly bimodal. This surprising issue, however, can be explained if only the complex trigonometric relation between course and track angle is taken into account (FIG. 1). Accuracy in determining the wind direction is a function of the wind speed: The scatter of the course will be larger if the wind is weak, and this in turn will compensate for the smaller drift. The apparent constant track angle within a large range of windspeeds, therefore, is the result of a combination of pure physics (winddrift) and the moth's inability to fly precisely upwind ("noise" superimposed on "basic orientation").

The question of how the wind direction is detected could be answered by manipulating the ground pattern itself (FIG. 2). When only stripes transverse to the wind were offered to the moth, course angles of $+55°$ or $-55°$ were stable, whereas with longitudinal stripes only, the due upwind or downwind direction was held. Drift causes the images of ground pattern to flow obliquely over the retina. This motion can be resolved in its longitudinal (L) and its transversal (T) component. The moths presented with any of this special pattern types chose a direction in which T was zero. The results, therefore, do not support the assumption that $\sqrt{(T^2 + L^2)}$, or any other function combining T and L, is kept constant. Rather, upwind flight is achieved by two independent control circuits, one of which minimizes the transverse component according to a common optomotor response, while the other keeps the longitudinal component at small positive values.

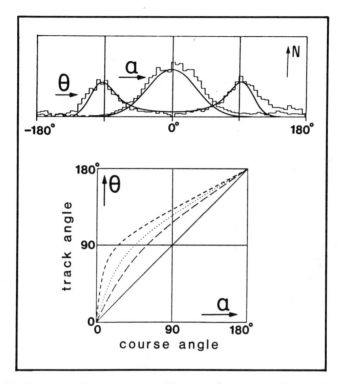

FIGURE 1. Density distribution of course angles (α) and track angles (θ) of a 10-minute flight of a male gypsy moth (bins 5° wide). Superimposed: theoretical distribution of track angles (θ), if the course angles (α) scatter around due upwind according to a Gaussian density distribution and if only the slope function of α versus θ (lower part) is taken into account. This function depends on the ratio of the moth's airspeed to windspeed (long dashes 2.0; dots 1.4; short dashes 1.1). In the depicted sample this ratio was 1.1 for both the theoretical and experimental cases.

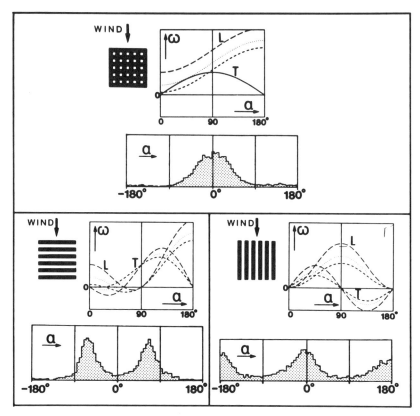

FIGURE 2. Transversal (T) and longitudinal component (L) of the apparent ground pattern motion plotted as a function of course angle (α) at different patterns offered to the moth: complete ground pattern (top), only stripes transverse to the wind (lower left), only stripes in line with the wind (lower right). Density distributions of course angles (α) as measured during 10-minute flights of male gypsy moths are depicted below the corresponding slope functions.

REFERENCES

1. KENNEDY, J. S. 1983. Zigzagging and casting as a programmed response to wind-borne odour: A review. Physiol. Entomol. **8:** 109-120.
2. KUENEN, L. P. S. & T. C. BAKER. 1983. A non-anemotactic mechanism used in pheromone source location by flying moths. Physiol. Entomol. **8:** 277-289.
3. PREISS, R. & E. KRAMER. 1986. Pheromone-induced anemotaxis in simulated free flight. *In* Mechanisms in Insect Olfaction. T. L. Payne, M. C. Birch & C. E. J. Kennedy, Eds.: 69-79. Clarendon Press. Oxford, England.
4. PREISS, R. & E. KRAMER. 1986. Mechanism of pheromone orientation in flying moths. Naturwissenschaften **7:** 387-389.

Putative Glutamergic and Aspartergic Cells in the Olfactory Bulb of the Rat

J. L. PRICE AND T. A. FULLER

Departments of Anatomy/Neurobiology and Psychiatry
Washington University School of Medicine
St. Louis, Missouri 63110

Glutamate and aspartate are widely used as excitatory neurotransmitters, especially in the forebrain. There have been several suggestions that one or both of these amino acids are used by the efferent fibers from the olfactory bulb in the lateral olfactory tract, although this has been questioned.[1]

To investigate whether any cells in the olfactory bulb can be identified as glutamergic and/or aspartergic, several experiments have been done using the "neurotransmitter-specific" retrograde axonal tracer [³H]D-aspartate ([³H]D-Asp). This tracer is taken up and retrogradely transported only by cells that possess a high-affinity uptake mechanism for glutamate and/or aspartate, and therefore may be putatively considered to be glutamergic and/or aspartergic.[2] Small injections of [³H]D-Asp (5-20 μCi in 25-175 nl), or of the "nonspecific" tracer WGA-HRP, were made into the layer I of the anterior piriform cortex in rats, near the lateral edge of the lateral olfactory tract. After a survival period of one day, the rats were killed by perfusion with 5% glutaraldehyde in 0.1 phosphate buffer, and their brains were prepared for autoradiography.

The injections of [³H]D-Asp labeled only a few cells in the mitral and external plexiform layers of the main olfactory bulb. In contrast, a large number of mitral cells are labeled in the accessory olfactory bulb and the anterior olfactory nucleus (FIG. 1). Futhermore, the labeled cells in the mitral cell layer of the main olfactory bulb are substantially smaller than other mitral cells (FIG. 2). Comparable injections of WGA-HRP label most cells in the mitral cell layer and many tufted cells in the external plexiform layer of the olfactory bulb, as well many cells in the accessory olfactory bulb and anterior olfactory nucleus. These experiments suggest that the majority of mitral and tufted cells do not use glutamate or aspartate as neurotransmitters, although a subpopulation of small mitral and tufted-like cells in the main olfactory bulb, and most of the mitral cells of the accessory olfactory bulb, may be glutamergic or aspartergic.

Injections of [³H]D-Asp directly into the main olfactory bulb fail to label mitral or tufted cells, but a few cells in the periglomerular region and superficial external plexiform layer take up the tracer. In at least one experiment, labeled cells were also found in the olfactory nerve layer, superficial to the periglomerular layer. These experiments indicate that in addition to the relay cells, there may be one or more groups of glutamergic or aspartergic short-axon cells within the olfactory bulb.

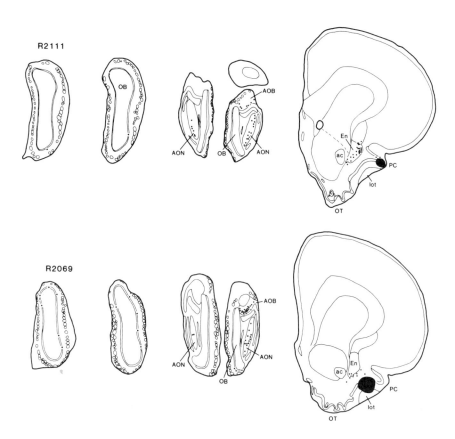

FIGURE 1. The distribution of retrogradely labeled cells in the main and accessory olfactory bulbs following injections [³H]D-Asp into the anterior piriform cortex. Each dot represents one cell.

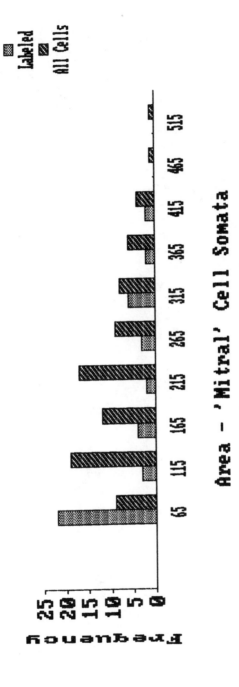

FIGURE 2. Histograms comparing the size of cells in the mitral cell layer of the olfactory bulb that were labeled with [^{3}H]D-Asp following an injection into the anterior piriform cortex ($n = 44$, mean $= 161$) with the size of all other mitral cells ($n = 86$, mean $= 211$). The difference between the two populations is significant ($p < 0.02$).

REFERENCES

1. FFRENCH-MULLEN, J. M. H., K. KOLLER, R. ZACZEK, J. T. COYLE & D. O. CARPENTER. 1982. *N*-Acetylaspartylglutamate: Possible role as the neurotransmitter of the lateral olfactory tract. Proc. Natl. Acad. Sci. USA **82:** 3897-3900.
2. CUENOD, M. & P. STREIT. 1983. Neuronal tracing using retrograde migration of labeled transmitter-related compounds. *In* Methods in Neuroanatomy. A. Bjorkjlund & T. Hokfelt, Eds.: 365-397. Elsevier. Amsterdam, the Netherlands.

Benzaldehyde Binding Protein from Dog Olfactory Epithelium[a]

STEVEN PRICE AND AMY WILLEY

Department of Physiology and Biophysics
Medical College of Virginia
Virginia Commonwealth University
Richmond, Virginia 23298

A protein with a high affinity for *o*-methyl phenols, of which the odorant anisole is the simplest compound, has been extracted from dog olfactory epithelium.[1] Called anisole binding protein, it was hypothesized to be a receptor for odorants of the anisole type. Further studies on it and isolation of a benzaldehyde binding protein are reported here.

"Anisole" and "benzaldehyde" columns were prepared by covalently coupling *p*-anisic acid and *p*-carboxybenzaldehyde, respectively, to ω-aminobutyl agarose via the carboxyl groups. Olfactory epithelium from dog nasal septum was extracted with 0.1 *M* Tris buffer, pH 7.2, and the residue was reextracted with the same buffer to which 0.1% sodium dodecyl sulfate had been added. The resulting extract was subjected to affinity chromatography on the two columns consecutively, and the anisole and benzaldehyde binding proteins were displaced from the columns by addition of 1 m*M* *p*-anisic acid or *p*-carboxybenzaldehyde, respectively. The eluates from the columns were subjected to polyacrylamide gel slab electrophoresis.[2] Protein on the slabs was visualized by silver staining.[3] Rabbits were immunized with anisole or benzaldehyde binding protein, and sera from them were tested for their effects on electroolfactograms (EOGs) of sagitally bissected heads of male ICR mice.[4] Except for the method of EOG recording and the use of slab rather than disc electrophoresis, the methods used in this study were essentially the same as those described previously.[1]

The electrophoretic patterns obtained from elution of the columns are shown in FIGURE 1. Nearly all the protein passes through the columns freely, eluting with Tris buffer (tracks B and E). After the buffer-elutable protein has cleared the column (tracks C and F), *p*-anisic acid and *p*-carboxybenzaldehyde displace additional protein from the anisole and benzaldehyde columns, respectively (tracks D and G). These are the anisole and benzaldehyde binding proteins. From their mobilities relative to those of the standards, their molecular masses are estimated to be about 61,000 daltons.

Effects of antisera and control sera (sera from the same rabbits, taken before immunization) on EOG responses to anisole, benzaldehyde and *iso*amyl acetate are shown in TABLE 1. The control sera have no significant effects. The serum from the rabbit immunized with anisole binding protein abolished the response to anisole and reduced the responses to the other odorants to about one-third. Serum from the rabbit immunized with benzaldehyde binding protein reduced responses to all three odorants to about one-third.

[a] Supported by the United States Army Research Office.

561

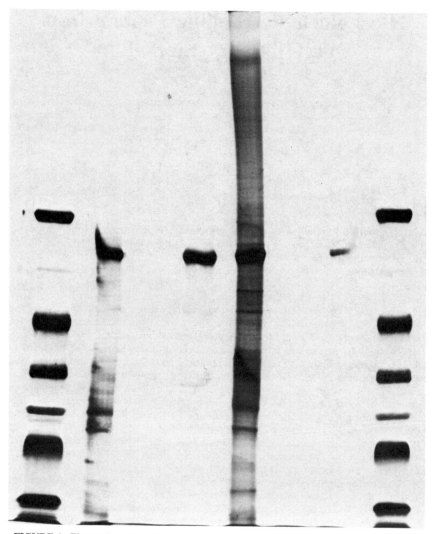

FIGURE 1. Electrophoresis of effluents from "anisole" and "benzaldehyde" columns on 10%
polyacrylamide gels, silver stained. The first and last tracks are protein standards of molecular
masses 66,000, 45,000, 36,000, 29,000, 24,000, 20,100, and 14,200 daltons (top to bottom). The
second through fourth tracks are eluates from the anisole column. The second is pooled material
eluted with Tris buffer (0.1 M, pH 7.2) in which there was significant absorbance at 280 nm.
The third is from the eluate after the absorbance had returned to baseline. The fourth is the
material displaced from the column with buffer including 1 mM p-anisic acid. The fifth through
seventh tracks are eluates from the benzaldehyde column after loading it with material that had
passed through the anisole column. The conditions for elution of the fifth through seventh tracks
are identical to those for elution of the second through fourth, respectively, except that p-
carboxybenzaldehyde is substituted for p-anisic acid. All eluates were concentrated by ultrafil-
tration before electrophoresis. The third, fourth, sixth, and seventh were concentrated about 50
times as much as the second and fifth tracks, and this should be borne in mind in comparing
the intensity of the bands.

Like anisole binding protein, benzaldehyde binding protein is present in olfactory, but not adjacent respiratory, epithelium. Each has a high affinity for a compound related to an odorant type. Each is extracted with a detergent, suggesting origin in a membranous component of the cells. Each induces the formation of antibodies that are inhibitors of responses to all odorants, although the antibodies induced by anisole binding protein in this rabbit, unlike those reported earlier,[1] are more effective against responses to anisole than to other odorants. The two proteins are not identical; although they have the same molecular weights, each can be completely removed from the extracts before isolating the other.

It has been hypothesized that the olfactory receptor proteins comprise a class with similar structures.[1,5] This is in accord with the existence of much protein in olfactory

TABLE 1. Effects of Sera on EOG Responses to Odorants[a]

| | Odorant (1/100 saturation in air) | | |
Sera	Anisole	Benzaldehyde	isoAmyl Acetate
Sera from rabbit immunized with anisole binding protein			
a. Control serum	$-6.8 \pm 7.9\%$ (35)	$5.8 \pm 9.9\%$ (10)	$5.4 \pm 14.5\%$ (10)
b. Immune serum	$-100.00 \pm 0\%$ (9)	$-61.6 \pm 7.5\%$ (10)	$-70.6 \pm 3.7\%$ (10)
Sera from rabbit immunized with benzaldehyde binding protein			
a. Control serum	$5.1 \pm 27.6\%$ (7)	$9.4 \pm 16.6\%$ (8)	$-1.4 \pm 26.9\%$ (5)
b. Immune serum	$-66.3 \pm 10.8\%$ (8)	$-56.4 \pm 9.7\%$ (9)	$-61.4 \pm 9.3\%$ (9)

[a] Figures are changes in EOG amplitude (mean $\pm$ SEM). Numbers of olfactory tissue samples are shown in parentheses. None of the results with control sera differ significantly from zero. All reductions with immune sera are significant ($p > 0.05$). The reduction in response to anisole with the serum against anisole binding protein is significantly greater than any other.

epithelium that cross-reacts with antisera against anisole binding protein,[5] and with the generalized inhibition of responses to odorants resulting from topical application of antisera against anisole or benzaldehyde binding proteins. The generalized inhibition probably results from antibodies directed against the region of the protein common to all, and the specific inhibition (seen with anisole binding protein antiserum, TABLE 1) probably represents actions of antibodies directed against the "active site."

ACKNOWLEDGMENTS

We thank Drs. Stephen J. Goldberg and Krishna Persaud for their generous advice and assistance.

REFERENCES

1. GOLDBERG, S. J., J. TURPIN & S. PRICE. 1979. Chem. Senses **4:** 207-214.
2. WEBER, K., J. R. PRINGLE & M. OSBORN. 1972. *In* Methods in Enzymology. Vol. XXVI. Enzyme Structure, Part C. C. H. W. Hirs & S. Timasheff, Eds.: 3-27. Academic Press. New York.
3. MERRIL, C. R., R. C. SWITZER & M. L. VAN KEUREN. 1980. Anal. Biochem. **105:** 361-368.
4. SHIRLEY, S., E. POLAK & G. H. DODD. 1983. Eur. J. Biochem. **132:** 485-494.
5. PRICE, S. & J. TURPIN. 1980. *In* Olfaction and Taste. VII. M. G. J. Beets, Ed.: 65-68. IRL Press. London.

Neural Coding of Gustatory Information in the Thalamus of an Awake Primate[a]

THOMAS C. PRITCHARD, ROBERT B. HAMILTON,
AND RALPH NORGREN

Department of Behavioral Science
The Milton S. Hershey Medical Center
Hershey, Pennsylvania 17033

The anesthetized rodent preparation, which has served gustatory neurophysiology so well at the peripheral, medullary, and pontine levels, has not been an appropriate model to investigate the role of taste neurons in the forebrain. General anesthetics, especially barbiturates and halogenated compounds, produce severe response depression and obstinate spindling of forebrain neurons. The use of an awake, behaving preparation not only circumvents problems associated with general anesthetics, but also allows observation of gustatory activity associated with ingestive behavior.

One week before data collection, a chronic recording chamber, indwelling indifferent electrodes, restraint bolts, and ground wires were surgically attached to the cranium of a 4-kg female rhesus monkey under general anesthesia. The monkey sat in a primate chair during data collection with its head secured but otherwise free to move its arms and legs. The activity of 85 single neurons from the ventral posterior thalamus was recorded with glass-coated tungsten microelectrodes, using standard transdural recording techniques. Gustatory neurons ($n = 26$) were tested with 0.3 ml of each of the following stimuli: 1.0 M sucrose, 0.1 M NaCl, 0.003 M HCl, 0.001 M QHCl, and distilled water. Additional concentrations of these stimuli, as well as thermal and tactile stimulation, were included when possible. Two additional neurons responded during consumption of a commercial blend of fruit juices. A response was considered significant if the activity during the three-second period following stimulus onset exceeded the mean distilled water response by 1.96 standard deviations. Water responses were compared in the same way to random samples of spontaneous activity.

RESULTS AND DISCUSSION

The spontaneous activity of gustatory neurons in the primate thalamus ranged from 1.6 to 32.6 spikes/sec ($\overline{X} = 10.5$). Sucrose-best neurons comprised 16 of the 28 gustatory neurons in the sample and, overall, produced the most vigorous responses

[a]This work was supported by Public Health Service Grant NS20518.

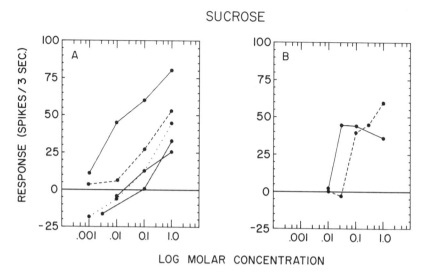

FIGURE 1. Response magnitude of seven gustatory neurons recorded from the thalamic taste area, plotted as a function of log sucrose concentration. The seven intensity response functions were plotted on two graphs only to make their presentation clearer. Six of the seven neurons responded to a sucrose concentration series with a monotonic response function. Two neurons (Panel A) showed inhibitory responses when the monkey was presented with low concentrations of sucrose, but for higher concentrations, these same cells increased their activity.

of the four standard taste solutions (15.8 spikes/sec.). Six of seven neurons tested with a sucrose concentration series showed a monotonic response function that included several inhibitory responses to concentrations below 0.01 M (see FIG. 1). The incidence of inhibitory responses in the entire sample, however, was low. Only 4% of the responses to the four basic stimuli at the probe concentrations were inhibitory. Although nine of the 28 gustatory neurons responded during tactile stimulation of the oral cavity, only four of the 37 neurons that responded exclusively to tactile stimulation were located in the thalamic taste area. Interestingly, the neurons that received convergent gustatory and tactile input often had nonoverlapping receptive fields. Water was an effective stimulus for nine (32%) of the gustatory neurons. Other nongustatory neurons ($n = 18$) located in the thalamic taste area fired in bursts spontaneously, but became quiescent just before or during fluid stimulation. Two other neurons showed excitatory responses under the same conditions. These "nongustatory" neurons could not be driven with either thermal or tactile stimulation and did not appear to be related to motor activity.

The high incidence of sucrose-sensitive neurons in the thalamus, as shown in these data, is consistent with previous reports that describe excellent sucrose responsivity in the primate chorda tympani nerve.[1-4] The monkey differs in this regard from the rat, whose chorda tympani nerve is more responsive to electrolytes than sugars.[5] The spontaneous rates and evoked responses observed in this experiment were more robust than those reported previously for either barbiturate-anesthetized[6] or paralyzed[7] rats.

REFERENCES

1. GORDON, G., R. KITCHELL, L. STROM & Y. ZOTTERMAN. 1959. The response pattern of taste fibers in the chorda tympani of the monkey. Acta Physiol. Scand. **46:** 119-132.
2. BROUWER, J. N., G. HELLEKANT, Y. KASAHARA, H. VAN DER WEL & Y. ZOTTERMAN. 1973. Electrophysiological study of the gustatory effects of the sweet proteins monellin and thaumatin in monkey, guinea pig and rat. Acta Physiol. Scand. **89:** 550-557.
3. PFAFFMANN, C., M. FRANK, L. M. BARTOSHUK & T. C. SNELL. 1976. Coding gustatory information in the squirrel monkey chorda tympani. *In* Progress in Psychobiology and Physiological Psychology. J. Sprague & A. Epstein, Eds. Vol. **6:** 1-27. Academic Press. New York.
4. SATO, M., H. OGAWA & S. YAMASHITA. 1975. Response properties of macaque monkey chorda tympani fibers. J. Gen. Physiol. **66**(6): 781-810.
5. PFAFFMAN, C. 1955. Gustatory nerve impulses in rat, cat, and rabbit. J. Neurophysiol. **18:** 429-440.
6. NOMURA, T. & H. OGAWA. 1985. The taste and mechanical response properties of neurons in the parvicellular part of the thalamic posteromedial ventral nucleus of the rat. Neurosci. Res. **3:** 91-105.
7. SCOTT, T. R. & M. S. YALOWITZ. 1978. Thalamic taste responses to changing stimulus concentration. Chem. Senses **3**(2): 167-175.

Rodenticide Flavor Profiles Identified through Generalization of Conditioned Flavor Avoidance[a]

RUSSELL F. REIDINGER,[b,c] J. RUSSELL MASON,[b]
AND CHARLES N. STEWARD [d]

[b]Monell Chemical Senses Center and
Biology Department
University of Pennsylvania
Philadelphia, Pennsylvania 19104

[c]United States Department of Agriculture
Animal and Plant Inspection Service
Animal Damage Control
Denver Wildlife Research Center
Denver Federal Center
Denver, Colorado 80225-0266

[d]Psychology Department
Franklin and Marshall College
Lancaster, Pennsylvania 17604

In these experiments, we investigated whether generalization of conditioned flavor avoidance (CFA) could be used to profile the components of flavors that we believed to be complex. Our studies also were designed to provide data of practical importance, in that five rodenticides (α-chlorohydrin [1.6 mg/ml], α-napthylthiourea [ANTU: 150mg/ml], calciferol [100 mg/ml], strychnine [100 mg/ml], Na warfarin [27mg/ml]) were used as conditioned stimuli.

With one exception, experimental groups were presented with a rodenticide conditioned stimulus (CS) in aqueous solution on the day of treatment. For strychnine, experimental groups were presented with the CS either in aqueous solution or in agar. In all cases, ingestion of the CS was followed by an intraperitoneal injection of LiCl. Control groups were given water (or plain agar) followed by LiCl injections. Generalization of CFA to four nontoxic flavors was then assessed. Additional conditioning and generalization trials followed until 24 flavors had been presented. CFA was exhibited toward all rodenticides, and avoidance generalized in every case to a subset of the 24 flavors. Generalization of CFA to aqueous strychnine was exhibited toward "bitter" flavors (0.2 M, 0.04 M Na saccharin, 0.41 M Na_2SO_4, 0.1 M $(NH_4)_2CO_3$, 0.1 M $MgSO_4$, 0.1 M L-phenylalanine, 3.0 M urea, 0.001 M SOA, $ps < 0.05$). Generalization of CFA to strychnine in agar was similar, although relatively fewer test flavors were avoided by the experimental groups (i.e., 0.2 M and 0.04 M Na

[a]Supported by Army Contract No. DAAG-29-82-K-0192.

saccharin, 0.0001 M QHCl, and 0.001 M SOA; ps < 0.05). Warfarin CFA was relatively weak, but generalization was exhibited toward "bitter," "sweet," and "salty" (0.1 M MgSO$_4$, 3.0 M urea, 0.1 M sucrose, 0.1 M KNO$_3$, 0.15 M L-phenylalanine, 0.001 M SOA, 0.1 M NaCl, ps < 0.05). Calciferol CFA generalized to "sweet" and "bitter" (0.1 M sucrose, 0.2% quassia, ps < 0.05), ANTU CFA primarily to "sour" (0.3 M NH$_4$Cl, 0.01 M acetic acid, 0.003 M citric acid, 0.1 M HCl, ps < 0.05), and α-chlorohydrin CFA to "sour" and "bitter" (0.3 M H$_4$Cl, 0.01 M acetic acid, 0.2% quassia, 0.2% gentain, 0.15 M L-phenylalanine, ps < 0.05).

These results demonstrate that rats can recognize the components of a complex flavor. However, we cannot as yet identify the specific sensory systems mediating either learning or response generalization. We also cannot discount the possibility that avoidance-mediated neophobia influenced consumption as well as flavor avoidance learning per se. Taste, texture, pH, and/or odor may have been involved (hence our use of the term flavor). Regardless, our results support the notion that flavor avoidance learning could be useful in the empirical development of rodenticide baits and prebait formulations. Moreover, the similarity of strychnine flavor profiles in drinking and feeding contexts suggests that the results of these flavor profiling experiments may have broad validity. An important cautionary note, however, is that differences in strychnine flavor profiling in feeding and drinking were sufficiently great to suggest that aqueous solutions of the rodenticides and flavors used in the experiments may not wholly represent the flavors of rodenticides incorporated into solid baits. Further experiments are required in which solid stimuli or commercial bait formulations are used.

Specialized Receptor Villi and Basal Cells within the Taste Bud of the European Silurid Fish, *Silurus glanis* (Teleostei)

KLAUS REUTTER

Anatomical Institute
University of Tuebingen
Oesterbergstrasse 3, 74 Tuebingen
Federal Republic of Germany

The ultrastructure of a teleost's taste bud (TB) is well known (e.g., Kapoor *et al.,*[1] Reutter[2]), and most of the work was done on American silurids belonging to the genus *Ameiurus* (= *Ictalurus*) (review: Reutter[3]). But up to now there is no information about the ultrastructure of the TB of the European silurid, *Silurus glanis*. The reason may be that this relatively rare and big fish (up to 2.5 m in length) is not commonly kept in the laboratory. We have now got two 60-cm animals, and we investigated the TBs of the barbels light and electron microscopically. Among others we found two striking cytological peculiarities that concern the light sensory cell's large receptor villus and the basal cells of a TB and which are dealt with here. These details have been occasionally mentioned in the literature but not discussed extensively.

RESULTS AND COMMENTS

Taste Bud Structure

As in other silurids, a *Silurus* TB is composed of light and dark sensory cells, which extend apically to the surface of the epidermis and where their receptor villi form the receptor field, the basally situated basal cells, and the TB nerve fiber plexus, which lies between the divided bases of the sensory cells and the basal cells.

FIGURE 1. Taste bud of *Silurus glanis,* ultrastructural details. (**a-c**) Receptor field region: ▶ Large receptor villi belonging to light sensory cells and small receptor villi of dark ones. Preparations: (**a**) glutaraldehyde (GA)/osmium tetroxide (OT)/lead citrate-uranyl acetate. (**b**) GA/OT/ruthenium red, uncontrasted. (**c**) GA/OT/lanthanum chloride, uncontrasted. (**d-e**) Region between basal cell (BC) and nerve fiber (NF) plexus of the taste bud. Spikelike processes of BCs are marked by arrowheads, synaptic regions by open arrows. Basal processes of dark sensory cells from desmosomes (D) to BCs. Preparations: GA/OT/lead citrate-uranyl acetate. Bars: In each case 1 μm.

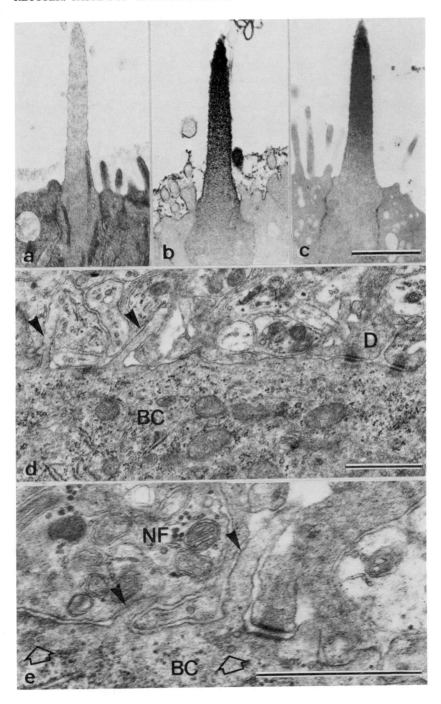

Large Receptor Villi of Light Sensory Cells

These villi contain longitudinally arranged tubular profiles of about 50 nm in diameter and seem to be directly fused to the plasmalemma of the villus (FIG. 1a). The content of the tubules is somewhat electron dense, but ruthenium red (marker of polyanions, such as occurring in mucous substances) and lanthanum chloride (marker of the extracellular space) do not mark the tubules' content additionally (FIG. 1b,c). This finding agrees with data of Crisp et al.[4] In view of the ruthenium red evidence, it is unlikely that the tubules, which seem to be part of the endoplasmic reticulum, will be correlated to mucus secretion. In contrast, both substances completely invade the body of a large receptor villus down to the apical cytoplasm of its light sensory cell (FIG. 1b,c). This result is in certain contradiction to the results of Fujimoto and Yamamoto,[5] who found ruthenium red as well as lanthanum nitrate within the tubules. It is still unclear how the tracers penetrate a villus, either via the (open?) tubular structures, or via the (easily penetrated?) villus's membrane, or both. Nevertheless, this phenomenon may suppose that membranes of large receptor villi belonging to light sensory cells are penetrated by chemicals, probably also by taste substances, and therefore they may be the favored chemoreceptive structures of a TB.

Basal Cells

It is of special interest that these disc-like cells, transversally orientated to the axis of the TB, possess spike-like processes that penetrate between the structures of the nerve fiber plexus. In the literature, these spikes are occasionally depicted and especially mentioned by Desgranges,[6] Ezeasor,[7] and Jakubowski.[8] In Silurus, one finds in close proximity to these spikes the sensory cell processes that form desmosomes with the basal cell and near to this the basal cell regularly synapses to a nerve fiber of the TB plexus. This structural arrangement resembles a Merkel cell-axon complex (Hartschuh et al.[9]) and leads, together with the data of Jakubowski[8] and former cytochemical tests and developmental investigations,[10,11] to the conclusion that the basal cell of the fish's TB is similar in structure and possibly in function to a Merkel cell. Our hypothesis is that the basal cells in a teleost's TB may not only have coordinative functions with respect to chemoreception and neurotransmission, but may also be a mechanoreceptor.

REFERENCES

1. KAPOOR, B. G., H. E. EVANS & R. A. PEVZNER. 1975. Adv. Mar. Biol. 13: 53-108.
2. REUTTER, K. 1986. Chemoreceptors. In Biology of the Integument. J. Bereiter-Hahn, A. G. Matoltsy & K. S. Richards, Eds. Vol. 2: 586-604. Springer. Berlin, Heidelberg, New York, Tokyo.
3. REUTTER, K. 1982. Taste organ in the barbel of the bullhead. In Chemoreception in Fishes. T. J. Hara, Ed.: 77-91. Elsevier. Amsterdam, the Netherlands.
4. CRISP, M., G. A. LOWE & M. S. LAVERACK. 1975. Tissue Cell 7: 191-202.
5. FUJIMOTO, S. & K. YAMAMOTO. 1980. Anat. Rec. 197: 133-141.
6. DESGRANGES, J. C. 1972. C. R. Acad. Sci. Ser. D (Paris) 274: 1814-1817.

7. EZEASOR, D. N. 1982. J. Fish Biol. **20:** 53-68.
8. JAKUBOWSKI, M. 1983. Z. Mikrosk.-Anat. Forsch. **97:** 849-862.
9. HARTSCHUH, W., E. WEIHE & M. REINECKE. 1986. The Merkel cell. *In* Biology of the Integument. J. Bereiter-Hahn, A. G. Matoltsy & K. S. Richards, Eds. Vol. **2:** 605-620. Springer. Berlin, Heidelberg, New York, Tokyo.
10. REUTTER, K. 1971. Z. Zellforsch. **120:** 280-308.
11. REUTTER, K. 1978. Adv. Anat. Embryol. Cell Biol. **55:** 1-98.

Hyperinnervation Produces Inhibitory Interactions between Two Taste Nerves[a]

D. R. RIDDLE, B. OAKLEY, S. E. HUGHES, C. L.
deSIBOUR, AND C. R. BELCZYNSKI

Department of Biology
University of Michigan
Ann Arbor, Michigan 48109

Several researchers have presented evidence that suggests that the processing of gustatory information at the level of the taste bud can involve interactions among taste fibers, for example, Bernard[1] and Miller.[2] We report here that two chorda tympani nerves in the same peripheral field may interact to inhibit gustatory impulse traffic to the brain.

METHODS

The proximal portion of the right chorda-lingual nerve was sutured to the distal stump of the left lingual nerve in 44 mongolian gerbils. Nine to fifteen months after the initial surgery, a second operation was performed on six gerbils. The native chorda tympani was removed in four gerbils and the foreign chorda tympani was removed in two. In 10 control animals the left lingual nerve and the right chorda-lingual nerve were transected. The latter was reconnected with an 11-0 suture to allow reinnervation of the right side of the tongue. Five to twenty months after the initial surgery, the chorda tympani nerves were exposed in the left and right middle ears and impulse activity simultaneously recorded from the two nerves while stimulating the tongue with a variety of taste solutions. It was also possible to electrically stimulate one chorda tympani while simultaneously recording from the other. Each animal was sacrificed at the end of the recording session. The tongue was removed and processed for light microscopy. Fungiform taste buds were counted and their positions on the tongue noted. A subset of taste buds was analyzed morphometrically. Counts of taste cells were made from plastic sections of fungiform taste buds from three experimental and four normal animals.

[a] Supported in part by United States Public Health Service Grant NS-07072.

RESULTS AND DISCUSSION

Punctate taste stimulation of the tip of the left side of the tongue in experimental animals elicited responses in both chorda tympani nerves. Stimulation of the right side was effective only after the taste solutions spread. Both nerves must have innervated the left side of the tongue. Histological analysis showed a near absence of fungiform taste buds on the right side of these tongues, with the exception of an occasional taste bud near the tip or along the midline. The eventual removal of the native or foreign chorda tympani showed that the native nerve alone supported all of the fungiform taste buds on the left side of the tongue, and that the foreign nerve would maintain about half of those taste buds. It appears then, that at least half of the fungiform taste buds on the left side of the tongue were dually innervated. Both the native and foreign chorda tympani nerves responded vigorously to taste stimulation and the concentration response functions were similar. However, simultaneous recording from the nerves revealed a number of response characteristics indicative of inhibition that were most profound in the foreign nerve. In 75% of the animals, responses of the foreign nerve to a variety of taste stimuli were both slower to rise and slower to return to baseline compared to responses in the native nerve. A 5-10-fold increase in the time required to reach 90% of peak height was not uncommon. Inhibition was observed in approximately three-fourths of the animals and was always much more pronounced in the foreign nerve. The vigor of inhibitory interactions varied unsystematically among animals and the taste solutions used as stimuli. In some instances a brief transient response was followed by a slow, delayed increase in activity. In other instances a transient, but unsustained, response was present. In the most dramatic cases, taste stimulation failed to elicit even a transient response and actually depressed spontaneous activity. With more concentrated taste solutions, this initial depression of spontaneous activity was slowly overcome and a neural discharge emerged. We observed none of these inhibitory interactions in control or normal animals.

Typically, in the same experimental animals whose foreign chorda tympani nerve displayed the inhibitory responses described above, we found that electrical stimulation of the native nerve depressed subsequent taste responses in the foreign nerve. Responses to both sucrose and NaCl were often suppressed by at least 35%. The effect was usually maintained for several minutes and, like the inhibition described above, was still observed after the native nerve was disconnected from the brain. It seems unlikely that the inhibition resulted from general autonomic effects, such as a change in blood flow through the papillae, since no change was seen in the response of the native nerve following electrical stimulation of either the native or foreign nerve. Since both nerves innervated many of the same papillae, one would expect them to respond similarly to autonomic influences. We observed no inhibitory effects of electrical stimulation in normal or control animals.

The anatomical substrate underlying these inhibitory interactions remains to be determined. Nor is it clear that inhibition by electrical stimulation operates through the same mechanisms as the inhibition observed during taste stimulation. Light microscopy revealed no anatomical changes that might account for the unusual physiological effects. Hyperinnervation failed to increase the number of taste buds, the size of taste buds, or the number of cells within taste buds. This is somewhat surprising given the clear trophic influence of taste neurons upon taste buds, and suggests that it is the character of the gustatory epithelium, not the density of innervation, that controls the number and size of fungiform taste buds. Electron microscopic studies now being conducted (Kinnamon *et al.,* unpublished) may reveal ultrastructural changes that could be responsible for the pronounced inhibitory interactions observed.

REFERENCES

1. BERNARD, R. A. 1972. Antidromic inhibition: A new model of taste receptor function. *In* Olfaction and Taste IV. D. Schneider, Ed.: 301-307. Wissenschaftliche Verlagsgesellschaft MBH. Stuttgart, Germany.
2. MILLER, I. J. 1975. Mechanisms of lateral interactions in rat fungiform taste receptors. *In* Olfaction and Taste V. D. A. Denton & J. P. Coghlan, Eds.: 217-221. Academic Press, Inc. New York.

Interactions between Taste Cells and Nerve Fibers in Murine Foliate Taste Buds[a]

SUZANNE M. ROYER AND JOHN C. KINNAMON

Laboratory for High Voltage
Electron Microscopy
Department of Molecular, Cellular
and Developmental Biology
University of Colorado
Boulder, Colorado 80309

In the mouse there are four or five rows of foliate papillae on the lateral surface of the tongue opposite the molar teeth. Clusters of taste buds, innervated by branches of the glossopharyngeal nerve and the facial nerve via the chorda tympani, are found in the epithelium on the sides of these papillae. Using both conventional transmission electron microscopy (CTEM) and high-voltage electron microscopy (HVEM), we have examined the morphology of murine foliate taste buds and have studied the interactions between taste cells and nerve fibers within these taste buds.

MATERIALS AND METHODS

Anesthetized adult white mice were fixed by perfusion through the left ventricle with fixative containing 2.5% glutaraldehyde and 1% paraformaldehyde in 0.05 M cacodylate buffer at pH 7.3. Small pieces of lingual tissue containing the foliate papillae were excised and processed according to standard methods. Ultrathin (60-90 nm) sections were examined using a JEOL JEM-100 CX electron microscope at 100 kV. Semithin (0.12-0.25 μm) sections were examined with the JEOL JEM-1000 high-voltage electron microscope at 750 kV.

RESULTS AND DISCUSSION

The taste buds of the mouse foliate papilla (FIG. 1A) are generally similar to those of the mouse vallate papilla.[1] We have observed three types of cytoplasmic speciali-

[a]This project was supported by Grants NS21688 and RR00592 from the NIH and a grant from the Procter & Gamble Co.

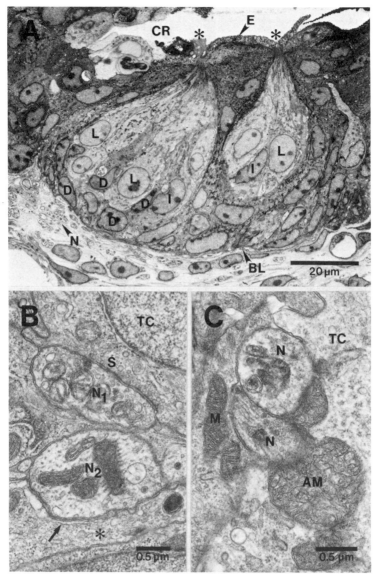

FIGURE 1. (A) Low-power HVEM of two foliate taste buds. The taste buds extend throughout the depth of the epithelium, from the basal lamina (BL) to the epithelial surface (E). In this plane of section, it is possible to see the taste pores (*) of both buds opening into the crypt (CR) between the papillae. Dark (D), light (L), and intermediate (I) cells are shown within the taste buds. Bundles of nerve fibers (N) are present in the connective tissue beneath the basal lamina. (B) CTEM of a synapse (S) from a taste cell (TC) onto a nerve fiber (N₁). There is a subsurface cisterna (arrow) in a nearby taste cell process (*) adjacent to its apposition with a second nerve fiber (N₂). (C) An atypical mitochondrion (AM) at the site of apposition between a taste cell (TC) and two small nerve fibers (N). The latter are partially enveloped by the large mitochondrion. Compare the atypical mitochondrion with the typical mitochondrion (M) in an adjacent taste cell process.

zations of foliate taste cells at points of close apposition between taste cells and nerve fibers: (1) afferent synapses from taste cells onto nerve fibers, (2) subsurface cisternae, and (3) atypical mitochondria. Subsurface cisternae and atypical mitochondria are sometimes closely associated with synapses or with each other.

As in murine vallate taste buds, both macular and fingerlike synapses are present in foliate taste buds[1] (FIG. 1B). Narrow subsurface cisternae, previously described in mouse vallate taste buds,[2] sometimes occur adjacent to the cytoplasmic leaflet of the taste cell membrane at sites of contact with nerve processes (FIG. 1B). The outer surfaces of these cisternae closely parallel the taste cell membrane at a distance of approximately 15 nm.

Atypical mitochondria, which are usually much larger than other taste cell mitochondria and which have expanded vesicular or tubular cristae (FIG. 1C), often occur in regions of close apposition between taste cells and nerve fibers. Such mitochondria are located immediately inside the taste cell membrane and are separated from it by a 20-nm space that often contains a discontinuous central line of electron-dense material. We have also observed such atypical mitochondria within the taste cells of both mouse vallate and fungiform papillae.

Presently, the functions of subsurface cisternae and atypical mitochondria in taste cells are unknown, but their occurrence in areas of contact between taste cells and nerve fibers and near synapses suggests that they might be involved in the formation of synaptic connections or in trophic interactions between taste cells and nerve fibers. Mitochondria resembling these atypical mitochondria have been observed in other tissues having elevated rates of respiration and in tumor cells,[3] suggesting the possibility of energy-requiring biochemical processes at sites of taste cell-nerve fiber apposition.

REFERENCES

1. KINNAMON, J. C., B. J. TAYLOR, R. J. DELAY & S. D. ROPER. 1985. Ultrastructure of mouse vallate taste buds. I. Taste cells and their associated synapses. J. Comp. Neurol. **235:** 48-60.
2. TAKEDA, M. 1976. An electron microscopic study on the innervation in the taste buds of the mouse circumvallate papillae. Arch. Histol. Jpn. **39:** 257-269.
3. ERLANDSON, R. A., B. TANDLER, P. H. LEIBERMAN & N. L. HIGINBOTHAM. 1968. Ultrastructure of human chordoma. Cancer Res. **28:** 2115-2125.

Postnatal Development of Cholinergic Enzymes and Acetylcholine Content in the Rat Olfactory Bulb

R. SAFAEI, R. MOUSSAVI, AND E. MEISAMI [a]

Department of Physiology-Anatomy
University of California
Berkeley, California 94720

Although newborn rats are capable of smell, during the postnatal period the olfactory bulb (OB) undergoes profound growth and development. The neurochemical aspects of OB development are still poorly understood.[1] Here we report on the postnatal changes in acetylcholine (ACh) content and the activity of acetylcholinesterase (AChE) and cholineacetyl transferase (ChAT), neuronal enzymes involved in ACh hydrolysis and synthesis; changes in pseudocholinesterase (BuChE), a nonneuronal enzyme with possible functions in the olfactory system, were also determined.[2]

Postnatal changes in ChAT activity, weight, ACh, and protein content of OB are shown in TABLE 1; the changes in specific and total (per OB) activities of AChE and BuChE are depicted in FIGURE 1. The data indicate that ChAT activity is present at birth but at a very low level (< 10% of the adult). The growth spurt in activity occurs only after day 10, reaching near-adult level by day 30. The increase in ACh content follows a time course similar to that of ChAT. The specific activity of AChE at birth is about 20% of the adult level while that of BuChE is already at its highest level. Thus the results indicate an earlier schedule of maturation for the hydrolytic enzymes, compared to ChAT.

Postnatally BuChE specific activity markedly *decreases* to reach adult level by day 15. AChE activity, however, increases markedly (growth spurt) but mainly after day 10, reaching adult levels by day 25. However, in terms of total activity (per whole OB), both AChE and BuChE show continuous increase postnatally except that the magnitude of increase for AChE is several times higher than for BuChE. Thus the decline in BuChE activity is not real, but due to a higher increase in the tissue mass relative to the enzyme (FIG. 1, TABLE 1).

The slow growth pattern of BuChE is consistent with a nonneuronal (glial-endothelial) localization and function for BuChE in the OB whereas the prodigious, late postnatal proliferation of ChAT, AChE, and ACh is compatible with a neuronal-synaptic localization and function for these cholinergic components. These results on AChE are in agreement with our previous studies on this enzyme.[3,4] Histochemical studies have associated the OB cholinergic activity with the centrifugal

[a] Address for correspondence: Dr. E. Meisami, Department of Physiology and Biophysics, University of Illinois, Urbana, IL 61801.

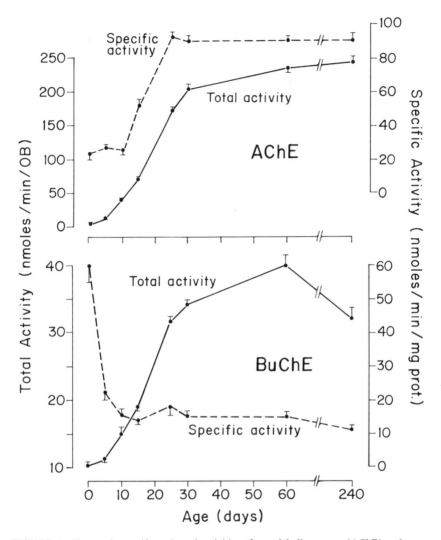

FIGURE 1. Changes in specific and total activities of acetylcholinesterase (AChE) and pseudocholinesterase (BuChE) of rat olfactory bulbs during the postnatal period. Each point is the mean of at least four duplicate determinations from the same number of albino rats. Enzyme activities were measured colorimetrically (Ellman *et al.*;[8] see also Meisami[9] and Meisami & Moussavi[3] and Meisami & Firoozi[4] for further technical details).

afferents to OB (from forebrain reticular nuclei, e.g., the horizontal limb of the nucleus of diagonal band.[4,5] Therefore the postnatal growth "spurt" in AChE, ChAT, and ACh activity, which occurs simultaneously after day 10, must reflect mainly the proliferation and maturation of fibers and synapses of the centrifugal cholinergic afferents within the OB. The earlier maturation of AChE, compared to ChAT and ACh, may represent the earlier development of intrabulbar postsynaptic compartments.

TABLE 1. Postnatal Changes in Weight and Protein Content and in Cholineacetyl Transferase Activity and Acetylcholine Content in the OBs of Albino Rat[a]

Age (days)	Weight (mg/OB)	Protein (mg/g wet wt)	ChAT Activity	ACh Content
			(% of adult level)[b]	
1	9	46	4	6
5	18	50	6	—
10	34	57	8	10
15	56	72	40	—
30	62	75	90	95
60	68	80	98	100

[a] n: at least three OB pairs per age group for weight, protein determination, and ChAT assay; for ACh assays, two to three assays per age group each using three to five pooled OB pairs.

Assay methods: ChAT, radiometrically (Rand and Johnson[6]); ACh: leech dorsal muscle bioassay; protein: Lowry et al.[7]

[b] Adult (6-month-old rat OB, weighing 74 mg/OB pair) levels (means): ChAT, 3 μmoles ACh/g wet wt/h; ACh, 60 nmoles ACh/g wet wt.

The developmental pattern of cholinergic elements in the OB resembles those seen in the cerebral cortex, which also receives its cholinergic input mainly from the forebrain reticular nuclei. The functional and behavioral significance of these neurochemical findings in the process of olfactory development remains to be elucidated.

ACKNOWLEDGMENTS

Authors are grateful to Dr. P. S. Timiras for support and to Dr. Hal Sternberg for help in ChAT assay. Part of this work was conducted at the Institute of Biochemistry and Biophysics, Tehran University, Tehran, Iran.

REFERENCES

1. MEISAMI, E. 1979. The developing rat olfactory bulb: Prospects of a new model system for developmental neurobiology. In Neural Growth and Differentiation. E. Meisami & M. A. Brazier, Eds.: 183-206. Raven Press. New York.
2. BUNDMAN, M. C., J. L. BRUCE, J. M. FRIGO, R. T. ROBERTSON & C. GORENSTEIN. 1984. Localization and characterization of pseudocholinesterase in rat brain. Neurosci. Abstr. 10: 683.
3. MEISAMI, E. & R. MOUSSAVI. 1982. Lasting effects of early olfactory deprivation on growth, DNA, RNA and protein content and Na-K-ATPase and AChE activity of the rat olfactory bulb. Dev. Brain Res. 2: 217-229.
4. MEISAMI, E. & M. FIROOZI. 1985. Acetylcholinesterase activity in the developing olfactory bulb: A biochemical study on normal maturation and the influence of peripheral and central connections. Dev. Brain Res. 21: 115-124.
5. MACRIDES, F. & B. J. DAVIS. 1983. The olfactory bulb. In Chem. Neuroanat. P. C. Emson, Ed.: 391-426. Raven Press. New York.

6. RAND, J. B. & C. D. JOHNSON. 1981. Anal. Biochem. **116:** 361.
7. LOWRY, O. H. *et al.* 1951. J. Biol. Chem. **192:** 265.
8. ELLMAN, G. L. *et al.* 1961. Biochem. Pharmacol. **7:** 88.
9. MEISAMI, E. 1984. J. Neurochem. **42:** 883.

Bretylium Tosylate Enhances Salt Taste via Amiloride-Sensitive Pathway[a]

SUSAN S. SCHIFFMAN,[b] SIDNEY A. SIMON,[c]
JAMES M. GILL,[b] AND TIMOTHY G. BEEKER[b]

[b]Department of Psychiatry
[c]Departments of Physiology and Anesthesiology
Duke University Medical Center
Durham, North Carolina 27710

Bretylium tosylate (BT) is an antifibrillary drug that has been shown to increase sodium transport through amiloride-sensitive pathways in frog skin.[1,2] Amiloride-sensitive sodium pathways have recently been found to mediate certain components of taste in both humans and rodents. Application of amiloride to the dorsal surface of the tongue diminished the taste of NaCl, LiCl, and sweeteners in humans;[3] reduced electrophysiological gustatory responses to NaCl and LiCl in rats and gerbils;[3-7] and blocked increases in short-circuit current across rat and canine lingual epithelium induced by NaCl, LiCl, and sugars.[5,6,8-11]

The purpose of the experiments reported here was to determine (1) if BT amplifies amiloride-sensitive components of taste (i.e., NaCl, LiCl, and sweeteners) and (2) if it does indeed amplify taste, does the enhancement involve an increase in epithelial transport similar to that observed in frog skin?

In order to determine the effect of BT on taste, three types of experiments were performed (see Schiffman et al.[12]). These included human psychophysical taste measurements, electrophysiological gustatory recordings in rats, and electrical measurements from isolated canine lingual epithelium. In the human experiments, half of the tongue was adapted to a drug (BT alone, BT and amiloride applied simultaneously, or amiloride alone). The other half of the tongue was adapted to a water control. A standard stimulus (0.2 M NaCl) dissolved in deionized water was applied to the drug-treated side of the tongue. Test stimuli were applied simultaneously to the nondrug side, and the concentrations were adjusted to match the perceived intensity of the standard. The results at pH 6.3 are shown in FIGURE 1. The striped bar represents the 0.2 M NaCl standard applied to the drug-treated side of the tongue. The white bar indicates the concentration perceived to match the standard. It can be seen that the 1 mM BT alone enhanced the perceived taste of 0.2 M NaCl by 33.5%. When amiloride (0.01 mM, 0.1 mM, or 1 mM) was applied simultaneously with BT, the enhancement was reduced or eliminated. The presence of BT protected against the reduction in the perceived intensity of NaCl by amiloride applied alone.

Both the multiunit and single-unit recordings from the NTS (nucleus tractus

[a]This research was supported in part by a grant to S. S. Schiffman (AG00443) and a grant to S. A. Simon (NS-20669).

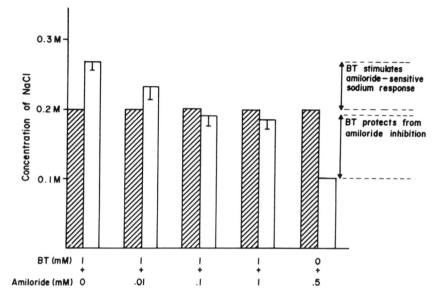

FIGURE 1. The striped bar represents the 0.2 M NaCl standard applied to the side of the tongue adapted to BT alone, mixtures of BT plus amiloride, or amiloride alone. The white bar represents the concentration perceived to match the standard.

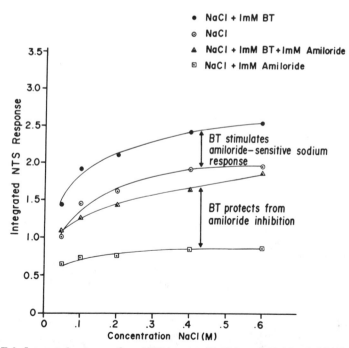

FIGURE 2. Integrated responses from NTS in rat to NaCl from 0.05 M to 0.6 M. Responses are given for NaCl alone, NaCl after adaptation to BT, NaCl after adaptation to a mixture of 1 mM BT, and 1 mM amiloride, and NaCl after adaptation to 1 mM amiloride.

solitarius) in rat parallel the human findings. Integrated responses from multiple units to NaCl from 0.05 M to 0.6 M are shown in FIGURE 2. Responses to NaCl are given for adaptation to 1 mM BT alone, 1 mM BT in combination with 1 mM amiloride, and 1 mM amiloride alone. BT enhanced the neural response to NaCl from 30-40% and, in addition, protected against the strong inhibition of NaCl found by amiloride alone. BT had no effect on the short-circuit current in isolated dog lingual epithelium. The fact that BT did not enhance short-circuit current suggests that BT does not affect the lingual epithelium in the same manner as a sodium-transporting epithelium such as frog skin.

BT was also found to enhance taste responses to LiCl in addition to NaCl in both humans and rats. Additional tastants, including other salts (KC1, CaCl$_2$), sweet, acid, or bitter compounds were not affected by BT.

In summary, BT was found to exhibit two distinct effects on taste transduction mechanisms.[12] First, it potentiated gustatory responses to NaCl and LiCl in humans and rats, and second, it protected against the inhibition of these responses by amiloride.

REFERENCES

1. BACANER, M. B. 1966. Bretylium tosylate for suppression of induced ventricular fibrillation. Am. J. Cardiol. **17:** 528-534.
2. BACANER, M. B. 1983. Prophylaxis and therapy of ventricular fibrillation: Bretylium reviewed and lidocaine refuted. Int. J. Cardiol. **4:** 133-152.
3. SCHIFFMAN, S. S., E. LOCKHEAD & F. W. MAES. 1983. Amiloride reduces the taste intensity of Na$^+$ and Li$^+$ salts and sweeteners. Proc. Natl. Acad. Sci. USA **80:** 6136-6140.
4. BRAND, J. G., J. H. TEETER & W. L. SILVER. 1985. Inhibition by amiloride of chorda tympani responses evoked by monovalent salts. Brain Res. **334:** 207-214.
5. DESIMONE, J. A. & F. FERRELL. 1985. Analysis of amiloride inhibition of chorda tympani responses. Am. J. Physiol. **249:** R52-R61.
6. HECK, G. L., S. MIERSON & J. A DESIMONE. 1984. Salt taste transduction occurs through an amiloride-sensitive sodium transport pathway. Science **223:** 403-405.
7. TEETER, J. H., W. L. SILVER & J. G. BRAND. 1983. Effect of amiloride on chorda tympani responses to salts. Soc. Neurosci. Abstr. **9:** 1022.
8. DESIMONE, J. A., G. L. HECK & S. K. DESIMONE. 1981. Active ion transport in dog tongue: A possible role in taste. Science **214:** 1039-1041.
9. DESIMONE, J. A., G. L. HECK, S. MIERSON & S. K. DESIMONE. 1984. The active ion transport properties of canine lingual epithelia *in vitro:* Implications for gustatory transduction. J. Gen. Physiol. **83:** 633-656.
10. MIERSON, S., G. L. HECK, S. K. DESIMONE, T. U. L. BIBER & J. A. DESIMONE. 1985. The identity of the current carriers in canine lingual epithelium *in vitro.* Biochim. Biophys. Acta **816:** 283-293.
11. SIMON, S. A. & J. L. GARVIN. 1985. Salt and acid studies on canine lingual epithelium. Am. J. Physiol. **249:** C393-C408.
12. SCHIFFMAN, S. S., S. A. SIMON, J. M. GILL & T. G. BEEKER. 1986. Bretylium tosylate enhances salt taste. Physiol. Behav. **36:** 1129-1137.

Some Effects of Olfactory Tract Cooling on Mitral Cell Response Patterns during Repetitive Natural Stimulation

D. SCHILD, T. FISCHER, AND H. P. ZIPPEL

Physiologisches Institut
Humboldtallee 23
Universität D-3400, Göttingen
Federal Republic of Germany

The characteristics of mitral cell discharge patterns in goldfish often change when the stimulus is repeatedly applied. This finding, which has statistical as well as functional (adaptation, plasticity) implications, has been obtained in the following way: Responses of goldfish mitral cells were recorded extracellularly with electrolytically sharpened insulated steel electrodes. With a recently developed computer-controlled device for odor application,[1] square pulses of odor (natural Tubifex food extracts, 10^{-4} g/l, and amyl acetate, 10^{-4} M, diluted in tap water) of at least 30 seconds' duration were applied. Thereafter, the olfactory mucosa was washed with tap water for at least 30 seconds. Each cell was stimulated 40 times with the same stimulus in order to investigate changes of the responses in the course of the stimulus repetitions. The results of one experiment were thus forty consecutive responses to the same stimulus.

The most common response to a stimulus pulse was an initial increase or decrease of activity (about 2-5 sec) followed by a fairly constant mean activity maintained during the pulse. The activity during the stimulus-free period (at least 30 sec), that is, the period between two stimulus presentations, usually began with a phasic reaction (about 2-5 sec) followed by an interval of fairly constant activity. The temporal structure of responses did not depend on what stimulus was applied.

In 49 of 69 experiments the responses to single stimuli were reproducible over all runs of the experiment. In 20 recordings the single-run responses were not reproducible. They rather showed (a) abrupt changes from the first to the second run or (b) slow changes that extended over up to 30 runs. Every pattern of this kind became, however, reproducible after a certain number of runs. The activity changes during the first runs can therefore be attributed to a process that leads either to a more pronounced response to a repeated stimulus or to an indifferent one.

In order to study the influence of the efferent fibers entering the olfactory bulb from the brain, 18 experiments were repeated under identical experimental conditions except that the olfactory tract was cooled to 1°C. One of the results was that in 12 out of these 18 experiments the responses were reproducible and had the same temporal structure as described above (phasic-static); six recordings showed activity pattern changes in the course of the 40 runs. Pattern changes in an experiment were not necessarily associated with pattern changes in the twin experiment with blocked efferent influence. Though this finding does not exclude that the efferent innervation of the bulb might lead to changes in the temporal structure of mitral cell response patterns, it suggests the existence of other mechanisms bringing about this effect. Considering the fairly stereotypical response behavior of receptors, innerbulbar plasticity could be taken into consideration.

REFERENCE

1. SCHILD, D. 1985. A computer-controlled device for the application of odours to aquatic animals. J. Electrophysiol. Tech. **12:** 71-79.

Contact Chemoreceptors on the Walking Legs of the Shore Crab, *Carcinus maenas*[a]

MANFRED SCHMIDT AND WERNER GNATZY

Zoologisches Institut
Gruppe Sinnesphysiologie
J. W. Goethe-Universität, Siesmayerstr. 70
6000 Frankfurt/M. 11, West Germany

Many small sensilla without hair shafts, called funnel-canal organs, are located on the walking legs of the shore crab, *Carcinus maenas.* An electronmicroscopic study showed that these sensilla are innervated by two mechanosensory neurons and one to 22 other sensory cells.[1] From the presence of a terminal pore, it was concluded that these cells are most likely chemoreceptive.[2]

Since the funnel-canal organs are the only sensilla on the epicuticular tip of the dactyl, a selective chemical stimulation of these sensilla is possible; therefore, extracellular recordings from the leg nerve were carried out. The dactyl tip was inserted through a piece of thick silicone tubing into a small stimulating chamber. This chamber was continuously superfused by artificial seawater, which could be replaced by various stimulus solutions in about 200 msec. The leg artery was perfused with oxygenated crab saline in order to keep the sensory cells in good condition.

About 20 different compounds, which excite other chemoreceptors of crustaceans,[3] were tested in a concentration range from 10^{-7} to 10^{-2} M. The pH and the osmolarity of all stimuli was adjusted to seawater (pH 7.9/860 mOsmol).

As to the specificity of the sensory units, they were divided into those that are broadly or narrowly tuned. The broadly tuned cells respond best to taurine and glycine, but to a lesser degree they are also stimulated by betaine and all amino acids tested except L-glutamate. The threshold sensitivity of these cells is between 10^{-6} and 10^{-4} M for taurine and glycine with a working range of about two to three orders of magnitude. Most cells with a narrow reaction spectrum respond to L-glutamate and L-glutamine. The threshold sensitivity is below 10^{-5} M for L-glutamate and about 10^{-4} M for L-glutamine. Besides these, taurine-only and glycine-only cells are also present. The threshold sensitivity of the taurine cells is between 10^{-6} and 10^{-5} M, that of the glycine cells between 10^{-6} and 10^{-4} M.

In comparison with other chemoreceptors of crustaceans,[3] the funnel-canal organs are fairly insensitive and therefore should be regarded as contact chemoreceptors. The working range of these sensilla for the most stimulatory compounds and the concentration of these compounds in bivalves (the main bait of *Carcinus*) are very similar. Therefore it is most likely the funnel-canal organs are involved in the last step of food detection.

[a]Supported by the Deutsche Forschungs Gemeinschaft; SFB 45/Al.

REFERENCES

1. SCHMIDT, M. & W. GNATZY. 1984. Are the funnel-canal organs the "campaniform sensilla" of the shore crab, *Carcinus maenas* (Decapoda, Crustacea)? II. Ultrastructure. Cell. Tissue Res. **237:** 81-93.
2. GNATZY, W., M. SCHMIDT & J. ROEMBKE. 1984. Are the funnel-canal organs the "campaniform sensilla" of the shore crab, *Carcinus maenas* (Decapoda, Crustacea)? I. Topography, external structure and main organization. Zoomorphology **104:** 11-20.
3. ACHE, B. W. 1982. Chemoreception and thermoreception. *In* The Biology of Crustacea. H. L. Atwood & D. C. Sandeman, Eds. Vol. **3:** 369-398. Academic Press. New York.

Further Investigations of the Receptor Potential in Primary Chemosensory Neurons by Intracellular Recording[a]

INGRID SCHMIEDEL-JAKOB,[b] PETER A. V. ANDERSON, AND BARRY W. ACHE

C. V. Whitney Laboratory
University of Florida
St. Augustine, Florida 32086

Knowledge of the electrical events associated with olfactory transduction is limited.[1] Fortunately, new techniques such as patch electrode recording[2] and morphologically favorable preparations provide hope for rapid progress in this area. The olfactory organ (antennule) of one such organism, the spiny lobster, consists of tufts of hairlike sensilla (aesthetascs) that are innervated by bipolar sensory neurons. These neurons were exposed in a hemicylindrical preparation of the antennule and treated enzymatically with papain and trypsin for intracellular recording using patch pipettes in the whole-cell configuration. The exposed neurons retain both their electrical and chemical excitability.[3]

The input impedances for hyperpolarizing current pulses ranged from 120 to 580 Mohm; with depolarizing current steps this fell to 34 to 70 Mohm. Transiently depolarizing the membrane to -30 to 40 mV evoked fast action potentials that overshot zero by as much as $+33.3$ mV. The rising phase of the action potential had a maximum slope of 71 V/sec; its repolarizing phase 50 V/sec. The duration of the action potential at half-peak amplitude was 1.37 msec. The time constant (τ) of these cells was in the range of 19.5 to 42.5 msec. With some cells there were indications of multiple time constants, but this phenomenon was not examined in any detail.

When stimulated chemically by an extract of crab muscle or amino acids, these cells produced prolonged, transient depolarizations (receptor potential) on which were superimposed a train of TTX-sensitive action potentials that do not necessarily overshoot zero mV. Chemically evoked depolarizations were dose-dependent and their amplitude was linearly related to the logarithm of the stimulus concentration. The depolarizations reached their peak amplitude in 96-300 msec and declined over 5-31 sec, depending on the strength of the stimulus. The amplitude of the receptor potential was dependent on the membrane potential of the cell and extrapolated to a mean reversal of $+22.3 \pm 7.3$ (SE) mV. All cells tested had similar extrapolated reversal

[a] Supported by National Science Foundation Award BNS 85-11256 and a grant from the Whitehall Foundation.

[b] Address for correspondence: Ingrid Schmiedel-Jakob, C. V. Whitney Laboratory, Rte. 1, Box 121, St. Augustine, FL 32086.

potentials, but actual reversal was never observed, even when these cells were depolarized to $+50$ mV. While these reversals may not accurately reflect the true reversal potential, they should provide useful information about the ionic basis of the transduction process in chemosensory neurons. Conductance changes occurred during chemically evoked depolarization. Surprisingly, a decrease as well as an increase in conductance of the membrane was observed. The membrane resistance changed during the rising phase and the plateau and gradually returned to its initial value. The conductance change for the increasing resistance was about 48% and was about 20% for the decreasing resistance. The underlying mechanisms were not studied in any detail. Further intracellular investigations are required in order to interpret these mechanisms in terms of a molecular view of olfactory reception.

REFERENCES

1. SHEPHERD, G. M. 1985. Welcome whiff of biochemistry. Nature **316:** 214-215.
2. HAMILL, O. P., A. MARTY, E. NEHER, B. SAKMANN & F. J. SIGWORTH. 1981. Improved patch-clamp techniques for high resolution current recording from cells and cell-free membrane patches. Pflugers Arch. **391:** 85-110.
3. ANDERSON, P. A. V. & B. W. ACHE. 1985. Voltage- and current-clamp recording of the receptor potential in olfactory receptor cells *in situ*. Brain Res. **338:** 273-280.

Progress on the Identification of the Folate Chemoreceptor of *Paramecium*

STEPHANIE SCHULZ, J. MICHAEL SASNER, AND
JUDITH VAN HOUTEN

Department of Zoology
University of Vermont
Burlington, Vermont 05405

The ciliated protozoan *Paramecium tetraurelia* has been used as a simple system in which to study chemical perception and transduction events. Folic acid is an essential vitamin for *P. tetraurelia* and, since it is secreted by the bacteria upon which paramecia feed, may serve as a food cue.[1] Paramecia are attracted to folate relative to chloride in behavioral assays. We have previously shown that it is the pterin moiety, and not the p-ABA or glutamate portions, of the folate molecule that is recognized by *Paramecium* in behavioral tests, and that this moiety is sufficient to elicit the observed chemoresponse. Detection of folate induces a membrane hyperpolarization that produces decreased turning frequency, resulting in the observed chemoresponse. A variety of biochemical approaches have been employed in attempting to identify a membrane protein that might serve as a chemoreceptor for folate.

[^{3}H]Folate binds to whole cells in a saturable, specific manner.[2] Binding studies using isolated cilia demonstrate that the ciliary contribution is < 1% of the total surface binding.[3] In electrophysiological studies, deciliated cells respond to folate in a manner identical to intact cells, supporting the observation that cilia contribute little to the cells' folate binding capacity. Recent studies of isolated cell body membranes (pellicles) without cilia confirm that the large majority of binding sites are not in the ciliary membrane (Schulz and Denaro, unpublished results). The pellicle folate binding site shows kinetic parameters consistent with multiple binding sites or a single cooperative binding site. The K_d is ~ 1 μM, with a total binding capacity of 1 nmole/mg pellicle protein. Binding to isolated pellicles shows much of the same specificity as the chemoresponse, but the relationship of this binding to chemoreception has not yet been established through the use of mutants, as it has been for whole-cell binding.

Folate-Sepharose affinity chromatography has been used to identify folate-binding proteins in Triton X-100 solubilized pellicles. Ten proteins, or groups of proteins, consistently elute specifically from folate-Sepharose columns with folate, but not with glutamate or *p*-ABA-glutamate. Methotrexate, a folate analogue that competes poorly with folate in behavioral assays, elutes these proteins with varying degrees of efficiency. Whole paramecia have been labeled with ^{125}I to identify which among the folate-binding proteins have externally exposed moieties. Nine individual polypetides within the ten groups of proteins can be labeled with ^{125}I. The folate-binding proteins have been further characterized by con A-Sepharose chromatography. Five of the folate-binding proteins elute specifically with α-methyl-D-mannoside, identifying them as glycoproteins. Four proteins are both labeled with ^{125}I and identified as glycoproteins; one of these proteins does not elute with methotrexate from a folate-Sepharose column. This protein (molecular mass 58 kDa) is a prominent folate-binding protein among soluble pellicle proteins but is a very minor component of cilia folate-binding soluble proteins and, therefore, is a good candidate for the *Paramecium* folate chemoreceptor.

We are circumventing potential problems of using affinity chromatography to identify relatively low-affinity binding proteins by using immunodetection of proteins cross-linked with folate. The chemoreceptor should be among the membrane proteins to which we can cross-link folate. Cross-linking folate onto whole cells specifically inhibits attraction to folate, but not to acetate.[4] Membrane proteins cross-linked with folate are identified by electroblotting proteins from polyacrylamide gels onto nitro-cellulose and immunodetection of the nitrocellulose with antifolate antibodies. Currently we are cataloguing the folate-binding proteins from normal cells and chemoreception mutants (Sasner, unpublished results).

REFERENCES

1. PAN, P., E. M. HALL & J. T. BONNER. 1972. Folic acid as second chemotactic substance in the cellular slime moulds. Nature (London) New Biol. **237:** 181-182.
2. SCHULZ, S., M. DENARO & J. VAN HOUTEN. 1984. Relationship of folate binding to chemoreception in *Paramecium*. J. Comp. Physiol. A **155:** 113-119.
3. VAN HOUTEN, J., S. SCHULZ & M. DENARO. 1983. Characterization and location of folate binding sites involved in *Paramecium* chemoreception. J. Cell Biol. **97:** 1797a.
4. VAN HOUTEN, J. & R. R. PRESTON. 1987. Chemoreception: *Paramecium* as a receptor cell. *In* Advances in Experimental Medicine and Biology. Y. Ehrlich, Ed. Plenum Press. New York. In press.

Perceptual Separability and Integrality in Odor Discrimination[a]

AMY L. SCHWARTZ, MICHAEL D. RABIN,[b] AND
WILLIAM S. CAIN

Department of Psychology Yale University, and
John B. Pierce Foundation Laboratory
New Haven, Connecticut 06519

Sensory characterization of commercial products often requires subjects to analyze odors of mixtures into components, insofar as possible. Subjects presumably accomplish this task more or less satisfactorily. Nevertheless, certain fundamental properties of odor perception (e.g., masking, assimilation, suppression in mixtures) would seem likely to limit success. The analysis of mixtures might therefore exhibit some characteristics of analytical processing and some of holistic processing. We explored this with a moderately challenging binary mixture that smelled like banana and crushed grass. The methodology entailed two-choice classification and the measurement of reaction time (RT). As expected, the outcome implied that the processing of odor mixtures falls somewhere between analytic and holistic.

METHODS

Four stimuli were created by orthogonally combining the banana-smelling odorant amyl acetate (A) with the crushed-grass-smelling odorant hexenal (H) at psychophysically matched levels of intensity: weak (W, 0.16%) and moderate (M, 4%). Orthogonal combinations of these two qualities and intensities resulted in the mixtures WA + WH, WA + MH, MA + WH, and MA + MH.

Eighteen subjects participated in three two-choice classification tasks. In all tasks, one stimulus was presented per trial. Subjects indicated one of two task-dependent responses by tapping a key.

In control tasks, subjects discriminated between pairs of mixtures in terms of one element in the mixture, while the other, irrelevant element was held constant, for example MA + MH versus WA + MH.

In orthogonal tasks, discrimination was between two elements of separate mixtures when the irrelevant element of each mixture varied orthogonally across stimuli. For example, one orthogonal task involved discriminating between WH and MH, while the other mixture element could be either WA or MA. Thus, subjects had to ignore

[a] Supported by Grant NS21644 from the National Institutes of Health.

[b] Address for correspondence: Michael D. Rabin, John B. Pierce Foundation, 290 Congress Avenue, New Haven, CT 06519.

TABLE 1. Reaction Time (sec) and Proportion of Errors for the Three Tasks

Task	RT	Error
Control	2.26	0.16
Orthogonal	2.69	0.21
Correlated	2.22	0.12

the irrelevant element in the mixture. If this type of task proves more difficult than the control task, it implies a failure of selective attention—and a lack of perceptual separability.

Correlated tasks were used to evaluate whether redundancy would aid the discrimination. The elements were correlated such that either element in the mixture defined the correct response. If the mixtures are separable, then this redundancy should not aid the subject.

The tasks, by definition, lead to the prediction that if the elements in the mixture are perceptually separable, there should be no orthogonal interference and no correlated benefit relative to the control. If the elements are integral, there should be both orthogonal interference and correlated benefit.

RESULTS

Paired t-tests showed (TABLE 1) that the orthogonal condition was more difficult than the control for both RT ($t(17) = 2.10$, $p = 0.05$) and errors ($t(17) = 2.09$, $p = 0.05$). The correlated and control conditions did not differ in either RT ($t(17) < 1$) or errors ($t(17) = -1.95$, $p > 0.05$).

These results support neither a separable nor integral pattern unequivocally. The failure of selective attention shown by the greater difficulty for the orthogonal conditions is consistent with integral (holistic) processing, but the equality of the correlated and control conditions is consistent with separable (analytic) processing. The data suggest that our subjects did not perceptually separate the elements in these mixtures thoroughly since they could not selectively attend to them. However, the lack of improved performance on the correlated task does not fit with a "classic" integral pattern. Hence, for this mixture and perhaps others, olfactory processing falls somewhere between analytic and holistic.

Mouse Monoclonal Antibody RB-8 That Distinguishes Chemically Distinct Zones in the Primary Olfactory Projection Recognizes a 125-Kilodalton Membrane-Associated Protein[a]

JAMES E. SCHWOB[b] AND DAVID I. GOTTLIEB

Department of Anatomy and Neurobiology
Washington University School of Medicine
St. Louis, Missouri 63110

Recently, we reported the isolation and immunohistochemical characterization of a novel mouse monoclonal IgG$_1$ designated RB-8, which identifies two chemically distinct zones in the primary olfactory projection of adult rats.[1] The axons from the ventrolateral part of the olfactory epithelium and their terminals in the glomeruli of the ventrolateral olfactory bulb are densely stained by RB-8 (termed RB-8 positive), while axons from the dorsomedial part of the epithelium and their terminals in the dorsomedial bulb are unstained or only lightly stained (termed RB-8 negative). RB-8 staining is nervous system specific. Other parts of the CNS and PNS stain non-homogenously. We report here biochemical characterization of the corresponding antigen in brain and in olfactory nerve.

A direct radioimmunoassay (RIA) based on the binding of [^{125}I]RB-8 IgG was used to analyze the RB-8 antigen in adult rat brain homogenates. The RB-8 antigen in whole-brain homogenates is membrane associated, since it is not detached by repeated washing with hypotonic buffer and is solubilized by detergent. RB-8 binds to brain membranes with high affinity. At saturation, 2.6 pm of labeled IgG binds per milligram of membrane protein. This value serves as a rough estimate of the concentration of antigen in brain membranes. The binding of RB-8 to membranes is immunologically specific, since [^{125}I]RB-8 is blocked by unlabeled RB-8 IgG but is not blocked by six other monoclonal antibodies (MAbs) or normal mouse serum. The RB-8 antigen is trypsin sensitive, indicating that it is a protein or protein-linked moiety.

Membrane proteins were harvested from the olfactory nerve layer of the bulb or from forebrain excluding the bulbs, separated by SDS-PAGE, and compared on

[a] Supported by National Institutes of Health Grants NS 12867, NS 07076, and NS 07057.

[b] Current address: Department of Anatomy & Cell Biology, SUNY Health Sciences Center, Syracuse, NY 13210.

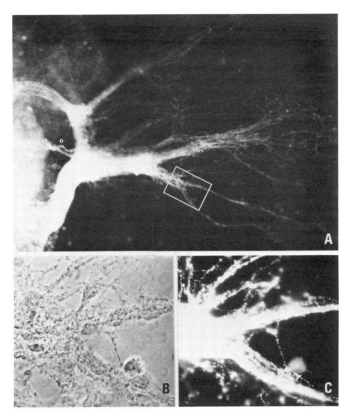

FIGURE 1. (A) Explant of fetal olfactory epithelium exposed to monoclonal antibody RB-8 while living, then fixed and stained for indirect immunofluorescence. The axonal plexus stains well with RB-8 in both the larger fasicles and smaller bundles.

(B) Phase-contrast photomicrograph of boxed area in (A).

(C) Higher power view of boxed area in (A). Comparing the RB-8 staining with the phase-contrast photograph suggests that most, if not all, the individual axons and small bundles of axons are labeled on their surface with RB-8.

immunoblots. Both olfactory nerve and forebrain contain immunoreactive proteins of 125 kDa that comigrate when electrophoresed together in a single lane. This protein is enriched more than 100-fold when brain membrane proteins are solubilized with octyl glucoside detergent and passed over an RB-8 immunoaffinity column. In comparison, a column made with MAb 224-1A6-A1, which does not bind to rat tissue, does not enrich the antigen. Thus, the binding to soluble antigen is immunologically specific.

The subcellular location of the RB-8 antigen was assessed in fetal olfactory epithelial explants that had substantial axonal outgrowths. In living cultures incubated with antibody before fixation, the plasma membrane prevents access to internal antigens. Under these conditions, axons are labeled by RB-8 as shown with indirect immunofluorescence (FIG. 1), and axonal membranes remain intact, since the axons fail to stain with antibody against vimentin, which is known to be an internal marker for olfactory axons.[2] After membrane permeabilization, these axons will also stain with anti-vimentin *in vitro.* We conclude that the RB-8 antigen is exposed on the cell surface.

In summary, the RB-8 antigen of olfactory axons is a membrane-associated, cell-surface protein of 125 kDa. The division of the primary olfactory projection into two zones based on the distribution of this antigen may be functionally important, given the evidence for some spatial organization in olfactory sensory coding and for a similar division of the rabbit olfactory system.[3] Furthermore, this dorsomedial-ventrolateral division clearly reflects the pattern of axonal connections between the epithelium and the bulb. Additional characterization of the RB-8 antigen will prove useful in understanding its functional role in the primary olfactory projection.

REFERENCES

1. SCHWOB, J. E. & D. I. GOTTLIEB. 1986. J. Neurosci. **6:** 3393-3404.
2. SCHWOB, J. E., N. B. FARBER & D. I. GOTTLIEB. 1986. Neurons of the olfactory epithelium in adult rats contain vimentin. J. Neurosci. **6:** 208-217.
3. MORI, K., S. C. FUJITA, K. IMAMURA & K. OBATA. 1985. Immunohistochemical study of subclasses of olfactory nerve fibers and their projections to the olfactory bulb in the rabbit. J. Comp. Neurol. **242:** 214-229.

The Janus Head of Taste

Department of Psychology and
Institute for Neuroscience
University of Delaware
Newark, Delaware 19716

Our approach to understanding the sense of taste has conformed closely to models developed for the nonchemical senses, with major topics being the existence of primary stimuli, of neuron types, of topographic organization and of underlying physical dimensions. This has not been fully satisfying, for as data accumulated and analyses became more sophisticated, definitive answers to this set of questions remained elusive. A possible shortcoming lies with the model. Taste (in rats) is not like the nonchemical senses in at least the following ways:

1. Anatomically: Taste is intermediate between the somatosensory and visceral systems. Many of its forebrain projections join somatosensory axons proceeding to the exteroceptive domain of the thalamocortical axis, but at least as many invade the ventral forebrain with visceral afferents.[1] In distinction to the nonchemical senses, then, taste sends its major projection to subcortical structures associated not with primary sensory evaluation but with feeding, reinforcement, motivation, emotion and autonomic processes.
2. Physiologically: Factors that affect the autonomic nervous system and the physiology of feeding can modify taste activity. This has been demonstrated in rats by manipulating sodium levels,[2] plasma concentrations of glucose[3] and insulin, degree of gastric distension,[4] and by development of conditioned taste aversions.[5] Both hypothalamic[6] and orbitofrontal cortical[7] taste-evoked activity is influenced by level of satiety in the macaque.
3. Functionally: The exteroceptive senses are constantly, and often dispassionately, evaluating the environment. Taste is called upon only sporadically, but its evaluation must activate the appropriate hedonic sensations to encourage consumption of nutrients and rejection of toxins. This information is somehow stored on a long-term basis appropriate to the time course of digestive rather than operant processes, so mediating the long CS-US intervals associated with taste aversion learning.

Thus taste represents a transition between the exteroceptive and visceral senses. Its receptors are at the interface between the external and internal milieux. It presents typical features of nonchemical sensory systems, such as identification of stimulus quality and intensity according to spatiotemporal codes, but its analysis is immediately related to, and is modifiable by, the internal condition of the organism. When interpreting gustatory neural data, it may be no less germane to ask what a rat's glucose utilization rate is than to ask about stimulus flow rate. To study taste realistically, we must address not only the questions associated with the nonchemical senses, but also those we would ask of a visceral sense: What nutrients or endogenous chemicals are in the digestive system, circulation, and CNS, and how might they affect the taste system directly or by vagal mediation?

Our neural and biochemical progress should come on two fronts: One, toward understanding the exteroceptive face of taste, particularly through the enlightening biophysical studies being conducted at the receptor level; the other, toward a knowledge of the interoceptive face through studies of the relationship between taste and feeding.

REFERENCES

1. NORGREN, R. 1976. Taste pathways to hypothalamus and amygdala. J. Comp. Neurol. **166:** 17-30.
2. CONTRERAS, R. 1977. Changes in gustatory nerve discharges with sodium deficiency: A single unit analysis. Brain Res. **121:** 373-378.
3. GIZA, B. K. & T. R. SCOTT. 1983. Blood glucose selectively affects taste-evoked activity in the rat nucleus tractus solitarius. Physiol. Behav. **31:** 643-650.
4. GLENN, J. F. & R. P. ERICKSON. 1976. Gastric modulation of afferent activity. Physiol. Behav. **16:** 651-658.
5. CHANG, F.-C. T. & T. R. SCOTT. 1984. Conditioned taste aversions modify neural responses in the rat nucleus tractus solitarius. J. Neurosci. **4:** 1850-1862.
6. BURTON, M. J., E. T. ROLLS & F. MORA. 1976. Effects of hunger on the responses of neurones in the lateral hypothalamus to the sight and taste of food. Exp. Neurol. **51:** 668-677.
7. ROLLS, E. T., Z. J. SIENKIEWICZ & S. YAXLEY. 1986. Hunger modulates the responses to gustatory stimuli of single neurons in the orbitofrontal cortex. In preparation.

Chemical Analyses of Hodulcin, the Sweetness-Suppressing Principle from *Hovenia dulcis* Leaves [a,b]

RICHARD SEFECKA[c]

IBM Instruments, Inc.
Danbury, Connecticut 06810

LINDA M. KENNEDY

Department of Biology
Clark University
Worcester, Massachusetts 01610

The chemistry of hodulcin (H), the selective sweetness-suppressing principle from *Hovenia dulcis*[1] is not known. However, there are known similarities and differences between compounds from *H. dulcis* and ziziphins (Zs) and gymnemic acids (GAs), the selective sweetness-suppressing compounds from *Ziziphus jujuba* and *Gymnema sylvestre,* respectively. Gymnemic acids are triterpene saponins, with a known genin structure (gymnemagenin).[2] Ziziphins are saponins, probably also triterpenes, yet not the same compounds as GAs.[3] *H. dulcis* contains saponins with the same genin structure (jujubogenin) as certain saponins from *Z. jujuba.*[4,5] However, taste activity of the structurally identified *H. dulcis* and *Z. jujuba* saponins has not been tested. To elucidate the chemistry of taste-active H and Zs, we analyzed a partially purified preparation of H and compared it with Zs and GAs.

Ziziphins and gymnemic acids were prepared and found active.[3,7] Hodulcin was prepared similarly and found to suppress magnitude estimates of the sweetness of 80 mM sucrose 50-100% in two humans. The preparations were analyzed by gradient elution reverse phase C18 HPLC (IBM Instr.; 40/60 MeOH/H$_2$O-100% MeOH gradient, 1 ml/min) with a unique, automatic computer-controlled column switching system for sample isolation (IBM Instr., Valco Instr.). Elution profiles (217-nm detection) were different for H and GAs and similar, but not the same for H and Zs (FIG. 1).

The H preparation was separated into three fractions, of which only one (8.5-13.5-min retention, FIG. 1) was active (80% suppression) in blind tests on L.M.K. The major component of the active fraction (12.5-min retention) was isolated, concentrated,

[a] This work was supported by Clark University Faculty Development and Biomedical Research Grants to L.M.K.

[b] The human tests were conducted at Clark University with prior approval of the Clark Committee on Rights of Human Participants in Research and Training Programs.

[c] Current address: IBM ISG Scientific and Technical Computing, Valhalla, NY 10595.

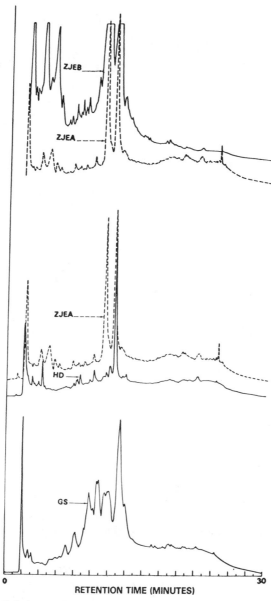

FIGURE 1. HPLC elution profiles for similarly extracted and partially purified samples of taste-active ziziphins (ZJEA), hodulcin (HD), and gymnemic acids (GS). The profile labeled ZJEB is for an inactive fraction from *Ziziphus jujuba* as in Kennedy and Halpern.[3] The GS sample is taste-active from 9-12 minutes (previous unpublished observations, R. Sefecka); the HD sample is taste-active from 8.5-13.5 minutes (text).

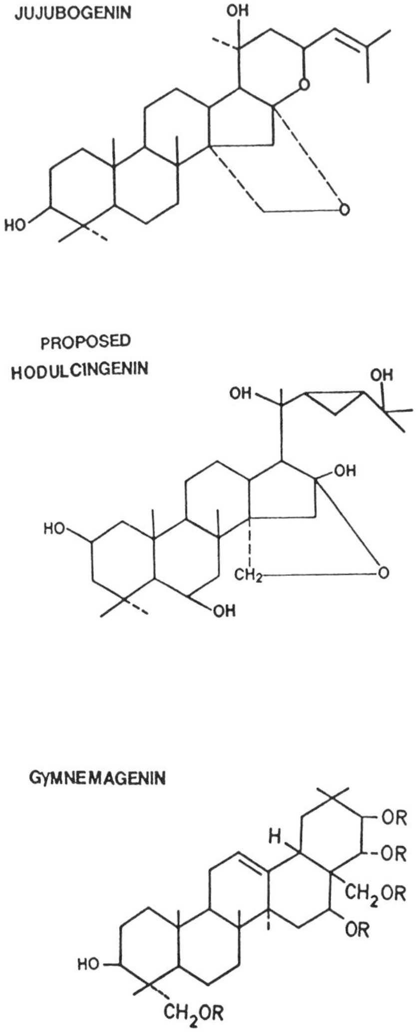

FIGURE 2. Triterpene aglycone structures for jujubogenin from *Ziziphus jujuba*,[4-6] gymnemagenin from *Gymnema sylvestre*,[2] and our proposed hodulcingenin from *Hovenia dulcis.*

collected, and then acid hydrolyzed as in Kimura *et al.*[4] for jujubogenin. The glycone fraction consisted of glucose and arabinose (IBM carbohydrate analysis HPLC; CH_3CN/H_2O, 80 : 20). Analysis by NMR (IBM NR300 AF Spectrometer) indicated a triterpene aglycone (genin) different from gymnemagenin and similar to, but not the same as jujubogenin (FIG. 2).

Thus we propose a genin structure for the major component of an active H fraction. Hodulcin is proposed to be a triterpene saponin, as are GAs and Zs; but our HPLC data for the similarly prepared H, GAs, and Zs taste-active samples indicate different compounds in the three. Also our NMR data indicate a "hodulcingenin" different from gymnemagenin[2] and similar but not the same as jujubogenin, reported elsewhere in this volume[6] to be the genin of taste-active Zs. Future comparisons of the chemical similarities and differences with the physiological taste effects of the three modifiers should be useful in elucidating sweetness transduction mechanisms. Further work on the minor components of the active H fraction, that is, their contributions to the taste activity and their structures, as well as work with individual components of each of the three modifiers, would also be useful.

ACKNOWLEDGMENTS

We thank G. Koller for assistance in obtaining leaves, D. Bachand and L. Wall for the figures, and J. Brink and R. Weihing for comments on the manuscript.

REFERENCES

1. SAUL, L. R., L. M. KENNEDY & D. A. STEVENS. 1985. Selective suppression of sweetness by an extract from *H. dulcis* leaves. Chem. Senses **10**(3): 445.
2. STOCKLIN, W. E. 1967. Gymnemagenin, proposed structure. Helv. Chim. Acta **50**: 491.
3. KENNEDY, L. M. & B. P. HALPERN. 1980. Extraction, purification and characterization of a sweetness-modifying component from *Ziziphus jujuba*. Chem. Senses **5**(2): 123-147.
4. KIMURA, Y., Y. KOBAYASHI, Y. TAKEDA & Y. OGIHARA. 1981. Three new saponins from the leaves of *Hovenia dulcis*. J.C.S. Perk. Trans. **I**: 1923-1927.
5. ADAMS, M. A. 1985. Substances that modify the perception of sweetness. *In* Characterization and Measurement of Flavor Compounds. D. D. Bills & C. J. Mussinan, Eds. ACS Symposium Series 289. Washington, D.C.
6. KURIHARA, Y., K. OOKUBO & B. P. HALPERN. 1987. Purification and chemical structure of taste modifiers: Taste-modifying protein and ziziphin. Ann. N.Y. Acad. Sci. This volume.
7. KENNEDY, L. M. & B. P. HALPERN. 1981. Actions of gymnemic acids and ziziphins: Dose-effect and time course relationships. Assoc. Chemorecept. Sci. **III**: 23.

Electron Microscopy of the Olfactory Epithelium in Zinc-Deficient Rats

SHUNTARO SHIGIHARA,[a] AKIHIRO IKUI, JUNKO
SHIGIHARA, AND HIROSHI TOMITA

Department of Otorhinolaryngology
Nihon University School of Medicine
Itabashi-ku Tokyo, Japan

MASAOMI OKANO

Department of Veterinary Anatomy
College of Agriculture and Veterinary Medicine
Nihon University
Fujisawa-shi Kanagawa, Japan

The role of zinc in living organisms is recognized as being important. In patients with olfactory disorders, it was found that the serum zinc level is lower than normal, that administration of zinc relieved symptoms. The purpose of this study was to demonstrate the relationship between zinc deficiency and olfactory disturbance.

MATERIALS AND METHODS

Wister strain rats were selected as subjects for the study. They were divided into the following six groups.

1. Young rat group in which a control diet was given. Male rats aged five weeks were fed for 15 weeks with the control diet, which contained 48.4 ppm zinc (five rats).

2. Young rat group in which a zinc-deficient diet was given. Male rats aged five weeks were fed for 9 to 17 weeks with the zinc-deficient diet containing 1.96 ppm zinc (eight rats).

3. Old rat group in which a control diet was given. Multipara female rats aged from 60 to 69 weeks were fed for 12 weeks with the control diet containing 53.0 ppm zinc (five rats).

[a] Address for correspondence: Dr. Shuntaro Shigihara, Department of Otorhinolaryngology, Nihon University School of Medicine, 1-32-16-203, Komone, Itabashi-ku Tokyo, Japan.

4. Old rat group in which a zinc-deficient diet was given. Multipara female rats aged 55 to 69 weeks were fed for 7 to 17 weeks with the zinc-deficient diet containing 8.6 ppm zinc (20 rats).

5. Young rat group in which zinc was readministered. Male rats aged five weeks were fed for 10 weeks with the zinc-deficient diet, and then each of them were fed for seven weeks either with the control diet containing 48.4 ppm zinc (one rat) or 33.3 ppm zinc (one rat).

6. Old rat group in which zinc was readministered. Multipara female rats aged 55 weeks were fed for 12 weeks with the zinc-deficient diet, and then for three weeks (one rat) and five weeks (two rats) with the control diet containing 53.0 ppm zinc.

They were anesthetized with Nembutal. After thorough perfusion with 2.5% glutaraldehyde and 2% paraformaldehyde, enucleated olfactory mucosa were then fixed in 2% osmic acid, dehydrated, embedded in Epon 812, sliced thinly, and double stained. Observations were then made under transmission-type electron microscope JEM-100CX (Nihon Denshi K.K.).

RESULTS

In the young zinc-deficient rat group, transmission electron microscopy revealed a marked widening of the intercellular space, shrinkage of the nucleus, irregularity

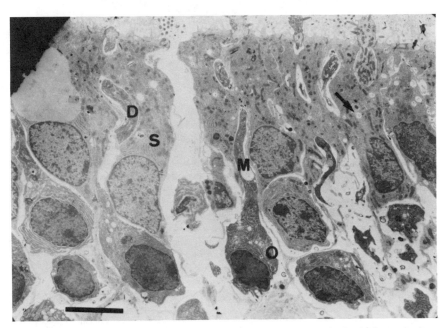

FIGURE 1. Olfactory mucosa of a young zinc-deficient rat. Multivesicular bodies and other granules in the supporting cells and receptor cells are increased. S, supporting cell; O, olfactory receptor cell; D, degenerating dendrite; M, multivesicular body. Arrow: various vesicles in a supporting cell. Scale bar: 5 μ.

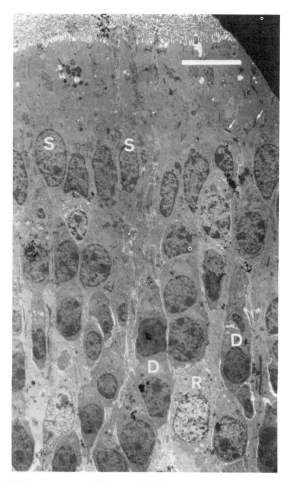

FIGURE 2. Olfactory mucosa of zinc-readministered young rats. Some groups of regenerating olfactory receptor cells were present, but there are also olfactory cells that have an electron-dense cytoplasm. S, supporting cell; D, degenerating olfactory receptor cell; R, regenerating olfactory receptor cell. Arrow: needlelike bodies in supporting cell. Scale bar: 10 μ.

of the chromosomes, and an increase in cytoplasmic small granules were observed. In addition, far more multivesicular bodies than normal were found. Similar degeneration was noted in the dendrites and olfactory vesicles. Immature cells containing many centrioles were observed as well as degenerating cells that included many membrane-bound bodies. In supporting cells the degenerative changes were not so marked as those in the olfactory receptor cells (FIG. 1).

In the old zinc-deficient rat group, degeneration of the olfactory receptor cells was found that was mild but similar to that in the young rat group.

In the young and old rat groups in which zinc was readministered, degenerative findings remained, but numerous groups of regenerated neurons were found (FIG. 2).

In the supporting cells of all rat groups that were fed with a zinc-deficient diet, many needlelike bodies, presumably attributable to degeneration, were observed.

The results of this study suggest that zinc deficiency causes degeneration in olfactory mucosa, and its changes are reversible.

Stereospecificity of the Alkyl Site for Optical Isomers of Dipeptides in the Labellar Sugar Receptor of the Fleshfly[a]

ICHIRO SHIMADA

Department of Biological Science
Tohoku University
Kawauchi, Sendai 980, Japan

YUJI MAKI

Department of Chemistry
Faculty of Science
Yamagata University
Yamagata 990, Japan

HIROSHI SUGIYAMA

Chemical Research Institute of Non-aqueous Solution
Tohoku University
Katahira, Sendai 980, Japan

There are four receptor sites in a single labellar sugar receptor of the fleshfly. They are the pyranose (P) site, the furanose (F) site, the alkyl (R) site, and the aryl (Ar) site.[4] Most of the stimulative dipeptides react with the R site.[2] L-Glu-L-Val is remarkably stimulative of them. We have already proposed the specific accessory site for the glutamyl moiety of Glu-Val(L-L), very close to the R site, based on the analysis of stimulating effectiveness of its analogues.[3]

We synthesized optical isomers of Glu-Val, Glu-norVal, Glu-Leu, and Glu-norLeu and examined their stimulating effectiveness. In case of Glu-Val, the order of effectiveness is L-L $\geq$ D-L > D-D > L-D. While Glu-Val (L-D) is almost ineffective, Glu-Val (D-D) is clearly effective, which is in contrast to the ineffectiveness of all the D-amino acids. This can be explained by considering the flexibility of the conformation of dipeptides and the strong interaction between the glutamyl moiety and the accessory site.

[a] Supported by a Grant-in-Aid of Special Research on Molecular Mechanisms of Bioelectrical Response (60115002) from the Japanese Ministry of Education, Science and Culture.

In other cases of dipeptides, the order of effectiveness of less stimulative L-D series is Glu-Val (L-D) < Glu-norVal (L-D) < Glu-norLeu (L-D) < Glu-norLeu (L-D). This may be due to the length of the side chain of COOH-terminal amino acids of dipeptides under the condition of specific binding of glutamyl moiety with the accessory site. Glu-norLeu (L-D), for example, is as stimulative as other optical isomers of Glu-norLeu. If the side chain becomes longer, it can bind more tightly with the subsite A of the R site. The interpretation was further supported by weak effectiveness of L-Glu-D-α-amino butyric acid that has a shorter side chain and of L-Val-D-Val that cannot bind the accessory site.

REFERENCES

1. SHIMADA, I. 1978. The stimulating effect of fatty acids and amino acid derivatives on the labellar sugar receptor of the fleshfly. J. Gen. Physiol. **71:** 19-36.
2. SHIMADA, I. & T. TANIMURA. 1981. Stereospecificity for amino acids and small peptides of multiple receptor sites in a labellar sugar receptor of the fleshfly. J. Gen. Physiol. **77:** 23-39.
3. SHIMADA, I., Y. MAKI & H. SUGIYAMA. 1983. Structure-taste relationship of glutamyl valine, the "sweet" peptide for the fleshfly: The specific accessory site for the glutamyl moiety in the sugar receptor. J. Insect Physiol. **29:** 255-258.
4. SHIMADA, I., H. HORIKI, H. OHRUI & H. MEGURO. 1985. Taste responses to 2,5-anhydro-D-hexitols; rigid stereospecificity of the furanose site in the sugar receptor of the fleshfly. J. Comp. Physiol. **157:** 477-488.

The Role of Taste in the Feeding Mechanism of the Carp (Cyprinidae)

FERDINAND A. SIBBING

Department of Experimental Animal Morphology and Cell Biology
Agricultural University
Marijkeweg 40
6709 PG Wageningen, the Netherlands

In many cyprinids and other fish, food is ingested with a flow of water, in case of the carp after detection by taste buds on its lips and barbels. In substratum feeders, such as carp and bream, the ingested water and waste are expelled while the palatable food is retained. Thus, besides the size-dependent selection of particles by the branchial sieve, a quality-dependent selective mechanism must also be available, sorting out food from nonfood. This paper deals with the sensor and effector system of such a refined "taste and sorting" mechanism in carps.

Light- and X-ray movies, combined with electromyography, showed that the subtle timing of upper jaw protrusion, opening of the mouth and the opercular valves, and

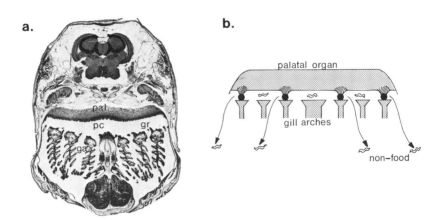

FIGURE 1. (a) Transverse section (7 μm) through the pharynx region of the carp. Note the muscular palatal organ (pal), unique for cyprinoid fish, and the perforated pharyngeal floor, composed by the gill arches (ga) and their rakers (gr). Unlike the orobuccal cavity, this pharyngeal cavity (pc) is slitlike, providing an extensive selection surface, and its lining is densely packed with taste buds (up to $820/mm^2$; from Sibbing *et al.*[3]).

(b) This scheme shows the mechanism of selective retention. Taste buds lining the palatal organ sense food particles and trigger a reflex mechanism, which activates the underlying palatal organ to produce local muscular projections. These clamp the sensed food particles locally against the floor, whereas waste is flushed with a flow of water through the branchial and opercular slits.

a. SELECTIVE RETENTION

(at oral compression)

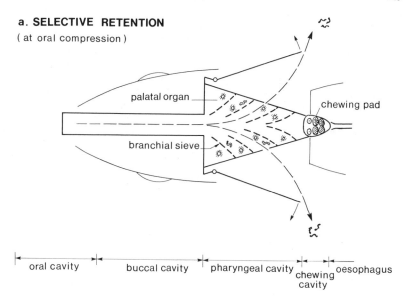

| oral cavity | buccal cavity | pharyngeal cavity | chewing cavity | oesophagus |

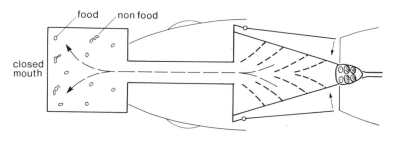

b. RESUSPENSION

(by closed protrusion)

FIGURE 2. Purification of a mixture of food and nonfood particles by a back-and-forth flow through the oropharyngeal cavity. During oral compression (**a**) the mixture is forced into the pharyngeal slit. Local bulging of the palatal organ clamps the food to be retained, whereas most waste is flushed with the water through the branchial and opercular slits (cf. FIG. 1b). Closed protrusion of the upper jaws (**b**) sucks the remaining mixture back to the expanded oral cavity where it is resuspended. Repetition of selective retention and resuspension effects a graded purification of the food (e.g., less than 5% of organic and inorganic debris).

of the volume changes in the oral, buccal, pharyngeal, and opercular cavities allows a versatile manipulation of the water and particles. Ten different movement patterns are employed for twelve types of feeding actions, dependent on the stage of processing and the type of food.[1] For example, toxic and large unpalatable or heavily contaminated food is ejected by spitting, less contaminated large food particles are purified by rinsing. Repetitive intake and spitting is employed for probing the substratum and for improving the grip on unmanageable lumps of food.

Histological and SEM techniques showed sensory structures in the oropharyngeal lining and permitted quantification of detailed distribution patterns of taste buds. Their densities vary from 40-50/mm^2 in the buccal lining to almost maximal densities of 820/mm^2 in the pharyngeal area, which permits very local spot measurement in gustation. Whether taste buds play a mechanoreceptive role also (cf. Reutter et al.[2]) is not clear. Mechanoreceptors monitoring the flow of water and particles are required for their adequate manipulation but have not been identified yet.

Oligovillous cells in the oropharyngeal lining[3] may belong to the common chemical sense and may react on toxic (cf. Whitear and Lane[4]) or feeding stimuli (cf. Silver & Finger[5]).

X-ray movies show that particles are slowed down and trapped in the gustatory area as the tubular buccal cavity widens into the slit-like pharynx (FIG. 1a). Taste buds in the pharyngeal roof lie on top of a muscular cushion, the palatal organ, which is active (electromyography) in selection and food transport.[1] Local contractions clamp food particles between roof and floor, that is, selective retention, whereas waste is flushed with the water through the branchial slits (FIG. 1b). Mucus is less and of a different type, less viscous, compared with the transport- and mastication area.[3] "Closed protrusion" and compressive movements of the snout manipulate the suspension repetitively back- and forwards, until the food is finally purified (FIG. 2). The slit-shaped cavity assures a large area for selection as well as close contact between roof and floor, whose gill rakers are almost equally densely packed with taste buds. The gill rakers are also movable and cooperate with the palatal organ.

Such a "taste and sorting" mechanism requires a very local steering of the effector apparatus and close coordination between roof and floor. These demands appear to be fulfilled anatomically by a viscerotopic mapping of the oropharyngeal lining in the vagal lobe, whose laminar structure and radially organized connections between sensory and motor neurons of both pharyngeal roof and floor[6,7] will allow very local reflexes in their opposed areas. The less punctuated mapping of the pharyngeal floor compared with its roof, which the latter authors found, can be explained by the overlap in working area of the gill rakers. This is due to their interdigitation between subsequent arches and their movement with respect to the palatal organ during feeding.

The integration of subsequent feeding actions[1] into an effective sequence and the remarkable plasticity in handling different types of food make us expect a highly organized substrate for the association of stimuli and for steering the feeding process in cyprinids. They help to explain the enormous success of carp all over the world.

REFERENCES

1. SIBBING, F. A., J. W. M. OSSE & A. TERLOUW. 1986. Food handling in the carp (*Cyprinus carpio* L.): Its movement patterns, mechanisms and limitations. J. Zool. London (A)**210**: 161-203.

2. REUTTER, K., W. BREIPOHL & G. J. BIJVANK. 1974. Taste buds types in fishes. II. Scanning electron microscopical investigations on *Xiphophorus helleri* Heckel (Poeciliidae, Cyprinodontiformes, Teleostei). Cell Tissue Res. **153**: 151-165.

3. SIBBING, F. A. & R. URIBE. 1985. Regional specializations in the oro-pharyngeal wall and food processing in the carp (*Cyprinus carpio* L.). Neth. J. Zool. **35**(3): 377-422.

4. WHITEAR, M. & E. B. LANE. 1983. Oligovillous cells of the epidermis: Sensory elements of the lamprey skin. J. Zool. London **199**: 359-384.

5. SILVER, W. L. & T. E. FINGER. 1984. Electrophysiological examination of non-olfactory, non-gustatory chemosense in the searobin, *Prionotus carolinus*. J. Comp. Physiol. (A) **154**: 167-174.

6. FINGER, T. E. 1981. Laminar and columnar organization of the vagal lobe in goldfish: Possible neural substrate for sorting food from gravel. Soc. Neurosci. Abstr. **7:** 665.
7. MORITA, Y. & T. E. FINGER. 1985. Topographic and laminar organization of the vagal gustatory system in the goldfish, *Carassius auratus.* J. Comp. Neurol. **238:** 187-201.

Trigeminal Chemoreceptors Cannot Discriminate between Equally Intense Odorants[a]

WAYNE L. SILVER

Department of Biology
Wake Forest University
Winston-Salem, North Carolina 27109

ADAM H. ARZT AND J. RUSSELL MASON

Monell Chemical Senses Center
Philadelphia, Pennsylvania 19104

Both electrophysiological[1] and psychophysical[2] evidence have demonstrated that trigeminal receptors in the nasal cavity respond to odorants. These free nerve endings constitute part of the "common chemical sense," whose major function is often purported to be the protection of the organism from noxious chemicals.[3] Despite these demonstrations of trigeminal chemoreception, it is not clear whether trigeminal receptors can discriminate between odorants matched for equal intensity. The present electrophysiological and behavioral experiments using tiger salamanders were designed to address this issue.

In electrophysiological experiments, integrated multiunit activity was recorded from the ophthalmic branch of the trigeminal nerve. Concentration-response curves were obtained for each of four odorants (amyl acetate [AA], cyclohexanone [CH], butanol [BU], and d-limonene [LI]) delivered to the nose via an air dilution olfactometer (also used in the behavioral experiments). The concentration of each compound necessary to produce an equivalent response (150% of a CH standard, $\approx$ 1,100 ppm) was then used as the background (cross-adapting) stimulus in cross-adaptation experiments pairing CH and AA or LI and BU.

A cross-adapting stream of one odorant severely reduced responsiveness to both itself and its partner (FIG. 1). At concentrations below background, responding was eliminated. Only concentrations above background increased neural activity above baseline.

For behavioral experiments, salamanders were trained to avoid AA (or LI) but not to respond to CH (or BU). In tests subsequent to concentration-response trials, both groups discriminated between odorant pairs that were matched for equal intensity, that is, concentrations that elicited 80% avoidance.

The animals were then given bilateral olfactory nerve cuts. After lesioning, higher concentrations ($\approx$ 1 log unit) were necessary to elicit 80% avoidance. In addition,

[a]Supported by National Science Foundation Grant #BNS-8310892.

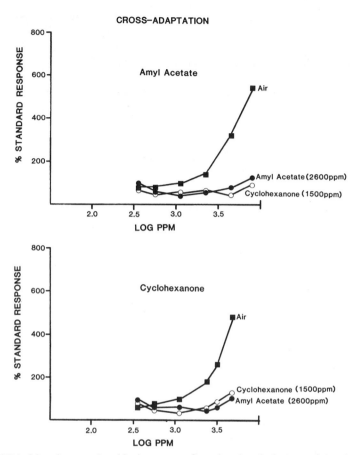

FIGURE 1. Mean integrated multiunit responses from the trigeminal nerves of six salamanders to increasing concentrations of AA and CH. Response magnitude is presented as a percent of the response to the standard stimulus, ≈1,100 ppm CH. Stimuli were presented against a background of either air, AA, or CH. The background concentrations of AA (≈2,600 ppm) and CH (≈1,500 ppm) elicited responses of equal magnitude as determined in earlier tests.

salamanders with severed olfactory nerves could no longer discriminate between odorants matched for equal intensity (concentrations that elicited either 80% or 90% avoidance) (FIG. 2).

On the basis of these electrophysiological and behavioral results, we conclude that trigeminal chemoreceptors are unable to discriminate between odorants matched for equal intensity. At least for the odorants used here, trigeminal chemoreceptors discriminate odorant quantity, not quality. We propose that qualitatively, all odorants are the same for the trigeminal system and speculate that the mechanism of stimulation is similar for all odorants.

POST-SURGICAL DISCRIMINATION PERFORMANCE

BEHAVIORALLY EQUAL CONCENTRATIONS

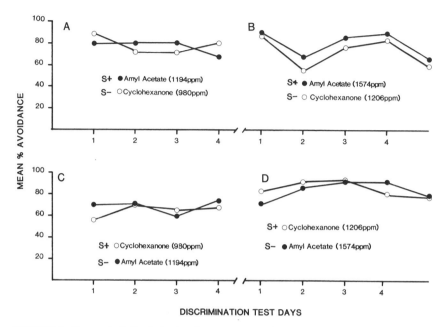

FIGURE 2. Postsurgical discrimination between AA and CH. Odorant concentrations were "equal," in that they elicited 80% avoidance (top panels) or 90% avoidance (bottom panels) during concentration-response tests.

REFERENCES

1. SILVER, W. L. & D. G. MOULTON. 1982. Chemosensitivity of rat nasal trigeminal receptors. Physiol. Behav. **28:** 927-931.
2. DOTY, R. L., W. E. BRUGGER, P. C. JURS, M. A. ORNDORFF, P. F. SNYDER & L. D. LOWRY. 1978. Intranasal trigeminal stimulation from odorous volatiles: Psychometric responses from anosmic and normal humans. **20:** 175-187.
3. SILVER, W. L. & J. A. MARUNIAK. 1981. Trigeminal chemoreception in the nasal and oral cavities. Chem. Senses **6:** 295-305.

Differences in Epithelial Responses of Rabbit Tongue to KCl and NaCl[a]

S. A. SIMON AND R. ROBB

Departments of Physiology and Anesthesiology
Duke University Medical Center
Durham, North Carolina 27710

J. L. GARVIN

Laboratory of Kidney and Electrolyte Metabolism
National Heart, Lung and Blood Institute
Bethesda, Maryland 20892

Recently, Contreras *et al.*[1] concluded from behavioral studies that " . . . rabbits seem to have a stronger perference for KCl than NaCl." They also measured the integrated chorda tympani responses (ICTR) of rabbits and showed that the responses are much greater for KCl than NaCl over a large concentration range. These ICTR recordings, which confirmed previous ones,[2,3] suggest a relation between ICTR and behavior. However both the above measurements are the consequences of events that occurred after these salts interacted with the tongue to produce an electrical response.

In a series of papers that characterized the epithelial properties of mammalian tongue, DeSimone and coworkers[4-7] and Simon and coworkers[8-10] have shown first, that tongues actively transport ions and second, that some of these transport processes are involved in taste transduction. In this report we will show that the epithelial responses of rabbit tongue are stimulated to a much greater extent by KCl than NaCl and that K^+ and Na^+ traverse rabbit tongue through different pathways. Our results are consistent with the ICTR and behavioral studies and, in addition, correlate the epithelial behavior of rabbit tongue to transduction events.

Methods for excising tongues of female New Zealand White rabbits and for measuring the open circuit potential (V_{oc}) and short circuit current (I_{sc}) are described elsewhere.[8-10]

In symmetric solutions of KH buffer, rabbit tongues exhibited the following electrical characteristics: $V_{oc} = 6.6 \pm 0.6$ mV, $I_{sc} = -26.7 \pm 2.4$ $\mu A/cm^2$, and Rm $= 245 \pm 18\Omega \cdot cm^2$ ($n = 26$). V_{oc} and I_{sc} are decreased above 40% or 95% when 1 mM ouabain or dinitrophenol are added to the serosal solution, respectively. These results show that: (a) the rabbit tongue actively transports ions, and (b) under these conditions about 40% of the active transport occurs through the Na-K-ATPase.

The dose-response curves of I_{sc} and V_{oc} to KCl and NaCl are seen in FIGURE 1. The response of V_{oc} to KCl and NaCl are slightly different when corrections for junction potentials are made. However, KCl stimulates I_{sc} to a much greater extent

[a] This research was supported by Grants NS-20669 and AG-00443 from the Institute of Sensory Disorders and Language, National Institutes of Health.

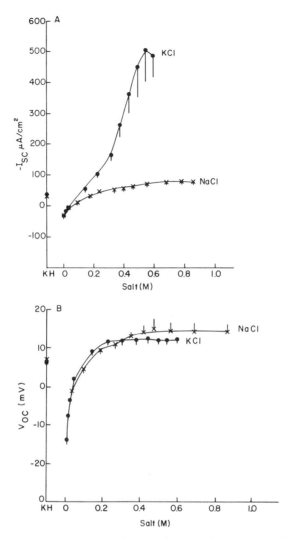

FIGURE 1. (A) Dose-response curves of short-circuit current (I_{sc}) versus KCl (open circles) and NaCl (x) on the mucosal side of rabbit tongue. Each mucosal salt solution also contained 2 mM Tris Cl and was bubbled with 100% O_2 to maintain its pH at 7.4. Temperature = 36°C. The serosal side was bathed in Krebs-Henseleit (KH) buffer. The composition of KH (in mM) was 121 NaCl, 6 KCl, 1.3 MgSO$_4$, 1.2 CaCl$_2$, 25 NaHCO$_3$ and 5.6 dextrose pH 7.4. On the extreme left-hand side, the mean values of I_{sc} in symmetric solutions of KH are given. Responses of I_{sc} from 0.01 to 0.1 M are not statistically significant. The I_{sc} is defined as negative when cations are following from the mucosal to serosal surface. Points are plotted as mean ± SE for six and three experiments for KCl and NaCl, respectively.

(B) Dose-response curves of open-circuit potential (V_{oc}) versus KCl (open circles) and NaCl (x) on the mucosal side of rabbit tongue. The conditions are the same as given in A. Under symmetric ionic conditions V_{oc} will be positive on the serosal side. Differences in V_{oc} between KCl and NaCl are slightly different when corrections for junction potentials are made.

TABLE 1. Comparison of Epithelial Changes with Respect to 0.5 M KCl for Primary Tastants with ICTR for Rabbit Tongue[a]

Stimulus M	Isc (x) / ISC(0.5 KCl)	Voc(x) / Voc(0.5 KCl)	ICTR(x) / ICTR(0.5 KCl)	
0.5 KCl	1.0	1.0	1.0[c]	1.0[d]
0.5 NH$_4$Cl	0.64	0.56		1.09
0.5 NaCl	0.17	1.16[b]	0.21	
0.1 NH$_4$Cl[f]	1.0	1.0		1.0
0.1 NaCl	0.31	1.0		0.25
0.3 KCl	1.0	1.0	1.0[e]	
0.3 NaCl	0.32	1.0	0.33	

[a] Isc, short-circuit current; Voc, open-circuit potential; ICTR, integrated chorda tympani responses.
[b] Significantly different than 1.0.
[c] Pfaffmann.[3]
[d] Beidler *et al.*[2]
[e] Contreras *et al.*[1]
[f] Ratios of two salts.

than NaCl. Therefore, the differences of I_{sc} in the presence of NaCl and KCl are due to the differences in the transport of the cations and not to differences in ionic strength (hence surface potential) or osmolality. The responses of I_{sc} to amiloride and ouabain in 500 mM KCl or NaCl were also different for KCl and NaCl. Amiloride (0.1 mM) added to the mucosal surface inhibited I_{sc} by 10% and 52% for KCl and NaCl, respectively, whereas 1 mM ouabain added to the serosal surface inhibited I_{sc} 10% and 55%. These pharamacological effects, taken together with the different shapes and magnitudes of the stimulation of I_{sc} by KCl and NaCl, suggest that (at concentrations greater than 150 mM) Na$^+$ and K$^+$ traverse the tongue through different pathways. At lower concentrations, no differences in I_{sc} were observed. A similar pattern was obtained by Contreras *et al.*[1] in measurements of ICTR. However, responses to NaCl and KCl were significantly different for concentrations greater than 50 mM. The ratio of the epithelial responses of V_{oc} and I_{sc} to 0.5 M KCl are given together with the integrated chorda tympani responses (ICTR) in TABLE 1. Also included are the epithelial and ICTR to NH$_4$Cl. There is good agreement between the relative stimulation of I_{sc} and ICTR, suggesting that epithelial responses are involved in the transduction process. The larger responses of KCl relative to NaCl may be reflected in the rabbits' diet, which consists of foods that have higher concentrations of KCl than NaCl. In comparison, for dogs, which are carnivores, the epithelial responses are slightly greater for NaCl relative to KCl.

REFERENCES

1. CONTRERAS, R. J., E. BIRD & D. J. WEISZ. 1985. Behavioral and neural gustatory responses in rabbit. Physiol. Behav. **34:** 761-768.

2. BEIDLER, L. M., I. Y. FISHMAN & K. W. HARDIMAN. 1955. Species differences in taste responses. Am. J. Physiol. **181**: 235-239.
3. PFAFFMAN, C. 1955. Gustatory nerve impules in rat, cat and rabbit. J. Neurophysiol. **18**: 429-440.
4. DeSIMONE, J. A. & G. L. HECK. 1980. An analysis of the effect of stimulus transport and membrane charge on the salt, acid and water-response of mammals. Chem. Senses **5**: 295-316.
5. DeSIMONE, J. A., G. L. HECK & S. K. DeSIMONE. 1981. Active ion transport in dog tongue: Possible role in taste. Science **214**: 1039-1041.
6. DeSIMONE, J. A., G. L. HECK, S. MIERSON & S. K. DeSIMONE. 1984. The active ion transport properties of canine lingual epithelia *in vitro*. J. Gen. Physiol. **83**: 633-656.
7. MIERSON, S., G. HECK, S. K. DeSIMONE, T. U. L. BIBER & J. A. DeSIMONE. 1985. The identity of the current carriers in canine lingual epithelium. Biochem. Biophys. Acta **816**: 283-293.
8. SIMON, S. A. & J. L. GARVIN. 1985. Salt and acid studies on canine lingual epithelium. Am. J. Physiol. **249**(Cell Physiol. **18**): C398-C408.
9. SIMON, S. A. & J. L. GARVIN. 1986. Different responses to salts and sugars in the anterior and posterior regions of dog tongue (abstract). Biophys. J. **49**: 22a.
10. SIMON, S. A., R. ROBB & J. L. GARVIN. 1986. Epithelial responses to rabbit tongue and their involvement in taste transduction. Am. J. Physiol. **251**(20): R598-R608.

The Odorant-Sensitive Adenylate Cyclase of Olfactory Receptor Cells

Differential Stimulation by Distinct Classes of Odorants

PAMELA B. SKLAR, ROBERT R. H. ANHOLT,[a] AND
SOLOMON H. SNYDER

*Departments of Neuroscience, Pharmacology and Experimental
Therapeutics, Psychiatry and Behavioral Sciences
The Johns Hopkins University School of Medicine
Baltimore, Maryland 21205*

We have characterized the odorant-stimulated adenylate cyclase in chemosensory cilia isolated from frog and rat olfactory epithelium.[1-3] In agreement with Pace *et al.,*[4] we find high levels of basal adenylate cyclase activity in frog olfactory cilia (2.0 ± 1.0 nmol/mg/min). In addition, we have demonstrated that rat olfactory cilia display comparably high levels of enzyme activity (3.6 ± 1.9 nmol/mg/min). Basal activity in both species is stimulated approximately twofold by GTP and approximately fivefold by guanosine-5'-0-(3-thiotriphosphate) (GTPγS) and forskolin. Calcium reduces GTP-stimulated activity with half-maximal effect at 10 μM. Odorants augment enzyme activity 30-65% above the basal level in a tissue-specific and GTP-dependent way, suggesting that odorant signals in the olfactory system may be transduced through regulatory GTP-binding proteins.[3,4]

We tested a series of odorants for their ability to stimulate adenylate cyclase in olfactory cilia (TABLE 1). Many floral, fruity, herbaceous, and minty odorants are potent stimulators of the enzyme. Citralva (3,7-dimethyl-2,6-octadienenitrile), a substituted terpenoid odorant having a fruity odor quality, is one of the most potent cyclase stimulators. We have, therefore, arbitrarily assigned a standard value of 100% stimulation to the cyclase activity elicited by 100 μM citralva. Maximal stimulation by citralva corresponds to a 55% increase in activity over the GTP-stimulated basal level. Many odorants that are structurally unrelated to citralva, such as menthone, D-carvone, L-carvone, 3-hexyl pyridine, 2-hexyl pyridine, hedione, helional, and coniferan, are also potent stimulators of the olfactory adenylate cyclase. We compared the concentration-response behavior of citralva, methone, the steroisomers D-carvone and L-carvone, and the potent bell-pepper odorant, 2-isobutyl-3-methoxypyrazine. These odorants all stimulate the enzyme in a concentration-dependent manner between 1 and 100 μM with similar potencies (FIG. 1A).

[a] Present address: Department of Physiology, Duke University Medical Center, Box 3709, Durham, North Carolina 27710.

TABLE 1. Stimulation by Odorants of the GTP-Dependent Adenylate Cyclase in Olfactory Cilia

Odorant (100 μM)	Stimulation (%)	Odorant (100 μM)	Stimulation (%)
Fruity		**Putrid**	
Citralva	100	Furfuryl mercaptan	29 ± 9(3)
Citral dimethyl acetal	69 ± 10(5)	Triethylamine	4 ± 7(5)
Citronellal	56 ± 5(3)	Phenylethylamine	0 ± 7(3)
β-Ionone	55 ± 5(5)	Isobutyric acid	−2 ± 7(3)
Citronellyl acetate	50 ± 9(4)	Pyrrolidine	−4 ± 6(2)
Isoamyl acetate	19 ± 11(8)	Isovaleric acid	−6 ± 8(5)
Limonene	5 ± 4(5)	**Odorous Chemical Solvents**	
Lyral	−4 ± 6(2)	Toluene	14 ± 8(3)
Floral		Ethanol	10 ± 7(3)
3-Hexyl pyridine	118 ± 10(3)	Chloroform	5 ± 2(3)
2-Hexyl pyridine	107 ± 8(3)	Butanol	4 ± 10(3)
Hedione	63 ± 4(3)	Pyridine	4 ± 10(4)
Coniferan	60 ± 20(2)	Xylene	−2 ± 3(2)
Geraniol	58 ± 3(3)	Acetic acid	−14 ± 33(3)
Helional	53 ± 6(5)	**Methoxy Pyrazines**	
Decanal	53 ± 8(2)	2-Isobutyl-3-methoxypyrazine	53 ± 4(10)
Amyl salicylate	40 ± 10(2)	2-Isopropyl-3-methoxypyrazine	36 ± 8(5)
Dimethyloctanol	33 ± 9(3)	2-Ethyl-3-methoxypyrazine	20 ± 8(5)
Acetophenone	30 ± 4(3)	2-Methyl-3-methoxypyrazine	9 ± 3(3)
α-Pinene	21 ± 12(3)	Methoxypyrazine	−5 ± 5(8)
Phenylethyl alcohol	19 ± 4(3)	**Alkyl Pyrazines**	
Lilial	−1 ± 4(2)	2,3-Diethyl-5-methylpyrazine	16 ± 9(3)
Minty		2-Methylpyrazine	7 ± 6(3)
Isomenthone	105 ± 10(3)	2,3,5,6-Tetramethylpyrazine	0 ± 11(3)
L-Carvone	74 ± 31(6)	Pyrazine	0
Menthone	71 ± 3(4)	2-Ethyl-3-methylpyrazine	−2 ± 7(3)
Eucalyptol	45 ± 8(4)	2,3,5-Trimethylpyrazine	−3 ± 1(3)
Herbaceous		2,3-Dimethylpyrazine	−6 ± 8(3)
D-Carvone	74 ± 4(7)	2-Ethylpyrazine	−7 ± 8(2)
Eugenol	47 ± 7(5)	**Thiazoles**	
Cinnamic aldehyde	34 ± 5(4)	2-Isobutylthiazole	59 ± 5(4)
Isoeugenol	31 ± 5(3)	2-Acetylthiazole	3 ± 7(5)
Benzaldehyde	27 ± 4(4)	2,4-Dimethyl-5-acetylthiazole	14 ± 6(2)
Ethyl vanillin	−3 ± 6(5)	Thiazole	−18

[a] Odorants were tested at 100 μM in the presence of 10 μM GTP. Data are expressed as a percentage of the activity observed in the presence of 100 μM citralva. Citralva stimulation was 17% and 22% of the stimulation by 2 μM forskolin and 10 μM GTPγS, respectively. Values are expressed as the mean ± SEM of (n) experiments.

Odorants structurally related to citralva, such as citral dimethyl acetal and citronellal, stimulate the enzyme to ∼69% and ∼59% of the citralva-stimulated level, respectively (TABLE 1). Several fruity, floral, minty, and herbaceous odorants stimulate activity to an intermediate level. Examples include amyl salicylate (40%), dimethyloctanol (33%), eucalyptol (45%), eugenol (47%) and cinnamic aldehyde (34%). Interestingly, the classes of fruity, floral, minty, and herbaceous odorants also contain some odorants that do not stimulate the enzyme such as limonene, lyral, phenylethylalcohol, lilial, and ethylvanillin (TABLE 1).

In contrast to the fruity, floral, minty, and herbaceous odorants, a number of putrid odorants and odorous chemical solvents do not stimulate adenylate cyclase activity (TABLE 1). Putrid odorants, such as isovaleric acid and triethylamine, and odorous chemical solvents, such as pyridine, do not enhance enzyme activity at concentrations up to 10 mM (FIG. 1B). The failure of certain groups of odorants to

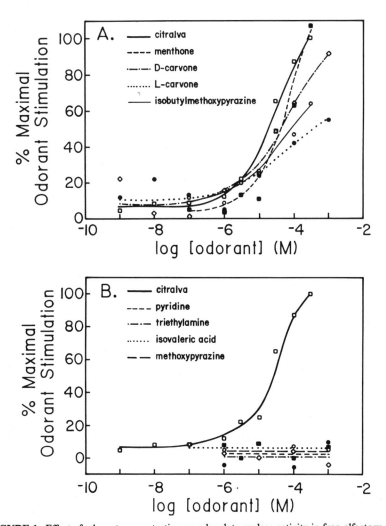

FIGURE 1. Effect of odorant concentration on adenylate cyclase activity in frog olfactory cilia. (**A**) Odorants that stimulate adenylate cyclase activity, (————) citralva, (- - -) menthone, (̄ ̄) 2-isobutyl-3-methoxy-pyrazine, (·—·—·) D-carvone, and (········) L-carvone. (**B**) Odorants that do not stimulate adenylate cylase activity, (— — —), pyridine, (·—·—·) triethylamine, (·····) isovaleric acid, and (— —) methoxy-pyrazine; (————) citralva is plotted for comparison. Data are expressed as a percentage of the enzyme activity observed in the presence of 100 μM citralva and are the mean of two to four experiments with a SEM $\leq$ 30% between different experiments.

stimulate the olfactory adenylate cyclase suggests that at least one additional transduction mechanism is involved in olfaction.[3]

To gain insight into the molecular parameters that determine the potency of an odorant as a cyclase stimulator, we investigated homologous series of structurally related odorants including the pyrazines, thiazoles, and pyridines. Stimulation of adenylate cyclase activity can be detected only when the parent compound, methoxypyrazine, thiazole, or pyridine has a hydrocarbon chain attached (TABLE 1). These results suggest that one factor that determines the potency of an odorant to activate the enzyme may be its hydrophobicity. Odorant hydrophobicity, as measured by a calculated octanol/water partition coefficient,[5] correlates with adenylate cyclase activity within the series of methoxypyrazine odorants ($r = 0.99$, $n = 5$). However, for odorants from varying odor classes, hydrophobicity is not closely correlated with cyclase activation ($r = 0.54$, $n = 56$). Thus, hydrophobicity alone is not the only factor that determines adenylate cyclase activation.

REFERENCES

1. PACE, U. & D. LANCET. 1986. Proc. Natl. Acad. Sci. USA 83: 4947-4951.
2. ANHOLT, R. R. H., U. AEBI & S. H. SNYDER. 1986. J. Neurosci. 6: 1962-1969.
3. SKLAR, P. B., R. R. H. ANHOLT & S. H. SNYDER. 1986. J. Biol. Chem. 261: 15538-15543.
4. PACE, U., E. HANSKI, Y. SALOMON & D. LANCET. 1985. Nature 316: 255-258.
5. CHOU, J. T. & P. C. JORS. 1979. J. Chem. Inf. Comput. Sci. 19: 172-178.

Failure of Rats to Acquire a Reversal Learning Set When Trained with Taste Cues

B. M. SLOTNICK, G. M. BROSVIC, [a] AND
S. R. PARKER

Department of Psychology
The American University
Washington, D.C.

If rats are trained on operant discrimination tasks with visual or auditory cues, they have slow acquisition and fail to acquire a learning set (fail to "learn to learn"). However, if they are trained with odor cues, their performance is exceptional. They rapidly learn simple odor detection and odor discriminations (often with few or no errors) and acquire reversal or multiple problem learning set tasks. They learn as well as monkeys do when trained on visual cues. In this study we ask whether this exceptional ability of rats to learn olfactory cues extends to other chemical sense stimuli such as taste.

METHODS

Eight rats were trained on a newly developed operant conditioning taste discrimination procedure (see Brosvic *et al.,* this volume). Briefly, a go, no-go discrimination procedure is used. The rat licks at a multiple-barrel stimulus delivery tube and is trained to switch its licking to a reinforcement tube when an S+ tastant is delivered but to continue licking at the stimulus delivery tube upon delivery of the S− tastant.

For four rats the stimuli were 0.9% NaCl and 0.1% saccharin (NaCl was the S+ for two rats). For the other four rats, the stimuli were 0.75% NaCl and 1% sucrose (NaCl was the S+ for two rats). When the rats learned the initial problem (criterion: 85% correct responding in two 20-trial blocks) the session was terminated and in the next session the significance of the stimuli was reversed and the animals were again trained to criterion. This procedure was continued until all rats were tested on seven reversal problems.

Note that rats trained in this manner on odors make fewer errors on the first reversal than in original learning and continue to improve until, after four to six reversal problems, each problem is solved in only a few errors. If lights or tones are used as discriminative cues, they make many more errors in the first reversal and do not acquire a learning set.

---

[a] Present address: Psychology Department, Glassboro State College, Glassboro, NJ 08028.

627

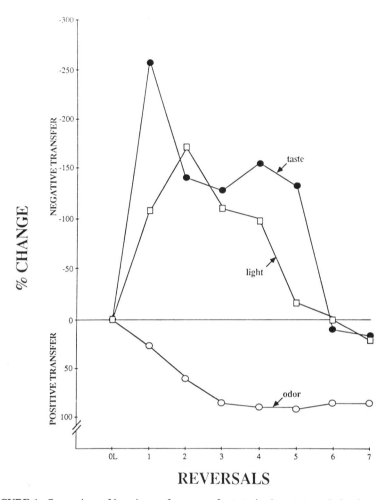

FIGURE 1. Comparison of learning performance of rats trained on taste and visual cues with rats trained on odor cues.

RESULTS

In all respects the performance of rats trained on taste cues was similar to that of rats trained on visual or tone cues. Each rat made many more errors in the first reversal than it did in initial learning and, while performance improved in subsequent reversals, none demonstrated acquisition of a learning set. A comparison of these data with those of rats trained on cues from other modalities (see FIG. 1) suggests that operant discrimination performance with taste cues closely resembles the relatively poor acquisition of a successive discrimination reversal shown by rats trained on odor cues. Thus, the advantage provided by odors for rapid learning of a discrimination reversal is not a property of chemical stimulation per se but may be specific to olfaction.

Perhaps it is not unexpected that a teleceptive modality like olfaction would play an important role in operant tasks involving the seeking out and avoiding of potential beneficial or noxious stimuli. In contrast, taste may play a critical role in identifying the consequences of ingestive behavior and thus provide a more potent stimulus for respondent conditioning in which illness serves as the unconditioned stimulus. Such conclusions are, in fact, predicted by Garcia, Hankins, and Rusiniak's model of the proposed roles of teleceptors and interoceptors in the control of learned behavior.

Odor Detection in Rats with Lesions of Olfactory Bulb Areas Identified Using 2-DG

B. M. SLOTNICK AND S. GRAHAM

Department of Psychology
The American University
Washington, D.C.

D. G. LAING AND G. A. BELL

CSIRO
Sidney, Australia

A number of studies using the 2-DG metabolic method have identified areas within the olfactory bulb glomerular layer that increase in activity with exposure of animals to specific odors. Because these results provide evidence for a spatial mechanism for coding odors, an examination of their functional significance is of interest. Recently, Laing *et al.*[1] found a distinct 2-DG focus centered in the glomerular layer of the dorsomedial quadrant of the rostral olfactory bulb in rats exposed to propionic acid vapor. We have examined the effects of destroying this quadrant of the bulb in rats on their ability to detect propionic acid and other odors.

METHODS

Eight rats were first trained to detect 0.5% (of vapor saturation) amyl acetate and then given sham lesions ($n = 3$) or lesions of the dorsomedial ($n = 3$) quadrant of the rostral olfactory bulbs. All tests used a positive pressure olfactometer and operant conditioning[2] in which an odor serves as the S+ stimulus and clean air serves as the S− stimulus. Postoperatively, all animals were tested for retention of the amyl acetate detection task, detection of propionic acid, propionic acid absolute threshold, and for detection of butanol, geranitorl, and acetic acid.

RESULTS

There were no appreciable differences in retention, acquisition, or threshold scores among the three groups. In particular, rats with dorsomedial lesions rapidly acquired the propionic detection task and had thresholds that were well within the range of controls.

These results indicate that an area of the olfactory bulb containing a major focus of metabolic activity induced by exposure to an odor can be removed without producing a deficit in the detection of that odor.

REFERENCES

1. LAING, D. G., G. A. BELL & A. PANHUBER. 1985. Human psychophysics and 2-DG reveal how and where suppression with odor mixture occurs. Chem. Senses **10:** 415.
2. SLOTNICK, B. M. & F. W. SCHOONOVER. 1984. Olfactory thresholds in unilaterally bulbectomized rats. Chem. Senses **4:** 325-340.

Taste Bud Development in Hamster Vallate and Foliate Papillae[a]

DAVID V. SMITH

Department of Otolaryngology and Maxillofacial Surgery
University of Cincinnati Medical Center
Cincinnati, Ohio 45267

INGLIS J. MILLER, JR.[b]

Department of Anatomy
Bowman Gray School of Medicine
Wake Forest University
Winston-Salem, North Carolina 27103

Our overall objective is to determine how multiple populations of taste receptors contribute to taste perception. In this project, we examined whether the taste buds in the hamster vallate and foliate papillae appear at the same time after birth and proliferate at similar or different rates to maturity. The date and time when a litter was delivered were recorded. Pups were overdosed in ether and fixed in AAF at ages from 1-120 days. Single vallate and foliate papillae were dissected from tongues, prepared in paraffin at 8 μm with serial sections, and stained with H & E. Each taste bud was marked on a diagram of the section, but only patent taste pores were counted. Foliate taste bud counts were obtained from 52 papillae: 19 bilateral and 14 unilateral. Vallate taste buds were counted in 25 hamsters. Foliate taste buds are shown at the top and vallate taste buds are at the bottom of the figure. The curves were fitted by least-squares to a polynomial equation.

No taste buds were present in either foliate or vallate papillae at birth. Developing taste buds were observed on the second postnatal day, but no open taste pores were observed until the third day. Pores were observed in both foliate and vallate papillae on day 4. The most rapid proliferation of taste buds occurred during the first two weeks in both kinds of papillae. Until about 10 days of age, only solitary taste buds are observed. After 10 days, pairs of closely oriented taste buds appeared, which gave the impression that taste buds may divide. From 15 to 30 days of age, the foliate taste buds slowed down their rate of appearance and then continued at a decreasing rate to reach a maximum at 60 days. The variation at each age could be observed in the pairs of foliate papillae from a single animal. Vallate taste buds increased in number

[a]This work was supported by Grants NS 23524 and NS 20101 from the National Institute of Neurological and Communicative Disorders and Stroke.

[b]Address for correspondence: Inglis J. Miller, Jr., Department of Anatomy, Bowman Gray School of Medicine, Wake Forest University, Winston-Salem, North Carolina 27103.

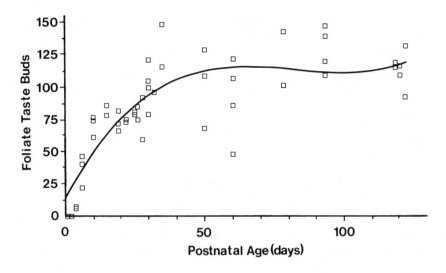

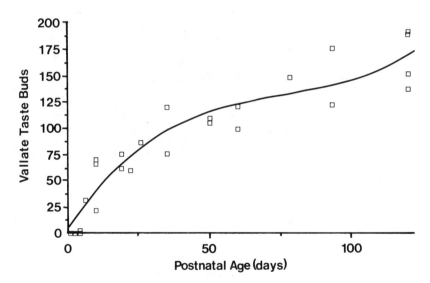

FIGURE 1. Taste bud development in hamster foliate and vallate papillae.

up to our oldest animals at 120 days of age. Thus, foliate and vallate taste pores appeared at the same age. After 60 days of age, the foliate papillae reached a maximum number of taste buds, while vallate taste buds increased in number to our oldest group at 120 days (FIG. 1).

Foliate and vallate taste buds appear functional during the first week of life. Although we have not determined the functional capacity of these neonate animals, we report elsewhere in this volume[1] the response characteristics of hamster vallate and foliate taste buds in adult animals. Quinine elicits a greater response in glossopharyngeal nerve (IXth) fibers that innervate foliate papillae than those that innervate vallate papillae. The IXth nerve also responds more to quinine and HCl than to sucrose and NaCl. Although these response characteristics yield no specific clues to the role of IXth nerve taste responses in the behavioral development of hamsters, important issues are involved. Previously, we[2] showed that vallate and foliate taste buds constitute about two-thirds of the total number of hamster taste buds. One of our objectives is to understand how taste information from the IXth nerve is processed in relation to the contribution from receptors innervated by the facial and vagus nerves. The relative contributions of these nerves at stages of the developmental processes could be useful in determining whether each nerve yields specific taste information or contributes complimentary responses to the others.

The hamster develops precociously in comparison to some other mammals. Young hamsters begin to eat solid food during the first week of life, and they are weaned from maternal suckling by about 20 days. The development of IXth nerve taste buds in relation to the other populations of taste receptors remains to be determined. The significance of variation in the bilateral number of folate taste buds is not clear. The apparent disparity in the rate of increase of total foliate and vallate taste buds may also have behavioral significance.

ACKNOWLEDGMENTS

We acknowledge the technical effort of Elise Williams and Steven Prough.

REFERENCES

1. HANAMORI, T., I. J. MILLER, JR. & D. V. SMITH. 1987. Taste responsiveness of hamster glossopharyngeal nerve fibers. Ann. N.Y. Acad. Sci. This volume.
2. MILLER, JR., I. J. & D. V. SMITH. 1984. Quantitative taste bud distribution in the hamster. Physiol. Behav. 32(2): 275-285.

Peripheral Olfactory Responses of Mature Male, Regressed, and Hypophysectomized Goldfish to a Steroidal Sex Pheromone, L-Serine, and Taurocholic Acid[a]

P. W. SORENSEN,[b] T. J. HARA,[c] AND N. E. STACEY[b]

[c]Department of Fisheries and Oceans
Freshwater Institute
Winnipeg, Manitoba R3T 2N6, Canada

[b]Department of Zoology
University of Alberta
Edmonton, Alberta T6G 2E9, Canada

Progress in understanding the actions of pheromones in teleost fish has been impeded by a lack of information on their chemical structure. Although several studies have suggested that sex steroids may function as behavioral pheromones in teleost fish,[1] olfactory sensitivity to sex steroids has not been demonstrated. Our studies of the goldfish indicate that 17α, 20β-dihydroxy-4-pregnen-3-one (17,20P) is released from periovulatory females and evokes rapid increases in serum gonadotropin (GtH) and sperm production in mature males.[2,3] Water-borne 17,20P does not increase GtH in females and gonadally regressed males. This study used electroolfactogram (EOG) recording to determine if the goldfish olfactory epithelium responds to sex steroids proposed to function as pheromones, and whether the sensitivity of the goldfish olfactory epithelium is influenced by sex or maturity. The olfactory sensitivity of mature spermiated males, regressed males and females, six-week hypophysectomized regressed fish, and sham-operated fish was measured to L-serine, taurocholic acid, and 17,20P. The EOG responses of males and regressed goldfish to these compounds and a variety of other L-amino acids and steroids have been reported elsewhere.[4]

The olfactory epithelia of mature males and regressed fish were acutely sensitive to 17,20P which had a detection threshold of 10^{-13} M, and at a concentration of 10^{-8} M evoked a response three times that of 10^{-5} M L-serine (FIG. 1). Previous studies have also demonstrated an extraordinary sensitivity to 17,20P, corroborating its role as a pheromone synchronizing spawning readiness between the sexes.[2,3] Although variable, the EOG responses of mature males, regressed fish and sham fish were all significantly larger ($p \leq 0.05$) than those of hypophysectomized fish to 10^{-5} M L-serine, 10^{-5} M taurocholic acid, and 10^{-8} M 17,20P (TABLE 1). Scanning electron microscopy confirmed a previous study using light microscopy that found hypophy-

[a]Funded by the Alberta Heritage Foundation for Medical Research.

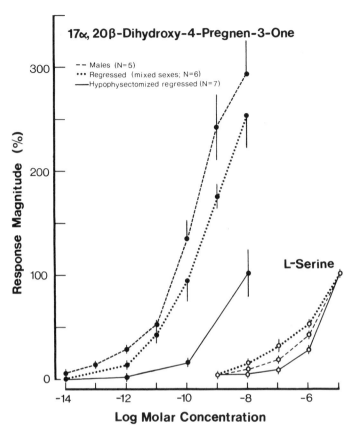

FIGURE 1. Semilogarithmic plot of the concentration-response relationship to 17α, 20β-dihydroxy-4-pregnen-3-one and L-serine of mature male, gonadally regressed, and hypophysectomized regressed goldfish. EOG response magnitude is represented as a percentage of that induced by the standard stimulant 10^{-5} M L-serine. Vertical bars indicate standard error.

TABLE 1. Average EOG Responses (mV $\pm$ SEM) Elicited by 10^{-5} M L-Serine, 10^{-5} M Taurocholic Acid (TCA), and 10^{-8} M 17α, 20β-dihydroxy-4-pregnen-3-one (17,20P) in Mature Males, Gonadally Regressed, Hypophysectomized, and Sham-Operated Goldfish[a]

Odorants	Males	Regressed	Hypophysectomized	Shams
L-Serine	0.53 ± 0.05 (13)	0.43 ± 0.04 (13)	0.18 ± 0.03 (8)	0.99 ± 0.9 (3)
TCA	1.12 ± 0.23 (5)	1.05 ± 0.19 (7)	0.32 ± 0.02 (7)	1.83 ± 0.26 (3)
17,20P	2.22 ± 0.39 (5)	1.16 ± 0.28 (6)	0.14 ± 0.15 (7)	2.21 ± 0.60 (2)

[a] Sample size is in parentheses.

sectomization to reduce the sensory olfactory epithelium of goldfish.[5] However, because hypophysectomization reduced EOG responses to all odorants, and neither sex nor maturity appeared to influence olfactory sensitivity, the fact that responsiveness to 17,20P pheromone is limited to mature males is not thought to be due to endocrine regulation of olfactory receptor function.

REFERENCES

1. STACEY, N. E., A. L. KYLE & N. R. LILEY. 1986. Fish reproductive pheromones. *In* Chemical Signals in Vertebrates. D. Duvall, D. Muller-Schwarze & R. M. Silverstein, Eds.: 117-133. Plenum Press. New York.
2. STACEY, N. E. & P. W. SORENSEN. 1986. 17α, 20β-dihydroxy-4-pregnen-3-one: A steroidal primer pheromone which increases milt volume in the goldfish, *Carassius auratus*. Can. J. Zool. **64:** 2412-2417.
3. DULKA, J. G., N. E. STACEY, P. W. SORENSEN & G. J. VAN DER KRAAK. 1987. Sex steroid pheromone synchronizes male-female spawning readiness in goldfish. Nature (London) **325:** 251-253.
4. SORENSEN, P. W., T. J. HARA & N. E. STACEY. 1986. Extreme olfactory sensitivity of mature and gonadally regressed goldfish to 17α, 20β-dihydroxy-4-pregnen-3-one, a potent steroidal pheromone. J. Comp. Physiol. **A160:** 305-313.
5. YAMAZAKI, F. & K. WATANABE. 1979. The role of steroid hormones in sex recognition during spawning behaviour of the goldfish, *Carassius auratus* L. Proc. Indian Natl. Sci. Acad. **B45:** 505-511.

The Electrolyte Distribution in Insect Olfactory Sensilla as Revealed by X-ray Microanalysis

R. A. STEINBRECHT

Max-Planck-Institut für Verhaltensphysiologie
8131 Seewiesen
Federal Republic of Germany

K. ZIEROLD

Max-Planck-Institut für Systemphysiologie
4600 Dortmund
Federal Republic of Germany

X-ray microanalysis allows the quantitative analysis of elemental concentrations *in situ* with the high lateral resolution of the electron microscope.[1] Thus, it is now possible to measure the electrolyte distribution directly in the various cellular and extracellular compartments of insect sensilla. Eventually, this information will provide a link between our understanding of the complex ultrastructure of these sensory organs and their functional interpretation as derived mainly from electrophysiological experiments.

Cryofixation and cryoultramicrotomy are essential prerequisites for X-ray microanalysis of electrolyte elements, because any solvent treatment might cause redistribution or even extraction of soluble elements. With antennae of the silkmoth, *Bombyx mori,* this has been achieved by quick immersion into liquid propane (at $-190°C$) and cryoembedding into heptane (at $-80°C$).[2] Ultrathin cryosections were cut in a Reichert FC4 cryoultramicrotome at $-110°C$, freeze dried and studied in a Siemens ST100F scanning transmission electron microscope. X-ray spectra were processed in an energy-dispersive spectrometer with multichannel analyzer (Link Systems). Dry weight concentrations of the analyzed elements were calculated according to the continuum method of Hall[3] after calibration with cryosections of salt solutions of known concentration (for a more detailed account of the procedure see Zierold[4]).

The ultrastructure of frozen, dried cryosections allows the unequivocal identification of all important sensillar compartments, such as dendrites and receptor lymph in cross sections of olfactory hairs or, in the antennal branches, the receptor-cell somata, dendrites, axons, the different auxiliary cells, as well as the extracellular compartments of the receptor-lymph and hemolymph spaces. The plasma membranes around the cellular compartments and the organelles within them are clearly discernible (FIG. 1).

X-ray microanalysis (TABLE 1) revealed a high concentration of potassium throughout the cellular compartments, but also in the extracellular receptor lymph, confirming earlier measurements on isolated receptor-lymph droplets.[5] It is assumed

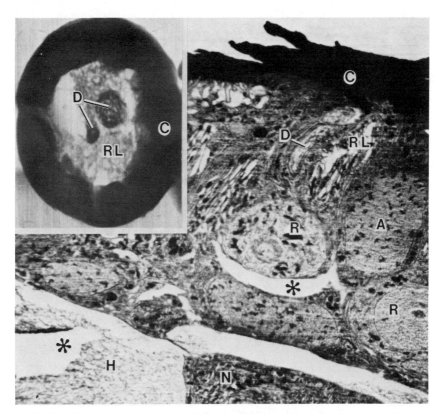

FIGURE 1 Unstained, ultrathin cryosection through antennal branch of *Bombyx mori* as seen after freeze drying in the scanning transmission electron microscope. The inset shows a cross-sectioned olfactory hair of a sensillum trichodeum at higher magnification. The asterisks mark spaces caused by sectioning and/or drying artefacts. A, auxiliary cell; C, cuticle; D, dendrites; H, hemolymph; N, nerve; R, receptor cell; RL, receptor lymph. Magnification × 4,000; inset × 30,000.

that receptor-lymph potassium in *Bombyx* is maintained high by an electrogenic potassium pump located in the apical membrane folds of one auxiliary cell, the trichogen cell.[6] Hemolymph potassium is low. Calcium is clearly present in the hemolymph, but below the detection limit in the cellular compartments and in the receptor lymph. Receptor lymph contains little chlorine; electroneutrality, therefore, most probably is established by organic polyanions, for example, sulfatized proteoglycans. This is also indicated by a fairly high sulfur content.

In preliminary experiments the potassium distribution was compared in stimulated and unstimulated sensilla. Antennae that received a strong, adapting bombykol stimulus before freezing showed a significant rise in receptor-lymph potassium (TABLE 1). This increase was unexpected because the receptor current should produce an ionic shift in the opposite direction and of a much lesser magnitude. Therefore, strong stimulation possibly activates additional ion movements, for example, the electrogenic

TABLE 1. Distribution of Electrolytes in Various Compartments of the Silkmoth Antenna[a]

Compartment	n	Potassium	Magnesium	Chlorine	Sulphur	Phosphorus
Hemolymph	15	96 (20)	137 (27)	97 (22)	157 (26)	454 (19)
Glia	5	195 (32)	41 (22)	28 (63)	213 (18)	371 (29)
Auxiliary cell cytoplasm	9	337 (22)	66 (32)	22 (49)	198 (15)	239 (20)
Auxiliary cell nucleus	5	491 (19)	72 (70)	32 (44)	344 (20)	536 (22)
Receptor cell cytoplasm	26	375 (20)	49 (41)	24 (23)	162 (16)	497 (20)
Receptor cell nucleus	8	539 (12)	43 (35)	36 (36)	220 (14)	440 (11)
Receptor cell axon	10	231 (40)	32 (52)	18 (50)	196 (19)	314 (29)
Dendrite unstimulated	11	283 (35)		51 (36)		
Dendrite stimulated	23	373 (37)		66 (60)		
Receptor lymph unstimulated	6	545 (45)		116 (48)		
Receptor lymph stimulated	17	664 (38)		158 (36)		

[a] Concentration in mmol/kg dry weight (in parenthesis: standard deviation as percent of mean).

potassium pump of auxiliary cells. The experiments are being repeated on a larger scale.

REFERENCES

1. HUTCHINSON, T. E. & A. P. SOMLYO. 1981. Microprobe Analysis of Biological Systems. Academic Press. New York and London.
2. STEINBRECHT, R. A. & K. ZIEROLD. 1984. A cryoembedding method for cutting ultrathin cryosections from small frozen specimens. J. Microsc. (Oxford) **136:** 69-75.
3. HALL, T. A. & B. L. GUPTA. 1982. Quantification for the X-ray microanalysis of cryosections. J. Microsc. (Oxford) **126:** 333-345.
4. ZIEROLD, K. 1981. Cryopreparation of mammalian tissue for X-ray microanalysis in STEM. J. Microsc. (Oxford) **125:** 149-156.
5. KAISSLING, K.-E. & J. THORSON. 1980. Insect olfactory sensilla: Structural, chemical and electrical aspects of the functional organization. *In* Receptors for Neurotransmitters, Hormones and Pheromones in Insects. D. B. Sattelle, L. M. Hall & J. G. Hildebrand, Eds.: 261-281. Elsevier/North-Holland. Amsterdam, the Netherlands.
6. STEINBRECHT, R. A. & W. GNATZY. 1984. Pheromone receptors in *Bombyx mori* and *Antheraea pernyi*. I. Reconstruction of the cellular organization of the sensilla trichodea. Cell Tissue Res. **235:** 25-34.

Sequential Interactions of Oral Chemical Irritants

DAVID A. STEVENS AND HARRY T. LAWLESS

Psychology Department
Clark University
Worcester, Massachusetts 01610

S. C. Johnson & Son, Inc.
Racine, Wisconsin 53403

In spite of its importance in contributing to overall impressions of flavor, the oral trigeminal system responsible for perceptions of oral chemical heat, irritation, and pungency remains poorly understood. One major unanswered question about this system concerns the breadth of tuning or specificity of receptors and fiber types for different oral chemical irritant compounds. A related perceptual question concerns whether qualitative differences exist among the sensations elicited by these different molecules. One psychophysical approach to studying this question of tuning is to stimulate sequentially, for example, in cross-adaptation studies.[1,2] In these paradigms, increases in response to the second-presented stimulus in a sequence (as compared with responses to repetition of the same stimulus) are viewed as evidence for specificity or narrowness of tuning. We investigated the pattern of responses to irritants presented in sequence, employing capsaicin and piperine, the irritant compounds from red and black pepper, respectively.

Twenty adult volunteers were paid for participation. Irritants were 20-ml samples of 1 mg/l capsaicin (Sigma Grade I) and 37.5 mg/l piperine (Sigma), which were approximately equal in perceived intensity, served at 37°C. The sequence of tasting was rinse with distilled water, expectorate, hold irritant sample in mouth (60 sec), expectorate, then hold the second irritant sample in the mouth (60 sec) and expectorate. Magnitude estimates of "burn" intensity were made at the 25th and 55th second during each irritant. The four possible irritant sequences were run on four different days. Ratings were normalized for each subject by a multiplier to set the mean rating of the first stimulus at the 25th second to the value of 100.

An analysis of variance showed that the intensity of the second sample was greater when the second irritant was different from the first (see FIG. 1), that there was an increase in intensity from the 25th to the 55th second, that this increase was enhanced when a different irritant was used second, and that this increase was greater for second samples (all F-ratios greater than 11.0, 1,19 df, $p < 0.01$). No differences between irritants were found.

The most notable result was a strong increment in response found when a different irritant was employed second, as compared with the magnitude of response when the same irritant was repeated. This result would be obtained if the second irritant recruited new receptors and/or fibers not stimulated by the first irritant. Such an explanation implies some specificity to oral trigeminal chemoreceptors, a narrowness of tuning. Some degree of specificity is in agreement with other results showing differences in

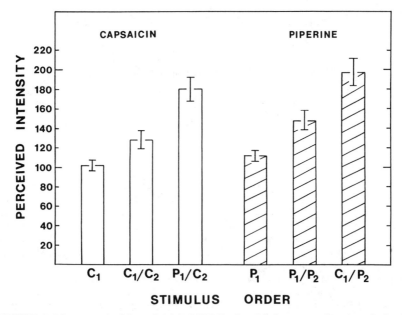

FIGURE 1. Mean perceived intensity ($\pm$ 1 SEM) of oral irritants as a function of stimulus order. Open bars indicate intensity of capsaicin, hatched bars intensity of piperine. C_1: intensity of capsaicin after a water rinse, as the first member of a pair. P_1: intensity of piperine after a water rinse, as the first member of a pair. C_1/C_2: intensity of capsaicin as the second member of a pair, following capsaicin. P_1/C_2: intensity of capsaicin as the second member of a pair, following piperine. P_1/P_2: intensity of piperine as the second member of a pair, following piperine. C_1/P_2: intensity of piperine as the second member of a pair, following capsaicin.

the spatial pattern of stimulation for these two irritants.[3,4] This is a narrower degree of tuning than that suggested by capsaicin-desensitization studies, in which capsaicin treatments were found to decrease or abolish responses to a broad array of irritant compounds.[5] However, desensitization effects, possibly acting through depletion of the neurotransmitter, substance P, may act more centrally than the cross-enhancement of response seen in our recruitment effect. Since it is possible to have different receptor mechanisms, but a common neurotransmitter, these effects are not necessarily in conflict.

REFERENCES

1. KOSTER, E. P. 1971. Adaptation and cross-adaptation in olfaction. Dissertation. University of Utrecht, the Netherlands.
2. MCBURNEY, D. H., D. V. SMITH & T. R. SHICK. 1972. Gustatory cross-adaptation: Sourness and bitterness. Percep. Psychophys. **11:** 228-232.
3. LAWLESS, H. 1984. Oral chemical irritation: Psychophysical properties. Chem. Senses **9:** 143-155.
4. STEVENS, D. A. & H. T. LAWLESS. Unpublished observations.
5. NAGY, J. I. 1982. Capsaicin as a probe for sensory neural mechanisms. *In* Handbook of Psychopharmacology. L. L. Iverson, S. D. Iverson & S. M. Snyder, Eds. Vol. **15:** 185-235. Plenum Publishing. New York.

Detecting Gas Odor in Old Age[a]

JOSEPH C. STEVENS AND WILLIAM S. CAIN

John B. Pierce Foundation Laboratory
New Haven, Connecticut 06519

Aging often impairs smelling, raising threshold, weakening strength, discrimination, and identification.[1] Few past 65-70 escape all loss. Elderly often cannot detect warning agents in gases. We address propane, which in the United States yearly causes some 6,500 explosions, 60 deaths, and 1,000 injuries, often disfiguring. Is the inability to smell ethyl mercaptan (EM), the usual warning agent, widespread enough to matter?

We first compared in our laboratory 21 old (70-85) and 21 young (18-25) persons for threshold and for suprathreshold strength of EM. Subjects were selected by sex, smoking abstinence, and nasal patency. The average old subject took for detection (force-choice, ascending limits task) 10 times greater concentration of EM (t = 5.94, 40 df, $p < 10^{-6}$) (FIG. 1). Compared to young subjects, old subjects judged that relative to saltiness of NaCl solutions, EM smelled weaker (FIG. 2), as have other odorants.[2]

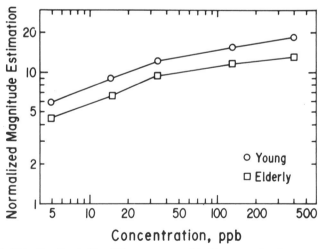

FIGURE 1. Detection thresholds of 21 young versus 21 old persons for the gas warning agent ethyl mercaptan (EM). "Standard level" is the concentration of EM (14 ppb) that would be achieved when a leak of propane from a freshly filed cylinder achieves a concentration of 0.47% (1/5 of the lower explosive limit).

[a] Supported by National Institutes of Health Grants AG-042897 and ES-00592.

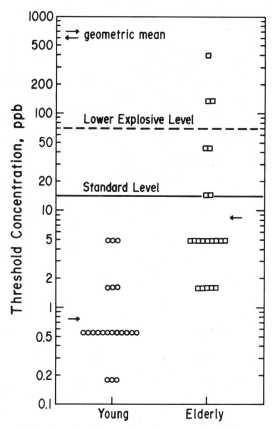

FIGURE 2. Mean magnitude estimations of the odor strength of EM vapor relative to salty taste strength as assessed by young and elderly subjects. The magnitude estimates of salts were used to normalize those of EM.

This result suggested a look at the "real" world, testing less select subjects under "everyday" conditions to help decide whether the problem is common enough to call for epidemiological scrutiny. The U.S. Department of Transportation specifies that LP gas have a distinct odor at or above concentration in air of 1/5th the lower explosive limit (LEL). One pound EM per 10,000 gallons of LP gas has been found to serve; 1/5 LEL occurs at 0.47% propane in air. Under these conditions a fresh cylinder of LP gas should yield about 14 ppb EM at 1/5 LEL. In the above study seven of 21 elderly failed to detect near or above 14 ppb and three near the LEL. Odorization of propane by EM often exceeds the minimum specified. As a source we used a "20-lb" commercial tank. Because the vapor pressures of the two gases differ, more EM is emitted as the tank empties. For realism we drew samples at five fullness levels: 100, 87, 63, 39, and 22%. By gas chromatography with a flame photometric detector, the EM levels were 6.5, 7.9, 13, 36, and 44 ppm. The samples were mixed with breathing air to produce 0.47% propane (1/5 LEL) and put in five-liter gas bags. The five samples thus made contained 30, 37, 59, 170, and 210 ppb, well above the "standard" 14 ppb. Even so, failure to detect was common—50 (45%) of 110 persons over 60 ($\bar{X} = 72$) and 6 (10%) of 52 persons under 40 ($\bar{X} = 23.5$).

For the tests, subjects chose five times which of three bags contained odorant (two contained air only). Contents were ejected at 3.4 l/min (± 0.6) 2.5 cm from subjects' nostrils. Reliable detection constituted four or five correct choices (binomial p, 0.045 and 0.0041). Six to eight young and 20 to 25 elderly subjects sampled at each of the five levels. Elderly subjects (87 female, 23 male) tested at two senior citizens centers, young subjects (26 of each sex) at the laboratory or a local college. Selection by sex, nasal patency, and smoking was intentionally avoided. Detection tended to improve with EM level emitted. Performance at one center was better than at the other, though both showed bad overall detection in elderly compared with young subjects. Epidemiological survey does seem called for.

REFERENCES

1. MURPHY, C. 1986. Taste and smell in the elderly. *In* Clinical Measurement of Taste and Smell. H. L. Meiselman and R. S. Rivlin, Eds. Macmillan. New York.
2. STEVENS, J. C. & W. S. CAIN. 1985. Age-related deficiency in the perceived strength of six odorants. Chem. Senses **10:** 517-529.

Voltage-Dependent Conductances in Solitary Olfactory Receptor Cells

NORIYO SUZUKI

Zoological Institute
Faculty of Science
Hokkaido University
Sapporo 060, Japan

To elucidate the chemoelectrical transduction mechanism in olfactory perception, the basic membrane properties of enzymatically isolated olfactory receptor cells from the bullfrog were studied in different ionic environments using the "Gigaseal" whole-cell clamp technique.[1] The receptor cells had resting potentials between -40 mV and -112 mV (mean -73.6 mV, $n = 14$). They had high input resistance (2-6 Gohm) and low input capacitance (3.7-5.0 pF, mean 4.5 pF, $n = 29$). The specific membrane capacitance was calculated as 0.8 $\mu F/cm^2$ from the estimated value of the surface area of receptor cells (573 μm). Repetitive action potentials were evoked in the receptor cells by the injection of depolarizing current. The frequency of the repetitive firing was a function of the magnitude of the injected current. Anode break spikes were observed on the cessation of hyperpolarizing current injection of appreciable strength.

The membrane currents of the receptor cells associated with depolarizing voltage pulses in normal saline were characterized by an initial inward current followed by a slowly developing outward current. No currents were activated by the hyperpolarizing voltage pulses. The inward current activated by the membrane depolarization consisted of two different current components; Na current (I_{Na}) and Ca current (I_{Ca}). I_{Na} was activated by membrane depolarization beyond -40 mV, was maximum at -10 mV and reversed its polarity at $+40$ mV. Most of I_{Na} was suppressed by the ordinary blocking dose of TTX for Na^+ channels, but about one-tenth of I_{Na} was left intact even with a high dose of TTX (20 $\mu g/ml$). This indicates that Na^+ channels of a different type are distributed in the receptor cells. I_{Ca} was activated by membrane depolarization beyond -20 mV, was maximal at $+10$ mV, and reversed its polarity at $+40$ mV. I_{Ca} showed a sustained response to depolarizing voltage pulses and no discernible inactivation was observed. Tail currents of I_{Ca} were observed on the cessation of the voltage pulses as in other tissue preparations.[2,3] The outward current activated by membrane depolarization consisted of at least three different current components; delayed rectifier K current (I_{Kdr}), Ca-activated K current ($I_{K(Ca)}$), and the transient outward current (I_{to}). Both I_{Kdr} and $I_{K(Ca)}$ were activated by membrane depolarization beyond -20 mV and were suppressed completely by the substitution of external K^+ and Ca^{2+} with TEA^+ and Co^{2+}, respectively. I_{to} was observed when the external NaCl and KCl were substituted with TEA-Cl, which suggests that anions are involved in this current.

[a] Supported by a Grant-in-Aid (No. 60115001) from the Japanese Ministry of Education, Science and Culture.

REFERENCES

1. HAMILL, O. P., M. E. MARTY, E. NEHER, B. SAKMANN & E. J. SIGWORTH. 1981. Improved patch-clamp techniques for high-resolution current recording from cells and cell-free membrane patches. Pflugers Arch. **391:** 85-100.
2. FENWICK, E. M., A. MARTY & E. NEHER. 1982. Sodium and calcium channels in bovine chromaffin cells. J. Physiol. **331:** 599-635.
3. HAGIWARA, S. & H. OHMORI. 1982. Studies of calcium channels in rat clonal pituitary cells with patch electrode voltage clamp. J. Physiol. **331:** 231-252.

Multimodal Neurons in the Lamb Solitary Nucleus

Responses to Chemical, Tactile, and Thermal Stimulation of the Caudal Oral Cavity and Epiglottis[a]

ROBERT D. SWEAZEY AND ROBERT M. BRADLEY

Department of Oral Biology
School of Dentistry
The University of Michigan
Ann Arbor, Michigan 48109

The receptor areas of the soft palate, caudal tongue, and epiglottis are important for the elicitation and control of ingestive behavior and upper airway reflexes. The afferent fibers that innervate these structures terminate primarily in the nucleus of the solitary tract (NST) or the spinal trigeminal nucleus.[1] To better understand the role of the NST in the integration of information from these separate receptor populations, we have recorded neural responses to mechanical, thermal and chemical stimulation.

METHODS

Experiments were performed on 31 anesthetized Suffolk lambs, aged 38-70 days, weighing 9-23 kg. The caudal tongue and palate were exposed by making an incision through the cheek and the laryngeal surface of the epiglottis was exposed by a midline incision from the first tracheal cartilage rostral to the base of the epiglottis. When a single NST neuron was isolated, the location of its receptive field(s) was mapped and responses to mechanical, thermal, and chemical stimuli recorded. The mechanical stimuli consisted of a moving soft brush and glass probe, and punctate stimulation applied using a modified Grass strain gauge. Punctate stimulation ranged from 0.5 to 10 grams. The thermal stimulation sequence was body temperature rinse (37-39°C), cool rinse (23-25°C), body temperature rinse, warm rinse (42-43°C), body temperature rinse. The chemical stimuli applied to the tongue and palate were 0.5 M KCl and NH$_4$Cl, 0.01 N HCl, and 0.154 M NaCl. These solutions were dissolved in distilled water, which also served as the rinse. Chemical stimuli applied to the epiglottis were

[a] Supported by National Institutes of Health Grant DE05728.

0.5 M KCl and NH$_4$Cl, 0.01 N HCl, and distilled water. The chemical solutions were dissolved in 0.154 M NaCl, which also served as the rinse for the epiglottis.

RESULTS

Responses were recorded from 53 single NST neurons. Generally, neurons responded to more than one of the stimulus modalities. Thirty-one percent responded to only one, 55% responded to two, and 14% responded to all three stimulus types. A greater proportion of neurons with receptive fields on the epiglottis (93%) responded to two or three of the stimulus modalities than neurons with oral cavity receptive fields (36%). Of the three stimulus modalities tested, mechanical stimuli were the most effective, eliciting responses in 91% of the neurons. Chemical stimuli produced responses in 60% of the samples while thermal stimulation was the least effective, eliciting responses from only 32% of NST neurons. Convergence of afferent information onto NST neurons occurred almost exclusively between the caudal tongue and palate. In these cases, the receptive fields on the tongue were aligned below those located on the palate. TABLE 1 summarizes the types of stimuli responded to on the basis of receptive field location.

Mapping the distribution of recording sites in the NST revealed that neurons with receptive fields located in the caudal oral cavity were generally located rostral, lateral, and ventral in the nucleus compared to those neurons having receptive fields located on the epiglottis.

DISCUSSION

The presence of large numbers of multimodal neurons suggests that the region of the NST investigated in this study is important in the integration of afferent information produced by complex stimuli in the caudal oral cavity and upper airway. Furthermore, these neurons may provide information important in the initiation of upper airway reflexes. The presence of neurons with opposing tongue and palate receptive fields,

TABLE 1. Percentage of Neurons Responding to Different Stimulus Combinations Classified by Receptive Field Location

Stimulus	Tongue Only ($n = 9$)	Palate Only ($n = 4$)	Epiglottis Only ($n = 28$)	Tongue and Palate ($n = 9$)	Other ($n = 3$)
Mechanical	67	50	0	56	33
Thermal	0	25	0	0	0
Chemical	0	0	7	0	0
Mechanical and thermal	11	25	4	33	0
Mechanical and chemical	0	0	60	11	67
Thermal and chemical	11	0	4	0	0
Mechanical, thermal, and chemical	11	0	25	0	0

which has been reported in other species,[2] would be particularly well suited for providing information relevant to the manipulation and subsequent swallow of a food bolus. The failure to find convergence in the NST between the widely separated receptor populations of the caudal oral cavity and epiglottis suggests that this peripheral separation is maintained at brainstem levels. This conclusion is further supported by the rough anatomical separation of neurons with oral cavity receptive fields from those having receptive fields on the epiglottis.

REFERENCES

1. SWEAZEY, R. D. & R. M. BRADLEY. 1986. Central connections of the lingual-tonsillar branch of the glossopharyngeal nerve and the superior laryngeal nerve in lamb. J. Comp. Neurol. **245:** 471-482.
2. HAYAMA, T., S. ITO & H. OGAWA. 1985. Responses of solitary tract nucleus neurons to taste and mechanical stimulation of the oral cavity in decerebrate rats. Exp. Brain Res. **60:** 235-242.

Quasi-Regenerative Responses to Chemical Stimuli in *in Vivo* Taste Cells of the Mudpuppy[a]

JOHN TEETER

Monell Chemical Senses Center and
Department of Physiology
University of Pennsylvania School of Medicine
Philadelphia, Pennsylvania 19104

Vertebrate taste receptor cells have commonly been considered to be electrically inexcitable, displaying a linear I-V relationship over a wide range of applied currents. Taste stimuli typically evoke slow changes in taste cell membrane potential that are graded in amplitude with stimulus concentration and are often, but not always, associated with changes in membrane resistance.[1] Recently, however, action potentials have been recorded from taste cells in isolated pieces of mudpuppy lingual epithelium in response to depolarizing currents[2] and chemical stimulation.[3] In addition, regen-

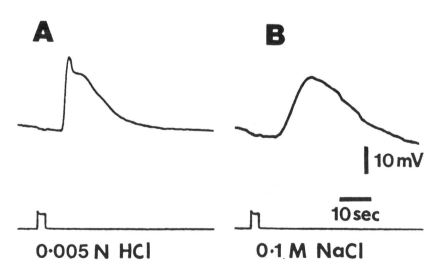

FIGURE 1. Slow potentials recorded from an *in vivo* mudpuppy taste cell in response to 0.005 N HCl (**A**) and 0.1 M NaCl (**B**). Vm = -55 mV, R_N = 120 MΩ.

[a]Supported by National Institutes of Health Grant NS-15804 and the Whitehall Foundation.

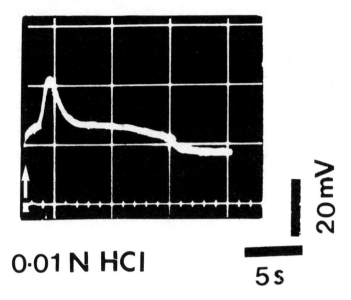

FIGURE 2. Response of a mudpuppy taste bud cell to a brief application (arrow) of 0.01 N HCl. A 20-mV depolarization, three seconds in duration, was superimposed upon a slower response, which appeared to follow the concentration profile of the stimulus. $Vm = -60$ mV, $R_N = 230$ MΩ.

erative anode-break potentials have also been recorded at the cessation of large hyperpolarizing currents in frog taste cells.[4] Voltage-dependent Na^+ and Ca^{2+} channels were shown to be involved in both types of responses. The role of voltage-gated channels in taste transduction, as well as their occurrence in species other than amphibians, however, remain uncertain. Action potentials have not been recorded in response to either applied current or taste stimuli from *in vivo* taste cells in any species, including mudpuppies.[5] Consequently, experiments were undertaken to reexamine the question of whether or not *in vivo* taste cells in mudpuppies generate action potentials in response to taste stimuli.

Mudpuppies were anesthetized either by immersion in MS-222 (1:8,000) or by an i.p. injection of 1-2 ml of 20% urethane. Anesthesia was maintained by adding MS-222 (1:10,000) to the aerated water used to perfuse the gills. Recordings were made from cells in taste buds on the dorsal surface of the tongue using 3 M KCl-filled microelectrodes (resistances of 60-150 MΩ). The indifferent electrode was inserted under the skin on the neck. Taste stimuli were applied at low flow rates (<5 ml/min) using a sample injection valve (0.1-0.5 ml stimulus sample) and more rapidly using a repeating dispenser or pressure injection system.

Acids elicited graded depolarizations in taste cells that had significantly faster rising times than those evoked by high concentrations of salts applied at the same flow rate (FIG. 1). Hyperpolarization of the cell with dc current resulted in acid and salt responses that were both faster rising and larger in amplitude. Taste cell responses to high concentrations of acids (0.005-0.01 N) usually had a distinct phasic component one to three seconds in duration (FIG. 1A), which was not observed in surface epithelial cells. Occasionally, acids produced slow potentials upon which a faster, apparently partially regenerative, depolarization was superimposed (FIG. 2). The amplitudes of both the slow and fast components of this type of response were increased slightly by hyperpolarization of the cell.

These results are consistent with the presence of voltage-dependent channels in mudpuppy taste cells and suggest that they contribute to the depolarizing responses elicited by taste stimuli. A phasic component to the responses elicited by concentrated acids and salts has also been observed in frog taste cells when the membrane potentials were increased by adaptation to water.[6] Action potentials comparable to those recorded from *in vitro* mudpuppy taste cells were not observed *in vivo,* either in response to depolarizing current or chemical stimulation, even in cells having resting potentials of -60 mV and input resistances of over 300 MΩ. It is possible that the slow depolarizations evoked by chemical stimuli result in inactivation of Na$^+$ channels to a level where a regenerative response cannot develop. The faster rising depolarizations evoked by concentrated acids (particularly at high resting potentials) may result in significantly less accommodation and, thus, a partially regenerative response.

REFERENCES

1. SATO, T. 1980. Recent advances in the physiology of taste cells. Prog. Neurobiol. **14:** 25-67.
2. ROPER, S. 1983. Regenerative impulses in taste cells. Science **220:** 1311-1312.
3. ROPER, S. & M. MCPHEETERS. 1984. Chemical stimulation evokes impulses in taste cells of the mudpuppy. Soc. Neurosci. Abstr. **10:** 656.
4. KASHIWAYANAGI, M., M. MIYAKE & K. KURIHARA. 1983. Voltage-dependent Ca^{2+} channel and Na$^+$ channel in frog taste cells. Am. J. Physiol. **244:** C82-C88.
5. WEST, C. H. K. & R. A. BERNARD. 1978. Intracellular characteristics and responses of taste bud and lingual cells of the mudpuppy. J. Gen. Physiol. **72:** 305-326.
6. SATO, T. 1977. An initial phasic depolarization exists in the receptor potential of taste cells. Experientia **33:** 1165-1167.

Effects of Short-Term Exposure to Lowered pH on the Behavioral Response of Crayfishes (*Procambarus acutus* and *Orconectes virilis*) to Chemical Stimuli[a]

ANN JANE TIERNEY AND JELLE ATEMA

Boston University Marine Program
Marine Biological Laboratory
Woods Hole, Massachusetts 02543

Amino acids are effective chemical stimuli for the receptors of a great many species. For crayfish, Hodgson[1] reported that glycine and glutamate stimulate antennular receptors and Hatt[2] found that many amino acids stimulate leg receptors. Behavioral responses of crayfish to amino acids have not yet been described and consequently the adaptive significance of the identified antennular and dactyl receptors is unknown. We conducted laboratory tests to determine how a mixture of eight amino acids affected feeding behavior, antennule movements, and walking in two crayfishes (*Orconectes virilis* and *Procambarus acutus*). We also assessed the effects of low pH on chemosensitivity of crayfish by testing animals at three different pH levels in the following order: 5.8, 4.5, 3.5, and 5.8.

The test stimulus was an equimolar mixture of L-alanine, L-serine, L-histidine, hydroxy-L-proline, glycine, taurine, L-phenylalanine, and L-asparagine. The mixture was injected into chambers (15 cm × 15 cm × 7 cm high; each chamber contained one crayfish) at three concentrations; $10^{-2} M$, $10^{-3} M$, $10^{-4} M$. At the end of testing at each pH level, food pellets were placed directly in front of the crayfish and the amount of food consumed by each animal was recorded.

At pH 5.8 animals spent significantly more time performing feeding movements, walking and (for *P. acutus*) lowering antennules in response to test stimuli compared to control stimuli (aged tap water). Response intensity generally increased with stimulus intensity. After 48 hours in acidified water (pH levels 4.5 and 3.5), both species showed a significant reduction in the amount of time spent performing feeding movements (FIG. 1). At pH 3.5, time with antennules lowered was also decreased in *P. acutus*. When the pH level was restored to 5.8, partial recovery of behavioral responsiveness was observed. *O. virilis* appeared to be more sensitive to acid exposure than *P. acutus*. At low pH levels feeding was more severely depressed in *O. virilis* than in *P. acutus* (FIG. 1), and showed less recovery during post-acidification trials.

[a] This study was supported by a grant from the United States Environmental Protection Agency (Dr. Clyde Bishop, Program Officer).

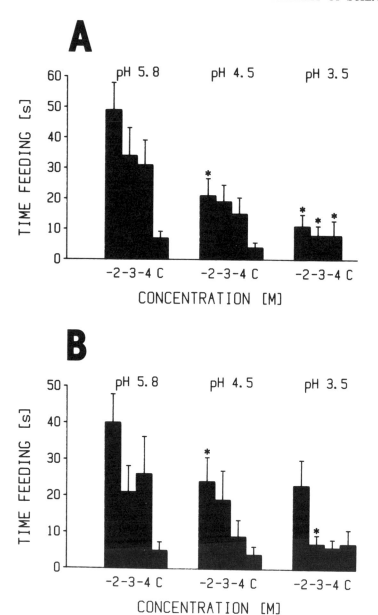

FIGURE 1. Feeding responses of *O. virilis* (**A**) and *P. acutus* (**B**) to three single-compound concentrations of an amino acid mixture ($10^{-2} M$, $10^{-3} M$, $10^{-4} M$) and a control at pH levels 5.8, 4.5, and 3.5. Bars indicate mean time (seconds) that 16 animals spent feeding; vertical lines indicate SEM. (*) Time spent feeding at pH 4.5 or 3.5 is significantly less than time spent feeding at pH 5.8 ($p < 0.025$; Wilcoxon's matched-pairs signed-ranks test). Concentrations were injected molarity.

General activity and amount of food actually consumed were not affected by acidification in either species. These results indicate that short-term acid exposure may interfere specifically with chemoreceptive processes at pH levels above those that cause failure of other organ systems.

Amino acid chemoreception in trout is also pH dependent.[3] Hara[3] suggested that a hypothetical amino acid receptor might contain two charged subsites, one negative and one positive, that interact with ionized α-amino and α-carboxyl groups of stimulus molecules. Amino acids are consequently expected to be most stimulating at their isoelectric points where dipolar forms are maximal. The amino acids we tested exist almost entirely in the dipolar form between pH levels 4 and 8. Thus, Hara's model may account for reduced responses at pH 3.5, but cannot account for reduced responses at pH 4.5. Changes in chemosensitivity due to pH probably result from changes in the state of ionization of the protein receptor, rather than of stimulus molecules.

REFERENCES

1. HODGSON, E. S. 1985. Electrophysiological studies of arthropod chemoreception. III. Chemoreception of terrestrial and freshwater arthropods. Biol. Bull. 115: 114-125.
2. HATT, H. 1984. Structural requirements of amino acids and related compounds for stimulation of receptors in crayfish walking leg. J. Comp. Physiol. A. 155: 219-231.
3. HARA, T. J. 1976. Effects of pH on the olfactory response to amino acids in rainbow trout, Salmo gairdneri. Comp. Biochem. Physiol. 54A: 37-39.

Response Characteristics of Olfactory Evoked Potentials Using Time-Varying Filtering

MITSUO TONOIKE

Electrotechnical Laboratory, Osaka Branch
11-46, 3-Chome, Nakouji
Amagasak: 661, Japan

Olfactory evoked potentials (OEPs) were recorded from the central region (C_z) of the intact human scalp with the referential derivation method.[1] Odorants were presented to the subject's nose for a duration of 200 milliseconds synchronized with the subject's respiration. The analogue wave forms of OEPs were digitized by converting every 8 milliseconds.

For the purpose of odor discrimination, time-varying filtering (TVF) developed by J.P.C.M. de Weerd,[2-4] was applied to the wave forms of OEPs.[5] The TVF method is able to handle the transient evoked potentials because it was developed for time-varying applications from the time-invariant wiener filtering. As OEPs are generally influenced by many factors such as effects of olfactory fatigue, arousal levels, and various other noises, the effectiveness of TVF was examined. The time-varying filtering function G (f,t) and the time-varying power density spectrum $\Phi_{ss}(f,t)$ of the signals were calculated for three odorants, amyl-acetate, vanillin, and dl-camphor by separating five banks of band pass filters (see FIG. 1a, b, c, d). From these results the G_1 (f,t) function and the $\Phi_{ss}^1(f,t)$ power density spectrum of the first band pass gave accurate estimates for the characteristic peak of OEPs. This showed that a frequency band of up to eight cycles is very important for the estimation of OEPs.

$G_1(f,t)$ and $\Phi_{ss}^1(f,t)$ showed a characteristic pattern for each odorant at each response time (see FIG. 2). The power density spectrum $\Phi_{nn}(f,t)$ of the various noise sources was also examined, and their characteristic patterns were shown in spectro-temporal representations. From the analysis of these experiments, it was shown that the characteristic frequency for the noise of OEPs was maximally about 20 cycles.

Finally, the TVF method was compared with the simple averaging (AV) and the a posteriori wiener filtering (APWF) methods which are the classical time-invariant filtering methods. These results show that the best estimates of OEPs were obtained with the TVF method. Response characteristics of OEPs for the above three odorants were drawn in the two-dimensional domain of frequencies and response times.

Estimated results for OEPs represent a model of human olfactory sensations using TVF function and power density spectra.

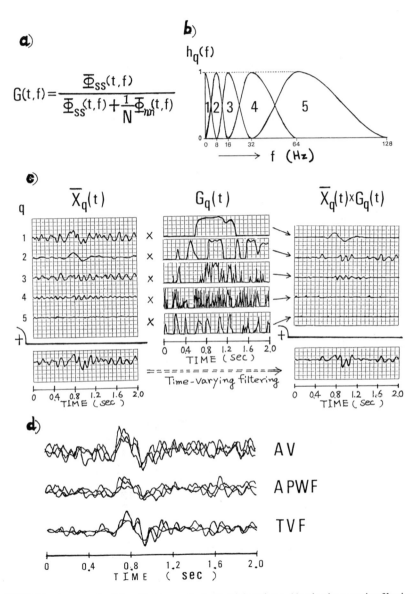

FIGURE 1. An example of the olfactory evoked potentials estimated by the time-varying filtering and its calculating procedure: (a) the TVF function given by time-varying filtering; (b) five bandpass areas and the low-pass filtering function used by the TVF method; (c) calculating process of the TVF method; and (d) comparisons among three filtering methods of the olfactory evoked potentials. AV: simple averaging; APWF: A posteriori wiener filtering; TVF: Time-Varying Filtering; odorant: amyl-acetate.

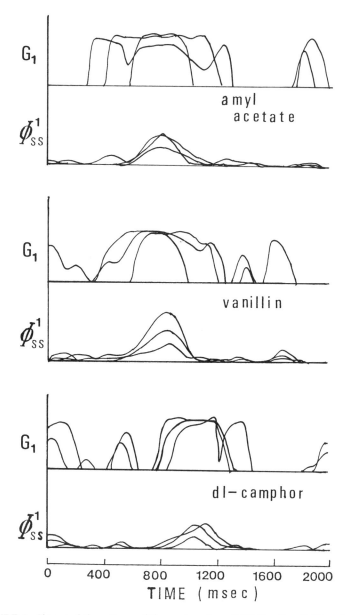

FIGURE 2. Characteristic patterns of three odorants on TVF function $G_1(t)$ and the power density spectrum $\Phi_{ss}^1(t)$ under the TVF method. Odorants: amyl-acetate, vanillin, dl-camphor.

REFERENCES

1. TONOIKE, M. & Y. KURIOKA. 1982. Precise measurements of human olfactory evoked potentials for odorant stimuli synchronized with respirations. Bul. Electrotech. Lab. 46(11): 622-633.
2. DE WEERD, J. P. C. & J. I. KAP. 1981. Spectro-temporal representations and time-varying spectra of evoked potentials. A methodological investigation. Biol. Cybern. 41: 101-117.
3. DE WEERD, J. P. C. 1981. A posteriori time-varying filtering of averaged potentials. 1. Introduction and conceptual basis. Biol. Cybern. 41: 211-222.
4. DE WEERD, J. P. C. & J. I. KAP. 1981. A posterior time-varying filtering of averaged evoked potentials. 2. Mathematical and computational aspects. Biol. Cybern. 41: 223-234.
5. TONOIKE, M. 1984. Analysis of olfactory evoked potentials using TVF method. Proceedings of The 18th Japanese Symposium on Taste and Smell. 9-12.

Effects of Injection of Calcium, EGTA, and Cyclic Nucleotides into the Mouse Taste Cell

KEIICHI TONOSAKI AND MASAYA FUNAKOSHI

Department of Oral Physiology
School of Dentistry
Asahi University
Gifu 501-02, Japan

Previously our studies[1] have shown that the sucrose depolarization response is not simply generated by the membrane depolarization and that the response does not depend entirely on the membrane resistance change. The sequence of events that couples stimulus-receptor interactions at the apical membrane of the taste cell must control the amplitude of receptor potential and the modulation of neurotransmitter release at synapses on the basolateral membrane of the taste cell. Complex intracellular taste transduction mechanisms are likely to exist in the taste cell. Recent experiments

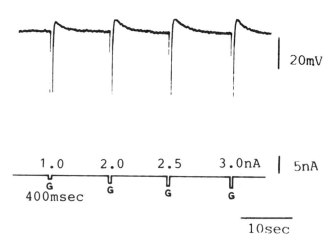

FIGURE 1. H-type taste cell membrane voltage response to 400-msec duration injections of 25 mM cGMP at 1.0 nA, 2.0 nA, 2.5 nA, and 3.0 nA. Titration of cGMP response is shown as a function of injection current. Bottom line indicates current stimulus monitor. Calibrations are shown at the end of each trace.

[a]Address for correspondence: Dr. Keiichi Tonosaki, Department of Oral Physiology, School of Dentistry, Asahi University, 1585-1 Hozumi, Hozumi-cho, Motosu-gun, Gifu 501-02, Japan.

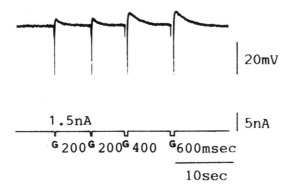

FIGURE 2. Another H-type taste cell membrane voltage response to injections of 1.5 nA of 25 mM cGMP at durations of 200 msec, 400 msec, and 600 msec. Titration of cGMP response is shown as a function of duration of injection time. Bottom line indicates current stimulus monitor. Calibrations are shown at the end of each trace.

have implicated intracellular messengers in controlling ion channels in membranes of photoreceptor cells.[2-6] Calcium ions and cyclic nucleotides can all act on the inner surface of the membrane to alter ion channel activity; therefore, calcium and cyclic nucleotides might be postulated as candidates for the role of intracellular transmitter in the taste cell. In order to determine which chemical serves as an internal transmitter in the process of taste transduction, we attempted to alter the internal ion concentration of taste cells by intracellular injections of specific chemicals. EGTA, cAMP, cGMP, calcium ion, sodium ion, potassium ion, chloride ion, or other ions were iontophoretically injected into the mouse taste cell from one or two barrels of a double- or triple-barreled microelectrode, while the other barrel was filled with Procion yellow dye solution and was monitoring membrane potential and resistance changes. After recording, Procion yellow was electrophoretically injected into the cell from the recording pipette by passing a steady current. Dye staining the recording cell is necessary in this type of experiment, because intracellular recording from the mouse taste cell does not eliminate uncertainty as to whether the recorded potential is obtained from the taste cell in the taste bud or from other surrounding tissue.[7,8] In this study, if a stained cell was not observed in the taste bud preparation, the responses were discarded. The experiments presented in this paper describe evidence about the H-type taste cell.[1] When an H-type taste cell is stimulated by sucrose solution, membrane potential depolarizes and membrane resistance of the taste cell to potassium and/or chloride ions increase.[1] Intracellular injected calcium or potassium ions induced hyperpolarization accompanied by a decrease of membrane resistance, while intracellular injected EGTA, cAMP, or cGMP induced depolarization accompanied by an increase of membrane resistance. The depolarization caused by cAMP injection was always smaller than that by cGMP. The effects of sodium and chloride ions were not clear. In the example shown in FIGURE 1, progressively larger pulses of cGMP cause graded transient depolarizations of progressively larger amplitudes, and in addition in FIGURE 2, progressively longer duration pulses of cGMP cause graded transient depolarizations of progressively larger amplitudes. The results suggest that calcium ions and cGMP are involved in the taste transduction process in mouse taste cells.

REFERENCES

1. TONOSAKI, K. & M. FUNAKOSHI. 1984. Effect of polarization of mouse taste cell. Chem. Senses **9:** 381-387.
2. BROWN, J. E., J. A. COLES & L. H. PINTO. 1977. Effects of injection of calcium and EGTA into the outer segments of retinal rods of bufo marinus. J. Physiol. **269:** 707-722.
3. MILLER, W. H. 1982. Physiological evidence that light-mediated decrease in cyclic GMP is an intermediary process in retinal rod transduction. J. Gen. Physiol. **80:** 103-123.
4. KAWAMURA, S. & M. MURAKAMI. 1983. Intracellular injection of cyclic-GMP increases sodium conductance in gecko photoreceptors. Jpn. J. Physiol. **33:** 789-800.
5. WALOGA, G. 1983. Effects of calcium and guanosine-3':5'-cyclic-monophosphoric acid on receptor potentials of toad rods. J. Physiol. **341:** 341-357.
6. SCHWARTZ, E. A. 1985. Phototransduction in vertebrate rods. Ann. Rev. Neurosci. **8:** 339-367.
7. TONOSAKI, K. & M. FUNAKOSHI. 1984. Intracellular taste cell responses of mouse. Comp. Biochem. Physiol. **78A:** 651-656.
8. TONOSAKI, K. & M. FUNAKOSHI. 1984. The mouse taste cell response to five sugar stimuli. Comp. Biochem. Physiol. **79A:** 625-630.

Effect of Dietary Protein on the Taste Preference for Amino Acids and Sodium Chloride in Rats

K. TORII, K. MAWATARI, AND Y. YUGARI

Life Science Laboratories
Central Research Laboratories
Ajinomoto Company, Inc.
Yokohama, Japan 244

Appetite and taste preference are affected by the nutritional status both within and outside normal nutritional limits. The major component of dietary protein is L-glutamate (Glu), which has the umami taste.[3] The synergistic enhancement of the umami taste of Glu by some 5'-ribonucleotides was recognized electrophysiologically[4] and psycophysically.[5,6] The biochemical mechanism of this phenomenon was suggested by the binding of Glu to the taste receptor membrane of the bovine circumvallate papillae. In this case, the maximal Glu binding was enhanced several-fold by certain 5'-ribonucleotides without any change in the dissociation constant.[7] Taste perception of L-amino acids (AA) and 5'-ribonucleotide seems to stimulate ingestion of nutrients. It is postulated that umami taste perception in animals may be a marker of dietary protein intake. Changes of taste preference for AA and NaCl in male, growing, Sprague-Dawley (SD), rats under various degrees of protein or some essential AA restriction were examined.

Fifteen kinds of AA or preferable taste solutions, 500 mM glycine (Gly), 150 mM NaCl, 150 mM Glu·Na (MSG) and 4.5 mM MSG + 4.5 mM guanosine 5'-monophosphate (5'-GMP), were offered to rats ($n = 6$, each group), in the choice of these solutions. Preference for NaCl with Gly and/or L-threonine (Thr) was induced under protein or some essential AA deficiency. Gly and Thr intake caused retention of the endogenous protein under the negative nitrogen balance. Umami taste preference was observed only when rats grew normally. The minimal level of dietary protein that induced a umami taste preference paralleled the protein requirement, which declined with age. The total Na intake in both normotensive SD rats and hypertensive rats (SHR) declined along with dietary protein increase. Preference for NaCl was sustained in SHR but disappeared in SD rats, reflecting a higher Na requirement in SHR.[8]

The quantitative ingestion of Lys in SD rats, fed Lys sufficient or deficient gluten diet (G+Lys or G−lys, respectively), in the choice from AA solutions was observed. The plasma and brain Lys alone altered under Lys deficiency, and then appetite for Lys was evoked. Subsequently, rats began to display a umami taste preference and normal appetite for food. Intakes of MSG, L-glutamine and L-arginine (Arg) occurred only when rats were fed either G+Lys or G−Lys concurrently with the choice of Lys solution. These results indicated that preference for umami taste and Arg is related

---

[a] Address for correspondence: Dr. K. Torii, Life Science Laboratories, Central Research Laboratories, Ajinomoto Company, Inc., 214, Maeda-cho, Totsuka-ku, Yokohama, Japan 244.

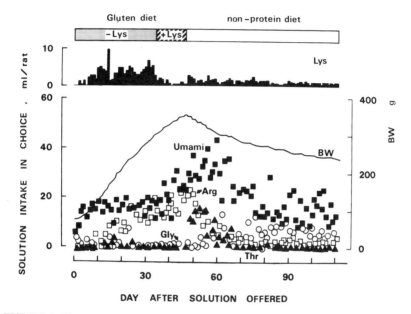

FIGURE 1. The pattern of preference for L-amino acids in growing rats subjected to a Lys deficiency. Rats were supplied with an isonitrogenous and isocaloric diet and 15 kinds of L-amino acid solutions as drinking solutions in a choice paradigm. Mean body weight is shown as a solid curve. The intake of each solution is shown as the total volume of (■) L-amino acid with umami taste quality; (□) Arg; (▲) Thr, and (○) Gly. Any other solutions were omitted from the figure because they were negligibly small.

to the nutritional status in which there is insufficient dietary protein for body need (FIG. 1).

In addition, experimental diets, G−Lys or G+Lys, with or without flavor, 0.2% saccharin Na, 0.17% MSG + 0.48% 5'-GMP, 1.8% NaCl, or 0.2% quinine sulfate, were offered to rats. G−Lys and G+Lys were given in alternate weeks in turn. The appetite for G−Lys was markedly suppressed and the palatability essentially unchanged regardless of any flavors added to diet (FIG. 2).

These data strongly suggest that taste preference and appetite are dependent on protein nutrition, which was defined by the amount and quality of dietary protein.

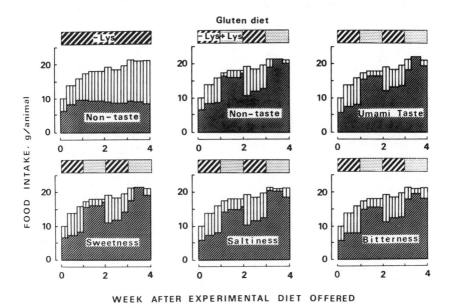

FIGURE 2. The effect of the addition of flavors on food intake in rats fed flavored diets with or without lysine. The experimental flavored diets, G−Lys and G+Lys, were given in alternate weeks in turn; controls were supplied with the unflavored G−Lys or G+Lys alone continuously. The mean values of food intake for each experimental group are shown in the shady zone (top left). Clear bars represent food intake values in controls given the G+Lys; for ease of comparison, these are contrasted against the shady zone for each experimental group.

REFERENCES

1. TORII, K., T. MIMURA & Y. YUGARI. 1986. Effects of dietary protein on the taste preference for amino acids in rats. *In* Interaction of the Chemical Senses with Nutrition. M. R. Kare & J. G. Brand, Eds.: 45-69. Academic Press. New York.
2. TORII, K., T. MIMURA & Y. YUGARI. 1986. Biochemical mechanism of umami taste perception, and effect of dietary protein on the taste preference for amino acids and sodium chloride in rats. *In* Umami: A Basic Taste. Y. Kawamura & M. R. Kare, Eds.: 513-563. Marcel Dekker. New York. In press.

3. GIACOMETTI, T. 1979. Free and bound glutamate in natural products. *In* Glutamic Acid: Advances in Biochemistry and Physiology. L. J. Filer, Jr., S. Garattini, M. R. Kare, W. A. Reynolds & R. J. Wurtman, Eds.: 25-34. Raven Press. New York.
4. SATO, M., S. YAMASHITA & H. OGAWA. 1970. Potentiation of gustatory response to monosodium glutamate in rat chorda tympani fibers by addition of 5'-ribonucleotides. Jpn. J. Physiol. **20:** 444-464.
5. YAMAGUCHI, S. 1967. The synergistic taste effect of monosodium glutamate and disodium 5'-inosinate. J. Food Sci. **32:** 473-478.
6. RIFKIN, B. & L. M. BARTOSHUK. 1980. Taste synergism between monosodium glutamate and disodium 5'-guanylate. Physiol. Behav. **24:** 1169-1172.
7. TORII, K. & R. H. CAGAN. 1980. Biochemical studies of taste sensation. IX. Enhancement of L-[^{3}H]glutamate binding to bovine taste papillae by 5'-ribonucleotides. Biochim. Biophys. Acta. **627:** 313-323.
8. TORII, K. 1980. Salt intake and hypertension in rats. *In* Biological and Behavioral Aspects of NaCl Intake. M. R. Kare, M. J. Fregly & R. A. Bernard, Eds.: 345-366. Academic Press. New York.

The Biochemistry of the Olfactory Purinergic System[a]

HENRY G. TRAPIDO-ROSENTHAL,[b] RICHARD A.
GLEESON, WILLIAM E. S. CARR, SCOTT M.
LAMBERT, AND MARSHA L. MILSTEAD

*C. V. Whitney Marine Laboratory and
Department of Zoology
University of Florida
St. Augustine, Florida 32086*

The olfactory system of the spiny lobster, *Panulirus argus,* is composed of aesthetasc sensilla present on the surface of the lateral antennular filament. These sensilla contain the dendritic terminals of primary chemosensory neurons. Among these neurons is a population that responds electrophysiologically to the purine nucleotide, AMP, a compound present at high concentrations in the prey of *P. argus.*[1,2] The response to AMP is attenuated by prolonged exposure to this nucleotide. Knowledge of the biochemical fate of stimulatory molecules in olfactory sensilla contributes to our understanding of how a chemosensory system operates.

METHODS

Studies of nucleotide dephosphorylation and nucleoside uptake were performed using sensilla-bearing sections of cuticle from antennular filaments. Sensilla were incubated with various compounds in 0.5- to 1-ml volumes of artificial sea water (ASW) at 22°C. Incubations were terminated by filtration and rinsing with ASW. Following the digestion of sensilla in NaOH, the amount of radioactivity was determined by means of liquid scintillation spectrophotometry.

RESULTS AND DISCUSSION

Experiments in which sensilla were incubated with [³H]Ado (from 0.05 to 10 μm), in the absence and presence of an excess of unlabeled Ado, revealed an uptake

[a] This work was supported by National Science Foundation Grants BNS-84-11693 and BNS-86-07513.

[b] Address for correspondence: Henry G. Trapido-Rosenthal, Whitney Marine Laboratory, Route 1, Box 121, St. Augustine, FL 32086.

system that internalizes this nucleoside. This uptake system has a K_m of $5.8\mu M$, and a V_{max} of 3.8 fmole/sensillum/min (133 pmole/mg protein/min). The uptake of tritiated adenosine is quite specific (FIG. 1). Unlabeled Ado is the most potent competitor for this process, followed in order by 2'-deoxyadenosine, inosine, and cyclohexyladenosine. The purine, adenine, and the amino acid, taurine, do not compete for the uptake of tritiated adenosine. The compounds nitrobenzylthioinosine and dipyridamole, potent competitors for vertebrate adenosine uptake systems,[3] are not particularly potent competitors for this uptake system.

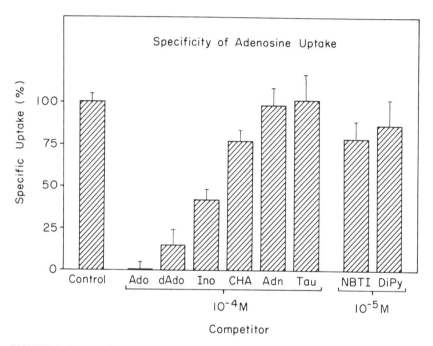

FIGURE 1. Competition by the indicated compounds for the specific uptake of [³H]Ado by aesthetasc sensilla. Sensilla were incubated for one hour in ASW containing 0.5 μM [³H]Ado plus competing compounds (dAdo = 2'-deoxyadenosine; Ino = inosine; CHA = cyclohexyladenosine; Adn = adenine; Tau = Taurine; NBTI = nitrobenzylthioinosine; DiPy = dipyridamole. Specific uptake is defined as that competed for by 100 μM unlabeled Ado.

The dephosphorylation of AMP before uptake was demonstrated by a double-label experiment in which sensilla were incubated with equimolar concentrations of ring-labeled [³H]AMP and AMP labeled with ³²P; ³H subsequently was found intracellularly, whereas ³²P was not (FIG. 2). The extracellular dephosphorylation of AMP (by definition a function of a 5'-ectonucleotidase) was very rapid; ³H from ring-labeled AMP appeared intracellularly at the same rate as ³H from adenosine (FIG. 2b). This ectonucleotidase activity is similar to that found in vertebrate systems, in that it is inhibited by ADP and the ADP analogue, α,β-methylene ADP (AMPCP).[4]

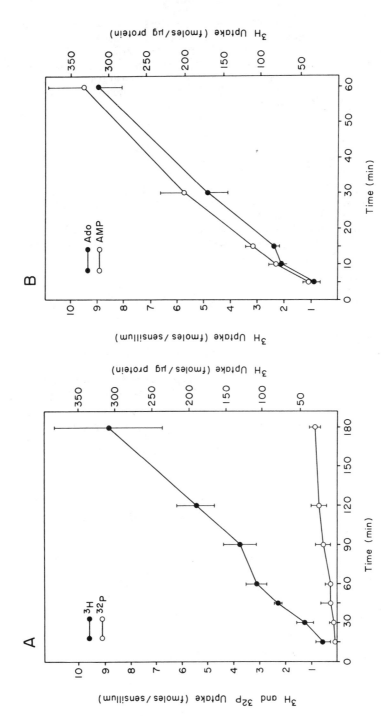

FIGURE 2. (**A**) Sensillar uptake of label from [³H]AMP (label in the adenosine ring) and [³²P]AMP (label in the phosphate group). Incubations contained both labeled forms of AMP, each present at a concentration of 0.25 μM, ± 500 μM unlabeled Ado. (**B**) Sensillar uptake of label from [³H]Ado and [³H]AMP. Incubations contained 0.5 μM [³H]Ado or 0.5 μM [³H]AMP, ± 500 μM unlabeled Ado.

The rapid dephosphorylation of AMP deactivates this signal molecule and the removal of the resulting Ado removes a possible inhibitory molecule from the receptor lymph of the sensillum. In the presence of ADP, another constituent of prey,[2] the dephosphorylation of AMP is inhibited. The ability of ADP to inhibit sensillar ectonucleotidase activity suggests that this nucleoside diphosphate could potentially modify the physiological response to AMP, by effectively lowering the threshold concentration at which the monophosphate can be detected.

SUMMARY

The olfactory system of the spiny lobster has chemosensory neurons that are excited by the purine nucleotide, adenosine 5'-monophosphate (AMP). We report here that this olfactory system also is capable of rapidly dephosphorylating AMP, and of internalizing the resulting nucleoside, adenosine. The roles of these processes in chemoreception by the lobster are discussed.

REFERENCES

1. DERBY, C. D., W. E. S. CARR & B. W. ACHE. 1984. Purinergic olfactory cells of crustaceans: Response characteristics and similarities to internal purinergic cells of vertebrates. J. Comp. Physiol. A 155: 341-349.
2. CARR, W. E. S. & C. D. DERBY. 1986. Behavioral chemoattractants for the shrimp, *Palaemonetes pugio:* Identification of active components in food extracts and evidence of synergistic mixture interactions. Chem. Senses 11: 49-64.
3. WU, P. H. & J. W. PHILLIS. Uptake by central nervous tissues as a mechanism for the regulation of extracellular adenosine concentrations. Neurochem. Int. 6: 613-632.
4. BURGER, R. M. & J. M. LOWENSTEIN. 1970. Preparation and properties of 5'-nucleotidase from smooth muscle of small intestine. J. Biol. Chem. 245: 6274-6280.

Responses of Neurons in the Nucleus of the Solitary Tract to Lingual and Palatal Stimulation with Preferred Chemicals

SUSAN P. TRAVERS[a] AND RALPH NORGREN

Department of Behavioral Science
College of Medicine
Pennsylvania State University
Hershey, Pennsylvania 17033

INTRODUCTION AND METHODS

Sucrose is a highly preferred taste stimulus for the rat. Many other chemicals preferred by this species, including both sugars and nonsugars, exhibit generalization to a taste aversion conditioned to sucrose, suggesting that there is a group of hedonically similar chemicals that are also qualitatively similar to one another for rats. Many of these same stimuli are described as "sweet" by humans. The rat's behavior toward sucrose and other psychophysically similar chemicals has been well investigated, in contrast to the neural responsiveness of this species to such stimuli. This lack of information is due to an unfortunate coincidence: The taste buds on the anterior tongue (AT) have been most frequently stimulated in neurophysiological studies but sucrose elicits poor neural responses from this receptor subpopulation in the rat.[1-3] Recent work from this laboratory,[3] however, has demonstrated that neurons in the rat nucleus of the solitary tract (NST) respond vigorously to sucrose when this chemical is applied to a different group of taste buds: those associated with the nasoincisor ducts (NID) of the hard palate. This study compares NST responses arising from AT stimulation to those arising from NID stimulation using a battery of preferred chemicals: 0.3 M sucrose, 0.3 M fructose, 0.3 M glucose, 0.3 M glycine, 0.02 M Na saccharin, 0.3 M maltose, and 0.1 M Polycose. Except for maltose and Polycose, these stimuli generalize to an aversion conditioned to sucrose in rats.[4,5,6] In addition, 0.3 M NaCl, 0.03 M HCl, 0.01 M QHCl were tested. Single-unit responses were recorded from the NST of acute, anesthetized rats using standard neurophysiological techniques. Rats were prepared using techniques that permitted free access to the oral cavity. Stimuli were applied *independently* to the AT and NID using camels' hair brushes (for details see Travers *et al.*[3]).

---

[a] Address for correspondence: Dr. Susan P. Travers, Department of Oral Biology, College of Dentistry, The Ohio State University, Columbus, OH 43210.

RESULTS AND DISCUSSION

As in a prior investigation[3] the effectiveness of the four standard taste stimuli differed for AT and NID stimulation: NaCl was most effective when applied to the tongue, sucrose to the palate. With the exception of Polycose, the other "preferred" stimuli tested were also more effective in stimulating the NID than the AT (FIG. 1). The order of effectiveness: sucrose > fructose > glucose for stimulating both taste receptor subpopulations parallels their relative efficacy for eliciting ingestion (summarized in Pfaffman[7]). On the other hand, maltose and Polycose are approximately as effective as sucrose in eliciting ingestion,[7,8] but much less effective than sucrose for NID stimulation, as is maltose for the AT (FIG. 1). The discrepancy between the

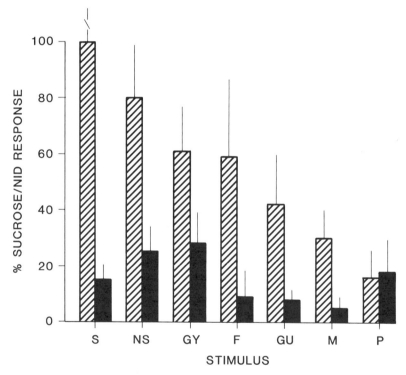

FIGURE 1. Mean responses ($+1$ SE) to 0.3 M sucrose (S), 0.02 M Na saccharin (NS), 0.3 M glycine (GY), 0.3 M fructose (F), 0.3 M glucose (GU), 0.3 M maltose, and 0.1 M Polycose (P) elicited by stimulating the NID (hatched bars) and AT (solid bars). Stimuli are ranked in order of effectiveness for NID stimulation. Magnitudes of responses are expressed relative to the response (impulses/5 sec) elicited by S applied to the NID. The AT responses to S, NS, GY and P are from 19 neurons that responded to both AT and NID stimulation and 16 other cells that received input only from the AT. The NID responses to these stimuli are from the same 19 AT-NID neurons plus three cells that responded only to NID stimulation. The averages for responses to F, GU and M are from a smaller number of neurons: AT responses: seven AT-NID neurons and five AT cells, NID responses: seven AT-NID neurons and three NID cells.

TABLE 1. Across-Neuron Correlations (Pearson's R)

Stimuli[a]	AT Only	NID Only	"Both"[b] Responses
Within electrolytes	0.68	0.96	0.65
Sweets X electrolytes	0.36	0.16	0.17
Within sweets	0.39	0.82	0.78

[a] Electrolytes = 0.3 M NaCl, 0.03 M HCl, 0.01 M QHCl, sweets = 0.3 M sucrose, fructose, glucose, glycine, and 0.02 M Na saccharin.

[b] Correlations in this category were calculated after selecting the most vigorous response to a given stimulus for each individual cell, regardless of which receptor subpopulation that response arose from. These selected responses were then used to calculate across-neuron correlations.

behavioral and neural efficacy of maltose has been noted previously with regard to the poor chorda tympani responses to this chemical.[7]

Across-neuron correlations between all possible pairs of the stimuli were calculated separately for AT and NID responses (TABLE 1). For AT responses, the mean correlation *among* the "sweet" stimuli was low, and nearly identical to the correlation *between* the electrolytes and the sweet stimuli. This suggests that the psychophysical similarity among sweet stimuli is not coded by AT responses. The correlations generated by NID responses were quite different; the correlation among the sweet stimuli increased, and was much higher than the correlation between the sweet stimuli and electrolytes, which decreased relative to the AT correlation. The NID responses, therefore, seem capable of coding the similarity among the sweet stimuli, and further separate these stimuli from the electrolytes. However, the correlation among the electrolytes is much higher than seen for AT responses, implying that the NID alone is less effective in differentiating among these different-tasting stimuli. A third set of correlations was generated, using information from both receptor subpopulations; the correlations among the sweet stimuli were high, these stimuli were sharply differentiated from the electrolytes, and the correlation among the electrolytes was lower than for NID responses alone. Therefore, the correlations for responses arising from both groups of receptors were more consistent with the psychophysical properties of the stimuli than were correlations derived from only the AT or NID.

REFERENCES

1. BEIDLER, L. M. 1953. Properties of chemoreceptors of tongue of rat. J. Neurophysiol. **16:** 595-607.
2. FRANK, M. E., R. J. CONTRERAS & T. P. HETTINGER. 1983. Nerve fibers sensitive to ionic taste stimuli in chorda tympani of the rat. J. Neurophysiol. **50:** 941-960.
3. TRAVERS, S. P., C. PFAFFMANN & R. NORGREN. 1986. Convergence of lingual and palatal gustatory neural activity in the nucleus of the solitary tract. Brain Res. **365:** 305-320.
4. NISSENBAUM, J. W. & A. SCLAFANI. 1987. Qualitative differences in polysaccharide and sugar tastes in the rat: A two-carbohydrate taste model. Neurosci. Biobehav. Rev. **11:** 187-196.
5. NOWLIS, G. H., M. E. FRANK & C. PFAFFMANN. 1980. Specificity of acquired aversions to taste qualities in hamsters and rats. J. Comp. Physiol. Psychol. **94:** 932-942.

6. PRITCHARD, T. C. & T. R. SCOTT. 1982. Amino acids as taste stimuli. II. Quality coding. Brain Res. **253:** 93-104.
7. PFAFFMANN, C. 1982. A model of incentive motivation. *In* The Physiological Mechanisms of Motivation. D. W. Pfaff, Ed.: 61-97. Springer-Verlag. New York.
8. FEIGIN, M. B. & A. SCLAFANI. 1987. Species differences in polysaccharide and sugar taste preferences. Neurosci. Biobehav. Rev. **11:** 231-240.

The Amplification Process in Olfactory Receptor Cells

DIDIER TROTIER AND PATRICK MacLEOD

Laboratoire de Neurobiologie Sensorielle
EPHE. 1 av. des Olympiades
91300 Massy, France

Many experimental data suggest that olfactory receptor cells are endowed with the utmost of sensitivity, which enables them, in favorable cases, to generate an axonal spike when they are stimulated by a single odorant molecule.[1] Does this amplifying process result purely from electrophysiological properties of the cell or does it imply a metabolic amplifying cascade by a second messenger? By applying patch-clamp recording techniques to isolated salamander olfactory receptor cells, we gathered new data which lead us to think of a comprehensive contribution of both processes in olfactory transduction.

Whole-cell recordings show voltage-dependent inward and outward currents well suited to generating axonal and possibly somatic action potentials. The input resistance

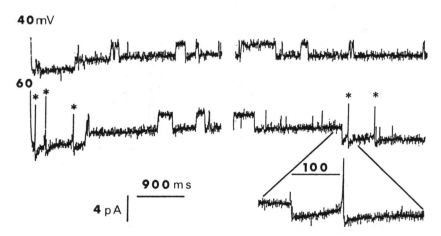

FIGURE 1. Cell-attached recordings from the somatic membrane of a salamander olfactory receptor cell. The pipette contained mainly 120 mM KCl. The inward current flowing through the channels could trigger action potentials (*) in the cell.

of a resting cell is high, usually well above 2 GΩ, and the time constant is about 30 msec.[2] Under these conditions, and provided they stay open long enough to charge the membrane capacitance, the opening of a few channels could elicit an action potential. A model is shown in FIG. 1. Cell attached recordings, with a K/low Ca-filled pipette, disclosed a large density of 27 ± 6 pS channels in the somatic membrane.

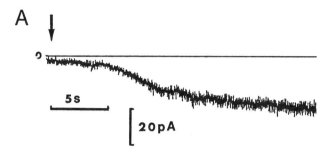

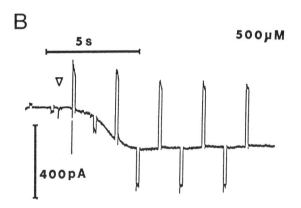

FIGURE 2. (A) Whole-cell recording of a salamander olfactory receptor cell. The electrode contained an intracellular-like solution. A bath application of forskolin (50 μM) induced an inward current and an increase of the membrane noise. Holding potential: -50 mV. Arrow: approximate time of arrival of forskolin.

(B) Whole-cell recording of a salamander olfactory receptor cell. The electrode contained an intracellular-like solution and 500 μM cAMP. After the onset of the WCR (triangle), cAMP diffused into the cell and elicited an inward current together with an increased membrane conductance. Holding potential: -50 mV. Test potentials: -100 and 0 mV.

By hyperpolarizing the patch membrane (positive pipette potentials), the inward current flowing through two channels was large enough to elicit an action potential in the cell. "Chemoreceptive" channels, with similar properties, located in the proximal part of the ciliary membrane would produce the same effect.

It has been proposed, based on electrophysiological[3] and biochemical[4] studies, that cAMP does play a role in olfactory transduction. We have observed, for the first time, that an increase of intracellular cAMP concentration opens depolarizing channels in olfactory receptor cells. Experiments were done in the whole cell recording (WCR) configuration, under normal ionic gradients, at -50 mV. Bath application (arrow in FIG. 2A) of an adenylate cyclase activator, forskolin, elicited an inward current and an increase of the membrane current noise. In other experiments, cAMP (50-500 μM) was injected by diffusion from the recording electrode. About 1 second after the onset of the WCR (triangle in FIG. 2B), an inward current resulted, associated with an increased membrane conductance, as indicated by test potentials to -100 and 0 mV. cAMP opens depolarizing channels in olfactory receptor cells. Together with the fact that some odorants activate adenylate cyclase,[4] this finding supports the idea of a second messenger participating in the olfactory transduction process.

REFERENCES

1. MOULTON, D. G. 1977. Minimum odorant concentrations detectable by the dog and their implications for olfactory receptor sensitivity. In Chemical Signals in Vertebrates, D. Müller-Schwarze & M. Mozell, Eds.: 455-464. Plenum Press. New York.
2. TROTIER, D. 1986. A patch-clamp analysis of membrane currents in salamander olfactory receptor cells. Pflügers Arch. 407: 589-595.
3. MINOR, A. V. & N. L. SAKINA. 1973. Role of cyclic adenosine-3',5' monophosphate in olfactory reception. Neurofysiologiya 5: 415-422.
4. PACE, U., E. HANSKI, Y. SALOMON & D. LANCET. 1985. Odorant-sensitive adenylate cyclase may mediate olfactory reception. Nature 316: 225-258.

Functional Morphology of the Olfactory Epithelium

KAZUYOSHI UENO, YUTAKA HANAMURE, JEUNG
GWEON LEE, AND MASARU OHYAMA

Department of Otolaryngology
Kagoshima University
Faculty of Medicine
Kagoshima 890, Japan

With the newly developed technique of scanning electron microscopy (SEM) using back-scattered electron imaging, three-dimensional structure and histochemical localization of argyrophil substances of the olfactory epithelium were observed. Additionally the three-dimensional structure of the basement membrane (BM) of the olfactory epithelium was studied with SEM, and laminin, which is an important component of the BM, was investigated using immunofluorescence.

MATERIALS AND METHODS

A total of five rabbits were used for the SEM study of the olfactory epithelium. After cracking the epithelia, they were stained with silver methenamine. Then we photographed the secondary electron image and back-scattered electron image, which were subsequently filtered and synthetically processed in color. For the SEM observation of BM, the specimens were immersed in 0.2% OsO_4 and the epithelial layer was removed. Additionally the localization and proliferation of laminin were studied by the immunofluorescence technique, using the olfactory epithelium of five laryngectomized dogs.

RESULTS AND DISCUSSION

The colored SEM of the olfactory epithelium demonstrated vividly that the nucleus of the olfactory cells and some of the basal cells had affinities to silver. These finding suggested that the basal cells, having affinity for silver, might differentiate into the olfactory cells.

SEM observation showed that the BM of the olfactory epithelium had many 20-μm diameter round holes, spaced at fairly regular intervals; mucous was secreted from

680

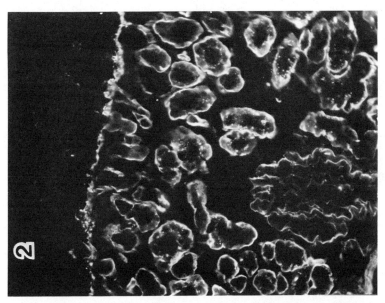

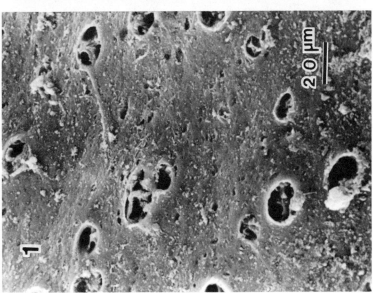

FIGURE 1. Olfactory epithelium with presumed openings of Bowman's glands.

FIGURE 2. Immunofluorescence stain shows laminin in basement membrane, olfactory epithelium, and Bowman's gland.

some of these holes. These holes were thought to be the openings of Bowman's glands (FIG. 1).

Laminin is a high molecular weight glycoprotein and an important component of BM. Using the immunofluorescence technique, laminin was clearly observed in the BM adjacent to the olfactory epithelium, and around vessels as well as Bowman's glands (FIG. 2). There were no significant differences of laminin staining between pre- and post-laryngectomized olfactory epithelium. It was also confirmed that the olfactory epithelium maintained its normal form. These findings suggest a role for laminin in the maintenance of proper tissue organization during olfactory epithelial regeneration.

REFERENCE

1. UENO, K. et al. 1985. Histochemical studies of glycoconjugates of basement membrane in the upper airway mucosa. In Basement Membranes. Seiichi Shibata, Ed.: 459-460. Elsevier. New York.

The Role of ATP and GTP in the Mediation of the Modulatory Effects of Odorant on Olfactory Receptor Sites Functionally Reconstituted into Planar Lipid Bilayers[a]

V. VODYANOY AND I. VODYANOY

Department of Physiology & Biophysics
University of California
Irvine, California 92717

The steady-state conductance of planar bimolecular lipid membranes (BLM) modified with rat olfactory homogenates becomes sensitive to very low concentrations of odorant in the presence of adenosine triphosphate (ATP) and guanosine triphosphate (GTP). The chemosensitivity is not observed when ATP and GTP are absent. Adenosine 3',5'-monophosphate (cyclic AMP) mimics the effect of the odorant. Effects of odorants and cyclic AMP are dose dependent. These data are consistent with the hypothesis that cyclic AMP is a second messenger in the initial steps of olfactory transduction.

Membrane fragments from rat olfactory epithelial homogenates that were incorporated into planar, essentially solvent-free BLM's have been used as a model system for initial events in mammalian olfactory transduction.[1] In the present work we use this to study the involvement of ATP, GTP, and cyclic AMP.

The conductivity of the BLM treated with olfactory epithelial homogenates increased in response to the odorant when the buffer solution used for the homogenate preparation contained ATP and GTP, or ATP and GTP are added to the media bathing the membrane before the odorant addition (FIG. 1A). Curve 1 of this figure shows an experiment in which vesicles prepared from the rat olfactory homogenate were pooled in a solution containing ATP and GTP before addition to the *cis*-compartment of the membrane chamber. Fifteen minutes after the membrane was treated with olfactory epithelial homogenate, diethyl sulfide at a final concentration of 10 nM was added to the same compartment. After the overshoot in the membrane conductance, a stationary value of conductance was reached. The effect of diethyl sulfide is dose dependent with a threshold of about.10 pM.[1] The curves 2 and 3 (FIG. 1A) illustrate the conductance upon two subsequent additions of 3',5'-cyclic AMP and 2',3'-cyclic AMP, respectively. The amplitude of the cyclic AMP effect appears to be dose dependent with a threshold in the region of 1 μM (FIG. 2).

[a]Supported by grants from the United States Army (DAAG29-85-K-0109), the National Institutes of Health (HL-30657), and the National Science Foundation (BNS-8508495).

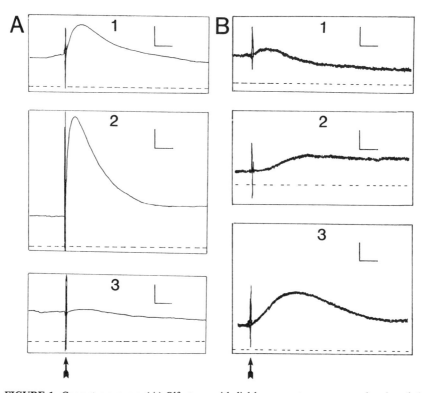

FIGURE 1. Current responses. (**A**) Olfactory epithelial homogenate was prepared and pooled in: 10 mM 3-N-morpholino propanesulfonic acid (MOPS), 50 mM sucrose, 15 μM ATP, 10 μM GTP, pH 7.4 with K$^+$ counterion. Membrane bathing solution: 30 mM KCl, 30 mM NaCl, 2 mM CaCl$_2$, pH 7.4 and 2.5 nM of 3,5-di-tert-butyl-4-hydroxybenzylide (SF-6847; see Vodyanoy and Vodyanoy[2]). The vertical base is equal to 5 pA, and the horizontal base is one minute. The membrane voltage was clamped at 10 mV. The dashed line is zero current. The current responses to: (1) diethyl sulfide of 10 nM; (2) 3'5'-cyclic AMP of 10 μM, and (3) 2'3'-cyclic AMP of 10 μM.

(**B**) Solution in the *cis*-compartment was continuously stirred during the measurement. (1) The current response to diethyl sulfide solution in ATP- and GTP-free conditions. Diethyl sulfide at a final concentration of 10 nM was added to the *cis*-compartment at the moment shown by the arrow. (2) ATP (final concentration of 20 μM) was added to *cis*-compartment at the arrow. (3) GTP (at a final concentration of 20 μM) was added at the arrow. Diethyl sulfide and ATP have been added already.

If neither ATP nor GTP was added to the reconstituted system, no conductivity increase was observed when diethyl sulfide was added (FIG. 1B, curve 1). When ATP was added to the system (FIG. 1B, curve 2) the conductance was restored to its previous value. Not until GTP was delivered to the system did the conductance increase (FIG. 2, curve 3) much like the conductance response shown in FIGURE 1A, curve 1.

If odorants act through a second messenger, then certain criteria of nucleotide involvement in this physiological process must be fulfilled.[3] For the olfactory system these criteria are as follows:[4] (1) The receptor cells should contain adenylate cyclase. (2) Odorants should modulate the activity of adenylate cyclase in a plasma membrane preparation from the receptor cells. (3) The system should not respond to odorant when messenger accumulation is prevented by blocking cyclic AMP synthesis. (4) Cyclic AMP should be able to simulate the action of odorant.

That criteria (1) and (2) are met for the olfactory receptors was clearly demonstrated for a cilia-enriched preparation.[5-7] The experiments reported in this work relate to the third and the fourth criteria[3] and give additional evidence for the possible involvement of cyclic AMP in olfactory transduction.[8]

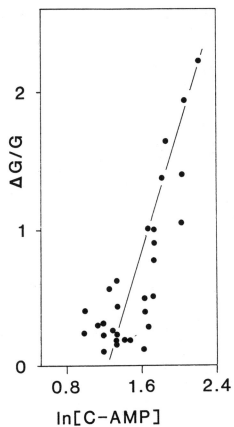

FIGURE 2. The planar BLM relative conductance dependence on cyclic AMP concentration. Data represent the results of 29 individual experiments. The planar BLM was pretreated with olfactory epithelial homogenate at the final concentration of about 10 μg/ml. Membrane conductance was measured under a clamp voltage of 10 mV.

REFERENCES

1. VODYANOY, V., R. B. MURPHY. 1983. Science **220:** 717-719.
2. VODYANOY, I., V. VODYANOY & J. K. LANYI. 1986. Biochim. Biophys. Acta **858:** 92-98.
3. ROBINSON, G. A., R. W. BUTCHER & E. W. SUTHERLAND. 1968. Ann. Rev. Biochem **37:** 149-174.
4. MENEVSE, A., G. DODD & T. M. POYNDER. 1977. Biochem. Biophys. Res. Commun. **77:** 671-677.
5. PACE, U., E. HANSKI, Y. SOLOMON & D. LANCET. 1985. Nature **316:** 255-258.
6. D. LANCET, J. HELDMAN & U. PACE. 1985. Soc. Neurosci. Abstr. **11:** 815.
7. CHEN, Z., M. GREENBERG, U. PACE & D. LANCET. 1985. Soc. Neurosci. Abstr. **11:** 970.
8. A. V. MINOR & N. L. SAKINA. 1973. Neurofiziologia **5:** 414.

Receptive Fields of Second-Order Taste Neurons in Sheep

Convergence of Afferent Input Increases During Development[a]

MARK B. VOGT AND CHARLOTTE M. MISTRETTA[b]

Department of Oral Biology
School of Dentistry and
Center for Human Growth and Development
University of Michigan
Ann Arbor, Michigan 48109

Previous neurophysiological studies have provided only a gross regional characterization of the lingual receptive fields of mammalian central taste neurons. Here we report the preliminary results of a quantitative analysis of the receptive fields of second-order taste neurons in terms of discrete receptor populations, and present evidence for an increase in afferent convergence on central taste neurons in sheep during development.

Afferent neurons in the chorda tympani nerve (CT) innervate taste buds located in fungiform papillae on the anterior tongue and synapse centrally on neurons in the nucleus of the solitary tract (NST) in the medulla. CT fibers[1] or NST neurons were studied in fetal sheep aged about 130 days gestation (term = 147 days) and lambs aged 40-50 days postnatal. Single neurons were isolated and responses to 0.5 M NH_4Cl, $NaCl$, and KCl were recorded. Then a fine platinum probe was used to electrically stimulate (10 microamps anodal current) individual fungiform papillae and the number and location of papillae in the receptive field of the neuron were recorded. The criterion for a chemical or electrical response was that impulse frequency should be greater than the mean plus two standard deviations of the spontaneous rate.

Most neurons had a receptive field located in the area of the highest concentration of fungiform papillae: the edge of the anterior one-third of the tongue extending about 1.5 cm on both the dorsal and ventral surfaces. In general, the fungiform papillae in the receptive field of a particular neuron were located together and were not scattered among other papillae. We also found there could be considerable overlap in the receptive fields of different neurons, that is, they might share many of the same papillae.

As presented in TABLE 1, the receptive field size of NST neurons was generally greater than that of CT neurons. Furthermore, the receptive field size of NST neurons was larger in lambs than fetuses, but for CT neurons receptive field size did not

[a] Supported by National Science Foundation Grant BNS 83-11497 to CMM.

[b] Address for correspondence: Charlotte M. Mistretta, Department of Oral Biology, School of Dentistry, University of Michigan, Ann Arbor, MI 48109.

TABLE 1. Receptive Field Size of CT and NST Neurons in Fetus and Lamb

Neuron Type	Format of Results	Number of Papillae in Receptive Field	
		Fetus	Lamb
NST neurons	$\overline{X}$ [a]	21	37
	n (SD)	8 (10)	30 (20)
	range	8-36	2-100
CT fibers	$\overline{X}$	13	11
	n (SD)	17 (9)	32 (7)
	range	2-40	2-29

[a] $\overline{X}$ = mean, n = number of fields, SD = standard deviation, range = smallest-largest receptive field.

increase (Age [fetus, lamb] by neurons [CT, NST] ANOVA; neurons and interaction Fs [1,83] > 6.0, ps < 0.02).

Although these data are preliminary, the larger receptive fields of NST neurons compared to peripheral neurons indicates a convergence of taste papilla input on central neurons. Furthermore, because the average receptive field size of NST neurons is larger in lambs than fetuses, the degree of convergence increases during development. Apparently this is due to increasing convergence of CT fibers onto NST neurons, since the total number of papillae innervated by a CT fiber does not increase between fetal and lamb ages.

REFERENCE

1. MISTRETTA, C. M., T. NAGAI & R. M. BRADLEY. Relation of receptive field size and salt taste responses in chorda tympani fibers during development. Ann. N.Y. Acad. Sci. This volume.

Variation in Olfactory Proteins

Evolvable Elements Encoding Insect Behavior

RICHARD G. VOGT AND GLENN D. PRESTWICH

Department of Chemistry
State University of New York at Stony Brook
Stony Brook, New York 11780

The pheromone-sensitive sensory hairs of the silk moth *Antheraea polyphemus* contain a pheromone-binding protein and a pheromone-degrading sensillar esterase.[1] Both proteins are unique to the extracellular lymph surrounding the sensory dendrites. Both proteins have been purified and their properties characterized.[2-4] Our data suggest that the binding protein functions as a pheromone carrier by phase partitioning the lipophillic pheromone molecules into solution, allowing them to move through the otherwise aqueous lymph space to the presumed membrane-bound receptor proteins. The esterase degrades pheromone rapidly (*in situ* half-life less than 15 msec), suggesting that it functions to maintain a low pheromone noise level within the hair. Together, the kinetic properties of these two proteins, along with those of the receptor, determine the efficiency of pheromone following during precopulatory flight. Thus, a part of this animal's precopulatory behavior is encoded in gene-coded kinetic properties of these proteins.

I have electrophoretically examined the pattern and activity of these two proteins from individual animals. FIGURE 1 shows a gel, with individual lanes corresponding to individual animals. The sensillar esterase (SE) appears as a closely spaced quartet to octet of bands. While all males possess multiple esterase bands, the presence of a particular identifiable band varies between individuals. Activity varies between bands from one individual and the same band from different individuals. Variation may be due to a combination of allelic differences, sequential modification (i.e. glycosilation) or polygenic expression. The binding protein (BP) appears as either a single fast or slow band.

We have recently sequenced the first thirty amino acids of the binding protein using combined techniques of electroblotting[5] and gas-phase microsequencing (Applied Biosystems 470A). This sequence is:

N-Ser Pro Glu Ile Met Lys Asn Leu Ser Leu Asn Phe Gly Lys Ala Met Asp Gln
Ser Lys Asp Glu Leu Asn/Ser Leu Pro Asp Ser Val Val

The sequence was confirmed from two individual animals, one from Long Island, New York, and one from Racine County, Wisconsin. The individual sequences matched exactly, except for residue 24 (Asn/NY, Ser/WI), reflecting evolved population differences.

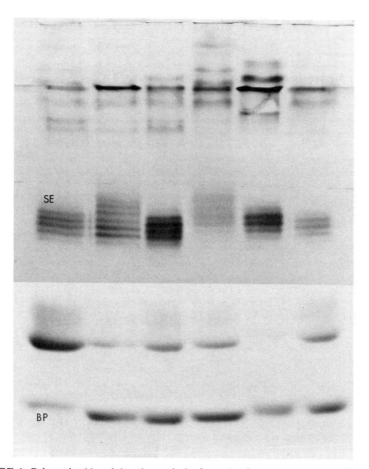

FIGURE 1. Polyacrylamide gel (nondenaturing) of proteins from several animals. Each lane contains a sample from a different animal (see text).

These patterns of variation suggest a model for the molecular basis of behavior that focuses on the evolvable elements of behavior. The temporal aspects of behavior are encoded in the kinetic properties of neural proteins. Behavioral variability can, in part, be ascribed to the allelic differences in these proteins. Evolution screens for "appropriate" behavior and selects for neural alleles with "appropriate" properties. The silk moth's sensory hair proteins provide a unique opportunity to study the molecular basis of sexual selection. Unlike the neural proteins of the CNS, these peripheral proteins have no shared function elsewhere in the animal to compromise their evolving design.

REFERENCES

1. VOGT, R. G. & L. M. RIDDIFORD. 1981. Nature **293:** 161-163.
2. VOGT, R. G., L. M. RIDDIFORD & G. D. PRESTWICH. 1985. Proc. Natl. Acad. Sci. USA **82:** 8827-8831.
3. VOGT, R. G. & RIDDIFORD. 1986. *In* Mechanisms in Insect Olfaction. T. L. Payne, M. C. Birch & D. E. J. Kennedy, Eds.: 201-208. Clarendon. Oxford, England.
4. VOGT, R. G. & L. M. RIDDIFORD. 1986. J. Chem Ecol. **12:** 469-482.
5. AEBERSOLD, R. H., D. B. TEPLOW, L. E. HOOD & S. B. H. KENT. 1986. J. Biol. Chem. **261:** 4229-4238.

Signal-to-Noise Ratios and Cumulative Self-Adaptation of Chemoreceptor Cells[a]

RAINER VOIGT AND JELLE ATEMA

Boston University Marine Program
Marine Biological Laboratory
Woods Hole, Massachusetts 02543

The taste receptor organs of the walking legs of the American lobster, *Homarus americanus,* are composed of mostly narrowly tuned populations of receptor cells.[1] Under natural conditions these receptor cells detect varying concentrations of their best stimulus against a chemically "noisy" background. Cell sensitivities change constantly due to: (1) self-adaptation to a constant background, (2) cumulative self-adaptation to repeated stimulus pulses, (3) cross-adaptation, and (4) mixture suppression. We have started to characterize adaptation and disadaptation time courses of

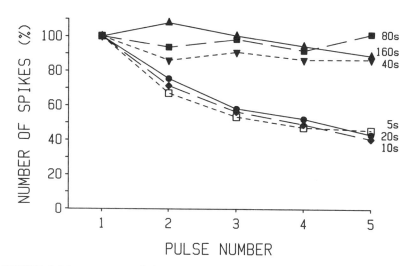

FIGURE 1. Mean responses of glutamate-best cells to five successive one-second glutamate pulses ($3 \times 10^{-4} M$ in ASW background) separated by either 5, 10, 20, 40, 80, or 160 seconds. For each cell responses are standardized to the first response ($=100\%$). $n = 5$ for each of the six pulse intervals.

[a] Supported by National Science Foundation Grant BNS 8512585.

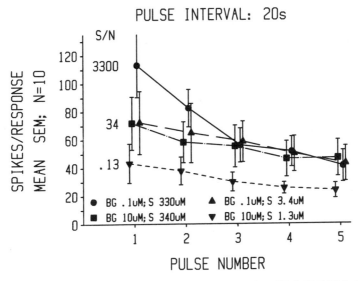

FIGURE 2. Effects of repetitive stimulation to five glutamate pulses (10^{-2}, $10^{-4}M$) in $10^{-7}M$ and $10^{-5}M$ glutamate background in 20-second interpulse intervals. BG: Glutamate background concentration. S: Applied glutamate stimulus concentration (peak). S/BG results in signal-to-noise (S/N) ratio: 3300 (●), 34 (■,▲), and 0.13 (▼).

glutamate receptor cells using a series of short (1-sec) standard glutamate pulses in backgrounds of different glutamate concentration and varying the interpulse intervals from 5 seconds to 160 seconds.

The temporal stimulus profile was calibrated by measuring the change of conductivity of flowing deionized water (20 ml/min) after injection of 50 μl 1.0 M NaCl solution. The stimulus chambers allowed 5-second interpulse intervals with less than 1% residual level of the previous stimulus pulse.

Single cells were identified in artificial seawater with $10^{-4}M$ glutamate pulses. Then, a series of five test pulses ($10^{-2}M$) was applied. Pulse intervals were 5, 10, 20, 40, 80, or 160 seconds. In the second experiment, a series of five pulses ($10^{-2}M$ or $10^{-3}M$) was applied in a low-glutamate background ($10^{-7}M$) in 20-second interpulse intervals. After three minutes of self-adaptation to an elevated glutamate background ($10^{-5}M$), the series was repeated.

Pulse intervals of 40 seconds and longer caused only slight variability in responses (both in number of spikes elicited and in duration of the response). Shorter interpulse intervals caused cumulative adaptation: There was no difference between 20-, 10-, and 5-second intervals (FIG. 1).

Adaptation to higher backgrounds reduced overall sensitivity and repetitive stimulation then caused only a slight further decrease in responsiveness. At two different background levels, the same signal-to-noise ratio resulted in similar responses, including similar cumulative adaptation (FIG. 2). The greater signal-to-noise ratio caused both stronger responses and greater cumulative adaptation, while the smaller ratio caused weaker responses and less cumulative adaptation.

Thus, (1) signal-to-noise ratios and not absolute stimulus levels were predictive of the responses of glutamate receptor cells and (2) as a population these cells do not fully recover from strong stimuli (high signal-to-noise ratio) until 40 seconds later. In addition, (3) there are pronounced differences between individual cells: Fast-adapting cells also disadapt faster than slow-adapting cells. These results may have consequences for animal behavior in turbulent odor clouds.[2]

REFERENCES

1. JOHNSON, B. R., R. VOIGT, P. F. BORRONI & J. ATEMA. 1984. Response properties of lobster chemoreceptors: Tuning of primary taste neurons in walking legs. J. Comp. Physiol. A **155:** 593-604.
2. ATEMA, J. 1987. Distribution of chemical stimuli. *In* Sensory Biology of Aquatic Animals. J. Atema, R. R. Fay, A. N. Popper & W. Tavolga, Eds.: 87-96. Springer Verlag. New York.

Long-Lasting Effects of Context on Sweetness Evaluation

TERESA ANNE VOLLMECKE[a]

Department of Psychology
University of Pennsylvania
Philadelphia, Pennsylvania 19104

Recent prior experience exerts a potent influence on current sensory-perceptual evaluations.[1,2] In the taste realm, the sweetness of beverages[3] and saltiness of soups[4] depend upon other beverages or soups experienced in the same session. Recent past experience, or context, long considered a nuisance variable in testing, may instead play a more important role in regulating the normal response to foods. Dietary experience such as exposure to a low-salt diet can alter both perceived intensity and preference of salt.[5] Part of this experience may be the operation of contextual effects. Context's influence on food evaluation is examined in this study using controlled exposure to a simple food—a colored sucrose beverage. Subjects evaluate sweetness in two separate contexts defined by different concentration ranges of sucrose.

METHODS

Forty individual subjects were tested twice, in sessions separated by one week (TABLE 1). In each session, they tasted seven samples each of three different sucrose beverages presented in counterbalanced blocks. All beverages were identically colored red and differed only in sucrose content. Subjects evaluated sweetness intensity using a nine-point category rating scale with verbal labels ranging from not at all sweet to extremely sweet.

TABLE 1. Experimental Design

Group	Session I Sucrose Conc.	Session II Sucrose Conc.
Low-high contexts	0.06, 0.14, 0.33	0.33, 0.77, 1.8
High-low contexts	0.33, 0.77, 1.8	0.06, 0.14, 0.33

[a] Current address: The Proctor & Gamble Company, 5299 Spring Grove Avenue, Cincinnati, OH 45217.

RESULTS

Both groups showed monotonically increasing sweetness evaluations for the three sucrose concentrations in low and high contexts (see FIG. 1). The difference in sweetness evaluation for the 0.33 M sucrose between contexts was examined with a two-way ANOVA with repeated measures. Significant main effects were found for groups ($F_{1,38} = 45.6$, $p < .01$) and context ($F_{1,38} = 147.8$, $p < .01$). Context and groups accounted for 57 and 12% of the variance, respectively (n^2). Subjects gave higher mean ratings for 0.33 M sucrose in the low context (mean low-high = 7.97, high-low = 5.62) than in the high context (mean low-high = 3.96, high-low = 3.46). Differences were significant at the 0.01 level (Tukey test). The two groups differed in their evaluations in the low, but not the high context. The high-low group gave significantly lower ratings for the 0.14 and 0.33 M sucrose beverages than the low-high group (Tukey test, $p < 0.01$). Context effects developed over the session. Ratings for 0.33 M sucrose in the low context increased, while in the high context they decreased with continued sampling (see FIG. 2).

SUMMARY AND DISCUSSION

The main results of this study are as follows:
1. Subjects show a clear effect of context, rating 0.33 M sucrose *sweeter* in the low context than the same concentration in the high context.

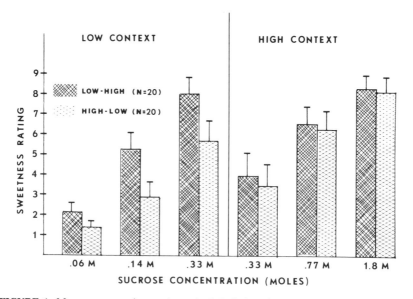

FIGURE 1. Mean sucrose ratings and standard deviations for five concentrations of sucrose presented in two separate contexts. The 0.33 M sucrose is rated in both contexts. Each bar represents the mean judgement of the last six samples of beverage for twenty subjects.

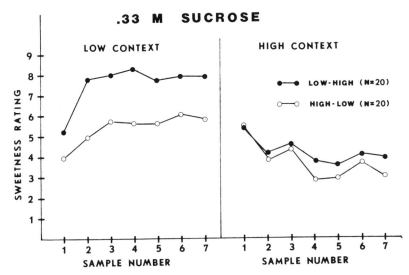

FIGURE 2. Mean sweetness ratings for seven samples of 0.33 M beverages in each context. Points represent the mean rating of twenty subjects.

2.   The context effect on sweetness evaluation develops quickly after initial exposure to the stimuli.

3.   Prior experience with the high context suppresses sweetness evaluations in the low context.

The first two effects are consistent with literature reports from other perceptual domains.[4] The third effect, the long-lasting impact of a high context, is unexpected. Context has previously been thought to exert only a short-acting influence restricted to other evaluations made in close temporal proximity. This study demonstrates that context can exert a more long-lasting influence. One possible reason a lasting effect follows the high, but not the low, context is that the former violates the normal expectation for sweetness in beverages (10% or less). This discrepancy between expectation and perceived sweetness leads to an enduring memory of the experience that is evoked the next time similar beverages are encountered.

REFERENCES

1.   HELSON, H. 1964. Adaptation Level Theory. Harper Row. New York.
2.   PARDUCCI, A. 1965. Category judgements: A range frequency model. Psychol. Rev. **72**(6): 407-418.
3.   RISKEY, D. R., A. PARDUCCI & G. K. BEAUCHAMP. 1979. Effects of context in judgements of sweetness and pleasantness. Percept. Psychophys. **26**(3): 171-176.
4.   LAWLESS, H. 1983. Contextual effects in category ratings. J. Test. Eval. **11**(5): 346-349.
5.   BERTINO, M., G. K. BEAUCHAMP & K. ENGELMAN. 1982. Long term reduction in dietary sodium alters the taste of salt. Am. J. Clin. Nutr. **36**: 1134-1144.

Taste and Salivary Gland Dysfunction

JAMES M. WEIFFENBACH, PHILIP C. FOX, AND
BRUCE J. BAUM

Clinical Investigations and Patient Care Branch
National Institute of Dental Research
National Institutes of Health
Bethesda, Maryland 20892

Salvia is normally present during tasting and is thought to be essential for the maintenance of oral structures subserving taste. We have recently described remarkably unimpaired taste perception in the complete and chronic absence of salivary gland function.[3] In the present study, we assess the taste function of persons with less severe salivary gland dysfunction.

The eight subjects of our earlier study were the only persons among 75 patients referred for evaluation of "dry mouth" to have no flow from either of the major glands under both resting and stimulated conditions. Their minor glands showed marked destruction. Deterioration of their teeth and oral soft tissues indicated chronic absence of saliva. One person had no taste complaints, denied any change in taste perception and demonstrated performance well within normal limits on threshold[1] and suprathreshold[2] tasks with stimuli representing each of the four basic taste qualities; thus, taste impairment is not a necessary consequence of salivary gland dysfunction. As a group, these subjects were essentially free from subjective complaints and were not impaired on objective measures of suprathreshold perception. The incidence of impairment on the threshold task was, however, elevated.

The 10 subjects of the present study were the only patients that lacked flow at rest but salivated in response to stimulation. On objective measures of taste function, they resembled the subjects reported earlier. Impairment of suprathreshold performance, relative to normal controls, was no more common than expected by chance. Three persons showed no impairment for either of two measures of suprathreshold perception for any of these four qualities. Similarly, threshold performance in this study resembled that observed in the earlier one. In each study, the incidence of impaired performance on the threshold task was elevated above that expected from normal subjects. However, as in the earlier case, a single person with normal thresholds for all four qualities demonstrated that impairment of threshold sensitivity is not inevitable.

While the objective findings obtained in the two studies are parallel, the subjective reports are strikingly different (see TABLE 1). Subjects with no salivary flow are less likely than those who salivate on stimulation to report decreased enjoyment of eating. Only one subject with complete salivary gland failure reported an overall decrease in

[a] Address for correspondence: James M. Weiffenbach, National Institutes of Health, Bldg. 10, Room 1A-05, Bethesda, Maryland 20892

TABLE 1. Salivary Flow and Subjective Reports[a]

Subjective Parameter	Subject Number									
	1	2	3	4	5	6	7	8	9	10
No Flow:										
Enjoyment of eating	↓	=	=	↓	=	↓	=	=		
Taste change	=	=	=	±	=	↓	=	=		
Smell change	=	=	↓	↑	=	±	=	=		
Some Flow:										
Enjoyment of eating	=	na	↓	=	↓	↓	↓	↓	=	↓
Taste change	↓	↓	↓	=	↓	↓	↓	↓	↓	↓
Smell change	=	=	ns	=	=	=	ns	=	=	↑

[a] Patient responses to a verbally administered questionnaire: ↓ = decrease; ↑ = increase; "=" = no change; ± = some qualities or odors increase and others decrease; ns = not sure; na = not available.

taste perception, whereas 9 of 10 persons with some residual gland function reported that their sense of taste was decreased.

It may be significant that the laboratory findings reflect performance after oral rinsing with distilled water. Thus, the oral milieu under which they are obtained is similar for those lacking gland function and those who produce saliva under stimulation. In contrast, subjective evaluations of perception and of the enjoyment of food reflect performance under conditions that could elicit salivation from persons with residual function. Speculation that differences in subjective evaluation depend on differences in stimulated salivation is supported by the observation that individuals with and without stimulated gland function make similar evaluations of their sense of smell.

REFERENCES

1. WEIFFENBACH, J. M., B. J. BAUM & R. BURGHAUSER. 1982. Taste thresholds: Quality specific variation with aging. J. Gerontol. 37(3): 372-377.
2. WEIFFENBACH, J. M., B. J. COWART & B. J. BAUM. 1986. Taste intensity perception in aging. J. Gerontol. 41(4): 460-468.
3. WEIFFENBACH, J. M., P. C. FOX & B. J. BAUM. 1986. Taste and salivary function. Proc. Natl. Acad. Sci. USA 83(16): 6103-6106.

Number, Size, and Density of Mitral Cells in the Olfactory Bulbs of the Northern Fulmar and Rock Dove

BERNICE M. WENZEL [a]

Department of Physiology
University of California
Los Angeles, California 90024

ESMAIL MEISAMI

Department of Physiology and Biophysics
University of Illinois
Urbana, Illinois 61801

The great variation in size of olfactory bulbs and conchae across avian species is well documented,[1] but detailed architectonic measurements have not been provided. Using morphometric and cell-counting methods, we have studied the olfactory bulbs of the Northern Fulmar (*Fulmarus glacialis*) and the Rock Dove (*Columba livia*). The former species represents procellariiform birds, which have very large bulbs, and the latter represents birds with bulbs of average size. The fulmar's use of olfaction in foraging has been described both anecdotally and experimentally,[2,3] and the pigeon's olfactory acuity has been studied as has its hypothesized reliance on odors in homing.[4]

The following measures were obtained from complete Nissl-stained serial coronal sections of olfactory bulbs: total volume, total number of mitral cells, mean diameter of mitral cells, total area of the mitral cell layer, and density of mitral cells per unit area of mitral cell layer. Mitral cell counts were based on the presence of a nucleolus. Values given are conservative and approximate, and include corrections for nucleolar overcounts due to binucleation and transection.

The data (TABLE 1) indicate that the relationship between bulb size and mitral cell number may be more complex than expected. The fulmar's bulb is 20 times larger than the pigeon's but has only six times more mitral cells. The total number of mitral cells in the fulmar is about twice that for rat[5] and rabbit[6] while the pigeon's total is in the range reported for mouse,[7] which has a bulb of similar size. The fulmar's bulb is larger because its mitral cells are bigger, twice the size of those in the pigeon, and the combined thickness of its functionally significant layers is greater (TABLE 1), also twice the pigeon's. Among these layers, the external plexiform and internal granular are very well developed.

[a] Address for correspondence: Dr. Bernice M. Wenzel, Department of Physiology, UCLA School of Medicine, Los Angeles, CA 90024.

The impressively large number of mitral cells in the Northern Fulmar, together with its very thick cortical bulbar sheet, should stimulate intensive olfactory research with the fulmars and related procellariiform species. Equally interesting is the relatively high number of mitral cells in the pigeon, in the same range as the mouse, implying that the pigeon may have some olfactory capacities comparable to those of such small rodents. To put these results in broader functional perspective, further counts are needed for the number of primary olfactory neurons and olfactory glomeruli and, thus, the convergence ratios in the primary afferent relay.

TABLE 1. Morphometry of Olfactory Bulb and Mitral Cell Number in Northern Fulmar and Rock Dove[a]

Structure	Northern Fulmar	Rock Dove
Bulb volume, mm^3	60	3
Area of mitral cell layer,[b] mm^2	35	5
Mitral cell:		
Diameter, μm	19	11
Number	120,000	20,000
Density, No./mm^2	3,500	4,000
Total thickness of bulb layers,[c] μm	900	450

[a] Each value is the mean of measurements obtained from olfactory bulbs of at least two animals.
[b] In the pigeon, the mitral cells occur in a narrow band rather than a monolayer.
[c] From glomerular to granular.

ACKNOWLEDGMENTS

Technical assistance was provided by H. Davidian, Dr. V. S. Fox, and A. Motamedi. A preliminary version of some of these results was reported to the Pacific Seabird Group in 1982 and to the Association for Chemoreception Sciences in 1983.

REFERENCES

1. BANG, B. G. 1971. Functional anatomy of the olfactory system in 23 orders of birds. Acta Anat. 79(Suppl.): 1-76.
2. HUTCHISON, L. H. & B. M. WENZEL. 1980. Olfactory guidance in foraging by procellariiforms. Condor 82: 314-319.
3. HUTCHISON, L. H., B. M. WENZEL, K. E. STAGER & B. L. TEDFORD. 1984. Further evidence for olfactory foraging by Sooty Shearwater and Northern Fulmars. In Marine Birds: Their Feeding Ecology and Commercial Fisheries Relationships. D. N. Nettleship, G. A. Sanger & P. F. Springer, Eds.: 72-77. Special Publication Canadian Wildlife Service. Ottawa, Canada.

4. WENZEL, B. M. 1983. The chemical senses. *In* Physiology and Behaviour of the Pigeon. M. Abs, Ed.: 149-167. Academic Press. London, England.
5. MEISAMI, E. & L. SAFARI. 1981. A quantitative study of the effects of early unilateral olfactory deprivation on the number of mitral and tufted cells in the rat olfactory bulb. Brain Res. **221:** 81-107.
6. ALLISON, A. C. & R. T. T. WARWICK. 1949. Quantitative observations on the olfactory system of the rabbit. Brain **72:** 186-197.
7. SCHÖNHEIT, B. 1970/71. Quantitative zytoarchitektonische Untersuchungen zur postnatalen Entwicklung des Bulbus olfactorius der Albinomaus (Mus musculus L. 1758, forma, alba), Stamm "Strong A". J. Hirnforsch. **12:** 375-388.

The Nervus Terminalis of the Shark

Influences on Ganglion Cell Activity[a]

JOEL WHITE AND MICHAEL MEREDITH

Department of Biological Science
Biology Unit I
Florida State University
Tallahassee, Florida 32306

The nervus terminalis (NT) is a ganglionated nerve that connects the brain and the peripheral olfactory structures. Although found in most vertebrates, including man, its function is unknown. Based on anatomical observations, the ganglion of the nerve could be merely a group of sensory cell bodies, analogous to a dorsal root ganglion, or it could function as a relay point in an efferent pathway, as in an autonomic ganglion. The data presented here suggest that the NT ganglion may be more complex than either of these possibilities.

FIGURE 1 shows the position of the NT on the ventral forebrain of the bonnethead shark, *Sphyrna tiburo*. Spontaneously occurring action potentials can be recorded in the NT (FIG. 2A, left side).[1] This spike activity is of efferent origin because spikes recorded either peripheral (at point B in FIG. 1) or central (at point A in FIG. 1) to the ganglion are eliminated by cutting the NT centrally (at point X in FIG. 1). Central disconnection of the NT has an additional consequence, not previously described. Approximately 0.5 to 0.8 seconds after the cut, there is an increase in multiunit activity in the nerve, resulting in a widening of the recorded baseline (FIG. 2A). This result is seen when recording either central (FIG. 2A) or peripheral to the ganglion.

Cooling the nerve can block action potential conduction reversibly and with less trauma than cutting. Cooling central to the ganglion (at point X in FIG. 1) results in elimination of the spontaneous efferent spikes recorded peripheral to the probe, and is followed by an increase in multiunit activity (FIG. 2B), confirming the results of the nerve-cut experiments. Upon warming to ambient temperature, the efferent action potentials resume and the multiunit activity diminishes, yielding a recording similar to that seen before cooling (FIG. 2B).

Following the central cut, the recorded multiunit activity is not eliminated by cutting the peripheral connections of the nerve (at point Y in FIG. 1), although its amplitude decreases slightly. If the nerve peripheral to the ganglion is cut first, no change in baseline is observed. A subsequent cut central to the ganglion results in an increase in baseline, similar to that seen when the central portion of an intact nerve is cut first. These experiments indicate that the majority of the multiunit activity, recorded after the nerve is cut centrally, is generated by cells in the main ganglion.

[a]Supported by National Science Foundation Grant BNS 841 21 41.

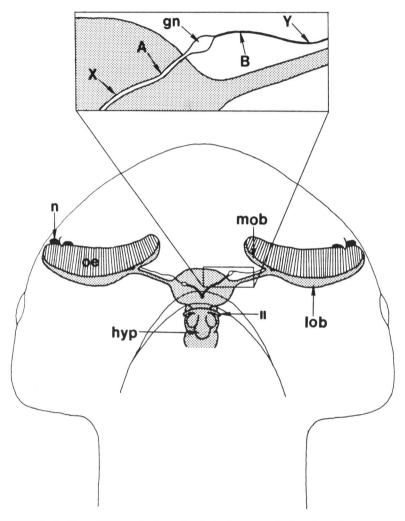

FIGURE 1. Position of NT in the bonnethead shark. Brain stippled. Rostral is toward top. Inset: A & X, points central to ganglion; B & Y, points peripheral to ganglion. Abbreviations: gn, ganglion of NT; hyp, hypophysis; lob, lateral olfactory bulb; mob, medial olfactory bulb; n, nostril; oe, olfactory epithelium; II, optic nerve.

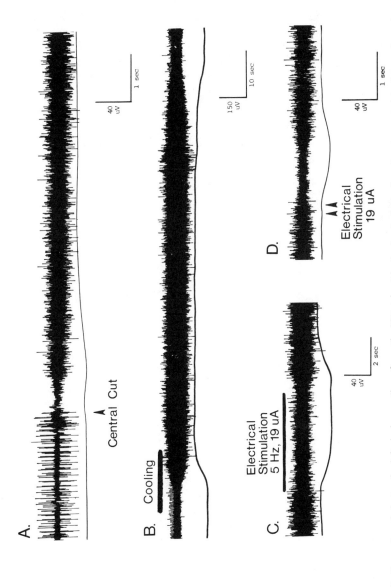

FIGURE 2. Examples of electrophysiological recordings from the NT using bipolar hook electrodes, showing raw data (upper traces) and integrated data (lower traces). (**A**) Cutting intact NT central to ganglion (FIG. 1, point X) while recording central to ganglion (FIG. 1, point A). (**B**) Cooling intact NT central to ganglion while recording peripherally (FIG. 1, point B). (**C**) Stimulation of central stump (FIG. 1, point A) while recording peripherally. (**D**) Stimulating peripheral to ganglion (FIG. 1, point B) while recording centrally.

The data from the nerve cutting and cooling experiments suggest that the efferent action potentials of the NT normally suppress the multiunit activity of the ganglion. The multiunit activity recorded peripheral to the ganglion can be suppressed by electrical stimulation central to the ganglion (FIG. 2C). Additionally, the multiunit activity can also be suppressed by stimulating peripheral to the ganglion (FIG. 2D).

This evidence, coupled with other evidence for synaptic inhibition[2] and reports of synaptic profiles in the ganglion[1] indicate that the ganglion is not simply a grouping of sensory cell bodies. Suppression of ganglion cell activity by efferent impulses also suggests that the terminalis ganglion is not just a relay point for efferent impulses. Rather than simply being relayed more peripherally, the efferent potentials in the intact nerve may influence peripherally or centrally directed spontaneous output from cells in the ganglion, or may modify afferent information from the olfactory epithelium.

REFERENCES

1. WHITE, J. & M. MEREDITH. 1985. Investigations of the nervus terminalis in elasmobranchs. Chem. Senses (Abstr.) **10:** 442.
2. FUJITA, I., M. SATOU & K. UEDA. 1985. Ganglion cells of the terminal nerve: Morphology and electrophysiology. Brain Res. **335:** 148-152.

The Solitary Nucleus
of the Hamster

Cytoarchitecture and Pontine Connections [a]

MARK C. WHITEHEAD AND LAWRENCE D. SAVOY

Department of BioStructure and Function
University of Connecticut Health Center
Farmington, Connecticut 06032

The solitary nuclear complex consists of a number of subdivisions with different cytoarchitectonic features.[2] These subdivisions differ also in the amounts of input they receive from lingual afferent axons. Chorda tympani, lingual (trigeminal), and glossopharyngeal (lingual branch) afferent axons synapse heavily throughout the central subdivision, less heavily in the lateral, ventral, and ventrolateral subdivisions, very lightly in the dorsal and laminar nuclei, and not at all in the medial, magnocellular and dorsolateral subdivisions.[2] In this study we determined which of these subdivisions contain cells that project rostrally to the pons. To reveal the pontine projection neurons, horseradish peroxidase injections were made that filled or nearly filled the parabrachial nucleus. Comparison of the morphologies of retrogradely labeled cells with Golgi-impregnated neurons permitted identification of cell types forming the pontine projection.

The vast majority of retrogradely labeled neurons were, in every case, located in the central subdivision at all rostrocaudal levels (FIG. 1). More cells were labeled in rostral than in caudal parts of this subnucleus. Moderate numbers of labeled cells were also observed in the lateral and ventral subdivisions, primarily at rostral levels, in the ventrolateral subdivision, and in the medial subdivision at all rostrocaudal levels. Few pontine projection neurons were seen in the dorsal and laminar subdivisions, and virtually none in the magnocellular and dorsolateral subdivisions. Thus, subdivisions receiving heavy input from the lingual periphery contain many pontine projection neurons, subdivisions receiving less input contain fewer projection cells, areas with no lingual input contain no projection cells. The medial subdivision is exceptional. It projects to the pons and probably receives general visceral sensory inputs (e.g. Gwyn et al.[1]).

Many of the retrogradely labeled cells of the central and lateral subdivisions had morphologies that were comparable to cell types previously identified in Golgi impregnations (FIG. 2). Most prevalent were elongate cells with ovoid-fusiform somata and dendrites oriented in the mediolateral plane parallel to axons entering from the solitary tract.[3] Also labeled were stellate cells that send three to four dendrites in all directions, although one or two dendrites often elongate mediolaterally.

[a] Supported by National Institutes of Health Grant NS16993.

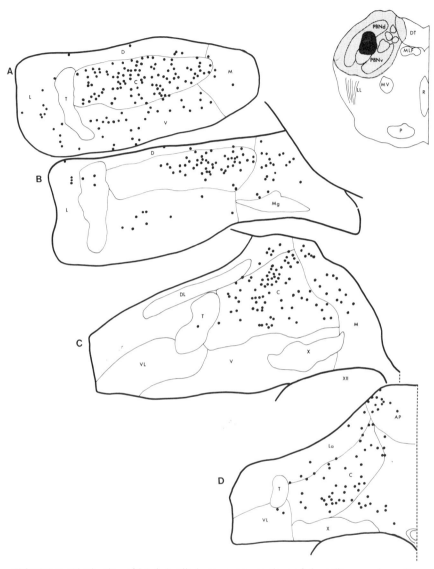

FIGURE 1. Distribution of labeled cells in transverse sections of the solitary nucleus after a large injection of horseradish peroxidase in the parabrachial nucleus (upper right). Distances of the sections from the obex in millimeters are: (**A**) +1.0; (**B**) +0.6; (**C**) +0.1; (**D**) −0.2. Abbreviations for FIGURES 1 and 2: AP, area postrema; C, central subdivision; D, dorsal subdivision; DL, dorsolateral subdivision; DT, dorsal tegmental nucleus; L, lateral subdivision; LL, lateral lemniscus; M, medial subdivision; Mg, magnocellular subdivision; MLF, medial longitudinal fasciculus; MV, motor trigeminal nucleus; P, pyramid; PBNd, v, parabrachial nucleus, pars dorsalis, and ventralis; R, raphe nuclei; T, solitary tract; V, ventral subdivision; VL, ventrolateral subdivision; X, vagal motor nucleus.

These results demonstrate that the cytoarchitectonic subdivisions of the solitary nucleus are distinguished by their afferent and efferent connections. Moreover, for each gustatory subdivision, there is a correlation between the density of lingual inputs it receives and the density of pontine projection neurons it contains. Several different types of neurons project to the pons from areas of the solitary nucleus that receive heavy gustatory inputs. The elongate cells in particular have a distinctive morphology and constitute a readily identifiable class of parabrachial projection neurons.

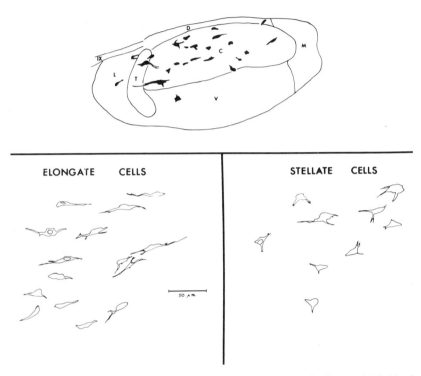

FIGURE 2. Elongate and stellate neurons comprise two categories of cell types labeled in the central subdivision. The drawings are from sections at a rostrocaudal level comparable to that depicted in 1A.

REFERENCES

1. GWYN, D. G., R. A. LESLIE & D. A. HOPKINS. 1985. Observations on the afferent and efferent organization of the vagus nerve and the innervation of the stomach in the squirrel monkey. J. Comp. Neurol. **239:** 163-175.
2. WHITEHEAD, M. C. 1985. The neuronal architecture of the solitary nucleus of the hamster. Chem. Senses. **10:** 436.
3. WHITEHEAD, M. C. 1986. Anatomy of the gustatory system of the hamster: Synaptology of facial afferent terminals in the solitary nucleus. J. Comp. Neurol. **244:** 72-85.

An Apparatus for the Detailed Analysis of Short-Term Taste Tests in Rats[a]

LAURA S. WILSON,[b] JAMES C. SMITH,[b] ROSS
HENDERSON,[b] JEFFREY SHAUGHNESSY,[b] MOHSSEN
S. NEJAD,[c] AND LLOYD M. BEIDLER[c]

[b] Department of Psychology
[c] Department of Biological Sciences
The Florida State University
Tallahassee, Florida 32306

Moment-by-moment analyses of feeding and drinking patterns in rats over 24-hour periods yield valuable information pertaining to taste quality and concentration discrimination that is not available from mere analysis of amounts consumed.[1] The purpose of the research reported here was to determine if such a detailed analysis of licking patterns would be equally revealing for short-term tests, where postingestional influences were minimized.

An apparatus was developed to test rats in single-bottle taste tests for up to 45 minutes. Licks are measured by tongue contact on the metal sipper tube. The number of licks each 500-msec period and a 1.0-msec inter-lick interval (ILI) histogram are measured by a microcomputer. Analysis programs allow for plotting a strip-chart record that shows the pattern of drinking over time. Using a selected criterion, the data are quantified in terms of bursts of licking, allowing for analyses of burst number, burst length, and interburst intervals. In addition, the ILI histogram is plotted and its mean and standard deviation values are determined. To evaluate the usefulness of this detailed analysis of short-term taste tests, examples involving different solution concentrations and quality differences were conducted.

TABLE 1. Mean Values for the Dependent Variables Indicated

Parameter	Water	0.25 M	0.5 M	1.0 M	F Value
Amount consumed (ml)	4.3	10.9	9.3	11.6	3.35
Total licks	763	1396	1435	1503	1.36
Number of bursts	17.2	13.3	10	8.2	1.61
Burst length (secs)	7.4	23.9	30.8	42.1	5.44[a]
Histogram St. Dev.	205	116	102	63	12.03[a]

[a] Significant beyond 0.05 level.

[a] Supported by a grant from NIA (1R01AG04932-02).

TABLE 2. Mean Values for the Dependent Variables Indicated[a]

Parameter	Saccharin	Sucrose	t Value
Amount consumed (ml)	11.50	14.25	1.48
Number of bursts	26.20	12.00	3.24[b]
Burst length (sec)	22.00	71.70	6.39[c]
Total drinking time	25.30	17.30	2.69[b]

[a] Note: These are matched t-tests with five degrees of freedom.
[b] Significant beyond the 0.05 level.
[c] Significant beyond the 0.01 level.

EXAMPLE 1

Six rats were tested on water and three concentrations of sucrose (0.25 molar, 0.50 molar, and 1.0 molar) with 10-minute single-bottle tests. In TABLE 1 it can be seen that although the amount consumed and total number of licks for the sugars are similar, they are consumed in different ways. For example, as the concentration increases, the bursts simultaneously decrease in number and increase in length. The decrease in the ILI histogram standard deviation indicates that there is less variability and fewer pauses in the licking as the concentration increases.

EXAMPLE 2

A comparison was made of licking patterns on qualitatively different solutions. Single-bottle daily 30-minute tests with 0.2% sodium saccharin were followed by similar testing with 32% sucrose (see TABLE 2). Again the amounts of the solutions consumed are similar, while the detailed analysis shows significant differences in the drinking patterns. The saccharin was ingested in many short bursts throughout the session, compared to the few long sucrose bursts concentrated in the first half of the session.

EXAMPLE 3

An analysis was made of sucrose ingestion before and after bilateral sections were made of the greater superficial petrosal nerves. Following surgery, number of bursts decreased and the interburst interval increased. In all three examples, this type of detailed analysis of short-term licking patterns allowed inferences about taste qualities and intensities not available from intake levels alone. It will also be useful in making inferences about taste perceptions of the rat following various surgical manipulations.

REFERENCE

1. SPECTOR, A. C. & J. C. SMITH. 1984. A detailed analysis of sucrose drinking in the rat. Physiol. Behav. **33:** 127-136.

Evidence for Participation by Calcium Ion and Cyclic AMP in Olfactory Transduction

BRUCE D. WINEGAR[a] AND ROLLIE SCHAFER

Department of Biological Sciences
North Texas State University
Denton, Texas 76203-5218

The roles of Ca^{2+} and cAMP in olfactory transduction were explored by aerosol application of inorganic cations, organic calcium channel antagonists, and cyclic nucleotide agonists onto the olfactory epithelium of urethane-anesthetized frogs (*Rana pipiens*). During extracellular recording, saturated vapors of isoamyl acetate were delivered every 100 seconds in 0.3-second pulses. Preliminary experiments with the tiger salamander (*Ambystoma tigrinum*) parallel the results reported below for the frog.

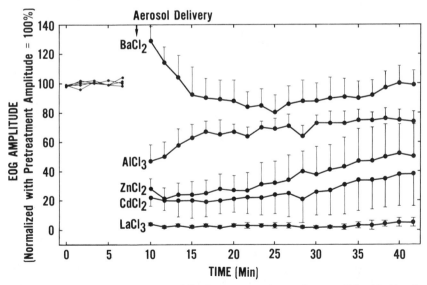

FIGURE 1. Averaged recovery curves following two-second aerosol sprays of the chloride salts of Ba^{2+} ($n = 5$), Al^{3+} ($n = 5$), Zn^{2+} ($n = 5$), Cd^{2+} ($n = 6$), and La^{3+} ($n = 5$). All were applied to give a calculated final concentration of 7.5 mM. Bars represent SD.

[a] Current address: Bruce D. Winegar, Ph.D., Department of Zoology, The University of Texas at Austin, Austin, TX 78712-1064.

Inorganic cations known to block inward calcium currents in other tissues inhibit electroolfactogram (EOG) responses. The rank order of inhibitory potency of the chloride salts is:

$$(La^{3+}) > (Zn^{2+}, Cd^{2+}) > (Al^{3+}, Ca^{2+}, Sr^{2+}) > (Co^{2+})$$

FIGURE 1 illustrates the effects of four of these salts on frog olfactory mucosa, applied in a two-second exposure that produces a calculated final concentration of 7.5 mM. Lanthanum ion virtually eradicates EOG responses, while Ba^{2+} (which in other tissues carries more current in calcium channels than Ca^{2+} itself) initially produces significant

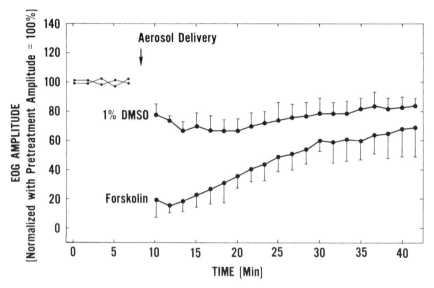

FIGURE 2. Averaged recovery curve following two-second aerosol sprays of forskolin to give a calculated final concentration of 1.5×10^{-5} M ($n = 6$). Forskolin was dissolved in 1% DMSO, and the DMSO curve is a control ($n = 7$). Bars represent SD.

enhancement (F = 43.04; $p < 0.001$; df = 19), followed by a slight inhibition. Magnesium ion has no inhibitory action at the same concentration while Ca^{2+} itself is significantly inhibitory (F = 5.74; $p = 0.0355$; df = 12) even at a final concentration of 1.5×10^{-4} M. A control two-second aerosol application of water depresses EOGs by an average of 5%.

The organic calcium channel antagonists diltiazem and verapamil, applied for two seconds to frog olfactory mucosa, reaching a calculated final concentration of 1.5×10^{-4} M, inhibit by 17% and 36%, respectively, averaged over 20 post-treatment EOGs. The effects of these agents at this concentration are completely reversible. Verapamil produces significant inhibition (F = 26.03; $p = 0.0003$; df = 12) after exposure to a calculated final concentration of 1.5×10^{-5} M.

Cyclic nucleotides, their mimics, and adenosine were tested for inhibition or enhancement. Dibutyryl cAMP (dbcAMP), a lipophilic mimic of cAMP, produces 54% inhibition following a two-second exposure to a calculated final concentration of 1.5 $\times$ 10^{-4} M. Dibutyryl cGMP, cGMP, cAMP, and adenosine all decrease EOGs by less than 15% compared to distilled water controls. Forskolin (FIG. 2), a reversible activator of adenylate cyclase, inhibits EOGs significantly beyond the 1% DMSO control (the solvent used for forskolin) at a calculated final concentration of 1.5 $\times$ 10^{-5} M (F = 17.17; p = 0.002; df = 11).

These data support the hypothesis that Ca^{2+} participates in olfactory transduction. The inorganic cations inhibit EOGs in rank order and concentrations consistent with calcium channel blockade in other tissues. Inhibition by organic calcium channel antagonists also suggests a role for calcium. The inhibitory effects of dbcAMP and forskolin indicate that cAMP may serve as a second messenger in olfactory transduction.

Identification of the Terminal Nerve in Two Amphibians Shown by the Localization of LHRH-like Material and AChE[a]

CELESTE R. WIRSIG [b] AND THOMAS V. GETCHELL

Department of Anatomy and Cell Biology
Wayne State University
Detroit Michigan 48201

The terminal nerve (TN) system is partially composed of a luteinizing hormone-releasing hormone-like immunoreactive (LHRH-ir) chain of neurons that has been shown to directly connect the nasal mucosa with central brain structures in fish[1-3] and mammals.[4-7] In addition, a subpopulation of TN neurons has been shown to contain acetylcholinesterase (AChE) in the hamster.[5] Some processes of the TN project to neural areas associated with the control of reproduction, for example, the preoptic area in fish[1] and the vomeronasal epithelium in mammals.[6] Because of these connections, it is thought that the TN may be involved with the pheromonal control of reproduction.

In this study, we have examined the TN of two amphibians, the tiger salamander and bullfrog, to gain additional information on TN peripheral and central connections. Procedures to label LHRH-ir material and AChE in the TN were used to map the nerve, as described in a previous paper.[8] In addition, the effects of hypophysectomy on LHRH-ir labeling of the salamander TN were examined.

FIGURE 1 illustrates the distribution of LHRH-ir TN neurons in the salamander brain and nasal mucosa. Labeled perikarya and smooth or varicose processes were found in the olfactory, vomeronasal, and trigeminal nerves. In trigeminal nerves projecting toward the vomeronasal organ, LHRH-ir fibers were found in conjunction with large, unlabeled neural perikarya. The greatest number of LHRH-ir perikarya was found in the olfactory nerves adjacent to the bulbs. The TN entered the brain through the olfactory bulbs where some LHRH-ir fibers terminated diffusely within olfactory nerve and glomerular layers. A condensed bundle of fibers continued along the ventral border of the telencephalon to terminate among a group of LHRH-ir neurons in the region of the lamina terminalis and preoptic recess. In the nasal mucosa the termination pattern of LHRH-ir fibers is less clear. A few fibers entered the olfactory and vomeronasal epithelium. A few labeled perikarya were seen in the olfactory epithelium or in ganglia in the lamina propria of the vomeronasal organ.

[a] Supported by National Institutes of Health Grant NS-16340.

[b] Address for correspondence: Department of Ophthalmology, Baylor College of Medicine, One Baylor Plaza, Houston, TX 77030.

interpretations are based on the proportions of responses that occur not on the main diagonal.

The OCM results from individual patients within each group were combined for comparison among the subject groups. The results from the PHP patients with normal Gs unit activity are shown in FIGURE 1. These results were not quantitatively different from 25 normal female controls whose $P(C) = 93.8\%$. Qualitatively, however, they differed from the normal controls in their decreased response to mint (L-Carvone) and orange (D-Limonene), the only substances in the OCM with stereochemical properties. There also is a weak tendency for a parosmic substitution of mint for orange, but not orange for mint. For the moment, the precise psychophysiological explanation for such confusions remains unclear.

RESPONSE

		AMMONIA	CINNAMON	LICORICE	MINT	MOTHBALLS	ORANGE	ROSE	RUB. ALCH.	VANILLA	VINEGAR	
		1	2	3	4	5	6	7	8	9	10	
AMMONIA	1	**.800**	—	—	—	—	—	—	.140	.020	.040	1
CINNAMON	2	.020	**.320**	.140	.120	—	.120	.120	.020	.100	.040	2
LICORICE	3	—	.220	**.380**	.140	.040	.020	—	.080	.100	.020	3
MINT	4	.020	.180	.220	**.160**	.040	.080	.060	.100	.040	.100	4
MOTHBALLS	5	.080	.020	.020	.040	**.420**	.040	.040	.160	.060	.120	5
ORANGE	6	.040	.140	.140	.120	.020	**.220**	.160	.040	.100	.020	6
ROSE	7	.040	.080	.040	.160	.120	—	**.200**	.080	.100	.180	7
RUB. ALCH.	8	.100	.060	.020	.060	.020	.020	.040	**.280**	.100	.300	8
VANILLA	9	—	.240	.040	.040	.020	.100	.120	.020	**.360**	.060	9
VINEGAR	10	.020	.160	.100	.080	.020	.080	.020	.060	.160	**.300**	10

(Left margin label: STIMULUS)

| 1 | 2 | 3 | 4 | 5 | 6 | 7 | 8 | 9 | 10 |

FIGURE 2. Odorant confusion matrix from five female pseudohypoparathyroid patients with deficient Gs unit activity. The rows represent the alternative odorants and the columns represent the proportion of responses to each odorant. The mean percent correct, calculated from the main diagonal, was 34.4%. Scores for the group ranged between 15% and 61%.

PHP patients with deficient Gs unit activity were markedly different both quantitatively and qualitatively from the normal controls and the PHP patients with normal Gs unit activity (FIG. 2). Quantitatively, there was no overlap between the percent correct between the two PHP groups. The difference was significant, $p = 0.002$ (Fisher Exact Test). Qualitatively, the pattern of responses from the PHP patients with deficient Gs unit activity was consistent with what has been observed in patients with known absence of olfactory nerve input resulting from trauma or surgical intervention.[2] Adequate responses are limited to the strong trigeminal stimulants (ammonia, isopropyl alcohol, and acetic acid), or the parosmic confusion of one trigeminal stimulant for another.

The foregoing findings confirm that normal Gs unit activity is necessary for human odorant perception. They also support the merit of the OCM as a modality to evaluate patients with presumed disturbances in their sense of smell.

REFERENCES

1. PACE, U., E. HANSKI, Y. SALOMON & D. LANCET. 1985. Odorant-sensitive adenylate cyclase may mediate olfactory reception. Nature **316:** 255-258.
2. WRIGHT, H. N. 1987. Characterization of olfactory dysfunction. Arch. Otolaryngol. **113:** 163-168.

Taste Cells in the Mudpuppy, *Necturus maculosus,* Are Electrically Coupled

JIAN YANG AND STEPHEN D. ROPER [a]

Department of Anatomy and Neurobiology
Colorado State University
Fort Collins, Colorado 80523

The functional organization of vertebrate taste buds is not well understood. Although individual taste cells may be functional units, it is also possible that a stimulus may activate a subset of taste cells that are electrically coupled and respond as a unit. We have investigated the existence of such electrically coupled units by injecting taste cells with a fluorescent dye, Lucifer yellow (LY), which crosses gap junctions in a number of epithelial and neural tissues. LY-coupling indicates the existence of electrical coupling between cells. We also injected LY into lingual epithelial cells to test for dye coupling between epithelial and taste cells.

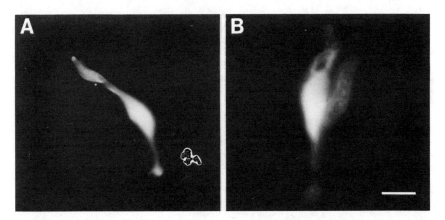

FIGURE 1. Morphology of Lucifer yellow stained taste cell and dye coupling between taste cells. (**A**) An elongated taste cell with one apical and one basal process. (**B**) Two taste cells were stained with Lucifer yellow by intracellular injection of a single cell. Note that the basal processes are out of the plane of focus. Calibration: 20 μm.

[a] Address for correspondence: Dr. Stephen D. Roper, Department of Anatomy and Neurobiology, Colorado State University, Fort Collins, CO 80523.

The lingual epithelium from mudpuppies (*Necturus maculosus*) was removed and placed in amphibian Ringer's solution. Taste cells were impaled with glass microelectrodes filled with 4% LY solution. Hyperpolarizing current was passed through the electrode for one to three minutes to inject LY into cells. Taste cells generate impulses[1]; thus, we confirmed stable impalements in taste buds by the presence of action potentials before, during, and after injecting LY.

A total of 91 taste cells in as many taste buds was injected with LY. LY-filled taste cells usually had an elongated shape with one apical process and one basal process (FIG. 1A). In a few cases, the processes were divided distally into two processes. The basal processes usually terminated in a number of finely branched, finger-like projections that extended to the base of the taste bud. This shape is similar to that revealed by histologic and ultrastructural studies on taste cells in the mudpuppy.[2,3]

In *taste buds,* dye coupling was observed in 19 cases. In 15 cases, two cells were stained with LY after a single intracellular injection. In four cases, three cells were stained. FIGURE 1B shows an example of dye coupling of taste cells. Larger subsets of coupled cells (> 3) were not observed. Dye-coupled pairs (or trios) of taste cells often appeared to be equally stained and cells in the pairs (or trios) had quite similar shapes. Although the site(s) of coupling could not be determined with certainty, we observed close apposition between cell bodies and/or apical processes.

In nontaste *epithelial cells,* we examined dye coupling in the superficial (mucosal) and basal (germinative) layers. Extensive LY coupling was found in the basal layer (15/15 cases), and in most instances the injected cell was coupled to three to five adjacent cells. In the superficial epithelium, 45 surface cells were injected with LY. Sixteen of these were immediately adjacent to taste pores. Only in a single case were surface epithelial cells dye coupled. No dye coupling was observed between epithelial cells and taste cells at the taste pore region.

Our results with LY show that some taste cells are dye coupled to one or two other taste cells in the mudpuppy taste buds. Since there is an average of about 100 cells in the mudpuppy taste bud, our data indicate that approximately 20% of the cells are electrically coupled, and in groups of two or three cells. It must be recognized, however, that weak electrical junctions may not be revealed by LY coupling. Thus, this figure represents a conservative estimate of the extent of electrical coupling between taste cells. West and Bernard[4] also reported electrical coupling in *Necturus* taste cells, but with their methodology they were unable to examine the number of cells to which any one cell was coupled. Furthermore, Teeter[5] has observed dye coupling between taste cells in the catfish.

REFERENCES

1. ROPER, S. D. 1983. Regenerative impulses in taste cells. Science **220:** 1311-1312.
2. FARBMAN, A. J. & J. D. YONKERS. 1971. Fine structure of the taste bud in the mudpuppy, *Necturus maculosus.* Am J. Anat. **131:** 353-370.
3. DELAY, R. & S. D. ROPER. In preparation.
4. WEST, C. H. K. & R. A. BERNARD. 1978. Intracellular characteristics and response of taste bud and lingual cells of the mudpuppy. J. Gen. Physiol. **72:** 305-326.
5. TEETER, J. 1985. Dye coupling in catfish taste buds. Proc. Jpn. Symp. Taste Smell **19:** 29-33.

Independence and Primacy of Umami as Compared with the Four Basic Tastes

SHIZUKO YAMAGUCHI AND YASUSHI KOMATA

Central Research Laboratories Ajinomoto Co., Inc.
1-1, Suzuki-cho, Kawasaki
Kanagawa 210, Japan

Umami is defined as the taste elicited by monosodium glutamate (MSG) or 5'-ribonucleotides such as IMP and GNP. In this paper, several lines of evidence indicate that umami is independent and "basic" as long as the so-called four basic tastes are regarded as basic.

In the first experiment, the similarities among the four tastes (sucrose, NaCl, tartaric acid, and quinine sulfate) and umami (MSG) in single and mixture solutions were examined. A multidimensional scaling yielded a spatial configuration reflecting the similarities among the stimuli.

All the stimuli were located within a four-dimensional regular polyhedron that has five vertices. The tastes composed of the four basic tastes were located within a three-dimensional tetrahedron, which was a subcomplex of the four-dimensional polyhedron with the four basic tastes located at four vertices. Umami was located at the other vertex, indicating that it constructs another dimension independent of the four basic tastes.

Another multidimensional scaling showed the dominance of umami in the tastes of natural foods. The tastes of broths made from meats (beef, pork, etc.) and fish fell outside the tetrahedron of the four basic tastes and were located close to umami. Those made from vegetables widely distributed around the five taste areas. However, when a small amount of IMP was added, the tastes approached umami due to the remarkable synergistic effect between IMP and glutamic acid contained naturally in the vegetable stocks. Thus the stocks examined were regarded to have dominant or potential umami, which is actualized or developed by a small amount of umami substances.

Umami did not enhance the four basic tastes, and vice versa, at threshold and at suprathreshold levels.

The hedonic properties of umami were examined in comparison with the four basic tastes. As far as simple aqueous systems were concerned, umami did not cause a pleasant sensation. In the selected flavored solutions or actual foods, umami clearly enhanced the hedonic tone. Umami increased the pleasantness of foods only within a certain range of concentration, and an excess of umami caused a rather unpleasant sensation by which the intake concentration became self-limited. These hedonic properties were similar to other basic tastes except for sweetness.

REFERENCES

1. KAWAMURA, Y. & M. R. KARE, Eds. 1986. Umami: A Basic Taste—Physiology, Bio-chemistry, Nutrition, Food Science. Marcel Dekker Inc. New York.
2. YAMAGUCHI, S. 1979. The umami taste. *In* Food Taste Chemistry. J. C. Boudreau, Ed.: 33-51. Am. Chem. Soc. Washington, D.C.

Gustatory Responses to Tetrodotoxin and Saxitoxin in Rainbow Trout (*Salmo gairdneri*) and Arctic Char (*Salvelinus alpinus*): A Possible Biological Defense Mechanism

KUNIO YAMAMORI AND MORITAKA NAKAMURA

School of Fisheries Sciences
Kitasato University
Sanriku, Iwate 022-01, Japan

TOSHIAKI J. HARA [a]

Department of Fisheries and Oceans
Freshwater Institute
Winnipeg, Manitoba, Canada R3T 2N6

Pufferfish toxin, tetrodotoxin (TTX), is a potent neurotoxin, with its lethal toxicity to humans 300 times that of KCN. TTX exerts its toxic action by specifically blocking the voltage-sensitive sodium channels in nerve and muscle membranes.[1] TTX is widely distributed in tetraodontid fishes, but its distribution seems much wider than originally thought. Although much information is available on the mechanism of its pharmacological action, little is known about its biological significance. The presence of the fish toxin presents some interesting questions: Do possessors have any adaptive significance, or do predators benefit as toxins act as warning signals emanating from prey? Some pufferfish release TTX into surrounding water when stimulated by electric shock or stressed by handling.[2] Our recent behavioral studies demonstrated that some fish species avoid food containing TTX, suggesting that fish may be able to detect TTX via the gustatory system (Yamamori *et al.*, unpublished). Saxitoxin (STX), a paralytic toxin, has also been found in some pufferfish species,[3] and its pharmacological action is identical to that of TTX. This study was designed to describe the gustatory sensitivity of rainbow trout and arctic char to TTX and STX by measuring the integrated electrical responses from the palatine nerve innervating the palate and inside upper lip.

[a] Address for correspondence: Toshiaki J. Hara, Department of Fisheries and Oceans, Freshwater Institute, 501 University Crescent, Winnipeg, Manitoba, Canada R3T 2N6.

The gustatory receptors of rainbow trout were extremely sensitive to TTX; it had a threshold concentration 10^{-7} M and at 10^{-5} M evoked a response three times that of 10^{-3} M L-proline, the most effective amino acid for this species (FIG. 1). The threshold for STX was lower, 10^{-8} M, but unlike TTX the response magnitude reached maximum at 10^{-6} M and saturation occurred with further increase in concentration. This situation was reversed in arctic char; lower threshold for TTX than STX (10^{-8} versus 10^{-7} M). The response magnitude was generally small, never exceeded that of 10^{-3} M L-proline at all concentrations tested. Cross-adaptation experiments indicated that receptor or fiber types for TTX are distinct from those that detect amino acids

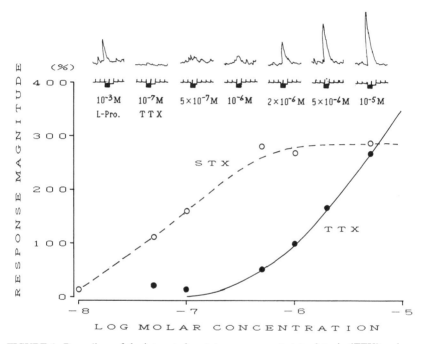

FIGURE 1. Recordings of the integrated gustatory responses to tetrodotoxin (TTX) and semilogarithmic plot of their concentration-response relationship, together with that for saxitoxin (STX), in rainbow trout. Response magnitude is represented as a percentage of that induced by the standard stimulant 10^{-3} M L-proline.

and bile salts, and that TTX and STX partially cross-react. Furthermore, the integrated gustatory responses to TTX or STX, a fast-adapting, phasic response, rapidly returned to the baseline even with continuous stimulation. Perfusion of the gustatory organs with these chemicals had no inhibitory effect. In frog and rat, TTX induced no electrical response in the gustatory systems, nor did it suppress the generator potentials in response to gustatory stimulation.[4]

These findings indicate the existence of a sensitive, specific gustatory receptor system for TTX and STX in rainbow trout and arctic char, suggesting a defense mechanism against poisonous prey in the aquatic environment. The results also suggest that TTX-sensitive sodium channels are not involved in the gustatory transduction.

REFERENCES

1. NARAHASHI, T. 1974. Chemicals as tools in the study of excitable membranes. Physiol. Rev. **54:** 813-889.
2. KODAMA, M., T. OGATA & S. SATO. 1985. External secretion of tetrodotoxin from puffer fishes stimulated by electric shock. Mar. Biol. **87:** 199-202.
3. NAKAMURA, M., Y. OSHIMA & T. YASUMOTO. 1984. Occurrence of saxitoxin in puffer fish. Toxicon **22:** 381-385.
4. OZEKI, M. & A. NOMA. 1972. The action of tetrodotoxin, procaine and acetylcholine on gustatory receptions in frog and rat. Jpn. J. Physiol. **22:** 467-475.

Influence of a Single Mutation on the Incidence of Pregnancy Block in Mice[a]

KUNIO YAMAZAKI, GARY K. BEAUCHAMP,
OSAMU MATSUZAKI, AND DONNA KUPNIEWSKI

Monell Chemical Senses Center
Philadelphia, Pennsylvania 19104

JUDY BARD, LEWIS THOMAS, AND
EDWARD A. BOYSE

Memorial Sloan-Kettering Cancer Center
New York, New York 10021

It has been shown that major histocompatibility complex (MHC) types affect the mating choices of mice[1] and that mice can be trained to distinguish arms of a Y maze scented by odors from MHC-congenic mice or their urines.[2,3] More recently, mice have been successfully trained in the Y maze to distinguish the scent of mice that differ genetically only by mutation of the H-2K gene, which belongs to the category of MHC genes known as class I, that plays a vital part in immunological recognition and response.[4] To investigate the relevance of H-2K genetic variation to reproductive behavior, we have now tested the effect of isolated H-2K genetic variation in the circumstances of "pregnancy block."

Pregnancy block (the Bruce effect)[5] refers to termination of pregnancy before implantation of the embryo (before day six of gestation in the mouse) after exposure of the female to an unfamiliar male, particularly of a strain different from the stud male.

All female mice in this study were of the inbred strain BALB, whose MHC type is H-2^d. The stud males were either B6/By (H-2^b) or B6.C-H-2^{bm1}, differing only by a mutation at the K^b gene. The second (test) male, to which the fertilized females were exposed, was either the same stud male, or a male genetically identical to the stud male (syngeneic male), or a male of the other H-2K type (B6.C-H-2^{bm1} if the stud male was B6/By, and vice versa). Females were monitored for return to estrus by visual inspection of the external genitalia. Return to estrus within seven days of stud mating was scored by blind testing as a blocked pregnancy or blocked pseudo-pregnancy.

[a] Supported by Grant GMCA-30296 from the National Institutes of Health.

730

The results showed that the incidence of pregnancy block was higher when the stud and unfamiliar males differed in the H-2K gene than when the stud and unfamiliar males were genetically identical (TABLE 1). Thus, the olfactory distinction of mice differing by a mutation of the H-2K gene can spontaneously influence neuroendocrine communication affecting reproduction.

TABLE 1. Incidence of Blocking of Pregnancy or Pseudopregnancy in Isolated BALB/c Females Exposed to Males Whose H-2K Type Differed or Did Not Differ from That of the Stud Male

Test Male	No. of Females	Females Returning to Estrus		
		No.	Percent	p
Stud male	142	25	18	
Syngeneic male[a]	164	31	19	
Congenic male[b]	183	73	40	<0.001

[a] A male genetically identical to the stud male.
[b] A male differing only at the H-2K gene.

REFERENCES

1. YAMAZAKI, K., E. A. BOYSE, V. MIKE, H. T. THALER, B. J. MATHIESON, J. ABBOTT, J. BOYSE, A. ZAYAS & L. THOMAS. 1976. Control of mating preferences in mice by genes in the major histocompatibility complex. J. Exp. Med. **144:** 1324-1335.
2. YAMAZAKI, K., M. YAMAGUCHI, L. BARANOSKI, J. BARD, E. A. BOYSE & L. THOMAS. 1979. Recognition among mice: Evidence from the use of a Y maze differentially scented by congenic mice of different major histocompatibility types. J. Exp. Med. **150:** 755-760.
3. YAMAGUCHI, M., K. YAMAZAKI, G. K. BEAUCHAMP, J. BARD, E. A. BOYSE & L. THOMAS. 1981. Distinctive urinary odors governed by the major histocompatibility locus of the mouse. Proc. Natl. Acad. Sci. USA **78:** 5817-5820.
4. YAMAZAKI, K., G. K. BEAUCHAMP, I. K. EGOROV., J. BARD, L. THOMAS & E. A. BOYSE. 1983. Sensory distinction between H-2^b and H-2^{bm1} mutant mice. Proc. Natl. Acad. Sci. USA **80:** 5685-5688.
5. BRUCE, H. M. 1959. An exteroceptive block to pregnancy in the mouse. Nature **184:** 105.

Monoclonal Antibodies Directed against Catfish Taste Receptors

Immunocytochemistry of Catfish Taste Buds[a]

JOAN YONCHEK AND THOMAS E. FINGER[b]

Department of Cellular and Structural Biology
University of Colorado Health Sciences Center
Denver, Colorado 80262

ROBERT H. CAGAN

Research and Development Division
Colgate-Palmolive Company
Piscataway, New Jersey 08854

BRUCE P. BRYANT

Monell Chemical Senses Center
Philadelphia, Pennsylvania 19104

Monoclonal antibodies were previously raised against a purified membrane fraction derived from the taste epithelium of channel catfish. One of these, antibody G-10, which was found to inhibit the binding of L-alanine to membrane fractions obtained from the taste epithelium, was used in these immunocytochemical studies. For immunocytochemistry, the culture supernatant from clone G-10 was purified on a protein-A column. The mandibular or maxillary barbels from channel catfish (*Ictalurus punctatus*) were fixed for one hour in 4% paraformaldehyde buffered with 0.1 M phosphate, pH 7.2, and then washed in phosphate buffer. One group of intact barbels was exposed to the antibody without permeabilization; thus the antibody had access only to sites on the external surface of the barbel. Another group of intact barbels was permeabilized by cryoprotection and two freeze-thaw cycles; also 0.3% Triton X-100 was added to the antibody solutions. In these cases, the antibody appeared to penetrate throughout the epithelium. Finally, a third group of barbels was cryoprotected in sucrose and

[a] Supported by National Institutes of Health Grants NS-15740, NS-00772, NS-15258, and NS-20486.

[b] Address for correspondence: Dr. Thomas E. Finger, Department of Cellular and Structural Biology, B-111, University of Colorado Medical School, 4200 E. 9th Ave., Denver, CO 80262.

sectioned at 12 μm either longitudinally or in cross sections on a cryostat. The sections then were thaw mounted onto gelatinized slides and exposed to antibody solutions containing Triton X-100.

The various tissues were exposed for 48 hours to the purified G-10 antibody, used at a 1 : 50 dilution in phosphate buffer. Following several buffer washes, the tissue was prepared with standard methods using peroxidase-antiperoxidase or avidin-biotin complexes to reveal the pattern of immunostaining. Tissue prepared for electron microscopy was osmicated, dehydrated, and embedded routinely following the immunoreaction. Control tissue was treated similarly to the experimental preparations

FIGURE 1. Immunoreactivity to antibody G-10 in an intact, permeabilized catfish barbel. (**A**) The arrow indicates a taste bud in which immunoreactive taste cells can be seen extending from the apical to the basal regions of the bud. (**B**) Superficial plane of focus showing the taste pore region of several taste buds (arrows). The typical annular pattern of immunoreactivity is clear. Original magnification ×430; reduced by 40%.

except that normal mouse serum, or a monoclonal antiserum directed against enkephalin, was substituted for the G-10 antibody. No specific staining as described below was observed in the control tissue.

The tissue treated with antibody G-10 exhibited clear immunoreactivity in both the sectioned and intact tissue. All taste buds contained some immunoreactive cells, but the number of immunoreactive cells varied among taste buds. In the permeabilized preparations, immunostaining was evident from the apical to the basal regions of each immunoreactive taste cell, although staining was heavier at the apical end (FIG. 1A). Furthermore, cells in the center of each taste bud were not immunoreactive, but were surrounded by immunoreactive taste cells (FIG. 1B). Electron microscopic analysis of nonpermeabilized barbels revealed specific immunoreactivity confined to the taste pore region with especially dense staining of the receptor microvilli. Little or no specific staining extended onto the flanking, supporting cells surrounding the apical pore of the taste bud (FIG. 2). Occasionally a more flocculent immunoreactivity was observed associated with the mucus of the taste pore region. Whether this represents artifactual diffusion of the G-10 antigen into the mucus, diffusion of the peroxidase product, or a biologically important phenomenon remains to be determined.

In summary, we find G-10 immunoreactivity selectively associated with the receptor cells of the barbel taste buds. Further, our results indicate that the G-10 antigen is present on the external face of the receptor microvilli. These results are consistent with the hypothesis that the G-10 antibody recognizes an antigen associated with taste receptor proteins of taste buds in catfish.

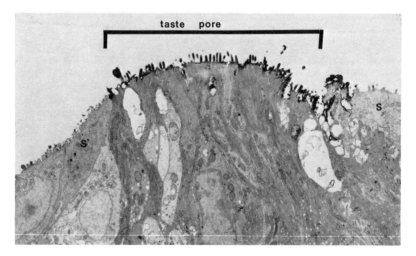

FIGURE 2. Electronmicrograph of a barbel that was exposed to G-10 antibody while intact and nonpermeabilized. Immunoreactivity is evident on the surface of the receptor microvilli in the taste pore region but is absent from the surface of the surrounding supporting cells (S). Original magnification ×8,300; reduced by 40%.

REFERENCE

1. GOLDSTEIN, N. I. & R. H. CAGAN. 1982. Biochemical studies of taste sensation: Monoclonal antibody against L-alanine binding activity of catfish taste epithelium. Proc. Natl. Acad. Sci. USA **79:** 7595-7597.

Crustacean Chemical Perception

Tuning to Energy and Nutrient Reward?

RICHARD K. ZIMMER-FAUST

Marine Sciences Institute
University of California
Santa Barbara, California 93106

A major problem for animals is how to exploit food resources efficiently. The ability to remotely sense both food quality and quantity and the effort needed for food capture is clearly advantageous, since this minimizes time in choosing food and maximizes net rate of energy or nutrient gain. Experiments in this study explored how crustacean chemical perception is tuned to odors that could signal energy and nutrient properties of food (see procedures of Zimmer-Faust *et al.*[1]). While amino acids are abundant in intact prey and provide usable nitrogen, ammonia is a nitrogenous waste product of protein catabolism. Elevated levels of ammonia might signal carrion of poor nutritional quality to marine predators. Biodegradatory bacteria selectively assimilate other low-molecular-weight compounds, including amino acids, while they release copious amounts of ammonia.[2] Consequently, interactions between these substances are proposed to signal the relative nutritional (nitrogen) quality of food. Supporting this hypothesis are data showing six littoral and bathypelagic species exhibiting probing and searching to amino acids, but not to ammonia. Amino acids cause forward ambulation (searching) in some species while ammonia induces only tail-flipping (fleeing) in others (see data for two representative species in FIG. 1A and B). Interactions between amino acids and ammonia are clearly antagonistic, since mixtures combining the two substances suppress both feeding and fleeing.

Adenosine 5'-triphosphate (ATP) is involved either directly or indirectly with all cellular chemical energy transfers. Additional experiments with the spiny lobster, *Panulirus interruptus,* show that chemical excitation by ATP > ADP > AMP = adenosine. This is the pathway followed by phosphoadenylates in autolytic degradation of animal flesh. The ratio of adenylate energy charge [AEC = (ATP + 0.5 ADP)/ (ATP + ADP + AMP)] is believed indicative of the total metabolic energy available to an organism;[3] consequently, its perception by a forager might assist in the recognition of prey energetic quality. Further experiments presented ATP, ADP, and AMP in all combinations while holding constituent concentrations constant. Both feeding and locomotory responses were found to be positively correlated with the AEC of solutions (FIG. 2, and Kendall's Tau: $T > 0.82, p < 0.02$, both comparisons), though data rejected hypotheses that lobsters were best stimulated by mixtures of highest total phosphoadenylate concentration, or by ATP (alone) independently of the other phosphoadenylates. Results indicate that crustacean chemical perception is tuned to odors that are likely to yield food of highest energy and nutrient payoff.

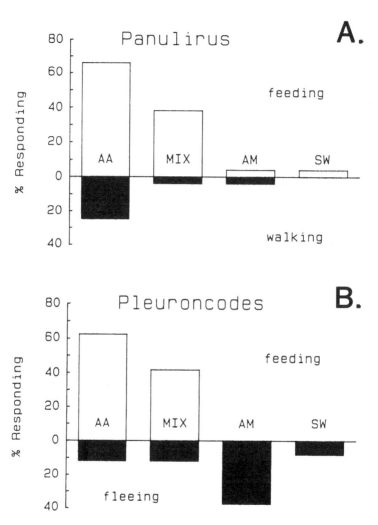

FIGURE 1. Feeding (□) and locomotory responses (■) by the spiny lobster, *Panulirus interruptus* (**A**) and by the pelagic red crab, *Pleuroncodes planipes* (**B**), to an amino acid (AA; glycine), ammonia (AM), amino acid + ammonia (MIX), and seawater (SW). Each chemical was presented at 10 μM, and 24 animals were tested with each solution.

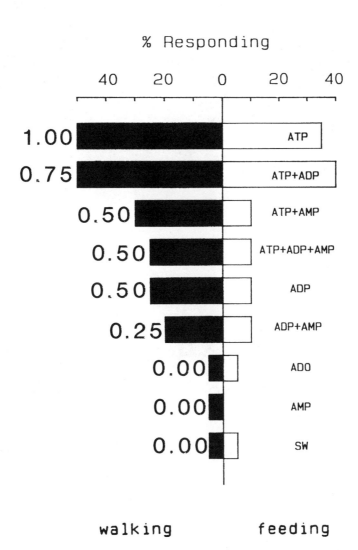

FIGURE 2. Feeding and locomotory responses by *Panulirus* are positively correlated with the adenylate energy charge of chemical solutions (0.00-1.00 scale). Each chemical was presented at 10 μM, and 20 animals were tested with each solution.

REFERENCES

1. ZIMMER-FAUST, R. K., J. E. TYRE, W. C. MICHEL & J. F. CASE. 1984. Chemical mediation of appetitive feeding in a marine decapod crustacean: The importance of suppression and synergism. Biol. Bull. **167:** 339-353.
2. HOLLIBAUGH, J. T. 1979. Metabolic adaptation in natural bacterial populations supplemented with selected amino acids. Estuarine Coastal Mar. Sci. **9:** 83-92.
3. ATKINSON, D. E. 1977. Cellular Energy Metabolism and Its Regulation. Academic Press. New York.

Behavioral Recovery and Morphological Alterations after Lesions in the Olfactory System in the Goldfish (*Carassius auratus*)

H. P. ZIPPEL

Physiologisches Institut
Humboldtallee 23
Universität D-3400, Göttingen
Federal Republic of Germany

D. L. MEYER

Institut für Anatomie
Kreuzbergring 36
Universität D-3400, Göttingen
Federal Republic of Germany

Lower vertebrates such as amphibia and fish are ideal organisms for the study of central nervous system (CNS) regeneration. In our initial studies we were able to demonstrate the functional and the morphological regeneration of fiber connections (receptor axotomy, dissection of the olfactory tracts, dissection of the anterior commissure) in the olfactory system.[1,2] From more recent experiments,[3] it is evident that after a total bilateral olfactory bulbectomy the receptors in the olfactory mucosa (OM) are present, but no connections were found between the OM and the telencephalon; behaviorally, no return to the preoperative behavioral threshold could be found even after a 12-month survival period.

In these investigations, after a partial bilateral ablation of the olfactory bulbs (OB), a rapid return to the preoperative behavior could be recorded at different postoperative time intervals; for this purpose the preoperative spontaneous and training behavior was compared with the postoperative reactions of the same animals. Immediately following surgery the animals fail to respond to low concentrations of olfactory stimuli but readily react to stimuli above the "taste" threshold (i.e., the animals respond positively to natural and synthetic odors used during training when applied in roughly 2 log units higher concentrations). Rostral bulbectomized (RBE) goldfish respond positively after a short regeneration period of about 7 to 10 days, whereas caudal bulbectomized (CBE) animals react positively after roughly six weeks (FIG. 1). From neuroanatomical investigations performed immediately after the behavioral test (RBE: 20 days, CBE: 9 weeks postop.), it is evident that shortly after the functional recovery a surprisingly high level of regeneration has been achieved: The spherical shape of the OB has been reestablished nearly completely in both the collectives. In RBE

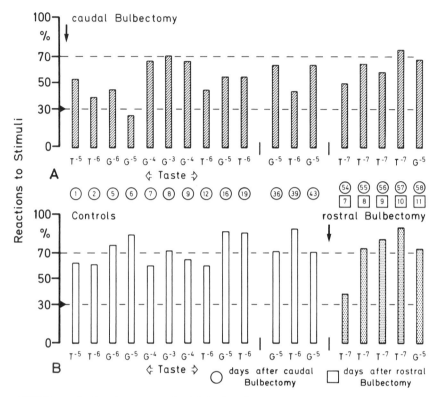

FIGURE 1. Behavior of 10 groups of goldfish after bilateral caudal (CBE: 5 groups) or rostral (RBE: 5 groups) bulbectomy. The animals were trained to Na-glutamate 10^{-5} molar (G^{-5}), and their behavior was tested with various concentrations of Tubifex food extracts (T; concentrations in g/l) before operation (not shown in figure). The columns represent reactions to the respective odor (orientation toward one out of two feeding funnels through which the stimulus is applied; for details see Zippel *et al.*[3]). Reactions close to or above the 70% level represent significant positive behavior. Up to the 36th postoperative day, CBE animals only react positively when stimulus concentrations were above the taste threshold (glutamate 10^{-4}, 10^{-3} molar, day 7-9), whereas the intact controls respond to low (olfactory) concentrations, especially the training stimulus (G^{-5}). From the 36th postoperative day onward in CBE-animals, a gradual return to the preoperative behavior (initially to the training stimulus [tests 36, 43 postop.] and finally to low concentrations of food odor [tests 54 to 57 postop.]) can be recorded. During the final test period, the CBE animals served as the controls for the RBE fish (former controls) in which from the eighth postoperative day onward, a rapid return to low-stimulus concentrations is evident.

animals the connections between the OM and the OB are not essentially different from those in intact animals, whereas in CBE animals the central connections between the OB and the telencephalon are evidently thinner and much more transparent. Light-microscopic investigations of the OM show no differences between postoperative and intact animals, and from HRP-studies it is evident that fiber connections between the OM and the OB exist, albeit in smaller numbers in comparison with intact or long-term regenerated fish. In a final series the olfactory mucosa was investigated in bilateral totally bulbectomized fish and after receptor axotomy[1] (at the olfactory bulb). After survival times of 6 h, 12 h, 24 h, and 2, 4, 6, and 9 days, no essential differences could be found between both the collectives: At no time after olfactory bulbectomy or after receptor axotomy could a significant amount of degenerating receptors be found. This finding might well explain the extremely short behavioral recovery of about 7-10 days after receptor axotomy and rostral bulbectomy in the goldfish, a time at which in many other vertebrates (e.g., trout[4]; salamander[5]; hamster[6]; rat[7]; monkey[8]) degenerative processes are at their maximum.

REFERENCES

1. ZIPPEL, H. P. & R. A. WESTERMAN. 1970. Geruchsdifferenzierungsvermögen der Karausche (*Carassius carassius*) nach funktioneller und histologischer Regeneration des Tractus olfactorius und der Commissura anterior. Z. Vgl. Physiol. **69:** 38-53.
2. ZIPPEL, H. P., R. V. BAUMGARTEN & R. A. WESTERMAN. 1970. Histologische, funktionelle und spezifische Regeneration nach Durchtrennung der Fila olfactoria beim Goldfisch (*Carassius auratus*). Z. Vgl. Physiol. **69:** 79-98.
3. ZIPPEL, H. P., W. BREIPOHL & H. SCHOON. 1981. Functional and morphological changes in fish chemoreception systems following ablation of the olfactory bulbs. *In* Proceedings in Life Sciences. Lesion-Induced Neuronal Plasticity in Sensorimotor Systems. H. Flohr & W. Precht, Eds.: 377-394. Springer Verlag. Berlin, Heidelberg, New York.
4. EVANS, R. E. & T. J. HARA. 1985. The characteristics of the elecro-olfactogram (EOG): Its loss and recovery following olfactory nerve section in rainbow trout (*Salmo gairdneri*). Brain Res. **330:** 65-75.
5. SIMMONS, P. A. & T. V. GETCHELL. 1981. Physiological activity of newly differentiated olfactory receptor neurons correlated with morphological recovery from olfactory nerve section in the salamander. J. Neurophysiol. **45:** 529-549.
6. CONSTANZO, R. M. 1985. Neural regeneration and functional reconnection following olfactory nerve transection in hamster. Brain Res. **361:** 258-266.
7. GRAZIADEI, P. P. C. & G. A. MONTI GRAZIADEI. 1980. Neurogenesis and neuron regeneration in the olfactory system of mammals. III. Deafferentation and reinnervation of the olfactory bulb following section of the fila olfactoria in rat. J. Neurocytol. **9:** 145-162.
8. MONTI GRAZIADEI, G. A., M. S. KARLAN, J. J. BERNSTEIN & P. P. C. GRAZIADEI. 1980. The olfactory nerve in monkey (*Saimiri sciureus*). Brain Res. **189:** 343-354.

Index of Contributors